Springer Geography

For further volumes:
http://www.springer.com/series/10180

Springer Geography

The Springer Geography series seeks to publish a broad portfolio of scientific books, aiming at researchers, students, and everyone interested in geographical research. The series includes peer-reviewed monographs, edited volumes, textbooks, and conference proceedings. It covers the entire research area of geography including, but not limited to, Economic Geography, Physical Geography, Quantitative Geography, and Regional/Urban Planning.

Ashraf Dewan • Robert Corner
Editors

Dhaka Megacity

Geospatial Perspectives on Urbanisation, Environment and Health

Springer

Editors
Ashraf Dewan
Department of Spatial Sciences
Curtin University
Western Australia

Robert Corner
Department of Spatial Sciences
Curtin University
Western Australia

ISBN 978-94-007-6734-8 ISBN 978-94-007-6735-5 (eBook)
DOI 10.1007/978-94-007-6735-5
Springer Dordrecht Heidelberg New York London

Library of Congress Control Number: 2013940776

Printed on acid-free paper

Springer is part of Springer Science+Business Media (www.springer.com)

This book is dedicated to our wives Shikha and Margaret who have given us their unstinting support during its genesis

Preface

This book is the result of a collaboration between two geographers – one an environmental geographer from a developed country, the other a physical and human geographer from a developing country. We share an interest in human–environment interactions that goes beyond the academic. We now find ourselves working in the same university where we are able to use geospatial tools to explore those human–environment interactions and the often deleterious effects that they have on not just the environment but on the human population themselves. We have assembled a talented collection of authors with expertise in many areas of human–environment interactions. Some of them are our own students, both past and present, and others are collaborators from a range of universities and institutes across the globe.

Humankind has been living in cities since the Chalcolithic era, but it was not until relatively modern times that cities began to approach the size and dominance that they currently possess. Two related phenomena occurred in the late twentieth century. The first was the expansion of existing cities to a new class of urban settlement known as the megacity, and the second is the fact that there are now more people living in the world's cities than in its rural areas. A primary driving factor for this has been rural–urban migration, which has been observed all over the world as the economies of countries transform from an agrarian base to an industrial base and in some "first world" countries to post-industrial service-based economies. The move to an urban lifestyle has always had positive and negative effects on those making that move. For many, it initially results in overcrowded conditions that are deleterious to health, but modern medicine is beginning to counteract that. The upshot of this is that rural–urban migrants frequently continue the high birth rate practices of their rural predecessors for at least a generation or two, without the moderating effect of higher infant mortality. This, in turn, fuels the population growth of the cities and megacities that these migrants inhabit.

South Asia, the home of some of the world's first cities in the Indus valley civilisation, is now the home of several megacities and a number of large conurbations that are on the threshold of becoming megacities. Unlike the megacities of the developed world, where population growth has slowed in recent years, the megacities of South Asia are bustling places where the demand for accommodation far outstrips the efforts of the civic administration either to control the supply of land for housing or to provide adequate infrastructure to service the burgeoning population. There are many reasons for this, and they are shared by the megacities of other developing countries. They include a basic lack of resources, a culture that sees no harm in taking bureaucratic shortcuts and poorly resourced planning and enforcement agencies whose underpaid staff may be prey to unscrupulous property developers. Regrettably, many of these developers care more about a rapid profit now than they do about the future environmental living conditions of their "clients".

Dhaka, the capital of Bangladesh, is a rapidly developing megacity. Whilst it has its own particular characteristics, it is a good example of the organic, often chaotic, development of megacities the world over. We hope that this book will be of interest to those who care about the future of our planet and its people and the way in which we accommodate our population as the whole world seeks to emulate the lifestyle of the so-called developed world.

We, finally, would like to thank all our authors who have taken their time to write their contributions in such a manner that they are widely accessible to readers of all levels, especially those who are seeking to understand the basic methodological and quantitative methods used.

Perth, Western Australia
February 2013

Robert Corner
Ashraf Dewan

Contents

1 **Introduction** 1
Robert J. Corner and Ashraf M. Dewan

2 **From a Town to a Megacity: 400 Years of Growth** 23
Sohel J. Ahmed, Kh. Md. Nahiduzzaman, and Glen Bramley

3 **Spatiotemporal Patterns of Population Distribution** 45
Robert J. Corner, Emmanuel T. Ongee, and Ashraf M. Dewan

4 **Climatic Variability** 61
Yusuke Yamane, Masashi Kiguchi, Toru Terao, Fumie Murata, and Taiichi Hayashi

5 **Monitoring and Prediction of Land-Use and Land-Cover (LULC) Change** 75
Robert J. Corner, Ashraf M. Dewan, and Salit Chakma

6 **Spatiotemporal Analysis of Urban Growth, Sprawl and Structure** 99
Ashraf M. Dewan and Robert J. Corner

7 **Key Driving Factors Influencing Urban Growth: Spatial-Statistical Modelling with CLUE-s** 123
Sohel J. Ahmed, Glen Bramley, and Peter H. Verburg

8 **Analysis of Urban Development Suitability** 147
Salit Chakma

9 **Impact of Land-Use Change on Flooding Patterns** 163
Takeo Onishi, Tahmina Khan, and Ken Hiramatsu

10 **Flood Vulnerability and Risk Assessment with Spatial Multi-criteria Evaluation** 177
Akiko Masuya

11 Supplementing Electrical Power Through Solar PV Systems 203
Md. Humayun Kabir and Wilfried Endlicher

12 Impact of Land Use and Land Cover Changes on Urban Land Surface Temperature 219
Ashraf M. Dewan and Robert J. Corner

13 Illustrating Quality of Life (QOL) 239
Ashraf M. Dewan, Kamrun Nahar, and Yohei Kawamura

14 Exploring Crime Statistics 257
Ashraf M. Dewan, Md. Rafiqul Haider, and Md. Ruhul Amin

15 Environmental Problems and Governance 283
Salim Momtaz and S.M. Zobaidul Kabir

16 Assessing Surface Water Quality Using Landsat TM and In Situ Data: An Exploratory Analysis 301
Razia A. Chowdhury, Towhida Rashid, and Sirajul Hoque

17 Emissions from the Brick Manufacturing Industry 319
Sarath K. Guttikunda

18 Rainfall Dependence of Hospital Visits of *Aeromonas*-Positive Diarrhoea 333
Masahiro Hashizume, Abu S.G. Faruque, and Ashraf M. Dewan

19 Modelling Spatiotemporal Patterns of Typhoid Cases Between 2005 and 2009 Using Spatial Statistics 345
Ashraf M. Dewan, Robert J. Corner, and Masahiro Hashizume

20 Spatiotemporal Analysis of Dengue Infection Between 2005 and 2010 367
Sarwa Ali, Robert J. Corner, and Masahiro Hashizume

Index 385

Contributors

Sohel J. Ahmed The Bartlett Development Planning Unit, University College London (UCL), London, UK

Sarwa Ali Department of Spatial Sciences, Curtin University, Bentley, Perth, WA, Australia

Md. Ruhul Amin Traffic North Division, Bangladesh Police, Dhaka, Bangladesh

Glen Bramley School of the Built Environment, Heriot-Watt University, Edinburgh, UK

Salit Chakma Department of Spatial Sciences, Curtin University, Bentley, Perth, WA, Australia

Razia A. Chowdhury Department of Geography, Faujderhat Cadet College, Faujderhat, Chittagong, Bangladesh

Robert J. Corner Department of Spatial Sciences, Curtin University, Bentley, Perth, WA, Australia

Ashraf M. Dewan Department of Spatial Sciences, Curtin University, Bentley, Perth, WA, Australia

Wilfried Endlicher Department of Geography, Humboldt-Universität zu Berlin, Berlin, Germany

Abu S.G. Faruque Department of Clinical Medicine, International Centre for Diarrhoeal Disease Research, Dhaka, Bangladesh

Sarath K. Guttikunda Division of Atmospheric Sciences, Desert Research Institute, Reno, NV, USA

Md. Rafiqul Haider Centre for Child and Adolescent Health, International Centre for Diarrhoeal Disease Research, Dhaka, Bangladesh

Masahiro Hashizume Institute of Tropical Medicine (NEKKEN), Nagasaki University, Nagasaki, Japan

Taiichi Hayashi Disaster Prevention Research Institute, Kyoto University, Uji City, Kyoto, Japan

Ken Hiramatsu Faculty of Applied Biological Sciences, Gifu University, Yanagido, Gifu, Japan

Sirajul Hoque Department of Soil, Water and Environment, University of Dhaka, Dhaka, Bangladesh

Md. Humayun Kabir Department of Geography and Environment, University of Dhaka, Dhaka, Bangladesh

S.M. Zobaidul Kabir Centre for Environmental Management, Central Queensland University, North Rockhampton, QLD, Australia

Yohei Kawamura Department of Mining Engineering, Curtin University, Kalgoorlie, WA, Australia

Tahmina Khan United Graduate School of Agricultural Science, Gifu University, Yanagido, Gifu, Japan

Masashi Kiguchi Institute of Industrial Science, The University of Tokyo, Tokyo, Japan

Akiko Masuya Department of Spatial Sciences, Curtin University, Bentley, Perth, WA, Australia

Salim Momtaz School of Environmental and Life Sciences, University of Newcastle, Ourimbah, NSW, Australia

Fumie Murata Faculty of Science, Kochi University, Kochi, Japan

Kamrun Nahar Department of Geography and Environment, University of Dhaka, Dhaka, Bangladesh

Kh. Md. Nahiduzzaman Department of City and Regional Planning, King Fahd University of Petroleum and Minerals, Dhahran, Kingdom of Saudi Arabia

Emmanuel T. Ongee Department of Spatial Sciences, Curtin University, Bentley, Perth, WA, Australia

Takeo Onishi Faculty of Applied Biological Sciences, Gifu University, Yanagido, Gifu, Japan

Towhida Rashid Department of Geography and Environment, University of Dhaka, Dhaka, Bangladesh

Toru Terao Faculty of Education, Kagawa University, Takamatsu, Kagawa, Japan

Peter H. Verburg Institute of Environmental Studies, VU University, Amsterdam, The Netherlands

Yusuke Yamane Faculty of Education, Tokoha Gakuen University, Shizuoka City, Shizuoka, Japan

Acronyms

ADB	Asian Development Bank
ADP	Annual Development Program
AEC	Atomic Energy Centre
AHP	Analytic hierarchy process
AIC	Akaike information criterion
AMIP	Atmospheric Model Intercomparison Project
AOI	Area of interest
AR	Assessment Report
ASCII	American Standard Code for Information Interchange
ASTER	Advanced Spaceborne Thermal Emission and Reflection Radiometer
ATMoS	Atmospheric Transport Modelling System
AVHRR	Advanced Very High Resolution Radiometer
BBS	Bangladesh Bureau of Statistics
BCAS	Bangladesh Centre for Advanced Studies
BDT	Bangladesh Taka (*name of Bangladeshi currency*)
BLD	Bangladesh Legal Digest
BLHI	Boundary layer heat island
BMD	Bangladesh Meteorological Department
BNBC	Bangladesh National Building Code
BOD	Biological oxygen demand
BP	Bangladesh Police
BPDB	Bangladesh Power Development Board
BTM	Bangladesh Transverse Mercator
BWDB	Bangladesh Water Development Board
CA	Cellular automata
CBD	Central business district
CCI	Coping Capacity Index
CEGIS	Centre for Environmental and Geographic Information Services
CETP	Common Effluent Treatment Plant
CLHI	Canopy layer heat island

CLUE	The Conversion of Land Use and Its Effects
CNG	Compressed natural gas
CO	Carbon monoxide
COPD	Chronic obstructive pulmonary disease
CVI	Composite Vulnerability Index
DAP	Detailed Area Plan
DCC	Dhaka City Corporation
DEM	Digital Elevation Model
DESA	Dhaka Electricity Supply Authority
DESCO	Dhaka Electric Supply Company
DIT	Dacca Improvement Trust
DL	Danger level
DM	Dhaka Megacity
DMA	Dhaka Metropolitan Area
DMAIUDP	Dhaka Metropolitan Area Integrated Urban Development Plan
DMB	Disaster Management Bureau
DMDP	Dhaka Metropolitan Development Plan
DMP	Dhaka Metropolitan Police
DN	Digital number
DND	Dhaka-Narayangonj-Demra
DNSC	Dhaka North City Corporation
DO	Dissolved oxygen
DOE	Department of Energy
DOE	Department of Environment
DPDC	Dhaka Power Distribution Company
DSCC	Dhaka South City Corporation
DSMA	Dhaka Statistical Metropolitan Area
ECA	Environmental Conservation Act
ECR	Environmental Conservation Rules
EIA	Environmental Impact Assessment
EIS	Environmental Impact Statement
EM-DAT	Emergency Events Database
ENSO	El Niño–Southern Oscillation
EO	Earth observation
EREC	European Renewable Energy Council
ERTS	Earth Resources Technology Satellite
ESDA	Exploratory spatial data analysis
ETM+	Enhanced Thematic Mapper Plus
FAP	Flood Action Plan
FCBTK	Fixed Chimney Bull Trench Kilns
FEM	Finite element method
GBM	Ganges-Brahmaputra-Meghna
GCP	Gross city products
GCP	Ground control points

GDP	Gross domestic product
GHI	Global Horizontal Irradiance
GIS	Geographic Information Science
GIS	Geographic Information Systems
GISTDA	Geo-Informatics and Space Technology Development Agency
GNI	Gross national income
GOB	Government of Bangladesh
GPS	Global Positioning System
GS	Grameen Shakti
GWR	Geographically weighted regression
HWL	Highest water level
ICDDR,B	International Centre for Diarrhoeal Disease Research, Bangladesh
IGBP	International Geosphere-Biosphere Project
IHDP	International Human Dimensions Program
IPCC	Intergovernmental Panel on Climate Change
IPPS	Independent Power Producers
IRS	Indian Remote Sensing
IRSO	Indian Remote Sensing Organization
ISDR	International Strategy for Disaster Reduction
ISODATA	Iterative Self-Organizing Data Analysis Technique
KMO	Kaiser-Meyer-Olkin
LDC	Less developed country
LIDAR	Light detection and ranging
LISA	Local indicators of spatial autocorrelation
LISS	Linear Imaging Self-Scanning Sensor
LST	Land surface temperature
LULC	Land use and land cover
MAUP	Modifiable area unit problem
MCA	Markov chain analysis
MCE	Multi-criteria evaluation
MESSR	Multispectral Electronic Self-Scanning Radiometer
MLC	Maximum Likelihood Classifier
MODIS	Moderate Resolution Imaging Spectroradiometer
MOEF	Ministry of Environment and Forest
MSS	Multispectral Scanner
MSW	Municipal solid waste
MW	Megawatt
NCEP	National Centers for Environmental Predictions
NDVI	Normalised Difference Vegetation Index
NDWI	Normalized Difference Water Index
NGO	Non-governmental organization
NOAA	National Oceanographic and Atmospheric Administration
NOC	No objection certificate

NO_x	Nitrogen oxides
OBIA	Object-based image analysis
OLS	Ordinary least square
OUA	Other urban area
PCA	Principle component analysis
PGCB	Power Grid Company of Bangladesh
PM	Particulate matter
PV	Photovoltaic
PVI	Physical Vulnerability Index
QOL	Quality of life
RADAR	Radio detection and ranging
RAJUK	Rajdhani Unnayan Katropakka (Capital Development Authority)
REB	Rural Electrification Board
RERC	Renewable Energy Research Centre
RMSE	Root mean square error
ROC	Relative operating characteristic
RS	Remote sensing
SAR	Synthetic aperture radar
SDD	Secchi disk depth
SDT	Secchi disk transparency
SEA	Strategic environmental assessment
SMA	Statistical Metropolitan Area
SO_2	Sulphur dioxide
SOB	Survey of Bangladesh
SPARRSO	Space Research and Remote Sensing Organization
SPSS	Statistical Package for the Social Sciences
SPOT	Système Probatoire d'Observation de la Terre
SPZ	Spatial planning zones
SRTM	Shuttle Radar Topography Mission
SUHI	Surface urban heat island
SVI	Social Vulnerability Index
SWERA	Solar and Wind Energy Assessment
TD	Transform divergence
TIGER	Topologically Integrated Geographic Encoding and Referencing
TIROS	Television Infrared Observation Satellite Program
TM	Thematic Mapper
TSS	Total suspended sediment
UGAT	Urban Growth Analysis Tool
UHI	Urban heat island
UN	United Nations
UNCHS	United Nations Centre for Human Settlements
UNDP	United Nations Development Programme
UNDRCO	United Nations Disaster Relief Coordinator
UNEP	United Nations Environment Programme

UNFCCC	United Nations Framework Convention on Climate Change
UN-HABITAT	United Nations Human Settlements Programme
UTM	Universal Transverse Mercator
VIF	Variance inflation factor
VOC	Volatile organic compounds
WASA	Water and Sewerage Authority
WB	World Bank
WFP	World Food Programme
WHO	World Health Organization
WiFS	Wide Field Sensors
WL	Water level
WLC	Weighted linear combination
WMO	World Meteorological Organization

List of Figures

Fig. 1.1 Spatial distribution of megacities and their population growth since 1970 ... 5
Fig. 1.2 Different spatial representations of Dhaka Megacity ... 7
Fig. 1.3 Location of the study area ... 14

Fig. 2.1 Evolution of Dhaka: (**a**) Pre-Mughal period, (**b**) Mughal period, (**c**) British period, (**d**) capital of East Bengal and Assam and (**e**) evolution of Dhaka at a glance, 1600–1980 ... 26
Fig. 2.2 DMDP Structure Plan 1995–2015 ... 34

Fig. 3.1 Density of population between 1991 and 2011, by distance category ... 53
Fig. 3.2 Population distribution and SDE in three census periods (**a**) 1991, (**b**) 2001, (**c**) 2011 ... 54
Fig. 3.3 Changes in population between 1991 and 2011, by census tracts ... 55
Fig. 3.4 Average change in population each year, 1991–2011 ... 57

Fig. 4.1 Annual temperature anomaly during 1953–2008 ... 65
Fig. 4.2 As in Fig. 4.1 for the pre-monsoon season (March–May) ... 66
Fig. 4.3 As in Fig. 4.1 for the monsoon season (June–Sept) ... 66
Fig. 4.4 As in Fig. 4.1 for the post-monsoon season (Oct–Nov) ... 67
Fig. 4.5 As in Fig. 4.1 for the winter season (Dec–Feb) ... 67
Fig. 4.6 Annual rainfall anomaly during 1953–2008 ... 70
Fig. 4.7 As in Fig. 4.6 for the pre-monsoon season (March–May) ... 71
Fig. 4.8 As in Fig. 4.6 for the monsoon season (June–Sept) ... 71
Fig. 4.9 As in Fig. 4.6 for the post-monsoon season (Oct–Nov) ... 72
Fig. 4.10 As in Fig. 4.6 for the winter season (Dec–Feb) ... 72

Fig. 5.1 Observed and simulated LULC between 1990 and 2011 ... 84
Fig. 5.2 Area percentage of actual versus simulated LULC categories in 2011 ... 89
Fig. 5.3 Simulated LULC maps of 2022 and 2033 ... 90

Fig. 6.1 False Colour Composite (*FCC*) of Landsat TM5 of April 2011 of the study area showing CBD and mean population centre in 2001 and 2011 ... 103
Fig. 6.2 Urban expansion between 2000 and 2011 with directional sectors ... 106
Fig. 6.3 Changes in built-up surface with ten annular buffers ... 107
Fig. 6.4 Direction of urban growth, 2000–2011 ... 108
Fig. 6.5 Urban growth types ... 109
Fig. 6.6 Density of urban areas between 2000 and 2011 ... 111
Fig. 6.7 Population density at census tracts in 2011 and change since 2001 ... 113
Fig. 6.8 Characteristics of the ten annular buffers in 2001 and 2011: (**a**) mean population density, (**b**) proportion of built-up surface, and (**c**) population per unit area of built-up surface ... 116

Fig. 7.1 Detailed modelling framework of the CLUE-s model ... 126
Fig. 7.2 An example of a network-based accessibility surface ... 131
Fig. 7.3 Restricted planning variables: (**a**) areas restricted for development, (**b**) areas promoted for development, and (**c**) fringe areas for accelerated development ... 132
Fig. 7.4 Size and weights for 21 × 21 cell neighbourhood ... 133
Fig. 7.5 Mean enrichment factors for built-up areas in 1999 ... 136
Fig. 7.6 Probability maps with neighbourhood effects: (**a**) t1–t2 and (**b**) t2–t3 ... 137
Fig. 7.7 Observed and simulated land-use for 1999 and 2005 ... 138

Fig. 8.1 Flowchart of urban suitability analysis ... 150
Fig. 8.2 Suitability potential for urban development in the study area ... 156

Fig. 9.1 Location of the study site ... 165
Fig. 9.2 Comparison of land-cover patterns between 1990 and 2011 ... 166
Fig. 9.3 Structure and parameters required to run tank model ... 167
Fig. 9.4 Comparison between calculated and observed river water levels at station 1 ... 170
Fig. 9.5 Change of water depth under conditions of 1990 land use and complete land-use conversion ... 171
Fig. 9.6 Comparison of inundated area between conditions of: (**a**) 1990 land use and (**b**) complete land-use conversion ... 172
Fig. 9.7 Comparison of inundated nodes between conditions of: (**a**) 2011 land use and (**b**) complete land-use conversion ... 173
Fig. 9.8 Change of water depth between conditions of 2011 land use and complete land use conversion ... 174

Fig. 10.1 Flood risk and vulnerability analysis model ... 191
Fig. 10.2 Spatial distribution of flood hazard ... 192

Fig. 10.3 Spatial patterns of vulnerability: (**a**) locational vulnerability, (**b**) socioeconomic vulnerability, (**c**) built environmental vulnerability, and (**d**) coping capacity, by community 194
Fig. 10.4 Spatial patterns of total vulnerability, by community 195
Fig. 10.5 Spatial patterns of flood risk, by community 197

Fig. 11.1 Well-illuminated roof areas 210
Fig. 11.2 A Zoomed-in well-illuminated roof areas in central Dhaka 211

Fig. 12.1 Location of the study area shown in natural colour composite of Landsat TM 2011 223
Fig. 12.2 Major LULC conversions in the study area 227
Fig. 12.3 Relative changes in LULC, 1990–2011 228
Fig. 12.4 Spatial patterns of LST (°C) and corresponding NDVI, 1990–2011 230
Fig. 12.5 Spatial distribution of temperature increase between 2000 and 2011 232
Fig. 12.6 Relationship between annual mean LST and F_r 233

Fig. 13.1 The first factor scores: environmental index 248
Fig. 13.2 The second factor scores: economic index 250
Fig. 13.3 The third factor scores: crowdedness 251
Fig. 13.4 Synthetic QOL index 252

Fig. 14.1 Seasonal variations in the occurrences of crime 268
Fig. 14.2 Number of crimes in DMA, by month 269
Fig. 14.3 Spatial variation of raw incidence rate of crimes 270
Fig. 14.4 Spatial distribution of the number of different crimes in DMA 272
Fig. 14.5 Spatial clusters (hotspots) of crime in DMA during August 2011–July 2012 273

Fig. 15.1 Legal framework of EIA in Bangladesh 288
Fig. 15.2 Major features of the Act and the Rules 289
Fig. 15.3 Screening and environmental clearance process of DOE 291
Fig. 15.4 Key strengths and weaknesses of legal and organisational arrangements of EIA in Bangladesh 292

Fig. 16.1 Distribution of Secchi disk transparency (SDT) in the rivers and lakes on 2 February 2011 312
Fig. 16.2 Distribution of TSS in the rivers and lakes on 2 February 2011 313

Fig. 17.1 Monthly average ambient $PM_{2.5}$ concentrations (mg/m^3) measured near Sangsad Bhaban (SB – traffic junction) and at the Atomic Energy Centre (AECD – campus); the average concentrations are segregated between brick manufacturing season (October to March) and the remaining months 320

Fig. 17.2 Modelling domain covering Dhaka region, along with the location of the brick kiln clusters and monitoring stations 322
Fig. 17.3 Wind speed and wind direction for year 2008; monthly precipitation totals (mm/h) and daily average harmonic mean mixing height (metres) estimated from the National Centers for Environmental Prediction (NCEP) reanalysis data for years 2004 and 2008 for the grid covering Dhaka region 325
Fig. 17.4 Modelled $PM_{2.5}$ (mg/m^3) concentrations averaged over the brick manufacturing season for Dhaka region 326
Fig. 17.5 Modelled $PM_{2.5}$ (mg/m^3) concentrations averaged over the brick manufacturing season for the emissions originating from individual brick kiln clusters in Dhaka region 328

Fig. 18.1 Seasonal variations in the number of cases of *Aeromonas*-positive diarrhoea per week and meteorological and river level data in Dhaka, 1966–2000. The river level data are missing from April 1997 to March 1998 338
Fig. 18.2 Relationship between the number of cases of *Aeromonas*-positive diarrhoea and average rainfall over lags of (**a**) 0–8 weeks, (**b**) 9–16 weeks and (**c**) 0–16 weeks (shown as a 3 df natural cubic spline) adjusted for seasonal variation, between-year variations and temperature. *RR* represents the relative risk of *Aeromonas*-positive diarrhoea. The centre line shows the estimated spline curve, and the upper and lower lines represent the 95 % CIs 339
Fig. 18.3 Percentage change (and 95 % CIs) in the number of cases of *Aeromonas*-positive diarrhoea for rainfall (per 10-mm increase) at each lag 340

Fig. 19.1 Location of hospitals studied 348
Fig. 19.2 Spatial clusters (*hotspots*) of typhoid cases in Dhaka (2005–2009). Hot and cold spots are measured by local GI* for each year as well as for the entire period 353
Fig. 19.3 The distribution of standardised residuals of typhoid cases from GWR as a function of demographic, socioeconomic and demographic + socioeconomic variables 357

Fig. 20.1 Frequency of dengue cases by census tract, 2005–2010 371
Fig. 20.2 Seasonal distribution of dengue occurrences 374
Fig. 20.3 Standard deviation ellipsoids of dengue occurrences, 2005–2009 376
Fig. 20.4 Moran's *I* scatterplots for total dengue frequency 377
Fig. 20.5 Spatial clustering of dengue, 2005–2010 379
Fig. 20.6 Dengue significance map 380

List of Plates

Plate 2.1 Dhalleshori temple in 1904 25
Plate 2.2 On the way to Tongi 27
Plate 2.3 Mughal ruins in the 1830s: (**a**) Bara Katra, (**b**) Tantibazar bridge, and (**c**) Saat Gumbad Masjid 29
Plate 2.4 Dhaka College in 1904 29
Plate 2.5 Slums located next to the railway line 38
Plate 2.6 Pedestrian congestion in the inner city area of Gulistan 39

List of Tables

Table 1.1 Demographic characteristics of Dhaka City Corporation (DCC) area 8
Table 1.2 Demographic characteristics of Dhaka Megacity 9
Table 1.3 Demographic characteristics of the study area, 2011 16

Table 2.1 Choice of travel mode in relation to distance to work place ... 37

Table 3.1 Descriptive statistics of population characteristics in the study area 51
Table 3.2 Monocentric density functions in the study area 52

Table 4.1 Annual and seasonal temperature variations from 1953 to 2008 64
Table 4.2 Trends of temperature anomalies 68
Table 4.3 Annual and seasonal rainfall variations from 1953 to 2008 69

Table 5.1 Land-use/land-cover classification scheme 79
Table 5.2 Accuracy assessment of LULC classification 82
Table 5.3 Results of LULC classification for 1990, 2000 and 2011 images showing area of each category, category percentage and area changed 86
Table 5.4 LULC transition probabilities, 1990–2000 87
Table 5.5 LULC transition probabilities, 2000–2011 88
Table 5.6 Area statistics of Markov-CA projected LULC for 2022 and 2033 90

Table 6.1 Urban growth types between 2000 and 2011 108
Table 6.2 Built-up area density and Shannon's entropy between 2000 and 2011 110
Table 6.3 Population distribution in annular buffer zones around the 2001 weighted population centre 115

Table 7.1 Logistic regression results for driving factors 134
Table 7.2 Conversion resistance factor values 136

Table 7.3 Conversion matrix for CLUE-s simulation 136
Table 7.4 Calibration and validation results compared to urban 1999 and urban 2005 139

Table 8.1 Database specifications 152
Table 8.2 Relative weighting of principal factors 153
Table 8.3 Relative weighting of sub-criteria 154
Table 8.4 Area and percentage suitability classes 157

Table 9.1 Optimised parameter values and Nash-Sutcliffe criteria 170

Table 10.1 Definitions of the key terminology used in this study 182
Table 10.2 Combinations of classes from maps of flood-affected frequency and floodwater depth, with assigned hazard ranks 185
Table 10.3 Decision hierarchy model and corresponding weights for physical vulnerability criteria 187
Table 10.4 Decision hierarchy model and corresponding weights for socioeconomic vulnerability criteria 188
Table 10.5 Decision hierarchy model and corresponding weights for built environmental vulnerability criteria 189
Table 10.6 Decision hierarchy model and corresponding weights for coping capacity criteria 190
Table 10.7 Area and percentage of population in each flood hazard zone 193
Table 10.8 Distribution of different parameters in relation to total vulnerability 196
Table 10.9 Distribution of population and other variables in different flood risk zones 198

Table 11.1 Calculation of well-illuminated roof area from the Quickbird image of 2006 211
Table 11.2 Solar PV-based power generation potential 212

Table 12.1 Major land use/cover conversions from 1990 to 2011 229
Table 12.2 Summary of retrieved LST from 1990 to 2011 229
Table 12.3 Mean normalised surface temperature by land use/cover type 231
Table 12.4 Pearson correlation coefficients between LST and NDVI by LULC type 233

Table 13.1 Correlation matrix 245
Table 13.2 Communality of 13 and 10 variables 247
Table 13.3 Rotated factor loading matrix 247
Table 13.4 QOL categories with different socioeconomic and environmental parameters 253
Table 13.5 Regression estimation of QOL models 253

Table 14.1 Crime statistics in Bangladesh, 2001–2010 261
Table 14.2 Crime statistics in DMA between August 2011 and July 2012, by crime type 268
Table 14.3 Coefficients of causal factors and crime incidence using linear regression 275

Table 16.1 Correlation between SDT, logarithmic transformation of SDT, and TSS with TM bands 308
Table 16.2 Coefficient of determination (r^2) and multiple regression equations for LN(SDT) and TSS 310

Table 17.1 Comparison of current and alternative brick manufacturing technologies available in Bangladesh 322
Table 17.2 Estimated emissions from brick kiln clusters in Dhaka region 324
Table 17.3 Model average $PM_{2.5}$ concentrations due to brick kiln emissions in Dhaka region 326

Table 18.1 Distribution of diarrhoea by pathogens by age in ICDDR,B Dhaka Hospital, 1996–2000 337
Table 18.2 Percentage change (and 95 % CIs) in weekly number of *Aeromonas*-positive diarrhoea for 10-mm increase in rainfall at lags of 0–16 weeks 341

Table 19.1 Explanatory variables used in the study 350
Table 19.2 Global autocorrelation analyses of typhoid data 352
Table 19.3 Ordinary least square (OLS) results 355
Table 19.4 Comparison of OLS and GWR results 356

Table 20.1 Annual dengue occurrences, 2005–2010 374
Table 20.2 Standardised age- and gender-specific incidences 375
Table 20.3 Spatial autocorrelation for dengue 378

Chapter 1
Introduction

Robert J. Corner and Ashraf M. Dewan

Abstract This book contains a series of chapters that describe various aspects of the megacity of Dhaka, the capital of Bangladesh. Bangladesh is a low-lying maritime nation in South Asia historically part of Bengal and governed until 1947 as part of the British colonial Indian Empire. When the constituent provinces of the Indian Empire were granted independence in 1947, Bengal was partitioned, with the eastern part becoming part of the new Dominion of Pakistan. Partition of the province caused considerable upheaval as the Dominion of Pakistan was intended as a homeland for Muslims on the Indian subcontinent and in time became the Islamic Republic of Pakistan. This resulted in a great deal of internal migration within the former provinces. Initially, Pakistan comprised two provinces, East and West Pakistan, but in 1971 East Pakistan obtained its independence as the separate sovereign nation of Bangladesh. Although the population of Bangladesh is predominately Muslim, the country currently has a secular constitution.

Keywords Megacities • Rural-urban migration • Population growth • Environmental degradation

1.1 Bangladesh: An Overview

Bangladesh lies between 20.34° and 26.38° N latitude and between 88.01° and 92.41° E longitude. It is bounded by India to the north, west and east, Myanmar to the southeast and the Bay of Bengal to the south. The total area of Bangladesh is 147,570 km^2 of which 119,624 km^2 (81.0 %) is effective land area, 8,236 km^2 (5.6 %) is occupied by river networks and the forest area consists of 19,710 km^2 (13.4 %) (BBS 2011a).

R.J. Corner (✉) • A.M. Dewan
Department of Spatial Sciences, Curtin University,
Kent Street, Bentley, Perth 6102, WA, Australia
e-mail: R.Corner@curtin.edu.au; A.Dewan@curtin.edu.au

A. Dewan and R. Corner (eds.), *Dhaka Megacity: Geospatial Perspectives on Urbanisation, Environment and Health*, Springer Geography,
DOI 10.1007/978-94-007-6735-5_1,

Bangladesh can be divided into three major physiographic regions: hills, terraces and floodplains. Floodplains constitute the lion's share of the country (80 %) and have been formed by the deposition of alluvium by three mighty rivers (the Ganges, Brahmaputra and the Meghna, collectively known as GBM) and their tributaries and distributaries; the northern and southern hilly areas of Tertiary and Quaternary age occupy about 12 % of the country, whilst only 8 % of country is made up of terrace areas (Brammer 2002). The country has a humid subtropical climate with wide seasonal variations in rainfall, generally warm temperatures and high humidity (Rashid 1991). There are three distinct seasons, being summer (March–May), the monsoon (June–October) and the winter or dry season (November–February). However, from a meteorological point of view, the year has been categorised into four seasons. They are pre-monsoon (March–May), monsoon (June–September), post-monsoon (October–November) and winter, which starts in December and ends in February (see Chap. 4). The pre-monsoon season is also known as summer and is the hottest part of the year, characterised by high evaporation rates. The mean temperature ranges from 27 to 31 °C during this period. Rainfall varies spatially with the frequent occurrence of thunderstorms which provide about 10–15 % of the annual total. Abundant rainfall occurs in the monsoon season with the influence of southerly or south-westerly winds. Temperature drops with the onset of the summer monsoon and approximately 75–80 % of total rainfall occurs during the rainy season. In contrast, the winter season is cool and dry with little rainfall. The mean winter temperature is 18 °C with the lowest being 3–5 °C in the northwest part of the country, and only 5 % of annual rainfall occurs during this season. This is the sunniest period in Bangladesh with high solar radiation. The hydrographic network of the country is very dense and more than 250 rivers crisscross the country (Chowdhury and Salehin 1987). These rivers act as lifelines in some of the areas and have supported agriculture, navigation, groundwater recharge, fish habitats and land-building activities since time immemorial.

The economy of Bangladesh is mostly agrarian with 48 % of the labour force engaged in agriculture and related activities (BBS 2010a). A provisional estimate shows that the agricultural sector, together with fishing and forestry, contributed 18.43 % to the GDP (Gross Domestic Product) in 2010–2011 at current market prices, whilst the contribution from the manufacturing sector including large-, medium- and small-scale industries was 18.17 % (BBS 2012a). Currently, the total GDP of the country is 110.6 billion US$ with the GDP *per capita* being 770 US$ (World Bank 2012). Based on the most recent Household Income and Expenditure Survey (HIES), the incidence of poverty is declining with time. For example, the rate of poverty was 40 % in 2005 and decreased to about 31.5 % in 2010 at national level. Similar results can be found for the rural and urban poverty rates (BBS 2011b). According to the most recent population and housing census (BBS 2012b), the literacy rate for males and females is 54.1 and 49.4 %, respectively, at the national level.

The ever-growing population in Bangladesh is currently placing enormous pressure on natural resources. Despite the fact that the annual growth rate has been reduced significantly, the size of population is still large when compared with the size of the country. As a result, both landlessness and environmental degradation have become pervasive in recent times (Choudhury 2008). The available statistics show that the population of Bangladesh was 71.4 million in 1974 with an annual growth rate of 3.5 % (BBS 1984) which increased to 87.1 million in 1981, 111.4 million in 1991 and 123.9 million in 2001 (BBS 1994, 2004). According to the 2011 population and housing census, the total population of the country is 149 million with an annual growth rate of 1.4 % (BBS 2012b). This shows that the total population of the country has increased by 72 million between 1974 and 2011. The density of population has also increased from 590/km^2 in 1981 to 976/km^2 in 2011. If the current trend continues, the population of Bangladesh is expected to reach to about 194 million in 2050 (UN 2012). Understandably, this population exerts tremendous pressure on a limited resource base. For instance, the total proportion of land holdings being farmed has been reduced from 72.7 % in 1983 to about 58.6 % in 2008 (BBS 2010a).

Along with national population growth, the percentage of people living in urban areas has also been rising since the country's independence from Pakistan in 1971 (Islam 2005a). For example, 1.8 million people were living in urban areas in 1951 which increased to 13.5 million in 1981, 22.5 million in 1991, 31 million in 2001 and 33.5 million in 2011 (BBS 2008, 2012a, b). This overwhelming growth in urban population is largely attributed to rural–urban migration, natural growth and the redefinition of urban areas (BBS 2008); however, it is asserted by many that rural–urban migration is the most dominant factor (Rouf and Jahan 2007; Islam 2005a). Because of such growth in urban population and associated infrastructure, every year Bangladesh loses 0.3 % of its cultivated land (BBS 2010a). Furthermore, the country faces enormous challenges in coping with the infrastructures and service requirements for its rapidly growing urban population (Rashid 2008). As a result, widespread environmental degradation has become a common feature in the urban areas of the country, particularly in the capital city – Dhaka (Dewan et al. 2012; Siddiqui et al. 2010; Islam 2005a).

Apart from rapid population growth, the probable effects of climate change have now become a matter of serious concern in Bangladesh in spite of its tiny contribution to the global carbon dioxide emission, at 0.3 metric tonne per capita per annum (World Bank 2012). Bangladesh's vulnerability to climate change is further aggravated by its geographical location, weak economy and low capacity to address the devastating impacts (Shahid 2010; Mirza 2009). Studies have demonstrated that the annual average rainfall of Bangladesh is increasing at a rate of 6.58 mm/year (Shahid 2011a) which has major implication for future flooding in the country. A 23–29 % increase in flooded areas is predicted for Bangladesh if global temperature rises by 2 °C (Mirza et al. 2003). In another study, Shahid (2012) predicted an average increase in temperature of 1.4 °C in 2050 and 2.4 °C in 2100. Rising temperature would seriously affect current agricultural practices (Shahid 2011b) and could severely disrupt food production in the region (Douglas 2009). A one-metre rise in sea level could inundate as much as 16 % of Bangladesh (Ahmed 2009) and might displace nearly 30 million people (BBS 2009 in Mirza 2009). Furthermore, global warming may increase the frequency of tropical cyclones and storm surges in the coast. A study by Islam and Peterson (2009) indicated that the rate of tropical storms making landfall has increased by 1.18 per year since 1950 which may be linked with global warming.

1.2 Definition of a Megacity

Half of the world's total population was living in urban areas in 2008 and this trend is expected to continue (Satterthwaite et al. 2010). It is projected that the world urban population will increase by 72 % by 2050, from 3.6 billion in 2011 to 6.3 billion in 2050 of which the majority of the population will be concentrated in the urban areas of the less developed regions (UN 2012). As a result, the number of so called 'megacities' will rise. Even though the concept of megacity evolved at the end of the twentieth century to describe the large urban agglomerations of the world, inconsistencies exist in literature with regard to the population threshold used to define a megacity. For example, Mitchell (1998) defines a megacity as having a population exceeding one million, whilst the UN (2003) defines a megacity as a conurbation that houses ten million people or more. Others have used a five million or an eight million population threshold to characterise megacities (Varis et al. 2006; Nicholls 1995). Nevertheless, in the1970s there were only two megacities with a population of more than ten million, which increased to 10 in 1990. Currently, 9.9 % of the world urban population lives in 23 megacities which is projected to increase to 37 in 2025 when they are expected to accommodate 13.6 % of the world urban population (UN 2012). Further estimates suggest that the number of people living in megacities has increased almost tenfold in the past 40 years, from 39.5 million in 1970 to 359.4 million in 2011, and could double again by 2025 (UN 2012). The largest increase in urban population between now and 2050 is expected to be concentrated in Asia and Africa (Seto et al. 2010). Figure 1.1 shows the development of urban populations in megacities of the world based on UN estimates (UN 2012). A distinct feature that can be observed from this figure is the fact that population growth in megacities such as London or New York in developed countries has been very slow between 1970 and 2011. However, an opposite picture can be seen for megacities in developing countries. India, for example, already has three megacities (Delhi, Mumbai and Kolkata) and with the addition of Bangalore, Chennai and Hyderabad is expected to have six by 2020 – meaning that India will have the largest concentration of megacities in the world (Chakrabati 2001). This gives some idea of the pressure of population growth in cities of developing countries.

1.3 Environmental Setting of Dhaka Megacity

Dhaka, the capital of the People's Republic of Bangladesh, first attracted attention when it became the provincial capital in 1905 (Rizvi 1969). After the partition of the Indian subcontinent in 1947 into India and Pakistan, Dhaka became the capital of the then East Pakistan. Bangladesh emerged as an independent state in 1971, and the country retained Dhaka as its capital. Since then, Dhaka has been the major focus of administrative, social, educational and cultural activities.

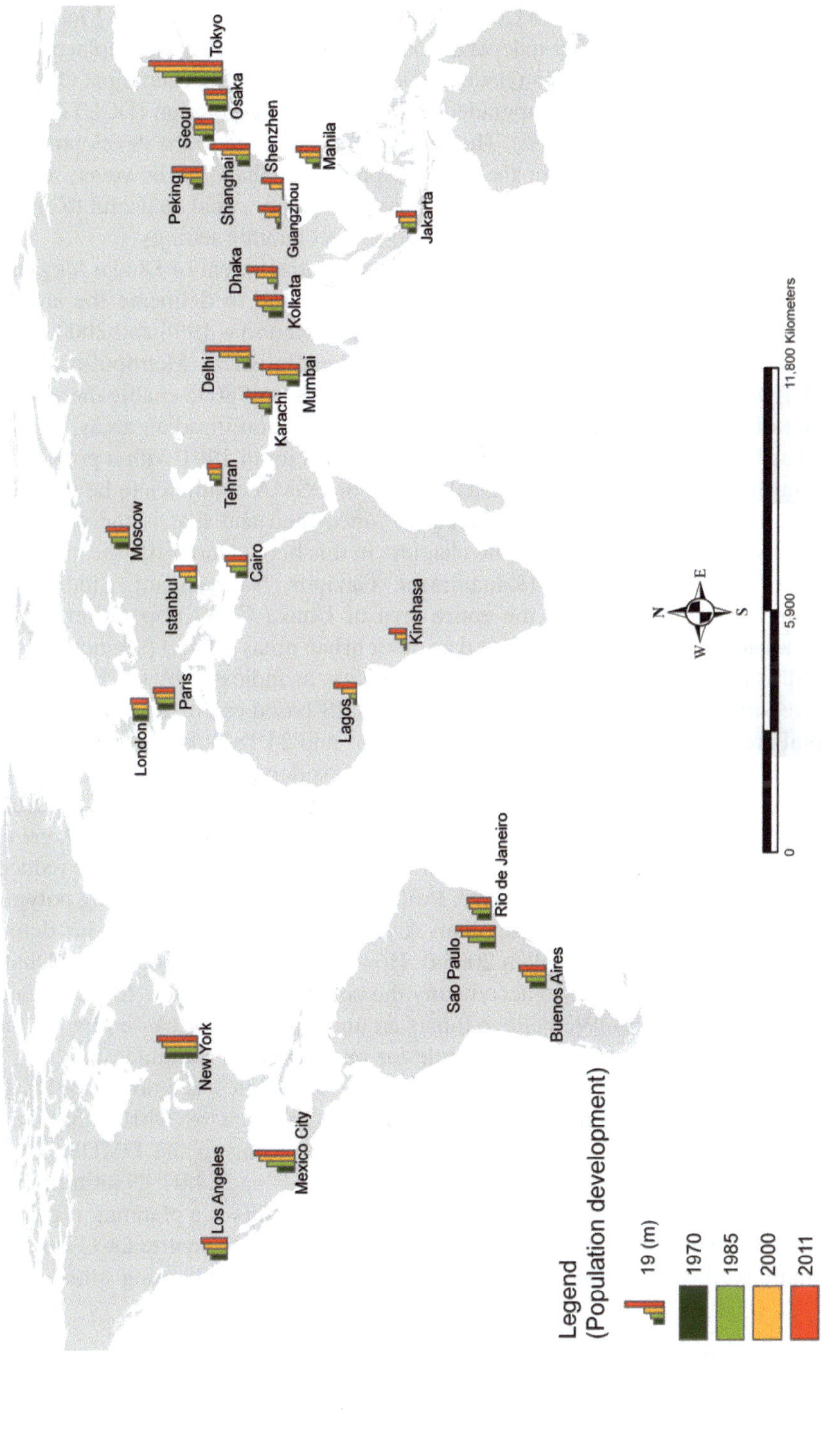

Fig. 1.1 Spatial distribution of megacities and their population growth since 1970 (Adapted from UN 2012)

The establishment of the municipality in Dhaka can be traced back to 1830. At that time, it was known as the 'Dhaka Committee' and later renamed to Dhaka Municipality in 1864 (Rizvi 1969). After independence, Dhaka Municipality was replaced with Dhaka Pourashava under the Bangladesh Local Council and the Municipal Order of 1972, which was subsequently upgraded to Dhaka City Corporation (DCC) in 1991 (Banglapedia 2006; Ahmed 1986). The historical background to the development of Dhaka Megacity is provided in the next chapter (see Chap. 2); however, a brief overview on the environmental setting of Dhaka Megacity could be useful to enable readers to grasp its spatial, environmental and socioeconomic settings.

There is uncertainty as to what constitutes the spatial extent of Dhaka Megacity because a number of different methods have been used to delineate the area of Dhaka (Islam 2005a). According to the Urban Area Report – 1998 and 2008 of the Bangladesh Bureau of Statistics (BBS) – Dhaka Statistical Metropolitan Area (DSMA) was created by the census authorities of BBS in 1980 to enable the smooth conduct of census. Because of rapid growth of population in urban areas, 'Dhaka SMA has been elevated to the status of Dhaka Megacity in 1991 with a population of 6.8 million' (BBS 1998, p. 78). Dhaka SMA or DSMA continues to be regarded as the territorial extent of Dhaka Megacity since then and that is the definition adopted by the authors of most of the chapters in this book. Currently, the megacity comprises six municipalities (Kadamrasul, Gazipur, Narayanganj, Siddirganj[1], Savar and Tongi) (Fig. 1.2a), the entire area of Dhaka City Cooperation (DCC) and 68 adjacent *unions* that are termed as other urban areas (OUA) [see notes at the end of this chapter]. The total area of the megacity, as indicated by the BBS Urban Area Report – 2008 – is 1,371 km^2; however, a GIS-based calculation revealed that the total area is 1,383 km^2. It lies between 23.55° and 24.18°N latitudes and 90.18° and 90.57° E longitudes (Fig. 1.2a). If one considers the boundary of Dhaka Megacity/DSMA defined by Professor Nazrul Islam, a distinguished urban researcher in Bangladesh, in his book *Dhaka Now: Contemporary Urban Development*, then the spatial extent of the area considered to be the megacity reduces substantially to a total area of 930 km^2. Professor Islam used the bounding polygon shown as Dhaka Statistical Metropolitan Area (DSMA) on Fig. 1.2b as his definition of Dhaka Megacity (see Islam 2005a). However, the DSMA polygon or Dhaka Megacity published by BBS to accompany the population census in 2001 indicates a larger area (Fig. 1.2b). We believe this is an unfortunate error caused by the fact that access to digital data is a great hurdle for researchers in Bangladesh.

Another spatial extent that is often used by academics is known as Dhaka Metropolitan Development Plan (DMDP). The total area of DMDP has been calculated with a GIS as being 1,472 km^2, but according to the DMDP (1997) report, the area is 1,528 km^2 and located between 23.54° and 24.04° N latitudes and 90.22° and 90.60° E longitudes (see Fig. 1.2b). In fact, this is a planning region of the Capital Development Authority (RAJUK) that includes the entire DCC area, the six municipalities mentioned previously and a number of adjoining other urban

[1] This municipality was declared in 2012. At the time of writing, its boundary had not been published digitally.

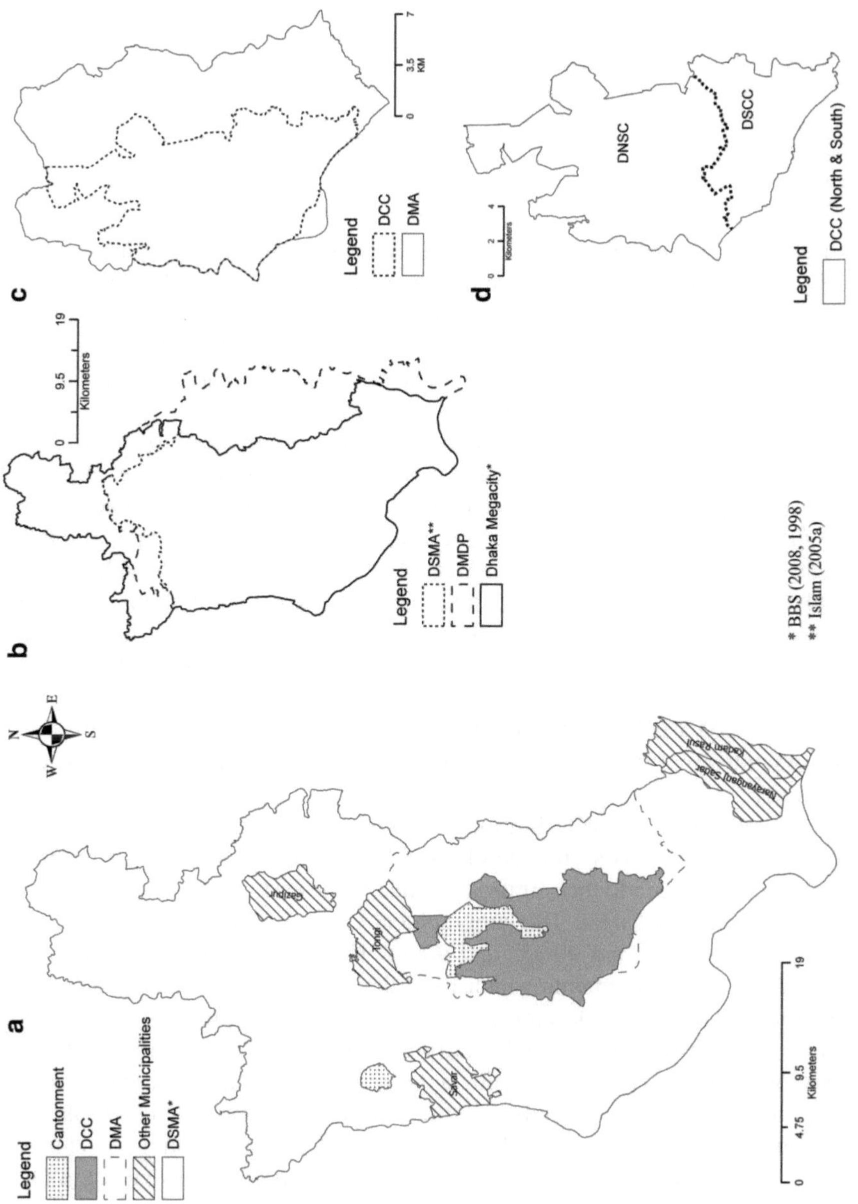

Fig. 1.2 Different spatial representations of Dhaka Megacity

Table 1.1 Demographic characteristics of Dhaka City Corporation (DCC) area

Year	Total households	Population	Density (km^2)	Sex ratio	Household size (average)
1961	61,983	368,575	10,840	145	5.9
1974	197,361	1,403,259	38,979	140	7.1
1981	367,513	2,475,710	18,755	141	6.1
1991	574,807	3,612,850	23,475	131	5.6
2001	1,109,514	5,327,306	34,629	131	4.7
2011	1,580,672[a]	7,033,075	49,182	131	4.5

(Adapted from BBS 1998, 2008, 2012b)
[a]Includes general + institutional households

areas (OUA). The DMDP plan does not include some outer urban areas of the megacity, namely, the entire Mirzapur *union*, part of Kalatia, Kashimpur and Konabari *unions* of Gazipur district to the north. It also excludes parts of Dhamsana *union* and all of Shimulia *union* of Dhaka district. However, DMDP includes many OUA in Narayanganj district and part of a municipality (Kaliganj) in Gazipur district to the east and southeast, respectively (DMDP 1997).

Dhaka Metropolitan Area (DMA), another term generally used to define Dhaka (Islam 2005a), is a police jurisdiction area that embraces the entire DCC and adjacent *unions* (Fig. 1.2c). It is located between 23.67° and 23.89° N and 90.32° and 90.51° E with an area of 303 km^2. Apart from the entire DCC area, a total of 11 neighbouring *unions* are within the DMA. These *unions* are Beraid, Bhatara, Satarkul, Dakshinkhan, Uttarkhan, Matuail, Saralia, Demra, Harirampur, Shyampur and Sultanganj (BBS 2008). It is bounded by four rivers that are flowing in DMA, namely, the Buriganga to the south, Balu to the east, Turag to the west and Tongi Khal to the north. According to the 2011 population and housing census, the total population of DMA is 8,906,039 and the density is 29,392 persons per km^2 (BBS 2012b). As noted by Professor Islam, DMA is commonly referred to as 'Dhaka' by most people.

The spatial extent of DCC is much clearer and has an area of 143 km^2 as estimated using GIS. It is surrounded by the Buriganga River to the south, Tongi Khal to the north, Jurain to the southeast, Badda to the east, Turag River to the west and Mirpur to the northwest. The geographical extent of DCC is between 23.69° and 23.89° N latitudes and 90.33° and 90.44° E longitudes. It also includes the Dhaka and Mirpur cantonments which have their own administrative unit, called the Cantonment Board. In December 2011, the entire DCC area was divided into two parts, namely, Dhaka South City Corporation (DSCC) and Dhaka North City Corporation (DNSC) (Fig. 1.2d) with a view to serving its inhabitants better; however, this decision was heavily criticised by policymakers and urban planners in the country. The major demographic characteristics of the DCC area are shown in Table 1.1 which indicates that both the population and population density in DCC area are increasing with time. According to the 1961 census, the total population of DCC was about 0.37 million which increased to more than 7 million in 2011, a net increase of 6.6 million people over 50 years. In 2001, the population density was 34,629 which increased to 49,182 per km^2 in 2011, showing that DCC is heavily crowded with many people living in small spaces (Table 1.1).

Table 1.2 Demographic characteristics of Dhaka Megacity

Year	Total HH	Population	Density	Sex ratio	Literacy rate	HH size	Growth rate (%)
1951	NA	411,279	4,815	165	–	6.4	–
1961	127,710	718,766	5,796	154	–	5.6	–
1974	341,167	2,068,353	6,156	137	–	6.1	11.15
1981	527,311	3,440,147	8,547	139	48.1	6	5.22
1991	1,088,378	6,487,459	4,795	126	57	5.4	6.55
2001	1,920,682	9,672,763	7,055	125	65.1	4.6	4.08
2011[a]	3,232,683	14,509,100	10,484	113	67.3	4.1	–

NA not available, *HH* households (Adapted from BBS 1998, 2008)
[a]Derived from BBS community series 2012

There are five major river systems flowing across the megacity, namely, the Buriganga–Dhaleshwari system to the south, Bansi–Dhaleshwari system to the west, Turag–Lubundha rivers to the north and the Lakhya–Balu system to the east and southeast. The megacity and its adjoining areas are composed of alluvial terraces of the southern part of the Pleistocene 'uplands' known as the Madhupur tract and low-lying areas around the junction of the Meghna and Lakhya rivers. In the course of time, this area has been merged with and dissected by the recent floodplains on its fringe to form the present landforms of Dhaka. The major geomorphic units of the area are the highlands (or the Dhaka Terrace), the lowlands or floodplains, depressions and abandoned channels (Miah and Bazlee 1968). Geomorphic classification reveals that a relatively young floodplain constitutes the largest area, followed by the higher terraces of the Pleistocene period (Kamal and Midorikawa 2004). Low-lying swamps and marshes located in and around the city are examples of other major geomorphic features. The elevation of Dhaka Megacity ranges from 0 to 16 m above the mean sea level (MSL) (FAP 8A 1991). With the exception of the Pleistocene terraces on the northern edge and further north, 71.5 % of the land is 0–5 m above mean sea level. Around 26.7 % of the land is 6–10 m, and the rest of the land (1.4 %) is 11–16 m above the mean sea level.

The climate of the megacity can be classified as tropical monsoon and is characterised by three distinct seasons: monsoon, summer and winter. The average annual rainfall is about 2,000 mm, of which 80 % occurs during the monsoon season. The temperature during the summer months ranges between 28 and 34 ° C. In winter, the temperature ranges between 10 and 21 ° C with little rainfall.

Dhaka Megacity has extensive administrative and infrastructure facilities, as well as extensive road and telecommunication networks. Hence, it has become the focus of urban expansion and the hub of economic activities. Due to its economic and socio-political significance, marginalised rural people are attracted to the area in the hope of better employment opportunities and improved lifestyles. As a result, Dhaka Megacity has become one of the fastest-growing cities in the world, primarily driven by explosive population growth. The population was 0.41 million in 1951 and 0.71 million in 1961. By 1974, it had risen to 2.06 million, an average annual growth rate of 11.15 % (BBS 2008; Islam 2005a). Interestingly, in the 1950s the total population of Dhaka was only two thirds that of Chittagong, another large city of Bangladesh (Ghosh 2007). In 1981, the population increased to 3.44 million and reached around 6.48 million by 1991 and 9.67 million in 2001 (BBS 2001, 2003) (Table 1.2).

Currently, the megacity's population is more than 14 million, with an average annual growth rate of 4.08 % between 1991 and 2001, which outpaced the country's annual growth rate of 1.3 % (BBS 2008, 2012a, b). Dhaka was one of the top ten megacities in the world in 2011; if the current rate of population growth continues, Dhaka will exceed Beijing in size by 2025, with a projected population of 22.9 million (UN 2012).

As can be seen in Table 1.2, the average household size and the sex ratio is declining, indicating an increase in both single-family households and the number of single females in the urban workforce (Siddiqui et al. 2004). Studies indicate that the rapid growth of the urban population is mainly driven by rural–urban migration. Islam (1991) reported that more than 60 % of people in Dhaka are migrants. A very recent study shows that Dhaka receives 300,000–400,000 migrants every year (Sanderson 2012). The observed age-specific population supports this estimation; the number of young people is higher than other age groups, suggesting that rural–urban migration is age-selective (Afsar 2000). In addition to rural–urban migration, another explanation exists for the phenomenal growth of Dhaka after the independence of Bangladesh. For example, Dutt and Noble (2004) noted that, prior to independence, urban growth in Bangladesh was very slow owing to the fact that most of the investment during that time took place in (West) Pakistan, and Bangladesh was something of a backwater, which may have resulted in lower urban expansion prior independence. Gaining national sovereignty in 1971 allowed the country to promote Dhaka as the heart of economic and social development, thereby providing better job opportunities, wages, infrastructure and other public services, which encouraged rural people to migrate. Understandably, these migrants exert tremendous pressure on land for housing and other urban services, which in turn has profound environmental implications for Dhaka Megacity.

Extreme inequality exists between the rich and the poor in Dhaka (Siddiqui et al. 2010; Begum 2007; Islam 2005a, b), as indicated by the per capita income of US $550, which is perhaps the lowest per capita income among megacities around the world (Islam 2005a). However, according to Islam (2005a), the megacity's gross city product (GCP) accounted for 17 % of the national GDP, suggesting the influence of its economy on national development. Rich people constitute only 3 % of the total population who enjoy a high standard of living, whilst the rest are in the middle- and lower-income groups (Hossain 2006). Further, as many as 56.63 % of the households do not own land, whilst 10.24 % own only 0.0330 acre (1 *katha*) of land or less (Islam 2005a). Consequently, a significant amount of the population lives below the poverty line and struggles to survive. The literacy rate is rising steadily, and it is higher among males. According to the 2011 population and housing census, the male literacy rate is 70.2 %, whilst the female literacy rate is 63 % (BBS 2012b).

As the growth of the megacity is outstripping its economic development, the large rural–urban migration, particularly by marginalised rural people, has resulted in a rapid increase in the number of slum dwellers (Rana 2011; Angeles et al. 2009; Islam 2005b). Although the incidence of poverty has recently been reduced (BBS 2011b), the percentage of urban poor in Dhaka is still as high as 45 %, of which 25 % are classified as extremely poor (Diaspora 2007 cited in Sanderson 2012).

Estimates suggest that the slum population in Dhaka has more than doubled in the past decade, reaching 3.4 million in 2005. The number of slum clusters (nearly 5,000 at present) increased by about 70 % during the same period (CUS et al. 2006), and the population density of the slums is much higher than that of non-slum urban areas (Sanderson 2012).

Sources of drinking water in the megacity vary according to the locality, although groundwater remains the major source (Ahmed et al. 2005). Fifty-three percent of households have access to piped water, whilst 43.1 % rely on tube wells for their potable water sources (BBS 2008). The remaining households depend on deep tube wells, ponds and other sources such as rivers. It may be noted that almost all of the inhabitants have access to pipe water in the areas governed by the various municipalities compared with other areas in the megacity. With the increasing dependency on groundwater extraction, the water table is dropping rapidly. The rate of decline has reached as high as 2.5 m per year in the vicinity of the Central Business District (CBD) (Hoque et al. 2007).

Due to increasing anthropogenic activities, the quality of surface water has been deteriorating in the rivers and other water bodies such as lakes and canals (Dewan et al. 2012). As Dhaka hosts more than 40 % of the country's industries (BBS 2005) and 75 % of the export-oriented garment industry (Islam 2005a), industrial effluents along with domestic wastes are directly discharged into the rivers, and are the primary cause of water contamination (Hossain and Rahman 2012; BBS 2010b; Rahman and Hossain 2008; Sohel et al. 2003; Karn and Harada 2001). Water quality parameters, such as dissolved oxygen (DO), have been found to be completely depleted in the Buriganga river during the dry season whilst biological oxygen demand (BOD) increases at the same time (Dewan et al. 2012; Sohel et al. 2003), indicating the existence of substantial amounts of inorganic and non-biodegradable components in the rivers (BBS 2010b; Karn and Harada 2001). Slum dwellers typically build open latrines on the roadside or construct hanging latrines over the water bodies. When natural and storm water run-off transports human waste to the surrounding water bodies, it results in severe water pollution (BBS 2010b; Hossain 2006; Hasan and Mulamoottil 1994).

Similar to water pollution, the level of air pollution in the megacity is critical and worsening (Begum et al. 2010; BBS 2010b). The major sources of air pollution are vehicle exhaust, resuspended road dust and manufacturing industries (Begum et al. 2011; Hasan and Mulamoottil 1994). Studies demonstrated that trace gases and aerosol particulate matter (PM_{10} and $PM_{2.5}$) in the air of Dhaka are higher than the permissible limit, particularly in the dry season (Guttikunda et al. 2012; Begum et al. 2006, 2010; Salam et al. 2008, 2013), which has serious implications for public health. For example, a recent study by Gurjar et al. (2010) suggests that the rate of morbidity in relation to chronic obstructive pulmonary disease (COPD), a disease that has links with NO_2, is significantly higher in Dhaka (2,100 cases per year) compared to other megacities in the region such as New Delhi in India.

The three primary sources of municipal solid waste (MSW) in Dhaka Megacity are residential, commercial and street sweeping. Solid waste discharge, which amounted to 1,040 t/day in 1985, rose by 300 % to 3,200 t/day in 2004 (JICA 2005). The solid waste of the megacity is characterised by high moisture content and low calorific value. One study shows that the percentage of organic material in domestic waste is decreasing, whereas the percentage of paper and plastic waste is increasing (Yousuf and Rahman 2007). However, the same study reported no proportional change in business and commercial wastes. Although medical waste contributes a small proportion (1 %) of the total MSW, its management remains somewhat unexplored (Hasan et al. 2006). Estimates suggest that approximately 10–25 % of medical wastes are hazardous and hence present a potential threat to public health (Ahmed 2000). Because of lack of resources and technology, the waste disposal method most usually practised in Dhaka Megacity is open dumping, which takes place in the low-lying areas both within and on the outskirts of the city, causing severe land and water pollution (UNEP 2006).

Inappropriate physical planning and piecemeal urban development in the megacity are also resulting in various environmental hazards of which floods and waterlogging are the most pervasive (Dewan 2013; Alam and Rabbani 2007). Although fluvial flooding is the most costly and remains one of the greatest threats of all natural hazards to people and property, pluvial flooding is becoming a grave concern for the inhabitants of Dhaka (Stalenberg and Vrijling 2009; Alam and Rabbani 2007; Tawhid 2004). Whilst Dhaka's peculiar geographical location makes it particularly vulnerable to flood (Bromley et al. 1989), unplanned urbanisation has accelerated the degree of vulnerability to flood, particularly in the recent past (Dewan and Yamaguchi 2008; Islam 2005b).

Apart from flooding, vulnerability to earthquake is also a matter of tremendous concern. According to the Bangladesh National Building Code (BNBC 1993), the megacity is located within an area whose seismic zoning coefficient is 0.15 g, which is ranked the second highest in terms of vulnerability. Experts fear that because of poor quality building materials and inappropriate construction techniques (Ansary 2003), severe damage is likely to occur in the event of an earthquake. Since the residents have no recent experience dealing with earthquake hazards, any tremor with a 7.0 magnitude would bring major human tragedy (Paul and Bhuiyan 2010). One study suggests that such a tremor could cost the lives of between 45,000 and 86,000 people (depending on the time of occurrence), and human injuries could be as high as 210,000, primarily owing to structural failure (Ansary 2004). Moreover, the construction of buildings and other urban infrastructures, which are now mainly through earth-filling on alluvial deposits, is elevating the potential of earthquake risks related to soil liquefaction (Kamal and Midorikawa 2004).

Other environmental hazards in the megacity include severe local storms such as tornados and nor'westers. Up to now, at least four violent tornadoes have hit the city and its suburbs. These tornadoes occurred on April 7, 1888, April 12, 1902, April 14, 1969, and April 5, 1972; they killed 118, 88, 562 and 75 people, respectively (Finch and Dewan 2004). Time series analysis of nor'wester occurrences (1954–2000) in Bangladesh revealed that Dhaka Megacity is highly exposed to local damaging storms, and it had the highest incidence of such events (on a weekly or fortnightly basis) resulting in severe property damage (Dewan and Peterson 2003; Peterson and Dewan 2002). Since urbanisation replaces natural land cover with asphalt and concrete infrastructures, rising temperatures due to the heat island effect during the summer are directly responsible for the formation of severe local storms in the city (Rana 2011). These weather-related hazards are likely to get worse in the context of global warming, which may have as yet unimagined impacts on people and property.

Due to high population density and extreme inequality in resources distribution, the law and order situation of the megacity is deteriorating at an alarming rate, making urban and social life increasingly insecure (Siddiqui et al. 2010; Islam 2005a). Available statistics suggest that the number of criminal offences has been increasing with time which is gradually becoming a serious social problem. For instance, about 60 % of the total crimes in the country are committed in Dhaka (Siddiqui et al. 2010), and out of 80 known organised criminal syndicates, 28 groups are actively operating in the city (Shafi 2010; Karzon 2006). The study by Siddiqui et al. (2010) further demonstrates that criminal offences including dacoity (gang robbery), robbery, violence against women, murder, burglary, theft cases, child abuse, kidnapping and smuggling have increased from 10,526 to 16,926 between 1985 and 2005. Of these, the incidence of murders was 15 per million in 1985 which increased to 19 per million by 2005. Another study found a total of 1,753 murder cases in 2005 whilst the number for 2001 was only 415 (Shafi 2010).

Many studies have demonstrated that Dhaka could be severely affected by climate change because of its high population density and poor capacity in dealing with the devastating effects (Roy 2009; Alam and Rabbani 2007; Senga 2004). The effects of climate change are projected to be most severe in relation to water-related disasters as widespread urban growth may intensify flooding in the coming years (Roy 2009). According to a study by Alam and Rabbani (2007), although long-term annual rainfall patterns show slight variation, monsoonal rainfall is showing an increasing trend in Dhaka. Low elevation, rapid population growth – particularly in slums – and limited adaptive capacity could put more people and property at risk in the event of climate change (Balica et al. 2012; Senga 2004) if non-climatic factors such as unplanned urban growth are not taken into account when developing appropriate countermeasures (Dewan 2013).

1.4 Description of the Study Area

As access to data (spatial and aspatial) is a serious hurdle in developing countries, particularly in Bangladesh, we could not consider the entire Dhaka Megacity for this book. Rather, we used a bounding polygon that includes part of both the Dhaka Megacity and DMDP areas. The study area is located between 23.61° and 23.97° N latitudes and 90.22° and 90.59° E longitudes and covers an area of 878 km^2 (Fig. 1.3). The study area comprises the entire DCC, DMA, Tongi and Savar municipalities with adjacent *unions* and other urban areas (OUA). The area is bounded by Tongi *Pourashava* to the north, Daudpur *union* to the east, Hazratpur *union* to the west and Kutubpur OUA to the south. Note that the territorial extent

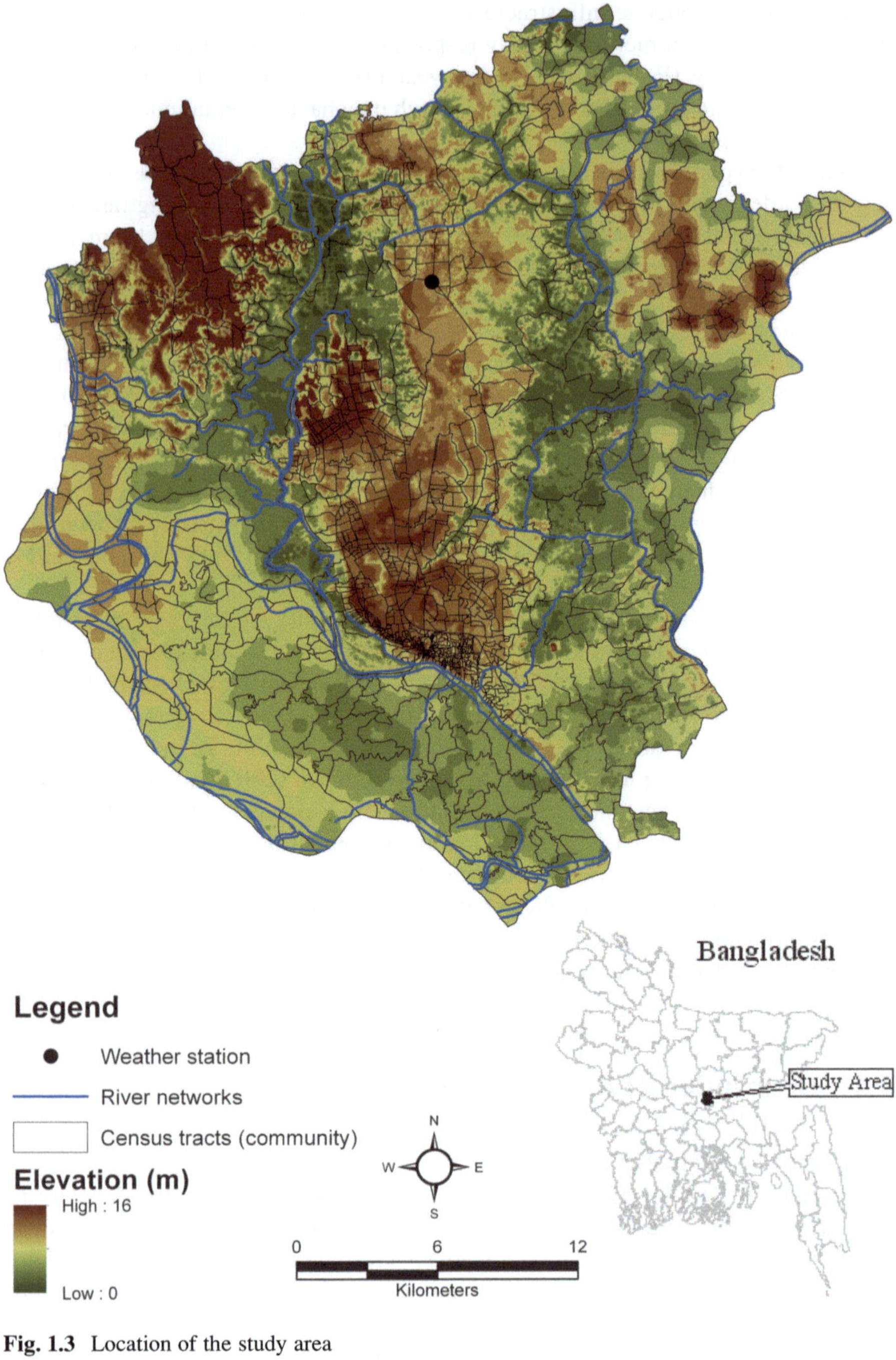

Fig. 1.3 Location of the study area

of OUA also includes a few unions of DMA. Whilst most of the studies presented in this book use this spatial extent, the study of emissions from the brick manufacturing industry (see Chap. 17) goes beyond this extent.

The total population of the study area according to last three censuses of 1991–2011 are 5.4 million in 1991, 8.1 million in 2001 and 11.3 million in 2011. The mean population density of these three periods is 25,106, 33,112 and 39,277 persons per km^2, respectively.

The demographic characteristics of the three municipalities in the study area, including Dhaka City Corporation (DCC), Savar, Tongi and DMA (including DCC) at the time of the 2011 population and housing census are shown in Table 1.3. The population composition of the study area as estimated by the 2011 population and housing census indicates that 16.5 % are in the age group of 0–9 years, 19.9 % are in 10–19 years, 52.1 % are within 20–49 years and 8.2 % are in 50–64 age groups, whilst 3.1 % are over 65+ age group (BBS 2012b). As shown previously, population density is much higher in the DCC area compared with the two municipalities and other spatial units in the study area. Similarly, the literacy rate is also higher owing to the fact that the DCC has the highest number of educational institutes and facilities in the megacity area. According to the 2011 census, the total population of DMA is 8.9 million with a density of 29,392 per km^2 (Table 1.3).

On average, 63 % of households in the study area have access to tap water, and 33 % use tube wells as the main source of drinking water, whilst 2.7 % rely on other sources such as ponds, wells and rivers. Approximately, 94 % of households are connected to electricity. On the other hand, only 87 % of households have access to safe sanitation (BBS 2012b).

A review of the housing characteristics of the study area shows that nearly 50 % of the houses are *pucca* [see notes at the end of this chapter] being constructed with brick walls and a cement roof; 28 % are *semi-pucca*; 19 % are *katcha*, constructed with fragile materials such as straw, bamboo and mud; and the rest are *jhupri* and are made from plastic materials and/or straw (BBS 2012b).

1.5 The Organisation of This Book

The chapters in this book can be placed in four broad categories or themes. The first of these looks at drivers of change such as population distribution. The second category deals with mapping and modelling of environmental issues. The third group looks at socio-environmental problems, whilst the final category employed spatial and temporal techniques to analyse health-related phenomena that the megacity of Dhaka is currently facing.

The historical development of Dhaka Megacity has been examined by Sohel J. Ahmed, Kh. Md. Nahiduzzaman and Glen Bramley and described in Chap. 2. They have specifically looked at the historical legacy of the various planning initiatives and master plans for Dhaka that were undertaken during the twentieth century. Population dynamics at the lowest available unit of census geography for

Table 1.3 Demographic characteristics of the study area, 2011

		Households	Population			Literacy rate (%)			
Locality	Area (km^2)	General	Total	Male	Female	Total	Male	Female	Density (km^2)
DCC	143	1,473,665	7,033,075	3,915,989	3,117,086	75.6	78.3	71.2	49,182
DMA	303	1,903,632	8,906,039	4,931,802	3,974,234	74.6	77.3	71.0	29,392
Savar PSA	28	72,201	286,008	286,008	148,958	73.6	76.5	70.5	10,214
Tongi PSA	33	116,387	476,350	476,350	255,236	65.3	68	59.7	14,343
Unions[a] and OUA	674	794,784	3,477,287	3,477,287	1,868,423	62.8	65.2	60.0	5,159

(Adapted from BBS 2012a, b)
PSA Pourashava/municipality
[a]OUA includes some *unions* of DMA

the last three census periods (e.g. 1991, 2001 and 2011) has been modelled and mapped by Robert J. Corner, Emmanuel T. Ongee and Ashraf M. Dewan and is presented in Chap. 3. Changes in the climate, particularly rainfall and temperature, in Dhaka using long-term meteorological records have been looked at by Yusuke Yamane, Masashi Kiguchi, Toru Terao, Fumie Murata and Taiichi Hayashi. The results of this study are presented in Chap. 4.

An analysis of land use and land cover (LULC) changes between 1990 and 2011 using geospatial techniques and a prediction of future land use changes are reported in Chap. 5 by Robert J. Corner, Ashraf M. Dewan and Salit Chakma. Using remotely sensed data, together with spatially referenced population data, Ashraf M. Dewan and Robert J. Corner investigated urban expansion, sprawl and urban structure. The results of that work are presented in Chap. 6. In Chap. 7, Sohel J. Ahmed, Glen Bramley and Peter H. Verburg demonstrate another approach to urban growth modelling using spatial–statistical techniques. Work carried out by Salit Chakma on the use of spatial multi-criteria techniques for assessing urban development suitability is presented in Chap. 8. In the light of many previous studies that suggest that flooding is likely to be a problem for major cities, particularly in developing countries, in the years ahead, two chapters are devoted to mapping and modelling its impact. In Chap. 9, the impact of land use change on flooding has been analysed with hydrological modelling and described by Takeo Onishi, Tahmina Khan and Ken Hiramatsu. A study on flood risk and vulnerability assessment at community level using multi-temporal remote sensing and other spatial data was conducted by Akiko Masuya, and the results of that study are presented in Chap. 10.

The third part of the book examines various social and environmental problems that the city is now encountering. Md. Humayun Kabir and Wilfried Endlicher discuss the current energy crisis in Dhaka and explore potential for the development of solar photovoltaic solutions. This is discussed in Chap. 11. In an attempt to understand the effects of land use and land cover change on land surface temperature (LST), Ashraf M. Dewan and Robert J. Corner made use of multi-date Landsat TM 5 data. This work is described in Chap. 12. The quality of life (QOL) in the Dhaka City Corporation (DCC) area has been investigated by Ashraf M. Dewan, Kamrun Nahar and Youhei Kawamura through an integration of remotely sensed images and census data, which is presented in Chap. 13. One year worth data on the incidence of crime in DMA forms the basis of a spatio-temporal analysis by Ashraf M. Dewan, Md. Rafiqul Haider and Md. Ruhul Amin. This is discussed in Chap. 14. Issues associated with the environmental governance of Dhaka city have been investigated by Salim Momtaz and S.M. Zobaidul Kabir. Their findings in Chap. 15 describe the environmental impact assessment (EIA) process and its shortcomings.

The last section of this book considers the health of both the environment and its inhabitants. A study by Razia A. Chowdhury, Towhida Rashid and Sirajul Hoque analysed water quality in the major rivers of DMA using Landsat TM 5 and in situ data. This is reported in Chap. 16. The spatial distribution patterns of air pollutants from brick kilns in the Dhaka region are analysed by Sarath K. Guttikunda and presented in Chap. 17. Masahiro Hashizume, Abu S. G. Faruque and Ashraf

M. Dewan analysed and described the variability of *Aeromonas*-positive diarrhoea with temporal techniques in Chap. 18. The prevalence of typhoid from 2005 to 2009 has been analysed through spatial statistics and is presented in Chap. 19 by Ashraf M. Dewan, Robert J. Corner and Masahiro Hashizume. Finally, Chap. 20 presents a study by Sarwa Ali, Robert J. Corner and Masahiro Hashizume using spatial clustering techniques to examine the incidence of dengue disease.

Notes

Katcha houses are generally made from mud, straw and bamboo with corrugated tin rooves.
Mahalla is the smallest urban geographic unit.
Mauza is the smallest rural geographic revenue unit having a Jurisdiction List Number (JLN).
Other Urban Area (OUA) includes *thana/upazila* headquarters and also development centres having urban characteristics.
Pourashava is a municipality incorporated and administered by local government under Pourashava Ordinance, 1977.
Pucca houses are constructed with solid walls and rooftops.
Semi-pucca houses are constructed with brick wall and corrugated tin roof.
Thana is a police jurisdiction area and comprising about 100 villages or city wards. A district consists of several *thanas*.
Union is a rural administrative geographic unit comprising of one or more *mauzas* and villages and governed by Union Parishad Institution.
Upazila is the second lowest tier of regional administration in Bangladesh.
Village is the lowest rural geographic unit either equivalent to a *mauza* or part of a *mauza*.
Ward is an urban administrative geographic unit comprising of one or more *mahallas* and governed by Ward Council.

References

Afsar R (2000) Rural-urban migration in Bangladesh: causes, consequences and challenges. University Press Limited, Dhaka
Ahmed SU (1986) Dhaka: a study in urban history and development 1840–1921. Curzon Press Limited, London
Ahmed A (2000) Hospital waste management: problems and strategic solutions. Paper presented at Independent University of Bangladesh, Dhaka
Ahmed MF (2009) Environment impacts of sea-level rise on the coastal areas of Bangladesh. In: Ahmed MF, Islam N, Matin MD (eds) Climate change and the tasks for Bangladesh. Bangladesh Poribesh Andolon and Bangladesh Environment Network, Dhaka, pp 88–100
Ahmed KM, Nishigaki M, Dewan AM (2005) Constraints and issues on sustainable groundwater exploitation in Bangladesh. J Groundw Hydrol 47(2):163–179
Alam M, Rabbani MG (2007) Vulnerabilities and responses to climate change for Dhaka. Environ Urbaniz 19(1):81–97
Angeles G, Lance P, Barden-O'Fallon J, Islam N, Mahbub AQM, Nazem NI (2009) The 2005 census and mapping slums in Bangladesh: design, select results and application. Int J Health Geogr 8:32

Ansary MA (2003) Status of earthquake risk mitigation in Bangladesh. Paper presented at the 3rd WSSI workshop on EQS, Bangkok
Ansary MA (2004) Seismic risk in urban areas of Bangladesh. Ministry of Environment and Forest, Dhaka
Balica SF, Wright NG, van der Muelen F (2012) A flood vulnerability index for coastal cities and its use in assessing climate change impacts. Nat Hazards 64(1):73–105
Bangladesh Bureau of Statistics (BBS) (1984) Bangladesh population census 1981, National series, analytical findings and national tables. Bangladesh Bureau of Statistics, Ministry of Planning, Dhaka
Bangladesh Bureau of Statistics (BBS) (1994) Bangladesh population census 1991, vol 1, Analytical findings. Bangladesh Bureau of Statistics, Ministry of Planning, Dhaka
Bangladesh Bureau of Statistics (BBS) (1998) Population census – 1991, vol 3, National series, Urban area report, Bangladesh Bureau of Statistics, Ministry of Planning, Dhaka
Bangladesh Bureau of Statistics (BBS) (2001) Population census – 2001, Zila series. Bangladesh Bureau of Statistics, Ministry of Planning, Dhaka
Bangladesh Bureau of Statistics (BBS) (2003) Population census – 2001, Community series. Bangladesh Bureau of Statistics, Ministry of Planning, Dhaka
Bangladesh Bureau of Statistics (BBS) (2004) Bangladesh population census 2001, National series. Bangladesh Bureau of Statistics, Ministry of Planning, Dhaka
Bangladesh Bureau of Statistics (BBS) (2005) Statistical yearbook of Bangladesh. Bangladesh Bureau of Statistics, Ministry of Planning, Dhaka
Bangladesh Bureau of Statistics (BBS) (2008) Population census – 2001, vol 3 National series, Urban area report, Bangladesh Bureau of Statistics, Ministry of Planning, Dhaka
Bangladesh Bureau of Statistics (BBS) (2010a) Census of agriculture 2008. Bangladesh Bureau of Statistics, Ministry of Planning, Dhaka
Bangladesh Bureau of Statistics (BBS) (2010b) Compendium of environment statistics of Bangladesh 2009. Bangladesh Bureau of Statistics, Ministry of Planning, Dhaka
Bangladesh Bureau of Statistics (BBS) (2011a) Monthly statistical bulletin – February 2011. Bangladesh Bureau of Statistics, Ministry of Planning, Dhaka
Bangladesh Bureau of Statistics (BBS) (2011b) Household Income and Expenditure Survey (HIES) – 2010. Bangladesh Bureau of Statistics, Ministry of Planning, Dhaka
Bangladesh Bureau of Statistics (BBS) (2012a) http://www.bbs.gov.bd/WebTestApplication/userfiles/Image/BBS/GDP_2011.pdf. Accessed 15 Sept 2012
Bangladesh Bureau of Statistics (BBS) (2012b) Population and housing census 2011, Community series – Dhaka, Gazipur, Narayanganj. Bangladesh Bureau of Statistics, Ministry of Planning, Dhaka
Banglapedia (2006) http://www.banglapedia.org/httpdocs/HT/D_0151.HTM. Accessed 5 June 2011
Begum A (2007) Urban housing as an issue of redistribution through planning? The case of Dhaka city. Soc Policy Adm 41(4):410–418
Begum BA, Biswas SK, Hopke PK (2006) Temporal variations and spatial distribution of ambient PM2.2 and PM10 concentrations in Dhaka, Bangladesh. Sci Total Environ 358(1–3):36–45
Begum BA, Biswas SK, Nasiruddin M (2010) Trend and spatial distribution of air particulate matter pollution in Dhaka city. J Bangladesh Acad Sci 34(1):33–48
Begum BA, Biswas SK, Hopke PK (2011) Key issues in controlling air pollutants in Dhaka, Bangladesh. Atmos Environ 45(40):7705–7713
BNBC (1993) Bangladesh national building code. Bangladesh National Building Code, HBRI-BSTI, Dhaka
Brammer H (2002) Land use and land use planning in Bangladesh. The University Press Limited, Dhaka
Bromley R, Hall P, Dutt A, Mookherjee D, Benhart JE (1989) Regional development and planning. In: Gaile GL, Willmott CJ (eds) Geography in America. Merrill Publishing Company, Columbus
Center for Urban Studies (CUS), National Institute of Population, Research and Training (NIPORT) and Measure Evaluation (2006) Slums in urban Bangladesh: mapping and census 2005. CUS, Dhaka

Chakrabati PGD (2001) Urban crisis in India: new initiatives for sustainable cities. Dev Pract 11 (2–3):260–272

Choudhury AKMK (2008) Land use planning in Bangladesh. A H Development Publishing House, Dhaka

Chowdhury JU, Salehin M (1987) Floods and their processes. In: Proceedings of the international seminar on evolution of scientific system of flood forecasting and warning in the Ganges, Brahmaputra and Meghna River Basins, Bangladesh, National Committee of International Commission for Irrigation and Drainage (ICID), Dhaka, pp 247–254

Dewan AM (2013) Floods in a megacity: geospatial techniques in assessing hazards, risk and vulnerability. Springer, Dordrecht

Dewan AM, Peterson RE (2003) Nor'wester damage and deaths in Bangladesh. Paper presented at eleventh wind engineering conference, Lubbock, 2–5 June 2003

Dewan AM, Yamaguchi Y (2008) Effects of land cover changes on flooding: example from Greater Dhaka of Bangladesh. Int J Geoinform 4(1):11–20

Dewan AM, Kabir MH, Nahar K, Rahman MZ (2012) Urbanization and environmental degradation in Dhaka Metropolitan Area of Bangladesh. Int J Environ Sustain Dev 11(2):118–146

Dhaka Metropolitan Development Plan (DMDP) (1997) Dhaka structure plan (1995–2015), vol I. Dhaka, RAJUK. Dhaka

Douglas I (2009) Climate change, flooding and food security in South Asia. Food Secur 1 (2):127–136

Dutt AK, Noble AG (2004) Urban development in South Asia. In: Dutt AK, Noble AG, Venugopal G, Subbiah H (eds) Challenges to Asian urbanization in the 21st century. Kluwer Academic Publishers, Dordrecht, pp 256–273

Finch JD, Dewan AM (2004) Tornadoes in Bangladesh and East India. http://bangladesh-tornadoes.org/climo/btorcli0.htm. Accessed 11 Apr 2011

Flood Action Plan (FAP) 8A (1991). Master plan study for Greater Dhaka Protection Project. Japan International Cooperation Agency, Dhaka

Ghosh S (2007) A farewell to urban planning: sequel to urbanization in South Asia. In: Jahan S, Maniruzzaman KM (eds) Urbanization in Bangladesh: patterns, issues and approaches to planning. Bangladesh Institute of Planners, Dhaka, pp 163–170

Gurjar BR, Jain A, Sharma A, Agarwal A, Gupta P, Nagpure AS, Lelievenld J (2010) Human health risks in megacities due to air pollution. Atmos Environ 44:4606–4613

Guttikunda S, Begum BA, Wadud Z (2012) Particulate pollution from brick kiln clusters in the Greater Dhaka region, Bangladesh. Air Qual Atmos Health. doi: 10.1007/s11869-012-0187-2

Hasan S, Mulamoottil G (1994) Environmental problems of Dhaka City. Cities 11(3):195–200

Hasan MM, Nahar A, Hira SMH, Arafat JB (2006) Sources and management of medical waste: a study on Dhaka. Geogr Environ J 5:185–206

Hoque MA, Hoque MM, Matin KM (2007) Declining groundwater level and aquifer dewatering in Dhaka metropolitan area, Bangladesh: causes and quantification. Hydrogeol J 15(8):1523–1534

Hossain S (2006) Social characteristics of a megacity: a case of Dhaka City: Bangladesh, TASA conference, Perth, 4–7 Dec 2006

Hossain AMMM, Rahman S (2012) Impact of land use and urbanization activities on water quality of the megacity, Dhaka. Asian J Water Environ Pollut 9(2):1–9

Islam N (1991) Dhaka in 2025 AD. In: Ahmed S (ed) Dhaka: past, present and future. The Asiatic Society of Bangladesh, Dhaka, pp 570–583

Islam N (2005a) Dhaka now: contemporary urban development, Golden jubilee series no. 2. Bangladesh Geographical Society (BGS), Dhaka

Islam N (2005b) Improving the lives of slum dwellers in Bangladesh, CUS bulletin on urbanization and development, no. 48. Center for Urban Studies, Dhaka

Islam T, Peterson RE (2009) Climatology of landfalling tropical cyclones in Bangladesh 1877–2003. Nat Hazards 48(1):115–135

Japan International Cooperation Agency (JICA) (2005) The study on solid waste management in Dhaka city, Clean Master Plan, Final report, JICA, Dhaka

Kamal ASMM, Midorikawa S (2004) GIS-based geomorphological mapping using remote sensing data and supplementary geoinformation: A case study of the Dhaka city area, Bangladesh. Int J Appl Earth Obs Geoinf 6(2):111–125

Karn SK, Harada H (2001) Surface water pollution in three urban territories of Nepal, India and Bangladesh. Environ Manag 28(4):483–496
Karzon SHR (2006) Theorising the crime and violence of Dhaka city. The Daily Star, Dhaka, 1 Apr 2006
Miah MM, Bazlee BL (1968) Some aspects of geomorphology of the Madhupur tract. Orient Geogr 12(1):39–48
Mirza MMQ (2009) Climate change risks in Bangladesh. In: Bhattacharya D, Ahmed MF, Islam N, Matin MD (eds) Climate change and the tasks for Bangladesh. Bangladesh Paribesh Andolon and Bangladesh Environment Network, Dhaka, pp 69–87
Mirza MMQ, Warrick RA, Ericksen NJ (2003) The implications of climate change on floods of the Ganges, Brahmaputra and Meghna Rivers in Bangladesh. Clim Chang 57(3):287–318
Mitchell JK (1998) Urban metabolism and disaster vulnerability in an era. In: Schellnhuber H-J, Wenzel V (eds) Earth system analysis: integrating science for sustainability. Springer, Berlin, pp 359–377
Nicholls RJ (1995) Coastal megacities and climate change. GeoJournal 37(3):369–379
Paul BK, Bhuiyan RH (2010) Urban earthquake hazard: perceived seismic risk and preparedness in Dhaka city, Bangladesh. Disasters 34(2):337–359
Peterson RE, Dewan AM (2002) Damaging Nor'westers in Bangladesh. In: 21st conference on severe local storms, American Meteorological Society, San Antonio
Rahman S, Hossain F (2008) Spatial assessment of water quality in peripheral rivers of Dhaka city for optimal location of water intake point. Water Resour Manag 22(3):377–391
Rana MMP (2011) Urbanization and sustainability: challenges and strategies for sustainable urban development in Bangladesh. Environ Dev Sustain 13(1):237–256
Rashid H (1991) Geography of Bangladesh, 2nd edn. The University Press Limited, Dhaka
Rashid KBS (2008) Bangladesh: resource and environmental profile. A H Development Publishing House, Dhaka
Rizvi SNH (1969) East Pakistan district gazetteers: Dacca. East Pakistan Government Press, Dacca
Roy M (2009) Planning for sustainable urbanization in fast growing cities: mitigation and adaptation issues addressed in Dhaka, Bangladesh. Habitat Int 33(3):276–286
Rouf MA, Jahan S (2007) Spatial and temporal patterns of urbanization in Bangladesh. In: Jahan S, Maniruzzaman KM (eds) Urbanization in Bangladesh: patterns, issues and approaches to planning. The Bangladesh Institute of Planners (BIP), Dhaka, pp 1–24
Salam A, Hossain T, Siddique MNA, Alam AMS (2008) Characteristics of atmospheric trace gases, particulate matter, and heavy metal pollution in Dhaka, Bangladesh. Air Qual Atmos Health 1(2):101–109
Salam A, Ullah MB, Islam MS, Salam MA, Ullah SU (2013) Carbonaceous species in total suspended particulate matters at different urban and suburban locations in the Greater Dhaka region, Bangladesh. Air Qual Atmos Health 6(1):239–245
Sanderson D (2012) Building livelihoods to reduce risk among the most marginalized in urban areas: strategic approaches from Dhaka. Environ Hazards 11(2):112–122
Satterthwaite D, McGranahan G, Tacoli C (2010) Urbanization and its implications for food and farming. Philos Trans R Soc B 365(1554):2809–2920
Senga R (2004) Natural and unnatural disasters: the relative vulnerabilities of Southeast Asian megacities to climate change, World Wide Fund for Nature (WWF), Gland, United Kingdom. www2.lse.ac.uk/IDEAS/publications/reports/pdf/SR004/WWF.pdf. Accessed 15 Mar 2012
Seto KC, Sánchez-Rodríguez R, Fragkias M (2010) The new geography of contemporary urbanization and the environment. Ann Rev Environ Resour 35:167–194
Shafi SA (2010) Urban crime and violence in Dhaka. The University Press Limited, Dhaka
Shahid S (2010) Recent trends in the climate of Bangladesh. Clim Res 42(3):185–193
Shahid S (2011a) Trends in extreme rainfall events of Bangladesh. Theor Appl Climatol 104 (3–4):489–499

Shahid S (2011b) Impact of climate change on irrigation water demand of dry season Boro rice in northwest Bangladesh. Clim Chang 105(3–4):433–453
Shahid S (2012) Vulnerability of the power sector of Bangladesh to climate change and extreme weather events. Reg Environ Change 12(3):595–606
Siddiqui K, Ghosh A, Bhowmik SK, Siddiqi SA, Mitra M, Kapuria S, Rajan N, Ahmed J (2004) Megacity governance in South Asia. The University Press Limited, Dhaka
Siddiqui K, Ahmed J, Siddiqui K, Huq S, Hossain A, Nazimud-Doula S, Rezawana N (2010) Social formation in Dhaka, 1985–2005: a longitudinal study of society in a third world megacity. Ashgate Publishing Group, Surrey
Sohel KMA, Chowdhury MAI, Ahmed MF (2003) Surface water quality in and around Dhaka city. J Water Supply Res Technol 52(2):141–153
Stalenberg B, Vrijling H (2009) The battle of Tokyo and Dhaka against floods. Built Environ 35(4):471–491
Tawhid KG (2004) Causes and effects of water logging in Dhaka city, Bangladesh. TRITA-LWR master thesis, Department of Land and Water Resource Engineering, Royal Institute of Technology, Stockholm
United Nations (UN) (2003) World urbanization prospects – the 2003 revision. UN, New York
United Nations (UN) (2012) World urbanization prospects – the 2011 revision. UN, New York
United Nations Environment Program (UNEP) (2006) Dhaka city state of environment 2005. UNEP, Dhaka
Varis O, Biswas AK, Tortajada C, Lundqvist J (2006) Megacities and water management. Water Resour Dev 22(2):377–394
World Bank (2012) http://data.worldbank.org/country/bangladesh#cp_cc. Accessed 17 Dec 2012
Yousuf TB, Rahman M (2007) Monitoring quantity and characteristics of municipal solid water in Dhaka city. Environ Monit Assess 135(3):3–11

Chapter 2
From a Town to a Megacity: 400 Years of Growth

Sohel J. Ahmed, Kh. Md. Nahiduzzaman, and Glen Bramley

Abstract The continued expansion of Dhaka means that within the next 20 years, it will become one of the most populated megacities in the world. Considering the importance, to its development, of understanding the nature of its pattern of growth and evolution, we examined that growth over a period of 400 years. The dynamics of Dhaka's urbanisation along with its trajectory over this period are discussed in this chapter. The pace of urbanisation was found to be increasingly rapid after independence from Pakistan in 1971. This rapid pace has caused a number of multifaceted problems, including severe environmental degradation. Issues associated with the urban planning process in Dhaka have been reviewed along with the major plans developed over the past 65 years. The chapter concludes with a critical and comparative analysis of those plans.

Keywords Development plans • Dhaka • Growth • Urbanization

S.J. Ahmed (✉)
The Bartlett Development Planning Unit, University College London (UCL), 34 Tavistock Square, WC1H 9EZ London, UK
e-mail: sohel.ahmed@ucl.ac.uk

Kh.Md. Nahiduzzaman
Department of City and Regional Planning, King Fahd University of Petroleum and Minerals, Dhahran 31261, Kingdom of Saudi Arabia
e-mail: nahid@kfupm.edu.sa

G. Bramley
School of the Built Environment, Heriot-Watt University, Edinburgh EH14 4AS, UK
e-mail: G.Bramley@hw.ac.uk

A. Dewan and R. Corner (eds.), *Dhaka Megacity: Geospatial Perspectives on Urbanisation, Environment and Health*, Springer Geography,
DOI 10.1007/978-94-007-6735-5_2,

2.1 Introduction

The speed of urbanisation is slower in the cities of developed world (UN 2011) than in the cities of the developing world where three million people were added to the global urban population each week during the last two decades (Moreno et al. 2008). This has already resulted in there being more urban dwellers than rural, a situation referred to as the *urban turn*[1] by some researchers (UNFPA 2007). These urban populations will continue to grow, and it is estimated that a further 20 % of the world's population will move to cities by 2050 (Butsch et al. 2009; World Bank 2009; Moreno et al. 2008; UN 2006, 2008). In coming years, Africa and Asia will have the largest urban agglomerations, with Africa projected to treble and Asia to double their present urban populations (UN 2012). Based on these projections, 70 % of the people of these two continents will be living in cities by 2030 (Kraas 2007; UN 2006). One phenomenon of this *urban turn* is the rise of a new category of human settlements – called megacities (Butsch et al. 2009).

Using the published literature, this chapter attempts to document the historical growth of Dhaka. It also critically discusses the dynamics of its urbanisation along with different trajectories of development. As the discussion proceeds, it highlights major development initiatives that have guided the growth of the city to its present form.

2.2 Trajectories of Development

Dhaka has grown from a small settlement confined by the rivers Buriganga and Dholai Khal. It has experienced many highs and lows throughout its history. This section briefly presents the five distinct phases of Dhaka's growth from a town to a megacity. The dominant features of these phases are described in the following sections.

2.2.1 Pre-Mughal Period (Before 1604)

It is not completely clear when and how a settlement was first formed at the site where the old core of Dhaka stands today, and its existence and development during pre-Mughal period is also subject to a great deal of debate. According to Chowdhury and Faruqui (2009) and Islam (1996), urban settlements have existed in the area where Dhaka is now back to seventh century CE. Some historians argue that it began as a trading post, serving the surrounding region by the rivers. This gave the site better locational advantages for development as a port or a trading town. Although Dhaka is now the capital city, there are three other locations adjacent to Dhaka which have

[1] In 2008, for the first time in human history, every second citizen lives in an urban settlement. Some researchers therefore labelled the crossing of the 50 % threshold in 2008 as the 'urban turn' (Butsch et al. 2009).

Plate 2.1 Dhalleshori temple in 1904 (Photograph taken by Fritz Kapp in 1904, part of an album of 30 prints from the Curzon Collection; courtesy to British Library photo collections)

enjoyed the status of capital during different regimes. Bikrampur is known to be the oldest settlement, dating back to the Vedic period, early days of Aryan migration into India at around 1300–1500 BCE (Chowdhury and Faruqui 2009; Islam 1996). During the Muslim dynasties of the Sultanate of Bengal, first Bhawal to the north and then Sonargaon to the south became the capitals of Bengal (late thirteenth and early fourteenth centuries). The use of these locations as capital cities suggests that the then rulers never decided to make Dhaka the capital.

A possible suggestion stems from its propensity to annual flooding from the surrounding rivers (Chowdhury and Faruqui 2009; Islam 1996). Many believe that the name of Dhaka was derived from one of the Hindu temples for the Goddess Dhaleshwari (Plate 2.1) established by a prominent king of the Sen Dynasty, Ballal Sen, during the pre-Islamic period. Along with this, the existence of Dhaka as a small town at this time is proved by two mosque inscriptions and other ancient literature. But this evidence is not unequivocal as there remains other versions[2] (Chowdhury and Faruqui 2009; Islam 1996, 2007).

From the pages of Baharistan-i-Ghaibi,[3] the name of various localities (e.g. Laksmibazar, Tantibazar, Shankharibazar, Kumartuli) denotes the prevalence of

[2] (1) The name might originate from Dak tree (*Butea frondosa*) which was believed to cover this area once; (2) it also could be the *dhak* or drum used when Islam Khan launched the city as capital; (3) the derivation could also be from a Prakrt dialect called Dhaka; (4) lastly, the name is also similar to 'Davaka' – a name found inscribed in pillar dated from Gupta Kingdom (Islam 2007).

[3] Written by a famous traveller named Alauddin Isfahan alias Mirza Nathan. This is a trusted source that provides sufficient foundation to Bengal history during the Mughal regime (Islam 2007).

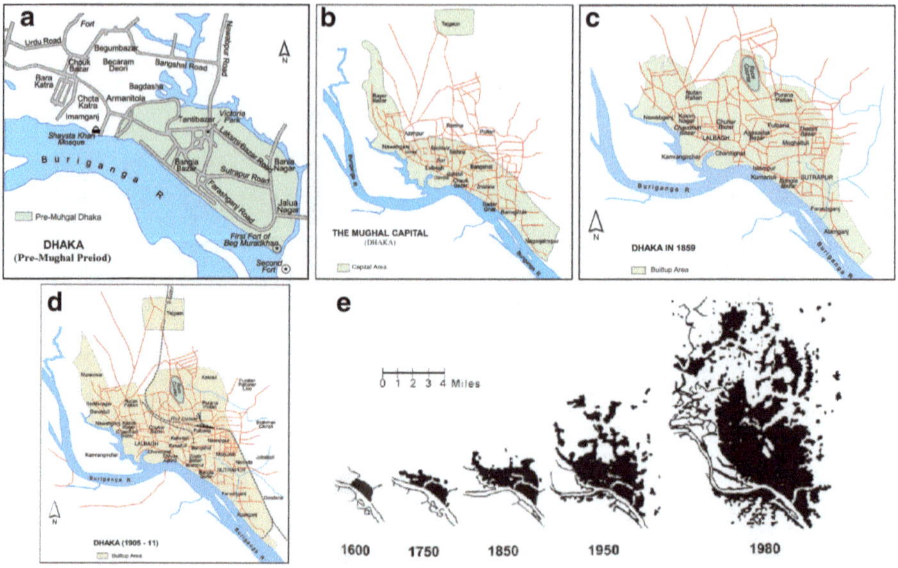

Fig. 2.1 Evolution of Dhaka: (**a**) Pre-Mughal period, (**b**) Mughal period, (**c**) British period, (**d**) capital of East Bengal and Assam (reconstructed by the authors), and (**e**) evolution of Dhaka at a glance, 1600–1980 (Adapted from Shankland Cox Partnership and Others 1981)

Hindu craftsmen in pre-Mughal Dhaka (Chowdhury and Faruqui 2009; Islam 1996). Little evidence about its size and population can be found before 1604. The sites close proximity to the then capital, Sonargaon, along with the major water transport networks of the Buriganga River and Dholai Khal, influenced its formation, with more growth occurring along the riverbanks. The area of Dhaka during this period has been estimated to be around 1 km^2, but the size of population is unknown (Islam 2007) (Fig. 2.1a).

2.2.2 *Mughal Period (1602–1764)*

When the Mughals invaded Bengal, they established their headquarters at Dhaka. The Mughal general, Man Singh, pitched his tent here, instead of in other capitals, due to its strategic locations in between rivers with two distinct purposes – to safeguard from enemies and to get easy access to surrounding regions (Hyder 1994). At first a *thana*[4] *(i.e. a fortified post)* was formed where pre-Mughal Dhaka was believed to have existed (Karim 2009). The location gained further importance when the Chandnighat fort was built in 1610 CE by Islam Khan Chisti. At this time, Dhaka

[4] *Thana* is a small administrative subdivision and is the lowest tier in police administration. It is also known as the third lowest tier in the hierarchy of local government. Currently, there are 522 *thanas* in Bangladesh of which 10 are located within the administrative confines of the Dhaka Metropolitan Statistical Area (BBS 2003).

Plate 2.2 On the way to Tongi (Sir Charles D'oyle's (Collector of Dhaka in early eighteenth century) sketches; courtesy to British Library photo collections)

was called Jahangirnagar and held the status of provincial capital for a little over a century (Islam 1996). Commercial activities and the needs of administration and defence led to the growth of Dhaka. The diaries of foreign explorers as well as many present place names are of help in identifying the boundaries of Dhaka at that time (Karim 2009; Islam 1996). From documentary evidence and Mughal sculptures in the 'old city', Mughal Dhaka seems to have included the present 'old Dhaka'. Commercial areas developed around the fort and along the river to the west, while residential areas mainly spread out in a northerly direction (up towards Tongi, Plate 2.2), and up to the fringes of Ramna near the Mir Jumla gate. Two roads leading to the north and to the east determined the city's growth at that time. During the administration of the subahdar Ibrahim Khan (1616–1620), its role as a trading centre extended to Southeast Asia, and it started to attract European traders (e.g. the Dutch, the Portuguese, the French and the British). It is argued that a greater amount of development took take place during the regime of Shaista Khan (1662–1679), when the city grew to an area of ca.96 mile^2 with about one million inhabitants (Karim 2009; Islam 1996). Records of the British East India Company mention that the boundary of Dhaka during this period was as follows: 'Buriganga to the south, Tongi to the north, Jafarabad-Mirpur to the west and Postogola to the east' (Islam 2007) (Fig. 2.1b).

The Chauk, close to the fort and Buriganga River, was the central business district(CBD). Cotton industries and other specialised industries (potters, weavers, shell cutters, jute painters etc.) flourished and clustered in between the Chauk and Bangla Bazar, another commercial centre which had existed before the Mughals

(Islam 1996). Tejgaon began its development as an industrial estate in the Mughal period as Dutch, English, Portuguese and French companies started to locate their factories in that area. Thus, during the Mughal regime, the growth of Dhaka was primarily driven by the Mughal's need for residences and administrative establishments. Professionals such as artisans and craftsmen and traders, including Europeans, added to this demand (Karim 2009). This expansion continued uninterrupted until the early eighteenth century when the capital was moved to the west Bengal town of Murshidabad and consequently Dhaka's short-lived glory began to fade (Chowdhury and Faruqui 2009; Islam 1996, 2007).

2.2.3 British Period (1764–1947)

Dhaka continued to suffer from a lack of political and commercial importance when the British East India Company took over political power of the region in 1757 and moved the capital of Bengal to Calcutta. This shift of power, and investment, led to a stalling of Dhaka's growth. The decline of the handcrafted cotton trade due to the promotion of cottons from the automated British industry added to the decline (Chowdhury and Faruqui 2009). By 1840, Dhaka had declined so much that 42 km^2 of its urban space had been vacated and 0.7 million of its inhabitants had left their homes (Ahsan 2009). The city was famously described by Bishop Heber in 1824 as '. . . .merely the wreck of its ancient grandeur' (Islam 1996, 2007) (Plate 2.3a–c).

However, the British rulers (particularly, the then collector, Charles Dawes) did not totally neglect this declining city and took some timely initiatives to reinvigorate its glorious past. The launch of the municipal committee in 1840 and the establishment of Dhaka College (Plate 2.4) in 1841 brought back life in different places, and the city started to modernise with metaled roads and sophisticated urban amenities like a piped water supply and street lights. In 1825, a racecourse was built in Ramna (later renamed to Suhrawardy Uddyan) and a large building complex and gardens were constructed adjoining Ramna (the present Shahbagh area) (Islam 1996, 2007).

Once the British Crown took over India from the East India Company, Dhaka started to see some substantial changes. Electricity was connected in 1878, and more roads were created and existing ones expanded. Drainage networks and the first planned residential area, Wari, were established. The construction of railway lines in 1885–1886 along the Narayanganj-Dhaka-Mymensingh route boosted the city's growth to the south and west side. In 1905 the city, once again, became the provincial capital of east Bengal and Assam and saw a 21 % increase in population in between 1901 and 1911 (Karim 2009; Islam 1996). However, the growth of Dhaka faced a huge setback when the proposed partition of Bengal, announced in 1905, was cancelled in 1911 – forcing Dhaka to become merely a town and the entire project of building a new capital was scrapped. During British Rule the history of urban development in Dhaka was a matter of both setbacks and a progress. Lack of financial resources remained the most important constraint to

Plate 2.3 Mughal ruins in the 1830s: (**a**) Bara Katra, (**b**) Tantibazar bridge, and (**c**) Saat Gumbad Masjid (Charles D'Oyle)

Plate 2.4 Dhaka College in 1904 (Photograph taken by Fritz Kapp in 1904, part of an album of 30 prints from the Curzon Collection; courtesy to British Library photo collections)

municipal urban development. Nevertheless, by the end of 1914, the government was contemplating a renewal programme for old Dhaka. The most noteworthy event of this time occurred in 1917, when the famous British town planner Sir Patrick Geddes visited Dhaka and recommended a detailed master plan for the city, which finally materialised over the next few decades. Dhaka saw some further growth during the Second World War due to its strategic location, and at that time the British built two airports and several hospitals. In the same period, the functional centre of the city around the fort (converted to a jail by the British) had gradually shifted to Victoria Park and the Ramna area which was the location of premier educational institutes such as Dhaka University and Ahsanullah Engineering School (later renamed to Bangladesh University of Engineering and Technology). Areas around Ramna were developed using garden city planning ideas with substantial provision of green spaces. Northward city expansion had been further facilitated with greater accessibility via roads and major thoroughfares from Ramna to Purana Paltan, Shantinagar and Segunbagicha (Fig. 2.1c) (Hyder 1994; Karim 2009; Islam 1996).

2.2.4 Pakistani Period (1947–1971)

When India and Pakistan were granted independence from Britain in 1947, the area previously known as East Bengal became East Pakistan, and Dhaka was declared the capital of this province. A new period of unprecedented growth began because the city had to accommodate people engaged in the new administration, business establishments and large number of people who had migrated from areas now in India. More commercial firms opened premises in Motijheel. The commercial area development around Ramna that experienced growth in the British period extended to the Motijheel area and was designated as the CBD in 1954 (Chowdhury and Faruqui 2009; Islam 1996, 2007). Dhanmondi, an area once covered predominantly by paddy fields, was planned as a residential area. Mirpur road, forming an axis to it, opened up Mohammadpur and Mirpur as additional residential areas in the 1960s to house Muslim migrants from India. The Tejgaon industrial belt, once used by foreign traders, remained officially designated as an industrial area for the newly structured city. An airport was also built at Tejgaon (currently known as the old airport). The Dacca Improvement Trust (DIT) was established in 1956 for planning and management of Dhaka and in 1959 prepared a master plan for an area of 320 $mile^2$. Under the Town Improvement Act of 1953, DIT undertook many city development projects, constructing roads, shopping malls, staff quarters, flats and developing industrial estates, commercial areas etc. Many planned residential areas were also part of this new master plan – notably the Gulshan Model Town, Banani in north-east and Uttara to the further north. The flood-free higher elevation areas along the Dhaka-Tongi axis were given priority as the sites for these model towns. In 1961, Dhaka City covered an area of 12 km^2 (Ahsan 2009) and had a population of less than one million (Chowdhury and Faruqui 2009; Ahsan 2009; Islam 1996, 2007) (Fig. 2.1d).

2.2.5 Bangladesh Period (Since 1971)

When East Pakistan became independent from Pakistan and was renamed as Bangladesh in 1971, Dhaka became the capital of the new country. This historical change in the status of the city also brought significant changes to the physical and socio-economic character of the city. Unprecedented population growth took place during the first decade after independence. This increase in population has led to an expansion of the city's urban footprint. There is no holistic approach to accommodate marginalised groups. The old 1959 Master Plan was used to guide and control growth long after its intended lifeline, and the proposed 1981 Dhaka Metropolitan Area Integrated Urban Development Plan (DMAIUDP) was never implemented. This meant that planners and those responsible for guiding growth or curbing development were inadequately equipped to manage its development. Natural topography in and around the city plays a large part in determining land value around planned residential and commercial areas, and private developers play a significant role in directing its development. Consequently, the old city's core areas had long been consolidated but yet continued to become denser (with a population density of 29,857/km^2) along with established residential areas in and around Dhanmondi, Gulshan, Banani, Baridhara and Uttara. The large number of people moving from rural areas in search of jobs in the city ended up in informal settlements (slums) which grew around these areas and reached the astonishing population density of 220,246 persons/km^2 (Angeles et al. 2009). By 2005 the number of slums was estimated as being 4,966 (Islam et al. 2009). The geographical dispersion of these slum populations is wide, and is caused by the wide occupation of public land by slums of different sizes, that are also in close proximity to areas with higher concentrations of formal jobs (that consequently create more informal employment opportunities) (Nahiduzzaman 2012; Islam et al. 2006, 2009; World Bank 2007). Under the sustained pressure of rural-urban migration, urban expansion has encroached on wetlands and productive farmlands in the fringe and peri-urban areas, even though these areas are subject to monsoonal flooding (Dewan 2013; Dewan et al. 2012; Dewan and Yamaguchi 2009a, b; 2008; Dewan et al. 2007; 2006; Tawhid 2004) (Fig. 2.1e: Dhaka in 1980).

2.3 Development Plans for Dhaka

Urban development planning in Bangladesh does not have a long history. The following sections provide a brief overview of the key planning initiatives that were used to guide Dhaka's urban planning and development pathways since the end of colonial rule.

2.3.1 Dacca[5] Master Plan 1959

The Dacca Master Plan 1959, prepared by DIT, was the first formal plan for Dhaka City. It was prepared for a 20-year time span expiring in 1979. The British

[5] The previous name for Dhaka.

consultants Minoprio Spensely and P.W. Macfarlane completed the project with British technical assistance under the Colombo Plan agreement (Chawdhury 1997). By 1960, the master plan had been drawn up to cover an area of 320 mile^2 that included the then Dhaka City and its adjacent municipalities such as Tongi and Narayangaj. A short report was accompanied by a map of the Dhaka City area at the scale of 1:40,000. The plan assumed a target population of about one million with the following assumptions:

(a) That the river Buriganga, as a major transport artery, will continue to play an important part in the economic life of Dacca.
(b) That the central area of Dacca (the old town) will be maintained as the principal business and shopping centre, particularly for small-scale enterprises.
(c) That the existing population will continue to increase at a rate of 1.75 % per annum.
(d) Expansion will be mainly northward onto flood-free land and will take the form of relatively self-contained new or satellite towns, with a view to reducing pressures on the existing urban areas and lowering the density of the core areas.

2.3.2 *Dhaka Metropolitan Area Integrated Urban Development Plan (DMAIUDP) 1981*

The Dhaka Metropolitan Area Integrated Urban Development Project (DMAIUDP) was the first ever attempt to prepare a *strategic* plan in Bangladesh (i.e. to prepare urban development strategy for the Metropolitan Dhaka with a more comprehensive system of planning). The project was prepared by Shankland Cox Partnership jointly funded by the Government of Bangladesh (GoB), the Asian Development Bank (ADB) and the United Nations Development Programme (UNDP) in 1981. Under this plan, the urban land use and the footprint characteristics of the city were analysed for the first time (Fig. 2.1e). In addition, the consultants came up with two options for review and selection. They were:

Option A: extensive land development immediately adjacent to the existing built-up areas by providing comprehensive flood protection
Option B: adoption of combined peripheral/northern expansion strategy as the basis of recommendations for urban development for future growth

Unlike the 1959 planners, those involved in the DMAIUDP did not underestimate population growth and allowed for a population increase to 6.6 million by 2000 (RAJUK 1993). The main difference was that they included the migration factor, which was missing in the 1959 Master Plan. Based largely on the experience of the 1970s, the DMAIUDP planners estimated almost four million migrants for the 1980–2000 period – 60 % of the total projected growth. There were some similarities between the 1959 Master Plan and DMAIUDP. They both proposed expansion to higher flood-free land – mainly to the north. Both followed strategies of dispersal to minimise pressure on older core areas. Both plans are highly directional, going against the natural trend for growth to be organic, and omnidirectional, and both were heavily dependent on public sector intervention for their implementation (RAJUK 1993).

The DMAIUDP plan, like many other project reports in Bangladesh, was not implemented. Possible reasons for this are that it was not a statutory plan and it was undertaken by the Planning Commission which was not empowered to execute a plan or policy. Therefore, the master plan (1959–1979) stayed in effect for almost double to its designed life, until the enforcement of new plan in 1995. However, many of the findings and deductions of the DMAIUDP study were accurate and were found useful in the preparation of the next development plan (1995–2015) and in the exploration of urban development issues during the last decades of the twenty-first century (Zaman and Lau 2000; RAJUK 1997).

2.3.3 Dhaka Metropolitan Development Plan 1995–2015

A new planning project was initiated in 1992 by the Government of Bangladesh (GoB) in collaboration with UNDP/UNCHS (HABITAT). The work was carried out by a consortium of international firms (Mott MacDonald Ltd. and Culpin Planning Ltd.) together with local consulting agencies (RAJUK 1997). The UN Centre for Human Settlements (UN-Habitat) acted as the executive agency for the overall project. The project was called 'Preparation of Structure Plan, Master Plan, and Detailed Area Plan–Metropolitan Development and Plans Preparation and Management in Dhaka', and its objectives were as follows:

(a) The preparation of integrated development plans and prioritised sectoral plans
(b) Increasing the capacities of RAJUK for making development plans at the metropolitan scale and implementing them with necessary regulatory functions

The project planning component was presented as a package of plan output which dealt urban development issues of Dhaka at a geographical hierarchy of three tiers: subregional, urban and suburban. The following three plans are advocated for dealing with the increasing physical expansion of Dhaka:

(a) Structure Plan (1995–2015)
(b) Urban Area Plan (1995–2005)
(c) Detailed Area Plan (2010)

These plans, collectively termed as Dhaka Metropolitan Development Plan (DMDP), offer development strategies, plans and programmes for 20 years for the entire RAJUK development area, consisting 1,528 km^2. Dhaka City Corporation and five adjoining municipalities (Savar, Narayanganj, Kadam Rasul, Gazipur and Tongi) are included in the DMDP planning area (RAJUK 1997).

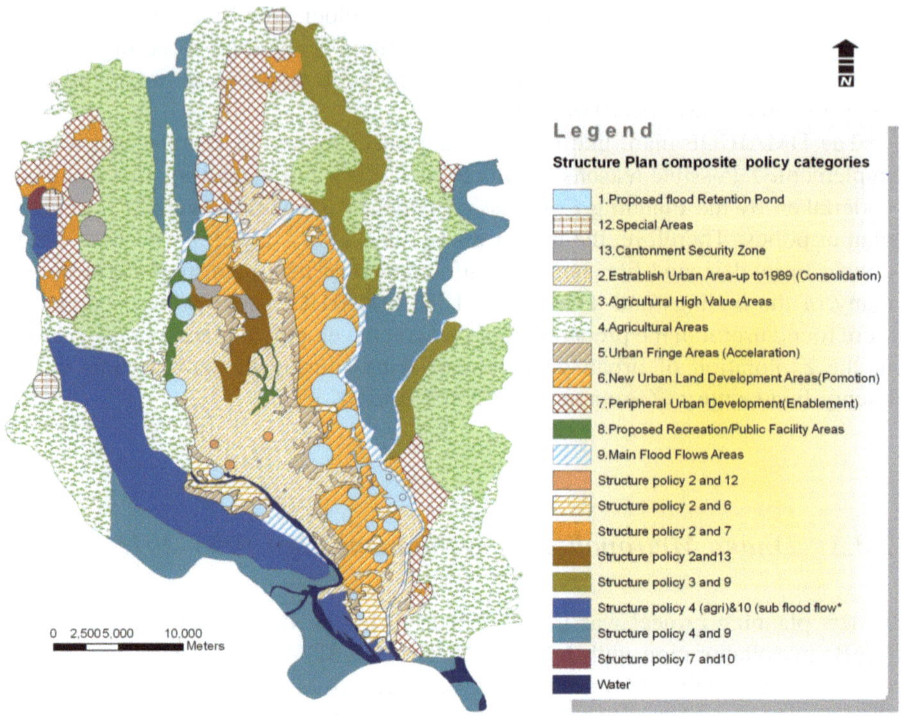

Fig. 2.2 DMDP Structure Plan 1995–2015 (Adapted from RAJUK 1997)

Dhaka Structure Plan (1995–2015)

The DMDP Structure Plan outlines the long-term development strategy for the period of 1995–2015 for the Dhaka Metropolitan Area allowing for a projected population of 15 million. The plan is contained in a report with the principle strategies shown on accompanying maps. A major summary plan embodied in the Structure Plan proposals is presented semi-diagrammatically at 1:50,000 scale. The plan defines a broad set of long-term spatial and sectoral policies for the entire development area under RAJUK's control. These policies are shown by a series of indicative maps, showing the sequence of anticipated development timelines (e.g. future road extensions) (RAJUK 1997) (Fig. 2.2).

The Structure Plan also focussed attention on the protection of the already degraded natural environment and on the effect of further human interventions. Its spatial plans and policies offered protection to high-quality low-lying areas, agricultural land, *khals* (ephemeral water-bodies) and river networks by restricting developments in these areas. In order to integrate these into the urban landscape and assist in beautifying the city, circular waterways round the city and a number of retention ponds in and around the city were also proposed. Additional proposals have been made to expand the city's open space ratio and to deter frequent water-logging and flooding (RAJUK 1997).

Urban Area Plan (1995–2005)

The Urban Area Plan (1995–2005) was proposed as a midterm development strategy for 10 years for the existing metropolitan area and areas that are to be brought under urban development over the next decade. The area covered by the plan was Dhaka City Corporation as well as Tongi-Gazipur and Savar-Dhamsona. The Urban Area Plan was a flexible interim development management document designed to eventually be superseded by the Detailed Area Plan (DAP) (RAJUK 1997).

According to RAJUK (1997), the full DMDP Urban Area Plan comprises an Explanatory Report which also describes the salient features for each of the 26 Spatial Planning Zones (SPZ), the resources map, depicting existing infrastructure locations, along with public and private sector development commitments. It also includes an interim Management Report, interim Planning Rules, a multi-sectoral Investment Programme and an Urban Area Plan, showing designated land-use management zones (RAJUK 1997).

The Detailed Area Plan (2010)

The Detailed Area Plan (DAP) provides detailed planning proposals for specific subareas in accordance with the broader policies and guidelines outlined in the Structure Plan and the Urban Area Plan, respectively. A DAP has been prepared for each of the 26 Strategic Planning Zones (SPZ). Each DAP provides a detailed analysis of the area and acts as a reference document, supported by detailed land-use maps and related information, for effective land management at *mauza* level (RAJUK 1997).

The DMDP, although following similar strategies to the previous plans, seems to have been able to avoid some of their shortcomings. It advocates the use of flood-free land in dealing with rising population growth and pressure. Different plans and strategies have been recommended for areas which are at different stages of urban growth. The strategies have been supplemented by a number of phased programmes and projects, especially the construction and extension of roads.

Unlike the plans in 1959 and 1981, it not only advocates growth in peripheral areas but also foresees growth in flood risk areas in the eastern fringes. This is facilitated by the construction of flood protection measures in areas with high demand for future urban development.

2.4 Assessment of the Success of the Plans

2.4.1 The Dacca Master Plan 1959

Although the Master Plan of 1959 advocated that the majority of the expansion take place in areas in the north and north eastern areas (where new planned residential areas were established at Gulshan and Banani), it also proposed large-scale wetland reclamation on the southern fringe as well. The plan was, in fact, a zoning plan suggesting broader planning guidelines for the city. The planners assumed a modest 40 % population increase over the 20 years, but it would be unrealistic to expect any plan to predict some of the events that took place in those two decades, notably the effects of independence from Pakistan in 1971. Nevertheless, many of the plan's recommendations have been followed and many of its provisions still apply. Although it is half a century old, the plan still remains as the basis for development control within the area.

The major assumptions (RAJUK 1993) that still hold good are:

(a) The continuing importance of the Buriganga River for transport
(b) The continuation of old Dhaka as a business core
(c) The new railway alignment
(d) The impossibility of substantial alleviation of annual flooding

Some of the major assumptions that have been proven invalid are:

(a) Annual growth rate of 1.75 % pa
(b) Non-extension of the cantonment (i.e. Headquarters of the Armed Forces)

There were a number of problems with the 1959 Master Plan. It is arguable that the main problem was that the plan did not foresee the rapid development that the core and surrounding area would experience. That is not surprising since the war of independence and its aftermath have fundamentally changed the sociopolitical-economic life of the city once Dhaka became the capital of an independent nation. Additionally, urban governance and management structures, in their infancy at that time, did not have the structure, manpower or vision to plan for the huge influx of people. The plan did not provide a suitable framework for inter-sectoral coordination of spatial planning, nor was it linked to the Annual Development Programme (ADP) or to the capital budgets of government development agencies. It also suffered from the generic deficiencies of the land-use-type 'master plan' approach which it principally adopted instead of a flexible development planning process. Apparently, it only identified 'where' general types of development are allowed but did not satisfactorily indicate 'what', 'how' or 'when' this development should take place and who should be responsible for it (Chowdhury 2003; RAJUK 1993).

While comparing the planned growth areas in the 1959 to the DMAIUDP 1981 Plan, there are certain geographical areas where urban expansion didn't occur (RAJUK 1993). The major reason for this is that the vast majority of the additional growth had been absorbed via densification in the core areas rather than by outward expansion. On the other hand, the designated areas for planned urban growth in the peripheral areas (i.e. the outer northerly and southerly areas such as Uttara, Tongi and Narayanganj) failed to attract significant growth.

From the DMDP Structure Plan report, it can be seen that the inner urban areas (i.e. areas categorised as urban in 1959) continued to absorb most of the growth within the Statistical Metropolitan Areas (SMA) in the 1990s and remain home to a large proportion of the total population (62 % in 1981; 61 % in 1991) (Ahsan 2009; RAJUK 1997). Over that period there was also a modest increase of the proportion

Table 2.1 Choice of travel mode in relation to distance to work place

	Travel mode			
Distance to work places (km)	On foot	By bus	On foot followed by bus	Total
0.5–2.6	35	–	–	35
2.6–5	34	1	–	35
5–10	–	12	8	20
10–12	–	2	8	10
Total	69	15	16	100

(Adapted from Nahiduzzaman 2012)

of the total population living in outer urban areas, mostly in the north and east, but there was a decline in Narayanganj and on the southern fringes. Population distribution and urban growth in Dhaka seems essentially centripetal in spite of plans and policies aimed at dispersal. This has largely to do with the spatial distribution of economic activity in Dhaka. According to BBS Labour Force Survey (1989), out of the ca. 2.2 million formal jobs that then existed in Dhaka metropolitan area, 42 % are located in the core areas and 90 % of people commuted less than 10 km to the core areas for economic activities (RAJUK 1993). Thus, there is an evidence of strong inertia imposed by the established urban structure with all types of work opportunities remained strongly centralised which has continued until recently, i.e. with majority of the jobs being within 5 km of the central core areas (Table 2.1) (Nahiduzzaman 2012). This explains why plans to disperse the population have only been partially successful and have largely affected the more mobile upper or middle income families only.

A recent study by Nahiduzzaman (2012), using the slum area of Jhilpar as a representative case, suggests that a majority (69 %) of informal jobs such as salesmen, shopkeepers, daily vendors (of vegetables, fish, clothes, recyclable products, etc.), day labourers, rickshaw pullers, CNG driven auto rickshaw drivers (popularly known as CNG driver), taxi cab drivers, house maids and beggars are located within 5 km distance from a slum, and that the primary mode of transport to get to these work locations is on foot (Table 2.1). Those who are employed in formal jobs such as garment industries, and private and public offices as clerks, office assistants responsible for refreshment, night guards, etc. do not commute more than 12 km. Thus, the study found that a typical journey time to work does not exceed 90 min although 30 % of work places are located at distances ranging from 5 to 12 km. As transport costs typically account for 4 % of urban poor's monthly income, travel to work on foot is the most preferred choice (i.e. 69 %) from their residences. This information suggests that the majority of the urban poor seek jobs within the walking distance in the inner city areas (Nahiduzzaman 2012) (Plate 2.5). This, therefore, increases the demand among the urban poor, and particularly newly arrived rural migrants, for accommodation in locations where low-skilled jobs are more easily found. A large proportion of the jobs are in the informal sector and usually located close to economic activities in the formal sector. This has led to the majority (85 %) of informal (slum) settlements being located in the inner part of the city, along the rail line or the flood protection embankment (Nahiduzzaman 2012; Hossain 2008; Syful 2007; World Bank 2007).

Plate 2.5 Slums located next to the railway line (Adapted from Yousuf 2009)

Thus, without a major dispersal of formal economic activities, basic urban amenities and/or policies to persuade new activities in peripheral locations, the movement of significant numbers of people away from existing built-up areas would remain a planner's dream.

Peripheral areas also have generally been less well provided with services and facilities compared to central Dhaka, which is another reason why they did not experience the urban growth, predicted by the 1959 Plan. Utility organisations (water, electricity, gas, sewerage) do not have long-term plans for these peripheral areas. This can be attributed to the lower population densities, which make the areas less attractive for short-term return on investment (ROI). The utilities, therefore, do not work coherently with development plans. This results in continuing poor access to utilities in peripheral areas, despite the potential for longer-term ROI (Zaman and Lau 2000; Islam 1999a, b) (Plate 2.6).

2.4.2 The Dhaka Metropolitan Development Plan 1995–2015

The DMDP was originally thought to have learned from previous failures; however, in reality it was not the case. On the other hand, the planning and management tasks for the DMDP Plan were more complex due to considerable changes in the social and political environment since 1959, and to the city having become more dynamic with very rapid change in both population and the urban footprint. Pressure on the natural environment had also increased. Even though the DMDP team formulated a series of plans at different spatio-temporal scales, growth has outpaced those plans. This is attributable to the fact that the plan-making process did not include any state-of-the art planning, forecasting or decision support tools, and thus, the plans did not benefit from any proper synthesis of trends in socio-economic data, location analysis, separate or integrated land-use-transport modelling (Choudhury 2008). The DAP was not completed until 5 years after its planned start date and became controversial due to there being more breaches and deviations from the plan than adherence (Bari and Efroymson 2009). Planned land-use changes have never been implemented, while much unplanned growth was not foreseen. It has been argued that these deviations from the plan served the vested interests of bureaucratic or

Plate 2.6 Pedestrian congestion in the inner city area of Gulistan (Adapted from Hasan 2013)

political entities, which once again highlights the severe planning and governance issues that are crippling the city's enormous potential (Bari and Efroymson 2009).

Over the years, different authorities have shared responsibility for the planning and development of Dhaka, including its administration, law and order and utilities and services. The present metropolitan governance of Dhaka involves three types of agencies – national, sectoral and local. A total of 22 ministries and 51 agencies are involved in the planning and development of Dhaka (Islam 2005; Siddiqui 2004; Islam 1999b). Even though RAJUK is the main agency for urban development control and planning, the administration of basic services to planning by DCC and other government agencies compounds the issue (Islam 1999a, b). The administrative capacity to manage the increasing size of the city, in terms of area and population, simply does not exist. Efforts to reform both RAJUK and Dhaka City Corporation (DCC) seem counterproductive. For example, a single window approach to provide building permits in 2007 by RAJUK did not work. It failed for a number of reasons such as the ingrained plurality in city planning and management, meaning that RAJUK has to share overlapping functions with other agencies, resulting in non-cooperation from other organisations that want to exert undue authority. In addition, continuous pressure from powerful vested groups militates against the transparency and efficiency of such a service. Although investment in training, more resources and automation is required, the fundamental shift in the culture of city governance is the foremost requirement. The recent dissolution of DCC into Dhaka north and south (Cox 2012) (see Chap. 1) can only further complicate city management since it will produce more planning and implementation agencies within each jurisdiction. The city needs to focus on better system thinking that includes better governance mechanisms, institutional and policy innovation while considering the city as a complex system. Once such innovative thinking can be infused into city governance to *inform* (in terms of indicators and performance measures at different spatial scales that can influence location decisions, individual and policy measures) and *to engage* different actors and institutions with different values, beliefs and interest for collaborative consensus building through discussion, dialogue and deliberation, only then can *clumsy*[6] or functional solutions beneficial *for all* (instead of *elegant* solutions beneficial *to a few* or detrimental to the system) can be achieved (Farooque 2011; Thompson 2008).

2.5 Concluding Remarks

Throughout its evolution from a town to a megacity, the growth and development of Dhaka always proceeded at an escalated pace, especially since the beginning of the British regime and after becoming a sovereign state capital in 1971. This primacy, coupled with the absence of an anti-urbanisation policy, has turned Dhaka into a megacity but without the support of necessary infrastructure facilities and urban amenities. Consequently, the city is facing a large number of problems and challenges that seem to exceed the capacity of the administrative authority to overcome them. Additional stresses are expected to be imposed by climate change, placing further pressure on the existing financial, technical, administrative and governing capacity of the various relevant city management authorities. This leads to an intriguing question: What kind of policies and actions would be required to make Dhaka resilient to both foreseeable and unforeseeable problems caused by urbanisation?

Rural-urban economic disparity is increasing, together with an increase in disparity between Dhaka and other major cities in Bangladesh. From the governance and planning perspective, there should be proper actor oriented regional and national policies and sustainable actions, involving relevant government agencies, development partners, and other stakeholders. These should embody the necessary

[6] The term *clumsy* is used by Thompson (2008) to describe a solution in which no single stakeholder gets all that they desire, but all stakeholders get some of what they desire.

rethinking of approaches while considering the city as a complex system and thus minimise such interurban and interregional differences. Arguably, while a deliberate attempt has been made to maintain Dhaka's status as being the primary city, there should be better policy and institutional innovation that can engage stakeholders and bring information and knowledge to effective city planning and management of Dhaka. The city also needs to learn and gain from research that is actively seeking for novel, tailored solutions, and evaluating current and past trends with similar international contexts and producing cutting-edge data that can assist in more effective and current decision and planning support. In this regard, plans with shorter periods can provide more room to accommodate change and cope better with rapidly changing structure of Dhaka Megacity.

References

Ahsan RM (2009) Changing pattern of the commercial area of Dhaka city. In: Ahmed SU (ed) Dhaka: past present future, 2nd edn. The Asiatic Society of Bangladesh, Dhaka, pp 431–450

Angeles G, Lance P, Barden-O'Fallon J, Islam N, Mahbub AQM, Nazem NI (2009) The 2005 census and mapping of slums in Bangladesh: design, selected results and application. Int J Health Geogr 8:32

Bangladesh Bureau of Statistics (BBS) (2003) Population census 2001: national report (provisional). Bangladesh Bureau of Statistics, Ministry of Planning, Dhaka

Bari M, Efroymson D (2009) Detailed Area Plan (DAP) for Dhaka Metropolitan Development Plan (DMDP): a critical review. WBB Trust, Dhaka

Butsch C, Etzold B, Sakdapolrak P (2009) The megacity resilience framework. Institute for Environment and Human Security, United Nations University, Bonn

Chawdhury A (1997) Evolution of urban planning in Bangladesh. National Geographical Association, Dhaka

Choudhury AKMK (2008) Land use planning in Bangladesh. A H Development Publishing House, Dhaka

Chowdhury IU (2003) The role of RAJUK in planned urban development. In: Islam N (ed) World Habitat Day 2003: souvenir on water and sanitation for cities. Bangladesh Institute of Planners and Center for Urban Studies, Dhaka, pp 88–91

Chowdhury A, Faruqui S (2009) Physical growth of Dhaka city. In: Ahmed SU (ed) Dhaka: past present future, 2nd edn. The Asiatic Society of Bangladesh, Dhaka, pp 56–76

Cox W (2012) Evolving urban form: Dhaka, New geography. Available at: http://www.newgeography.com/content/003004-evolving-urban-form-dhaka. Accessed 23 Sept 2012

Dewan AM (2013) Floods in a megacity: geospatial techniques in assessing hazards, risk and vulnerability. Springer, Dordrecht

Dewan AM, Yamaguchi Y (2008) Effects of land cover changes on flooding: example from Greater Dhaka of Bangladesh. Int J Geoinform 4(1):11–20

Dewan AM, Yamaguchi Y (2009a) Using remote sensing and GIS to detect and monitor land use and land cover change in Dhaka Metropolitan of Bangladesh during 1960–2005. Environ Monit Assess 150(1–4):237–249

Dewan AM, Yamaguchi Y (2009b) Land use and land cover change in Greater Dhaka, Bangladesh: using remote sensing to promote sustainable urbanization. Appl Geogr 29(3):390–401

Dewan AM, Yeboah KK, Nishigaki M (2006) Flood hazard delineation in Greater Dhaka, Bangladesh using integrated GIS and remote sensing approach. Geocarto Int 21(2):33–38

Dewan AM, Islam MM, Kumamoto T, Nishigaki M (2007) Evaluating Flood hazard for land-use planning in Greater Dhaka of Bangladesh using remote sensing and GIS techniques. Water Resour Manag 21(9):1601–1612

Dewan AM, Yamaguchi Y, Rahman MZ (2012) Dynamics of land use/cover changes and the analysis of landscape fragmentation in Dhaka Metropolitan, Bangladesh. GeoJournal 77(3):315–330

Farooque M (2011) Dhaka megacity: from reactive government to anticipatory governance, BAPA-BEN urbanization conference, January 8, BIM, Dhaka

Hasan M (2013) GFDL. http://www.gnu.org/copyleft/fdl.html, Retrieved from http://commons.wikimedia.org/wiki/File%3ADhaka_Crowd_in_Gulistan_Crossing.JPG on 1 Feb 2013

Hossain S (2008) Rapid urban growth and poverty in Dhaka city. Bangladesh e-J Sociol 5(1):1–25

Hyder Z (1994) Organic cities and the case of Patrick Geddes in Dhaka. Department of Architecture, Massachusetts Institute of Technology, Cambridge, MA

Islam N (1996) Dhaka from city to megacity: perspectives on people, places, planning, and development issues. Urban Studies Programme, Dhaka

Islam N (1999a) Urbanisation, migration and development in Bangladesh: recent trends and emerging issues, Centre for Policy Dialogue (CPD) and United Nations Population Fund (UNFPA), CPD-UNFPA publication series no 1, Dhaka, pp1–24

Islam N (1999b) Dhaka city: some general concerns. Asian cities in the 21st century: contemporary approaches to municipal management, vol 3, Reforming Dhaka city management. Asian Development Bank (ADB), Manila, pp 71–82

Islam N (2005) Dhaka now: contemporary urban development. The Bangladesh Geographical Scociety, Dhaka

Islam S (2007) Banglapedia: national encyclopaedia of Bangladesh. In: Miah S, Ahmed W, Chowdhury AM, Rahman SMM, Siddiqui K, Kabir SMH (eds), Asiatic Society of Bangladesh, Dhaka

Islam N, Angeles G, Mahbub A, Lance P, Nazem NI (2006) Slums of urban Bangladesh: mapping and census, 2005. Centre for Urban Studies, Dhaka

Islam N, Mahbub AQM, Nazem NI (2009) Urban slums of Bangladesh. Center for Urban Studies, Dhaka

Karim A (2009) Origin and development of Mughal Dhaka. In: Ahmed SU (ed) Past present future, 2nd edn. The Asiatic Society of Bangladesh, Dhaka, pp 34–55

Kraas F (2007) Megacities and global change in East, Southeast and South Asia. ASIEN 103 (4):9–22

Moreno EL, Bazoglu N, Mboup G, Warah R (2008) State of the world's cities 2008/2009: harmonious cities. United Nations Human Settlements Programme (UN-Habitat), Nairobi

Nahiduzzaman KM (2012) Housing the urban poor: an integrated governance perspective – a case study on Dhaka, Bangladesh. PhD dissertation, The Royal Institute of Technology, Stockholm

Shankland Cox Partnership and Others (1981) Dhaka metropolitan area integrated urban development project, Planning Commission, Government of Bangladesh, Dhaka

Rajdhani Unnayan Katripakko (RAJUK) (1993) Strategic growth options – Dhaka 2016. UNDP, UN-HABITAT, RAJUK, Dhaka

Rajdhani Unnayan Katripakko (RAJUK) (1997) Dhaka Metropolitan Development Plan, 1995–2015, Vol I & II. RAJUK, Dhaka

Siddiqui K (2004) Megacity governance in South Asia: a comparative study. The University Press Limited, Dhaka

Syful M (2007) Physical density and urban sprawl: a case of Dhaka City. Department of Urban Planning and Environment, Royal Institute of Technology (KTH), Stockholm

Tawhid KG (2004) Causes and effects of water logging in Dhaka city, Bangladesh. Department of Land and Water Resource Engineering, Royal Institute of Technology (KTH), Stockholm

Thompson M (2008) Organising and disorganising: a dynamic and non-linear theory of institutional emergence and its implication. Triarchy Press, Devon

United Nations (2012) World urbanization prospects: the 2011 revision-highlights. Department of Economic and Social Affairs, Population Division, New York

United Nations (UN) (2006) World urbanization prospects: the 2005 revision. Department of Economic and Social Affairs, Population Division, New York

United Nations (UN) (2008) World urbanization prospects: the 2007 revision. Department of Economic and Social Affairs, Population Division, New York

United Nations (UN) (2011) Population distribution, urbanization, internal migration and development: an international perspective . Department of Economic and Social Affairs, Population Division, New York

United Nations Population Fund (UNPFA) (2007) State of world population 2007: unleashing the potential of urban growth. In: Marshall A (eds), United Nations Population Fund, New York

World Bank (2007) Dhaka: improving living conditions for the urban poor, Bangladesh development series no. 17. The World Bank Office, Dhaka

World Bank (2009) World development report 2009: reshaping economic geography. The World Bank, Washington, DC

Yousuf BA (2009) Mysteries Bangladesh. Available at: http://asadbinyousuf.wordpress.com/2009/03/02/slums-situated-near-the-komlapur-rail-station/. Accessed 24 Jan 2013

Zaman QM, Lau SSY (2000) City expansion policy versus compact city demand: the case of Dhaka. In: Jenks M, Burgess R (eds) The compact cities: sustainable urban form for developing countries. Spon Press, London, pp 141–152

Chapter 3
Spatiotemporal Patterns of Population Distribution

Robert J. Corner, Emmanuel T. Ongee, and Ashraf M. Dewan

Abstract Population studies worldwide have suggested that urban population densities generally follow an exponential decay pattern as one travels outwards from the central business district (CBD). Dhaka has experienced phenomenal population growth over the past two decades. This chapter uses econometric and GIS techniques to map and model recent population dynamics using census data for three successive census years (1991, 2001 and 2011) aggregated at the lowest level of census geography. Linear and non-linear regression models were tested to examine urban density form. The study found that a negative exponential function was best suited for the study area since it produced the highest coefficient of determination (R^2). Additionally, temporal trends of the population density gradient for the study area revealed gradual flattening. Further, it was found that the *y*-axis intercept (an indicator of CBD density) did not drop over time as general theories for cities experiencing economic growth would suggest. The visualisation of population change was conducted through standard deviational ellipses and simple spatial analysis. The results revealed that, with the exception of a few census tracts, the magnitude of population change is (are) still high in the area, and that a suburbanisation trend has set in over the period since the penultimate census.

Keywords Population growth • Population density • Spatial analysis • Monocentric model • Demographic gradient

---

R.J. Corner (✉) • E.T. Ongee • A.M. Dewan
Department of Spatial Sciences, Curtin University,
Kent Street, Bentley, Perth 6102, WA, Australia
e-mail: R.Corner@curtin.edu.au; e.ongee@curtin.edu.au; A.Dewan@curtin.edu.au

A. Dewan and R. Corner (eds.), *Dhaka Megacity: Geospatial Perspectives on Urbanisation, Environment and Health*, Springer Geography,
DOI 10.1007/978-94-007-6735-5_3,

3.1 Introduction

The spatial distribution of human population is a key factor in assessing the causes and impacts of global environmental change (Tian et al. 2005; Elvidge et al. 1997). Accurate and timely information on the variation of population distribution both within and between urban areas is crucial for a number of social, economic and environmental applications (Liu et al. 2008; Sutton 1997). The analysis of population distribution can support planners and policy makers in identifying impacts related to social and environmental functionality of the city (Millward and Bunting 2008) and subsequently assist in formulating relevant policy measures. The density and spatial patterning of human population has, therefore, been a subject of academic research in a number of disciplines for many years (Fonseca and Wong 2000).

Geospatial techniques have been extensively used in estimating population in urban areas around the world, as is shown by a recent review of the topic (Patino and Duque 2013). Whilst remotely sensed data has been used to estimate population which can then be correlated with census or ground-based measurement through regression techniques (Wu et al. 2005), Geographic Information Systems (GIS), on the other hand, are used to illustrate the spatial patterning of population via spatial analysis. The major objective of GIS-based population study is to examine urban density gradients and to capture urban spatial structuring with econometric and spatial analyses following the pioneering work of Clark (1951). The basic premise of Clark's work is that density in urban areas declines, in a negative exponential manner, with the improvement of transportation and communication over time. Thus, urban population distribution reduces with distance from a central point and flattens with time as people become vehicle dependent. Inspired by Clark's work, others have tested and refined his classic work to examine the geographical distribution of population and employment density in urban areas over time. For example, Newling (1969) conceptualised four stages of density gradients as youth, early maturity, late maturity and old age. Batty and Kim (1992) argued that an inverse power function is the most appropriate to model urban population density, whilst Parr (1985) suggested that an inverse power function is more appropriate to the urban fringe and hinterland, with a negative exponential function being more suitable for explaining density in an urban area. Nonetheless, the goodness of fit varies according to the place studied (Wu et al. 2005).

Of the studies that use econometric and spatial analytical techniques to model and map urban population distribution, most are from developed countries with only a few from developing countries (Feng et al. 2009; Suárez and Delgado 2009; Millward and Bunting 2008; Griffith and Wong 2007; Mennis 2003; Wang and Zhou 1999; Wang and Meng 1999; Sutton 1997; Martin 1996, 1989; Gordon et al. 1986; Alperovich 1982; Griffiths 1981; Krakover 1985; Berry and Kasarda 1977; Newling 1969). The majority of studies look at one of two density function approaches, monocentric models and polycentric models. The monocentric model investigates how density varies with distance within an urban area with an assumption that a city has only one centre (e.g. central business district) in which population and economic activities are concentrated, whilst the latter model assumes that a city has a number of sub-centres other than the city centre and urban residents have access to all centres. Thus, population densities are functions of distances to all these centres (Small and Song 1994). As noted by Wang and Meng (1999), the polycentric model examines the influence of each sub-centre on the distribution of urban population in a given area. A somewhat differing result was reported by McMillen and McDonald (1997) in their study on Chicago's employment density, suggesting bicentric density patterns. By contrast, McDonald (1989) reviewed urban density studies and made two general findings: (1) that the density gradients flatten over time and (2) the rate of flattening of the density curve depends on the size of the urban area. Adam (1970) noted that the density pattern is related to the age of the city. In other words, older cities have a higher central density, whereas younger cities tend to be steeper in density gradients. Studies in developing countries yielded different results when compared to those from developed countries. For instance, Berry and Kasarda (1977) reported that, in Calcutta, the density gradient did not flatten over time and central density increased contemporaneously with expansion of urban areas. This result is contrary to findings from China which showed that Beijing's density gradient flattened over time (Wang and Zhou 1999), and attributed this to the polycentric nature of the city, which is partly the result of China's reforms towards market economy (Feng et al. 2009). These two examples from China and India suggest that there is no clear pattern to the trends of urban population distribution in developing countries, whilst the results from developed countries are almost identical.

Since its independence from Pakistan in 1971, Bangladesh has experienced rapid growth in human population (see Chap. 1). This growth has impacted upon some cities more than others due mainly to the economic disparity between rural and urban areas (BBS 2010). As a result, there has been considerable conspicuous growth in the urban population, severely affecting the social and environmental sustainability of major cities, particularly the megacity of Dhaka. Siddiqui and his colleagues (2010) suggest that as a result of this rapid population growth, most of Dhaka, except for a few high-class residential areas, appears to be either slums or near-slums.

Despite the fact that population increase is an ever-present phenomenon in Bangladesh, research with respect to its distribution is mainly confined to temporal analysis (Rouf and Jahan 2007; Islam 2002; Khan and Rahman 2000; Eusuf 1996). Based on population and the rank-order method, Eusuf (1996), for instance, demonstrated the changing patterns of urban centres in Bangladesh between 1872 and 1991 which showed that both the number and size of urban centres in the country had increased noticeably as a function of population growth. Similarly, other researchers have also used change in population count in urban areas through time to demonstrate population increase in the cities, particularly in Dhaka (Rana 2011; Siddiqui et al. 2010; Islam 2009, 2005). The major objective of these temporal studies is to quantify the rate of urbanisation or urban growth of cities. Though these studies are useful for understanding population growth in urban areas over time, they lack the spatial component that is essential to predicting the environmental consequences of population growth in urban areas (Shoshany and Goldshleger 2002; Elvidge et al. 1997). Moreover, temporal studies alone are unable to examine the pattern of increase or decrease in population distribution. Since population is the primary factor that is driving urban expansion and environmental changes in Dhaka (Dewan et al. 2012), mapping and modelling of population distribution is of great importance to urban planners and policy makers who are often charged with monitoring urban growth and developing adequate land-use policies.

Although many studies of population distribution over space have been carried out to date, particularly in post-industrial cities, work is required to determine whether the urban density models found to be applicable in developed countries can also be applied to less developed countries, especially to cities with totally different population dynamics such as in Dhaka. This chapter seeks to better understand the population dynamics of Dhaka megacity through quantifying temporal changes in population density and exploring its spatial distribution across the study area (see Chap. 1). Of particular interest is the investigation of the well-established theory that population density is generally inversely related to distance from a city's central business district (CBD).

3.2 Data and Methods

3.2.1 Data Preparation

The data used for the analysis consisted of both spatial and aspatial data. One of the biggest problems in working with spatially referenced data in Bangladesh is the unavailability of updated census boundary information, and even when available, they are often inaccessible to the public because of the so-called security concern of relevant organisations and the national government. Although, socioeconomic and demographic data from the census are available in tabular format, they are not accompanied by a spatial database such as the TIGER files in the USA. As a result, it is extremely difficult for anyone to study spatially referenced population distribution.

To overcome this, a spatial database of the census tracts was created by digitising the small area atlas of the Bangladesh Bureau of Statistics (BBS), with input from the digital database of SPARRSO, Center for Environmental and Geographic Information Services (CEGIS) and numerous field visits. It should be noted that neither of the databases contained up-to-date digital census tract boundaries, and around 2 % of the census tracts were missing. That is, 25 new census districts were created between the decennial censuses of 1991 and 2001. To overcome this problem, the names of the 1991 census tracts were first matched with the 2001 names using the BBS community series publications. The additional census polygons in the 2001 census are the result of subdivision, due to population increase, of polygons used in the previous census. To create the polygons for the 2001 census, the digitised polygon files were re-referenced to the street grid, which is an up-to-date street network from the Detailed Area Plan (DAP) of the Capital Development Authority (RAJUK). Census units created between 1991 and 2001 were then demarcated and mapped out in the field. This work was guided by a hard copy of a map which highlighted the road networks that were used to split the original census tracts into those use for the 2001 population census and was obtained from BBS. Additionally, a high-resolution mobile-mapping GPS device (Trimble Nomad 800GXE) was used to confirm the road locations. The editing of the census tracts was performed in ArcGIS (v. 10). The final census tract file consists of 1,212 polygon features, which include 441 rural communities (*mauza*) and 771 urban communities (*mahalla*) (Dewan 2013). Again between the 2001 and 2011 censuses, some of the census tracts were subdivided. Since no spatial data was available for these new tracts, the older geography (2001) was used and the demographics for the new constituent tracts were amalgamated.

Demographic attributes were obtained from the BBS community series (BBS 1993, 2003, 2012), which represent the population censuses of 1991, 2001 and 2011. These data were encoded in a spreadsheet because digital versions were not available for the 1991 and 2001 censuses, and then linked with the census tract boundary using a unique ID. Population density was calculated as the total number of people residing in a census tract divided by its total area (km^2).

3.2.2 Modelling Urban Population Density

The central business district (CBD) is critical to population density modelling. In order to test different models of urban population on the data for Dhaka, the CBD had to be identified beforehand. With reference to previous research, Motijheel was chosen as the CBD since it has featured prominently in previous population-related studies of Dhaka (Barter 2012; Islam et al. 2009). Whilst some studies have used mathematical techniques to choose CBD (see Alperovich 1982), this research has opted to rely on expert knowledge instead. Using the Euclidean distance function within a GIS, distance to other areas was calculated, and four regression models, namely, linear, reverse exponential, negative exponential and power, were constructed to identify whether the urban density gradient is similar or dissimilar to that of other cities in the world. To minimise the impact of outliers on density gradient analysis, statistical outliers were removed from the data using the method proposed by Iglewicz and Hoaglin (1993) which detects outliers based on modified Z-score values, M_i:

$$M_i = 0.6745 \frac{(x_i - \tilde{x})}{MAD} \tag{3.1}$$

where MAD = median absolute deviation and $\tilde{x}$ = median. In this method, modified Z-scores with an absolute value of greater than 3.5 are considered outliers. For this study, the observations (x_i) were the population densities for each of 3 years. Further, units with zero values of population density and the CBD itself were disregarded from the analysis. This was done to ensure that logarithmic models did not fail since the logarithm of zero is undefined. It was also thought unreasonable for an entire *mahalla* to have a population of zero. The following equations were used to derive an urban density gradient for the 3 years considered:

$$\rho(r) = a + br \tag{3.2}$$

where r = distance from centre of urban area (km), b = slope and a = intercept.

$$\rho(r) = ae^{-rb} \tag{3.3}$$

where a = maximum population density (km^{-2}), r = distance from centre of urban area and b = exponential decay coefficient or density gradient (km^{-1}).

$$\rho(r) = ar^{b} \tag{3.4}$$

where r = distance from centre of urban area, $\rho(r)$ = population density at distance r from CBD (km^{-2}), a = intercept (population density (km^{-2}) at distance zero from CBD) and b = growth/decay rate.

$$\rho(r) = a + b\ln(r) \tag{3.5}$$

where r = distance from centre of urban area (km), a is the intercept and b (ln) = slope.

The linear and logarithmic models were estimated using ordinary least squares regression (OLS), whilst the negative exponential and power models were estimated using non-linear least squares (NLS) regression. The NLS approach ensured that the dependent variable (population density) was not log-transformed and remained the same in all cases hence making the different models comparable (Feng et al. 2009; Wang and Meng 1999). This technique is an extension of OLS which has the advantage of being able to fit a broad range of functions and the disadvantage of requiring iterative optimisation procedures to compute parameter estimates (NIST/SEMATECH 2003).

3.2.3 Mapping Population Distribution

To determine the distribution of population in the three census years, we first used the standard deviational ellipse to gain a better understanding of the geographical aspects of population. The standard deviational ellipse (SDE) examines the standard deviation of the features from the mean centre separately from the x-coordinates and the y-coordinates to define the axes of the ellipse (Mitchell 2005), and it is very useful to assess the orientation of a dataset (Wong and Lee 2005). Using population density as a weight field for each year, three weighted ellipses with a radius of one standard deviation were derived to identify those areas with the highest concentrations of population.

To map the magnitude of population change between the three census years, we considered the population count in each census tract. We then computed percentage of change for each tract for the periods of 1991–2001, 2001–2011 and 1991–2011. Next, the direction of change was labelled as positive or negative (depending on the sign of percentage population change values). Standardised Z-scores of the percentage change values were computed for the three periods. The scores were subsequently classified on the basis of their standard deviation. Finally, a series of change maps were produced to show the amount and direction of population change for each spatial unit. The maps measured change direction as negative and positive and magnitude of change as scaled values (1–4): 1 (Z-score <-1.96), 2 (Z-score between -1.96 and 0), 3 (Z-score between 0 and 1.96) and 4 (Z-score >1.96).

3.3 Results and Discussion

Table 3.1 shows the population characteristics in the study area. Census tracts have an average area of 0.72 km^2, ranging from 0.001 to 11.05 km^2. The population shows an increasing trend, suggesting that the highest population of a census tract was 72,836 in 1991 which grew 174,048 in 2011, an increase of 101,212 within 20 years time. Population density is also extreme in the study area which may be attributed to vertical expansion of the city in recent years. Another explanation could be the existence of slums in the many census tracts which are usually places of high density (CUS et al. 2006). The average size of the population is also increasing so as the average density (Table 3.1).

We fitted the monocentric density function using both linear and non-linear least square models after excluding zero-density spatial units and outliers. The best model was chosen as one with the highest overall R^2 value (Wang and Zhou 1999). It can be seen from the estimated models in Table 3.2 that the negative exponential function was dominant and therefore provided the best fit in all 3 years relative

Table 3.1 Descriptive statistics of population characteristics in the study area

		1991		2001		2011	
	Area (km^2)	Population	Density (km^2)	Population	Density (km^2)	Population	Density (km^2)
Minimum	0.001	0	0	2	22	15	27
Maximum	11.05	72,836	170,692	143,208	177,434	174,048	179,215
Mean	0.72	4,513	25,106	6,718	33,112	9,554	39,277
Std. dev.	1.30	6,358	32,923	10,648	39,730	16,179	43,420

Table 3.2 Monocentric density functions in the study area

Model	Year	Equation	R^2
Linear	1991	PDEN[a] = 53,147.2–2,729.7 (distance)	0.3521
	2001	PDEN = 67,292–3,327 (distance)	0.3593
	2011	PDEN = 74,666–3,445 (distance)	0.3225
Exponential	1991	PDEN = 78,640 $e^{-0.1582\ (\text{distance})}$	0.4148
	2001	PDEN = 92,550 $e^{-0.1363\ (\text{distance})}$	0.4025
	2011	PDEN = 94,740 $e^{-0.1108\ (\text{distance})}$	0.3462
Power	1991	PDEN = 71,321.6613 $(\text{distance})^{-0.5553}$	0.3133
	2001	PDEN = 87,551.7432 $(\text{distance})^{-0.5168}$	0.3044
	2011	PDEN = 94,175.4143 $(\text{distance})^{-0.4648}$	0.2644
Logarithmic	1991	PDEN = 71,365.7–23,090.3 ln(distance)	0.3986
	2001	PDEN = 88,117–27,450 ln(distance)	0.3870
	2011	PDEN = 94,817–27,713 ln(distance)	0.3303

[a]PDEN = population density

to the others. This finding is in agreement with Clark's (1951) theory that urban population densities decay exponentially with increasing distance from the city centre. This finding is also in agreement with other studies in China (Feng et al. 2009; Wang an Meng 1999) but differs from the study of Calcutta (Berry and Kasarda 1977), another city in a developing country. As noted above, the model fit depends on the area being analysed. Additionally, the steady decline in the absolute value of the estimated density gradient parameter (*b*) of the negative exponential function in three census years confirms the econometric model that predicts flattening of population density gradients as cities grow and economies develop (McDonald 1989).

Whilst the steady decline in population density gradients over the three census years in the study area suggests the onset of suburbanisation, particularly very recently, it was also noticed that the model intercept (*a*) increased with every subsequent year. Since the model intercept predicts population density of the CBD, the data shows that the population density of the CBD has not yet peaked. Contrary to expectations, this research found that the y-axis intercept does not drop over time. This trend is peculiar since one would expect the density to fall in the presence of suburbanisation and economic growth. Since Dhaka is 400 years old, the increase in model intercept could be related to age of the city (Adam 1970). However, we do not have a method to justify this claim. Further study is, therefore, needed to examine this phenomenon.

As shown in Table 3.1, a noticeable feature is that the density per unit area is increasing with time. For instance, the highest density was 170,692 persons/km^2 in 1991 which rose to 177,434 and 179,215 in 2001 and 2011, respectively. Note that the density gradually reduces away from the city centre, suggesting horizontal dispersal since 2001, but the rate appears to be very slow. As low-lying flood-prone areas surround the main part of Dhaka, this may have influenced the geographic distribution of population. Further grouping of density by distance category shows that the population density is highest between 0 and 5 km from the CBD and it is still rising (Fig. 3.1). A reasonable explanation is that the concentration of

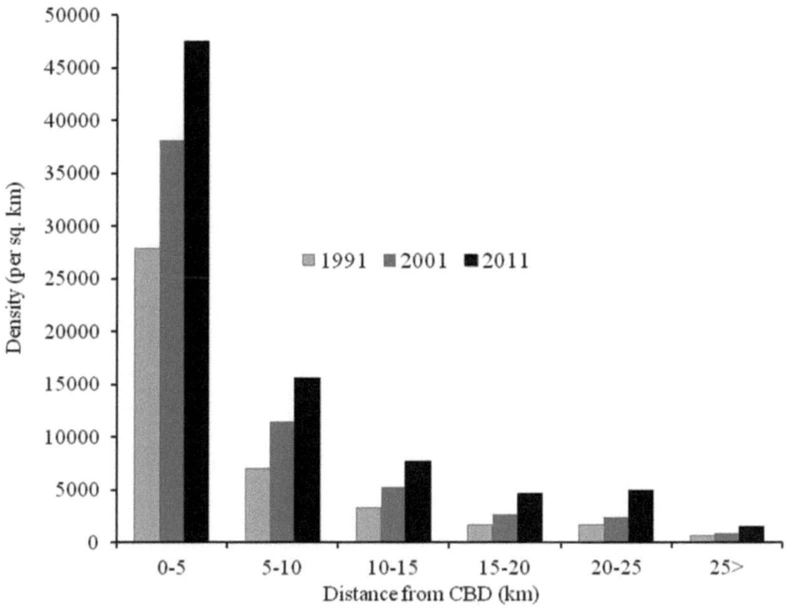

Fig. 3.1 Density of population between 1991 and 2011, by distance category

educational and administrative organisations and economic activities is very high in this zone compared to other distance categories. In 1991, the density of the 0–5 distance category was 27,912 persons/km^2 (aggregated) which increased to 38,094 and 47,522 in 2001 and 2011, respectively, revealing that the consolidation of old core areas (known as Old Dhaka) is still continuing (see Chap. 2).

The second highest density by distance measure can be found between 5 and 10 km from CBD (Fig. 3.1). Instead of decreasing in density, especially at the city centre, the result showed that the rate of increase in aggregated population density is increasing with each 5 km increment from the CBD. This trend became highly pronounced in 2011, which is clearly linked with the population pressure in Dhaka.

Three weighted standard deviational ellipses (SDE) have been derived for three census years and shown in Fig. 3.2a–c. These provide an overview of the standard deviation of the population distribution and the direction of the ellipses. It can be seen that the rotation of ellipses has increased between 1991 and 2011, suggesting the ellipses are becoming skewed towards the northwest-southeast. In addition, it gives an illustration of the location of the main population concentration which is in agreement with econometric analysis of population density function above. The same figure also shows three choropleth maps of total population according to census tracts, from which it is easily seen that the population concentration in the peripheral areas is increasing with time owing to the fact that there is little space left in the old core areas. However, this type of development is known to have serious implications for the environment as peripheral areas are mostly floodplain and wetlands in Dhaka (Dewan and Yamaguchi 2009a, b; 2008).

The spatial distribution of population change is shown in Fig. 3.3. This indicates that most of the spatial units in the study area have registered population growth,

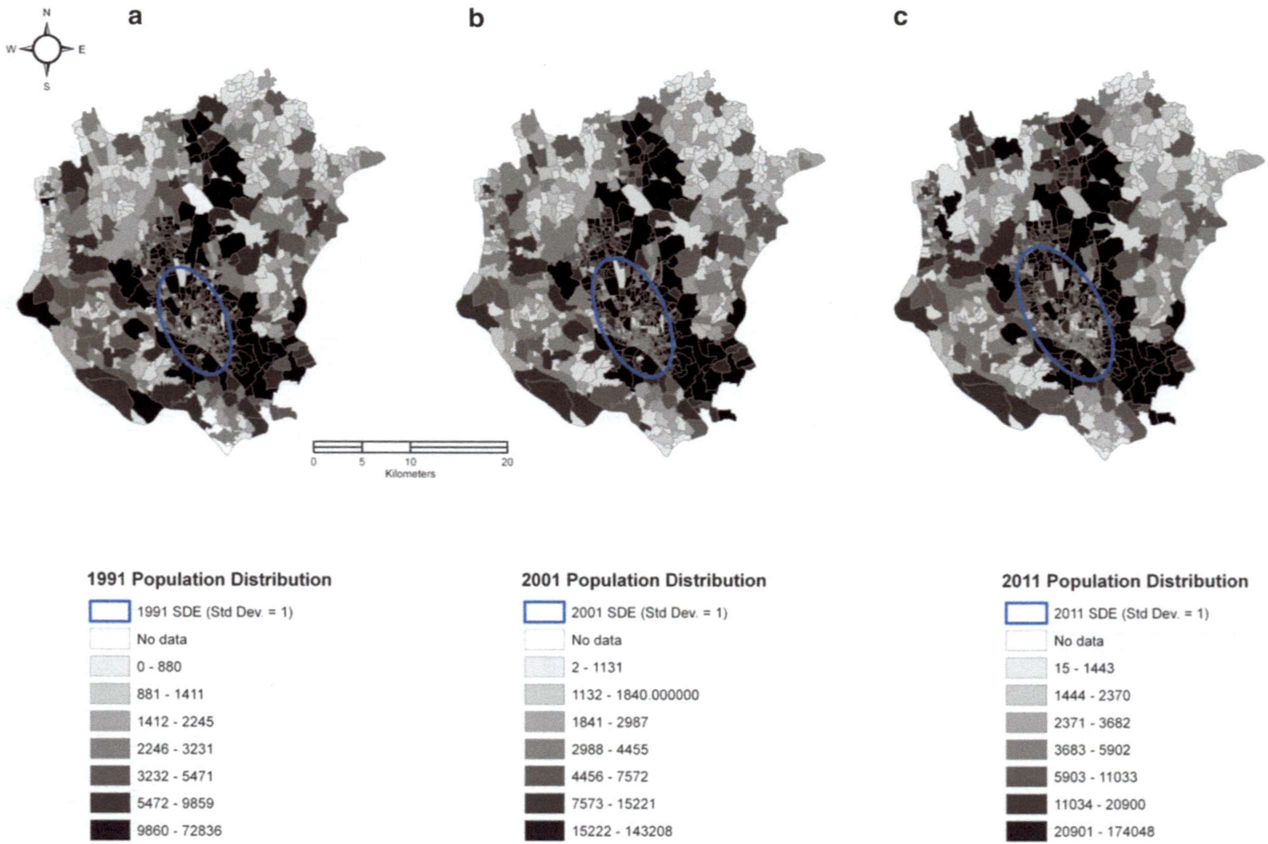

Fig. 3.2 Population distribution and SDE in three census periods (**a**) 1991, (**b**) 2001, (**c**) 2011

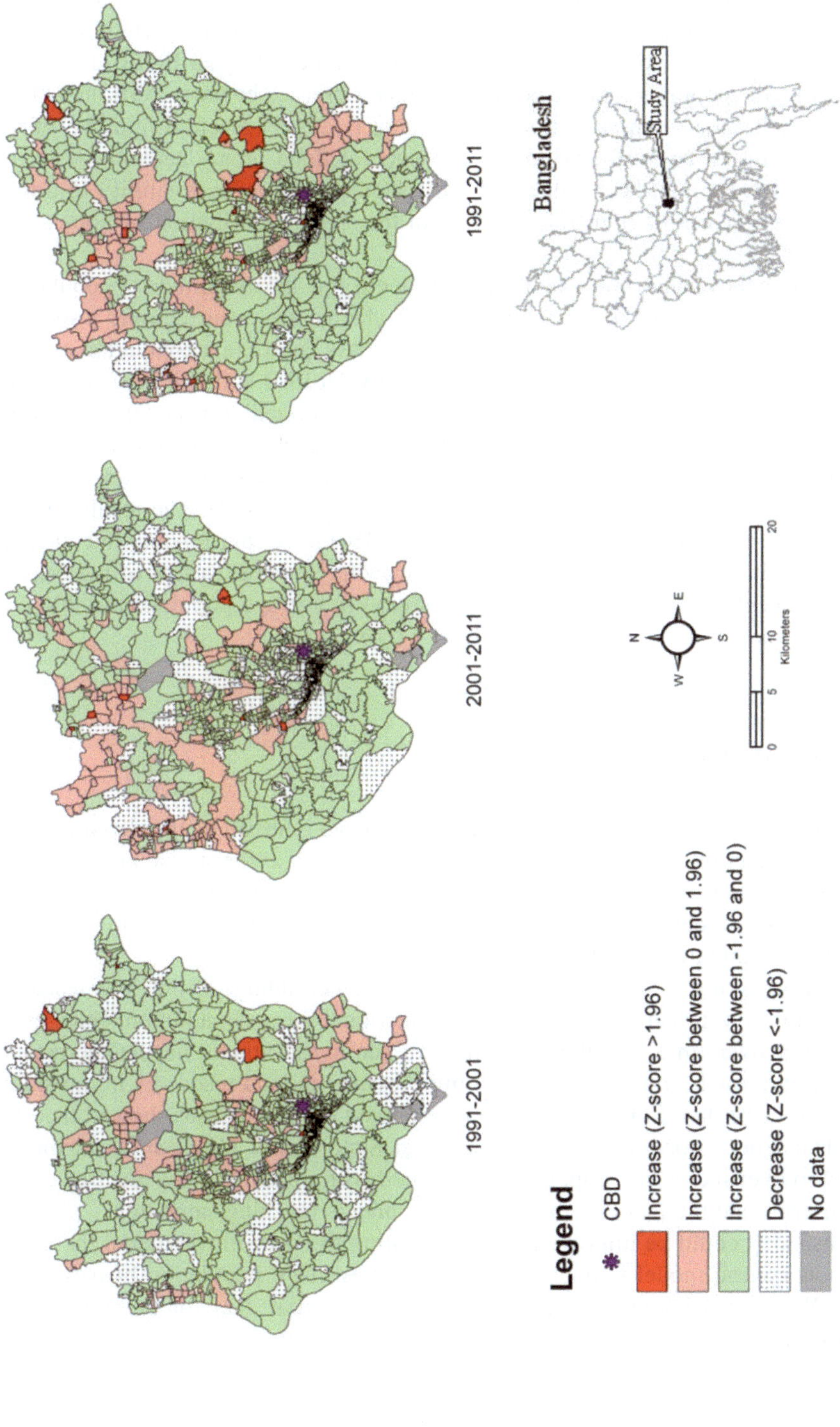

Fig. 3.3 Changes in population between 1991 and 2011, by census tracts

whereas a few of them showed a decrease in total population. For example, in 2001 a total of 247 census tracts experienced decline in population relative to their 1991 population. Whilst a few census tracts located in the central part of the city registered decline in population in 2001, this number reduced by 2011. At the same time, the population in the peripheral areas increased between 2001 and 2011, indicating that the suburbanisation process is of relatively recent origin, but fairly slow. This might be attributed to the existence of flood-prone lands in the study area. Apart from that, land value and other factors such as accessibility to urban centres may have important roles to play in the spatial distribution of population (Dewan 2013).

Although the change maps are useful for identifying the magnitude of population change, they are unable to show how fast this change has occurred over a specific period of time in a given spatial unit. To answer this question, a population change rate map was calculated for a 20-year period between 1991 and 2011 by subtracting the 1991 population in each census tract from the 2011 population and then dividing the derived value by 20 to get the population change each year. This map shows an average population change (decrease/increase) over 20-year period and useful to identify which areas grew rapidly and which slowly (Fig. 3.4). The result reveals that out of 1212 census tracts, 83.4 % tracts (1010) have registered an increase in population in terms of average population growth. However, a few of the census tracts adjacent to the CBD registered a decrease in average population in 20-year period as shown in the inset map of Fig. 3.4 (box).

3.4 Conclusions

The objective of this chapter was to analyse changes in urban density gradient and spatial distribution of population. Using the census tract as the spatial unit of study, population data from the last three census years have been considered and analysed. Econometric modelling and spatial analysis techniques were used to map and model the population distribution within a GIS framework.

The results of this study corroborated many of the findings of previous research on the subject of urban population density functional forms. In addition, changes in population over time revealed that most of the census tracts have experienced population growth, indicating the ongoing pressure on land resources in the study area. The latest data from the 2011 population and housing census showed that the suburbanisation process has started, albeit at a slow rate. This can be attributed jointly to ever-increasing population combined with recent improvement of transport arteries and the fact that the inner urban area is reaching saturation.

This study could be improved by investigating the polycentric model theory as the monocentric model is unable to accurately estimate population in peripheral areas. We also acknowledge that availability of spatial data for the subdivided census tracts used in 2011 would help to improve this study.

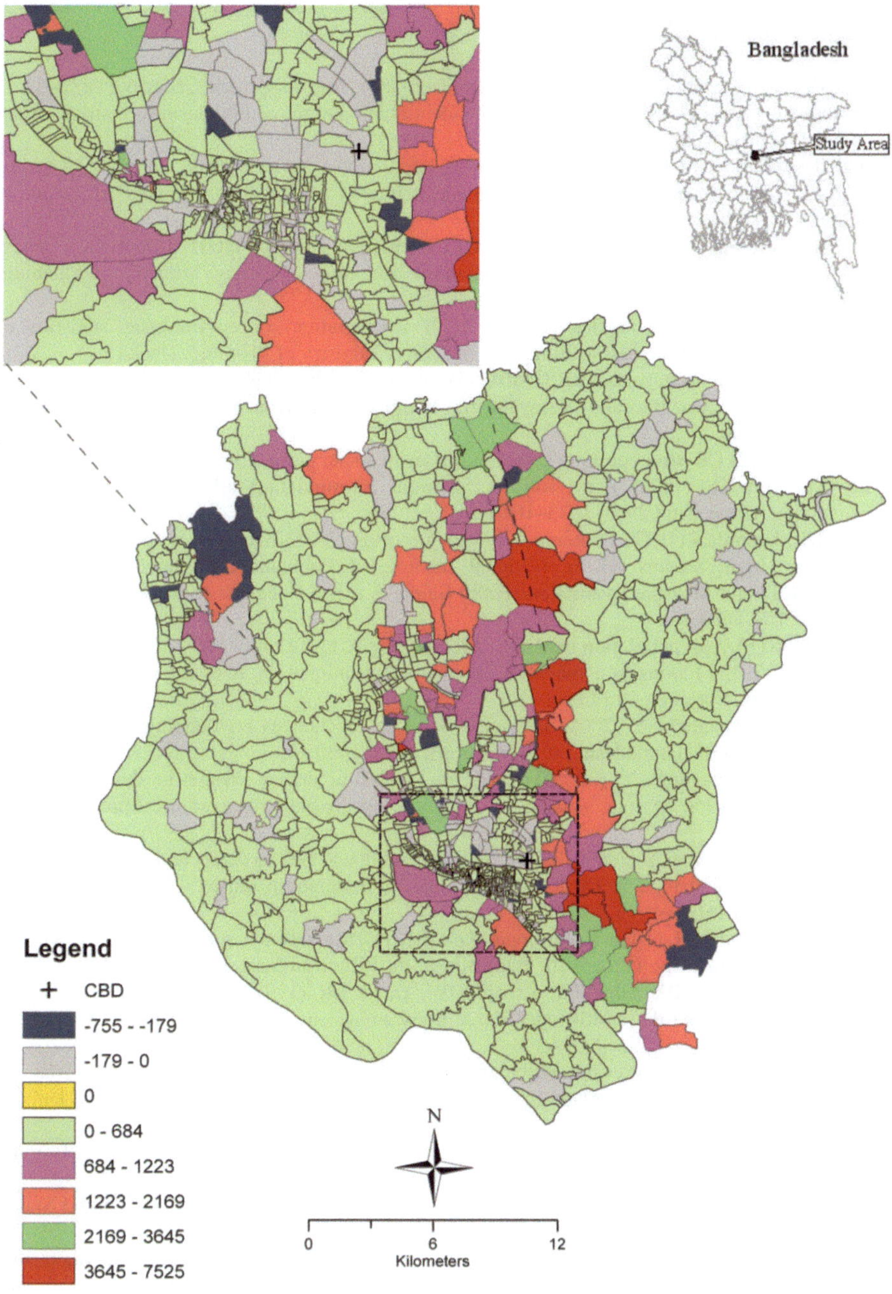

Fig. 3.4 Average change in population each year, 1991–2011

References

Adam JS (1970) Residential structure of Midwestern city. Ann Assoc Am Geogr 60(1):32–62
Alperovich G (1982) Density gradients and the identification of the central business district. Urban Stud 19(3):313–320
Bangladesh Bureau of Statistics (BBS) (1993) Population census – 2001, Community series, Dhaka, Gazipur and Narayanganj. Bangladesh Bureau of Statistics, Ministry of Planning, Dhaka
Bangladesh Bureau of Statistics (BBS) (2003) Population census – 2001, Community series, Dhaka, Gazipur and Narayanganj. Bangladesh Bureau of Statistics, Ministry of Planning, Dhaka
Bangladesh Bureau of Statistics (BBS) (2010) Compendium of environment statistics of Bangladesh 2009. Bangladesh Bureau of Statistics, Ministry of Planning, Dhaka
Bangladesh Bureau of Statistics (BBS) (2012) Population and housing census – 2011, Community series, Dhaka, Gazipur and Narayanganj. Bangladesh Bureau of Statistics, Ministry of Planning, Dhaka
Barter PA (2012) Off-street parking policy surprises in Asian cities. Cities 29(1):23–31
Batty M, Kim KS (1992) Form follows function: reformulating urban population density functions. Urban Stud 29(7):1043–1070
Berry BJL, Kasarda J (1977) Contemporary urban ecology. McMillan, New York
Center for Urban Studies (CUS), National Institute of Population, Research and Training (NIPORT) and Measure Evaluation (2006) Slums in urban Bangladesh: mapping and census 2005. CUS, Dhaka
Clark C (1951) Urban population densities. J R Stat Soc A 114(4):490–496
Dewan AM (2013) Floods in a megacity: geospatial techniques in assessing hazards, risk and vulnerability. Springer, Dordrecht
Dewan AM, Yamaguchi Y (2008) Effects of land cover changes on flooding: example from Greater Dhaka of Bangladesh. Int J Geoinform 4(1):11–20
Dewan AM, Yamaguchi Y (2009a) Land use and land cover change in Greater Dhaka, Bangladesh: using remote sensing to promote sustainable urbanization. Appl Geogr 29 (3):390–401
Dewan AM, Yamaguchi Y (2009b) Using remote sensing and GIS to detect and monitor land use and land cover change in Dhaka Metropolitan of Bangladesh during 1960–2005. Environ Monit Assess 150(1–4):237–249
Dewan AM, Kabir MH, Nahar K, Rahman MZ (2012) Urbanization and environmental degradation in Dhaka metropolitan area of Bangladesh. Int J Environ Sustain Dev 11(2):118–146
Elvidge CD, Baugh KE, Kihn EA, Kroehl HW, Davis ER (1997) Mapping city lights with nighttime data from the DMSP operational linescan system. Photogramm Eng Remote Sens 63(6):727–734
Eusuf AZ (1996) Urban centres in Bangladesh: their growth and change in rank-order. In: Islam N, Ahsan RM (eds) Urban Bangladesh: geographical studies. Department of Geography, University of Dhaka, Dhaka, pp 7–20
Feng J, Wang F, Zhou X (2009) The spatial restructuring of population in metropolitan Beijing: toward polycentricity in the post-reform era. Urban Geogr 30(7):779–802
Fonseca JW, Wong DWS (2000) Changing patterns of population density in the United States. Prof Geogr 52(3):504–517
Gordon P, Richardson H, Wong H (1986) The distribution of population and employment in a polycentric city: the case of Log Angeles. Environ Plann A 18(2):161–173
Griffith DA (1981) Modelling urban population density in a multi-centred city. J Urban Econ 9 (3):298–310
Griffith DA, Wong DW (2007) Modeling population density across major US cities: a polycentric spatial regression approach. J Geogr Syst 9(1):53–75

Iglewicz B, Hoaglin D (1993) How to detect and handle outliers. In: Mykytka EF (ed) The ASQC basic references in quality control: statistical techniques. ASQC Quality Press, Milwaukee
Islam N (2002) Urbanization in Bangladesh: current trends. In: Islam N, Baqee A (eds) Nagarayane Bangladesh. Department of Geography and Environment, University of Dhaka, Dhaka, pp 13–24 (in Benagli)
Islam N (2005) Dhaka now: contemporary urban development, Golden Jubilee series no. 2. The Bangladesh Geographical Society (BGS), Dhaka
Islam N (2009) Dhaka in 2025 AD. In: Ahmed S (ed) Dhaka: past, present and future, 2nd edn. The Asiatic Society of Bangladesh, Dhaka, pp 608–621
Islam I, Sharmin M, Masud F, Moniruzzaman M (2009) Urban consolidation approach for Dhaka City: prospects and constraints. Glob Built Environ Rev 7(1):50–68
Khan JR, Rahman MR (2000) Urbanisation and the trend and spread of urban poverty: a study on Rajshahi city. In: Khan JR, Rumi SRA, Shaikh AH (eds) Geography of Bangladesh: selected articles. Department of Geography and Environmental Studies, University of Rajshahi, Rajshahi, pp 185–214
Krakover S (1985) Spatial-temporal structure of population growth in urban regions: the cases of Tel-Aviv and Haifa, Israel. Urban Stud 22(4):31–328
Liu XH, Kyriakidis PC, Goodchild MF (2008) Population-density estimation from using regression and area-to-point residual kriging. Int J Remote Sens 22(4):431–447
Martin D (1989) Mapping population data from zone centroid locations. Trans Inst Br Geogr 14 (1):90–97
Martin D (1996) An assessment of surface and zonal models of population. Int J Geogr Inf Syst 10 (8):973–989
McDonald JF (1989) Econometric studies of urban population density. J Urban Econ 26 (3):361–385
McMillen DP, McDonald JF (1997) A nonparametric analysis of employment density in a polycentric city. J Reg Sci 37(4):592–612
Mennis J (2003) Generating surface models of population using dasymetric mapping. Prof Geogr 55(1):31–42
Millward H, Bunting T (2008) Patterning in urban population densities: a spatiotemporal model compared with Toronto 1971-2001. Environ Plann A 40(2):283–302
Mitchell A (2005) The Esri guide to GIS analysis, vol. 2: spatial measurements & statistics. Esri Press, Redlands
National Institute of Standards and Technology (NIST) and International SEMATECH (2003) Engineering statistics handbook, NIST/SEMATECH. Available at: NIST/SEMATECH e-Handbook of Statistical Methods, http://www.itl.nist.gov/div898/handbook/
Newling BE (1969) The spatial variation of urban population densities. Geogr Rev 59(2):242–252
Parr JB (1985) A population-density approach to regional spatial structure. Urban Stud 22 (4):289–303
Patino JE, Duque JC (2013) A review of regional science applications of satellite remote sensing in urban settings. Comput Environ Urban Syst 37:1–17
Rana MMP (2011) Urbanization and sustainability: challenges and strategies for sustainable urban development in Bangladesh. Environ Dev Sustain 13(1):237–256
Rouf MA, Jahan S (2007) Spatial and temporal patterns of urbanization in Bangladesh. In: Jahan S, Maniruzzaman KM (eds) Urbanization in Bangladesh: patterns, issues and approaches to planning. Bangladesh Institute of Planners, Dhaka, pp 1–24
Shoshany M, Goldshleger N (2002) Land-use and population density changes in Israel – 1950 to 1990: analysis of regional and local trends. Land Use Policy 19(2):123–133
Siddiqui K, Ahmed J, Siddiqui K, Huq S, Hossain A, Nazimud-Doula S, Rezawana N (2010) Social formation in Dhaka, 1985–2005: a longitudinal study of society in a third world megacity. Ashgate Publishing Group, Surrey
Small KA, Song S (1994) Population and employment densities: structure and change. J Urban Econ 36(3):292–313

Suárez M, Delgado J (2009) Is Mexico city polycentric? A trip attraction capacity approach. Urban Stud 46(10):2187–2211

Sutton P (1997) Modelling population density with night-time satellite imagery and GIS, Computers. Environ Urban Syst 21(3/4):227–244

Tian Y, Yue T, Zhu L, Clinton N (2005) Modeling population density using land cover data. Ecol Model 189(1–2):72–88

Wang F, Meng Y (1999) Analysing urban population change patterns in Shenyang, China 1982-90: density function and spatial association approaches. Geogr Inf Sci 5(2):121–130

Wang F, Zhou Y (1999) Modelling urban population densities in Beijing 1982–90: suburbanization and its causes. Urban Stud 36(2):271–287

Wong DWS, Lee J (2005) Statistical analysis of geographic information with ArcView GIS and ArcGIS. Wiley, Hoboken

Wu S-S, Qiu X, Wang L (2005) Population estimation methods in GIS and remote sensing: a review. GISci Remote Sens 42(1):58–74

Chapter 4
Climatic Variability

Yusuke Yamane, Masashi Kiguchi, Toru Terao, Fumie Murata, and Taiichi Hayashi

Abstract This chapter discusses the long-term variability of temperature and rainfall in Dhaka for the years 1953–2008 using surface meteorological data provided by the Bangladesh Meteorological Department (BMD). Baseline climate values are defined as the mean values between 1972 and 2000, and deviations from these means are examined as annual anomalies. The annual surface temperature anomaly has a positive trend with statistical significance and a rate of increase of 1.9 °C per 100 years. Positive statistically significant trends are also found for three out of four individual seasons. No significant trends were found for annual and seasonal variability in rainfall.

Keywords Climate variability • Dhaka • Temperature • Rainfall

Y. Yamane (✉)
Faculty of Education, Tokoha Gakuen University, 1-22-1, Sena, Aoi-ku, Shizuoka City, Shizuoka 420-0911, Japan
e-mail: yamane@tokoha-u.ac.jp

M. Kiguchi
Institute of Industrial Science, The University of Tokyo, 4-6-1, Komaba, Meguro-ku, Tokyo 153-8505, Japan
e-mail: kiguchi@rainbow.iis.u-tokyo.ac.jp

T. Terao
Faculty of Education, Kagawa University, 1-1, Saiwaicho, Takamatsu, Kagawa 760-8522, Japan
e-mail: terao@ed.kagawa-u.ac.jp

F. Murata
Faculty of Science, Kochi University, 2-5-1, Akebonocho, Kochi 781-8520, Japan
e-mail: fumie@kochi-u.ac.jp

T. Hayashi
Disaster Prevention Research Institute, Kyoto University, Gokasho, Uji City, Kyoto 611-0011, Japan
e-mail: hayashi@rcde.dpri.kyoto-u.ac.jp

A. Dewan and R. Corner (eds.), *Dhaka Megacity: Geospatial Perspectives on Urbanisation, Environment and Health*, Springer Geography,
DOI 10.1007/978-94-007-6735-5_4,

4.1 Introduction

Climate change is having significant impacts on the global environment and human society, and in order to predict changes in climate in the future, we need to comprehensively understand the history of climate change not only on a global scale but also regionally.

Various meteorological disturbances occur in Bangladesh throughout the year and greatly affect society, often in contrary ways. For example, while the heavy rainfall in the monsoon season (June–September) is beneficial for agriculture, it can also cause severe flood damage. Before and after the rainy season, cyclones strike Bangladesh and often cause devastation. In the pre-monsoon season (March–May), thunderstorms frequently occur and the accompanying lightning, hail and tornadoes cause extensive damage. Cold snaps and fog cause damage in the winter season. In short, Bangladesh is prone to meteorological disasters.

Climate change can affect the variability, frequency and magnitude of meteorological disasters. Therefore, it is especially important to understand the potential climate change in Bangladesh. In order to investigate future climate change, it is of primary importance to first identify the characteristics and trends of climate change in the past.

This study focuses on climate change in Dhaka, the capital of Bangladesh. Dhaka has experienced rapid urbanisation with significant additional concentrations of population, housing and industrial factories. Therefore, understanding the impact of climate change on human society in Dhaka is important. Although there have been some studies of climate change for the whole of Bangladesh (e.g. Pant and Rupa Kumar 1997), there has been little research focusing on climate change exclusively in the urban area of Dhaka.

The present study looks at the climate change in Dhaka using long-term data for the period between 1953 and 2008. We focus on temperature and rainfall, which are important meteorological factors especially for human society, and investigate the long-term variability of these factors. We believe that the present study promotes understanding of climate change in Dhaka and the progress of climate study in the future.

4.2 Data and Methods

The surface meteorological data used in this study were provided by the Bangladesh Meteorological Department (BMD) (see Chap. 1 (Fig. 1.3) for the location of weather monitoring station). Data used in this analysis are three hourly temperature and daily rainfall data between 1953 and 2008 in Dhaka. Temperature and rainfall data are not available for the year of 1974.

The methods to constructing annual data in this study are as follows. Daily mean temperature is calculated if the number of three hourly observations in a day is greater than or equal to seven. Monthly mean temperature is then calculated if less than 20 % of the mean daily temperature values in a month are missing. The annual average temperature is then calculated if all monthly mean temperature values are available for the year. Seasonal average temperature is calculated if all monthly mean temperature values are available for the season. Monthly mean rainfall is calculated if less than 20 % of the daily rainfall observations are missing. Annual and seasonal rainfall values are calculated in the same manner as the annual and seasonal mean temperatures.

Baseline climate values in the present study are defined as the mean values between 1972 and 2000. Anomaly data for each year is defined as the difference between the annual and baseline climate values.

The present study shows the variability of temperature and rainfall for each season. Bangladesh experiences four seasons: the pre-monsoon season from March to May, the monsoon season from June to September, the post-monsoon season from October to November and a winter season from December to February. The highest temperatures occur in the pre-monsoon season and the lowest in the winter season. The monsoon season sees the largest amount of rainfall with, on average, more than 75 % of annual rainfall occurring then.

4.3 Temperature Variability

Table 4.1 shows the year-to-year variability of temperature during 1953–2008 for the annual, pre-monsoon season, monsoon season, post-monsoon season and winter season. Long-term average annual temperatures are 25.5 °C, with averages for the pre-monsoon season, monsoon season, post-monsoon season and winter season being 27.6, 28.4, 25.2, and 19.7 °C, respectively. Maximum temperatures are 26.5, 28.9, 29.3, 26.8, and 21.6 °C for the annual, pre-monsoon season, monsoon season, post-monsoon season and winter season. Minimum temperatures are 24.8, 26.3, 27.3, 23.6, and 17.9 °C for the annual, pre-monsoon season, monsoon season, post-monsoon season and winter season. Figure 4.1 shows the variability of the temperature anomaly in Dhaka between 1953 and 2008, which indicates an underlying increasing trend. The rate of increase based on the liner trend line is 1.9 °C per 100 years. This value is about twice as large as that of surface temperature averaged across the globe (0.7 °C per 100 years) reported in the Fourth Assessment Report by the Intergovernmental Panel on Climate Change (IPCC) report (IPCC 2007). A Mann-Kendall rank statistical test was conducted in order to evaluate the statistical significance of the increasing trend in temperature anomaly shown in Fig. 4.1 and indicates that the increasing trend is statistically significant at the 99 % level.

Figures 4.2, 4.3, 4.4, and 4.5 show the variability of temperature anomalies for the pre-monsoon season, monsoon season, post-monsoon season and winter seasons, respectively, in Dhaka during 1953–2008. The results indicate an increasing trend for all seasons. The rates of increase for the four seasons derived from the linear trend lines are summarised in Table 4.2. The highest rate in the four seasons is

Table 4.1 Annual and seasonal temperature variations from 1953 to 2008

Year	Annual	Pre-mon.	Mon.	Post-mon.	Winter
1953	25.1	28.0	27.3	24.5	19.9
1954	25.1	28.1	28.0	23.6	19.3
1955	25.0	27.2	28.0	25.3	18.7
1956	24.8	27.3	27.9	23.9	18.9
1957	25.1	27.5	28.4	24.4	19.0
1958	25.7	28.1	28.6	25.3	19.8
1959	24.8	27.4	27.9	23.7	18.8
1960	25.3	27.9	28.2	24.2	19.6
1961	24.8	28.0	27.9	23.7	17.9
1962	25.0	27.5	28.2	24.1	18.7
1963	25.0	26.3	28.4	24.4	19.6
1964	25.2	27.5	28.0	25.2	19.3
1965	25.1	27.2	27.7	24.9	[a]
1966	[a]	[a]	28.1	24.7	[a]
1967	25.1	26.7	28.1	24.0	20.0
1968	25.1	26.7	28.2	24.9	19.3
1969	25.5	27.9	28.2	24.9	19.6
1970	25.4	28.0	28.3	24.7	19.5
1971	[a]	[a]	[a]	[a]	[a]
1972	25.2	27.6	28.3	25.3	18.6
1973	[a]	27.0	[a]	24.9	[a]
1974	[a]	[a]	[a]	[a]	[a]
1975	25.2	27.8	27.9	24.8	19.4
1976	25.3	27.6	27.8	25.6	19.3
1977	[a]	26.6	28.1	[a]	19.4
1978	25.1	26.3	28.2	25.7	19.3
1979	25.8	28.5	28.2	26.3	19.7
1980	25.7	27.8	28.5	25.2	20.0
1981	[a]	[a]	28.8	26.0	18.4
1982	25.4	27.3	28.4	24.7	20.0
1983	25.2	26.9	28.7	24.9	19.0
1984	25.6	27.9	28.2	25.9	19.7
1985	26.0	28.0	28.5	25.7	20.9
1986	25.9	27.7	28.7	25.3	20.7
1987	[a]	28.0	[a]	26.3	20.8
1988	26.3	27.9	29.0	26.1	21.2
1989	25.9	28.6	29.0	25.7	19.4
1990	25.7	26.5	28.8	26.0	20.8
1991	25.8	27.7	28.6	25.3	20.5
1992	25.8	28.4	28.9	25.5	19.2
1993	25.4	26.6	28.5	25.4	20.0
1994	25.8	27.7	29.1	25.5	19.5
1995	25.9	28.7	28.9	25.8	19.3
1996	25.7	28.3	28.5	25.0	19.9
1997	25.4	27.2	28.7	25.3	19.1
1998	25.9	27.0	29.3	26.8	19.7

(continued)

Table 4.1 (continued)

Year	Annual	Pre-mon.	Mon.	Post-mon.	Winter
1999	26.3	28.9	28.6	25.7	21.0
2000	25.7	27.1	29.0	26.0	19.9
2001	25.9	27.8	28.7	26.1	20.7
2002	25.8	27.2	28.6	25.7	19.6
2003	25.8	27.6	28.9	25.9	20.2
2004	25.9	28.5	28.5	25.1	21.1
2005	26.2	28.2	29.1	25.4	21.6
2006	26.5	28.4	29.0	26.1	20.0
2007	25.7	27.9	28.7	25.5	19.7
2008	25.9	28.4	28.7	25.4	
Average	25.5	27.6	28.4	25.2	19.7
Maximum	26.5	28.9	29.3	26.8	21.6
Minimum	24.8	26.3	27.3	23.6	17.9

[a]No data

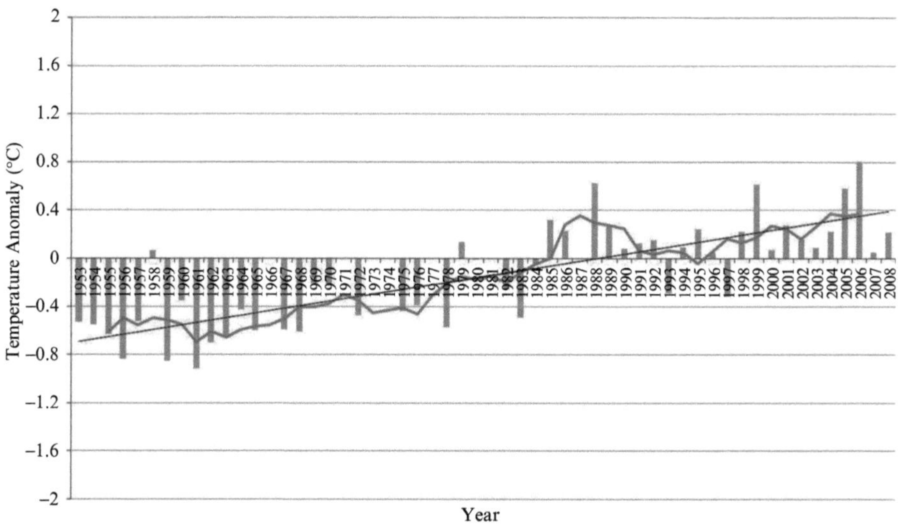

Fig. 4.1 Annual temperature anomaly during 1953–2008 (*solid curve* and *line* represent the 5-year moving average and linear trend line, respectively)

3.0 °C per 100 years of the post-monsoon season. The rates of increase for the winter season and post-monsoon season are 2.5 °C per 100 years and 1.7 °C per 100 years, respectively. The positive trends for the post-monsoon season, monsoon season and winter season are statistically significant at the 99 % level. The pre-monsoon season trend is positive and the rate of increase is 0.9 °C per 100 years, but this trend is not statistically significant. The positive trends of the post-monsoon season and winter season contribute greatly to the increasing trend of annual temperature anomaly.

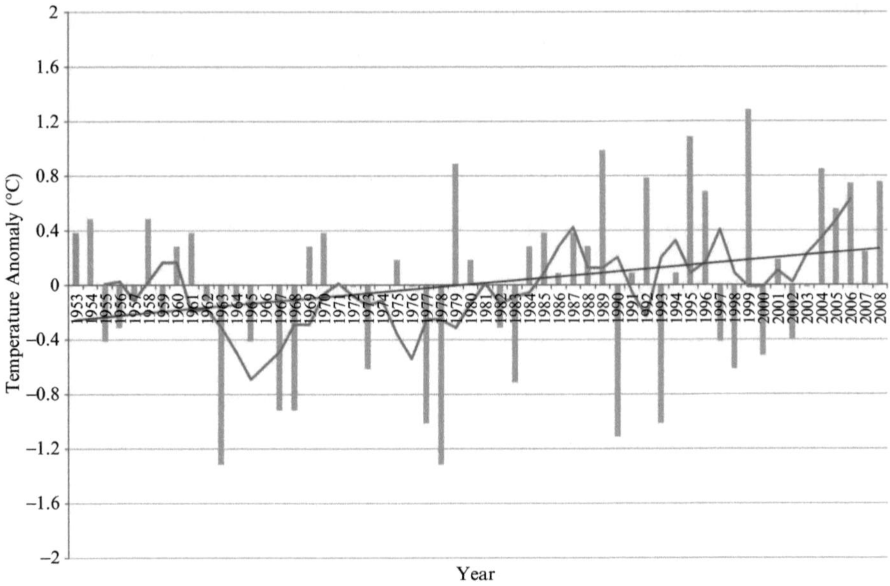

Fig. 4.2 As in Fig. 4.1 for the pre-monsoon season (March–May)

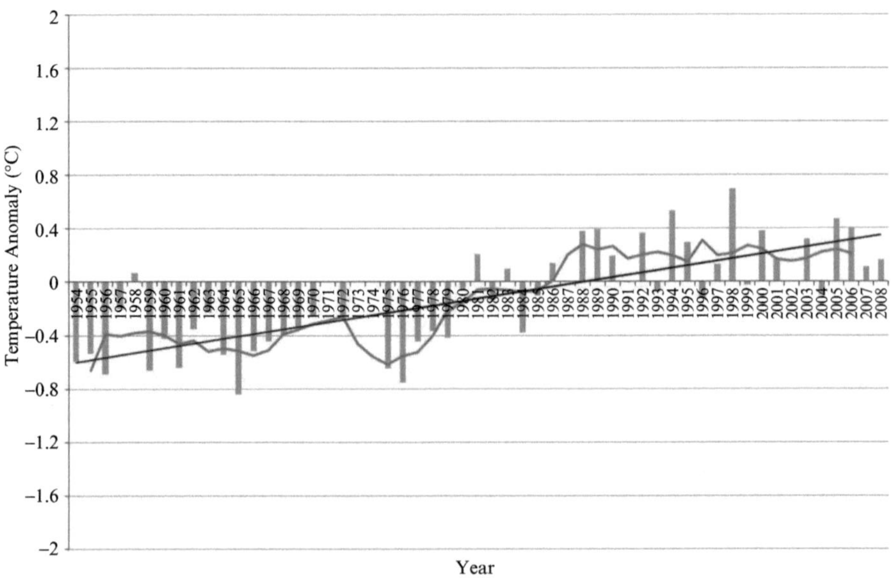

Fig. 4.3 As in Fig. 4.1 for the monsoon season (June–Sept)

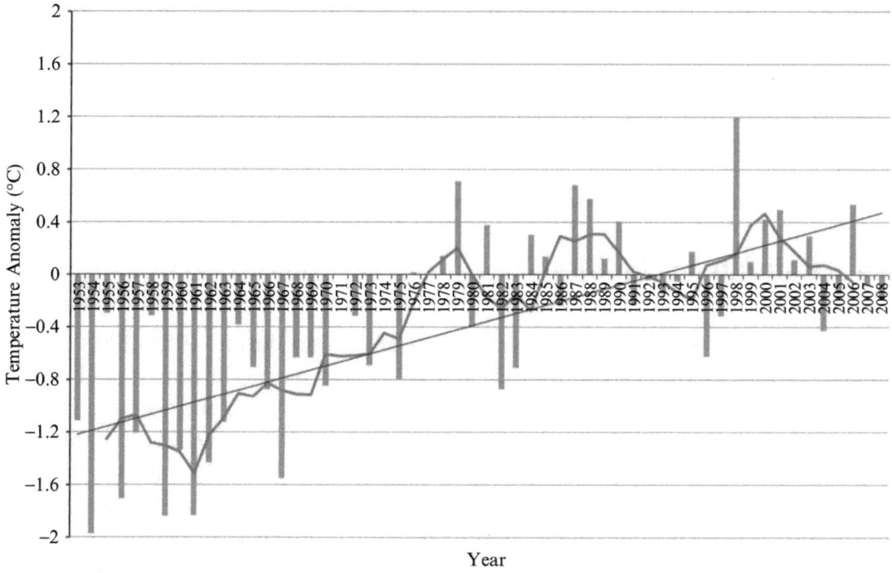

Fig. 4.4 As in Fig. 4.1 for the post-monsoon season (Oct–Nov)

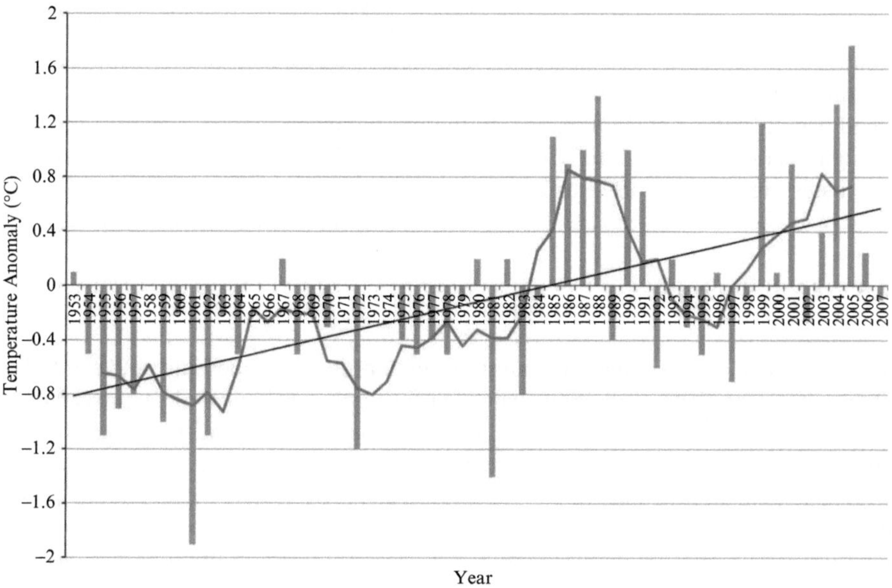

Fig. 4.5 As in Fig. 4.1 for the winter season (Dec–Feb)

Table 4.2 Trends of temperature anomalies

Annual	1.9 °C/100 year
Pre-monsoon	0.9 °C/100 year
Monsoon	1.7 °C/100 year
Post-monsoon	3.0 °C/100 year
Winter	2.5 °C/100 year

All trends are statistically significant (99 %) except for the pre-monsoon

4.4 Rainfall Variability

Table 4.3 shows the year-to-year variability of rainfall during 1953–2008 for the annual, pre-monsoon season, monsoon season, post-monsoon season and winter season. Average rainfall for these is 2,076, 475, 1,371, 207, and 39 mm, respectively. Maximum rainfall for the same periods is 3,028, 869, 2,120, 568, and 154 mm, with minimum rainfall being1169, 137, 703, 29, and 0 mm for the four seasons. Figure 4.6 shows the variability of the rainfall anomaly in Dhaka during 1953–2008. Although the linear trend line shows a positive trend, the Mann-Kendall test indicates that it is not statistically significant. From the middle of the 1980s, the magnitude of anomaly variability is relatively large and fluctuates considerably. On the contrary, from the 1960s to the early part of the 1980s, the magnitude of anomaly variability is relatively small.

Figures 4.7, 4.8, 4.9, and 4.10 show the variability of rainfall anomaly for the four seasons. Although positive trends for the pre-monsoon season, monsoon season and post-monsoon season can be found from the linear trend lines, the trends for these three seasons are not statistically significant. In the winter season, the linear trend line does not show a clear trend and the statistical test indicates the trend has no statistical significance. In the pre-monsoon season, negative anomaly is significant from 1953 to 1970 and after 2001, and the anomaly variability fluctuates greatly from 1972 to 2000.

4.5 Conclusions

The present study examined the long-term variability of temperature and rainfall in Dhaka between 1953 and 2008 using surface meteorological data provided by the BMD. Surface temperature has a positive trend with statistical significance and the rate of increase is 1.9 °C per 100 years. For each season, trends are positive with statistical significance in the monsoon season (from June to September), post-monsoon season (from October to November) and winter season (December to February). The rates of increase are 1.7 °C per 100 years in the monsoon season, 3.0 °C per 100 years in the post-monsoon season and 2.5 °C per 100 years in the winter season. The trend of the pre-monsoon season has no statistical significance. For rainfall, trends for annual and seasonal variability have no statistical significance.

Table 4.3 Annual and seasonal rainfall variations from 1953 to 2008

Year	Annual	Pre-mon.	Mon.	Post-mon.	Winter
1953	1,934	454	1,403	65	14
1954	2,269	297	1,730	218	10
1955	1,577	255	1,152	170	17
1956	2,495	393	1,898	174	150
1957	1,554	247	1,141	29	95
1958	1,258	269	703	191	19
1959	2,453	410	1,456	568	0
1960	1,834	393	1,379	62	13
1961	2,170	444	1,661	52	15
1962	1,786	377	1,214	180	0
1963	1,971	368	1,411	189	54
1964	2,332	550	1,407	324	28
1965	2,117	382	1,526	181	[a]
1966	[a]	[a]	1,363	275	50
1967	2,053	569	1,374	75	5
1968	1,900	342	1,410	143	1
1969	1,540	246	1,188	105	24
1970	1,995	260	1,252	459	31
1971	[a]	[a]	1,688	213	[a]
1972	1,808	600	1,092	105	21
1973	[a]	784	[a]	192	116
1974	[a]	[a]	[a]	[a]	[a]
1975	2,145	428	1,430	257	7
1976	2,238	610	1,499	122	66
1977	1,861	707	781	283	44
1978	2,251	666	1,467	98	16
1979	1,837	137	1,432	201	86
1980	2,218	615	1,268	300	52
1981	1,865	655	1,032	91	50
1982	1,805	339	1,254	197	75
1983	2,388	804	1,238	253	32
1984	3,028	836	2,120	58	9
1985	2,053	671	1,284	79	32
1986	2,500	461	1,605	409	7
1987	2,187	372	1,667	111	77
1988	2,482	869	1,200	366	35
1989	1,627	313	1,030	240	48
1990	2,103	507	1,270	284	41
1991	2,850	628	1,675	406	154
1992	1,169	178	858	85	52
1993	2,819	757	1,774	236	67
1994	1,540	570	834	69	39
1995	1,751	352	1,156	203	22
1996	2,044	461	1,205	357	9
1997	1,921	420	1,430	40	75
1998	2,312	666	1,410	183	0
1999	2,374	449	1,544	381	57

(continued)

Table 4.3 (continued)

Year	Annual	Pre-mon.	Mon.	Post-mon.	Winter
2000	2,121	832	960	272	1
2001	1,684	481	1,007	195	26
2002	1,875	434	1,247	168	25
2003	1,693	359	1,130	134	45
2004	2,347	338	1,801	208	4
2005	2,637	537	1,676	420	0
2006	1,919	366	1,487	66	30
2007	2,885	359	2,065	431	79
2008	2,385	341	1,738	227	
Average	2,076	475	1,371	207	39
Maximum	3,028	869	2,120	568	154
Minimum	1,169	137	703	29	0

[a]No data

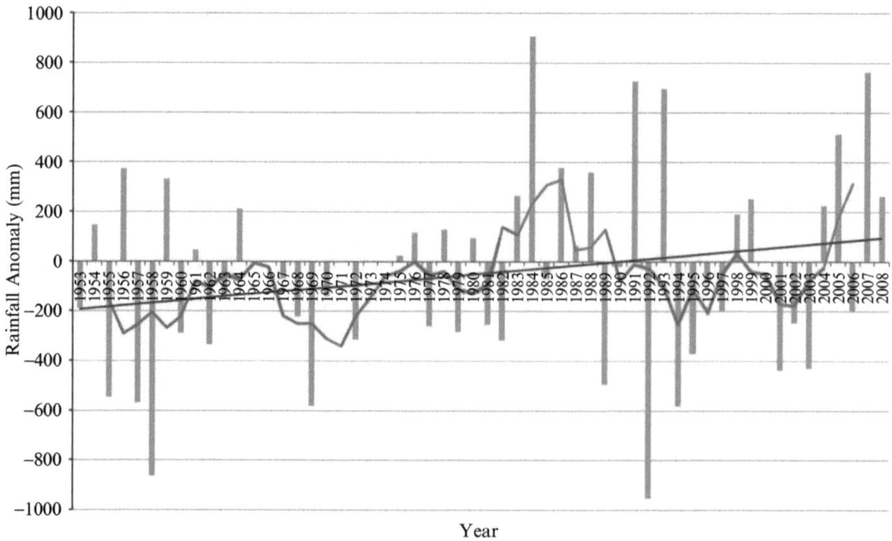

Fig. 4.6 Annual rainfall anomaly during 1953–2008 (*solid curve* and *line* represent the 5-year moving average and linear trend line, respectively)

For the sake of completeness, this study should be extended to investigate the relationship between climate change and urbanisation in Dhaka. To clarify the effect of urbanisation on climate change, it is necessary to compare climate change in Dhaka with that of a nonurbanisation area around Dhaka.

Although this study shows statistical aspects of the long-term changing of temperature and rainfall, no meteorological factor causing the variability is discussed. It is important that we conduct further analysis in order to clarify the

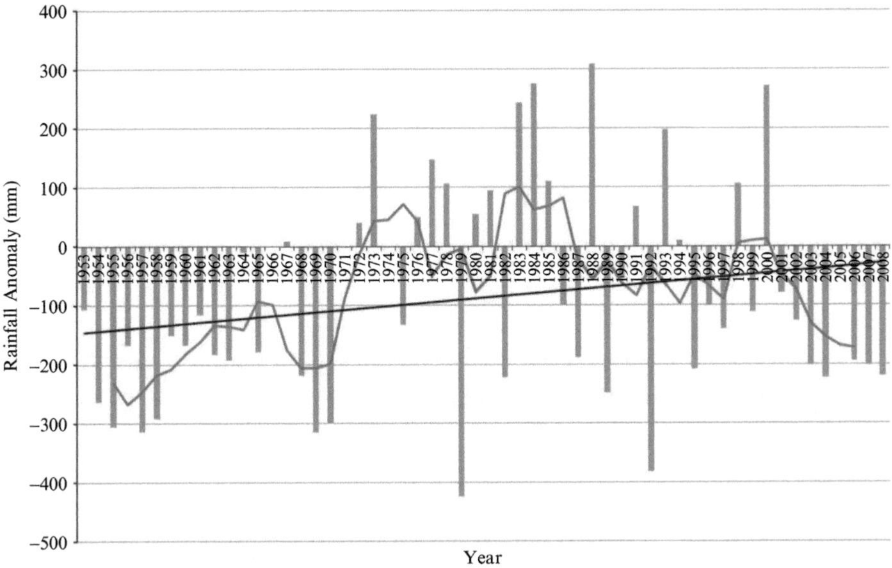

Fig. 4.7 As in Fig. 4.6 for the pre-monsoon season (March–May)

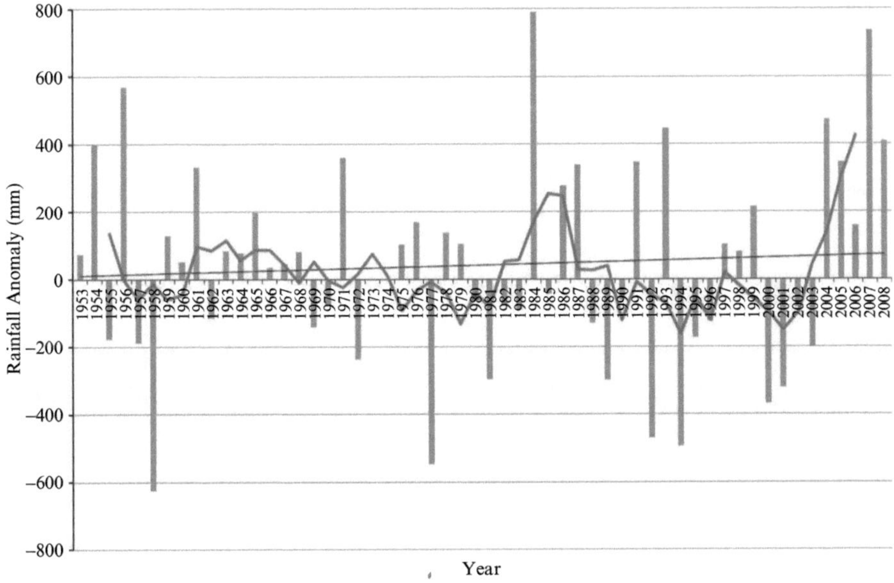

Fig. 4.8 As in Fig. 4.6 for the monsoon season (June–Sept)

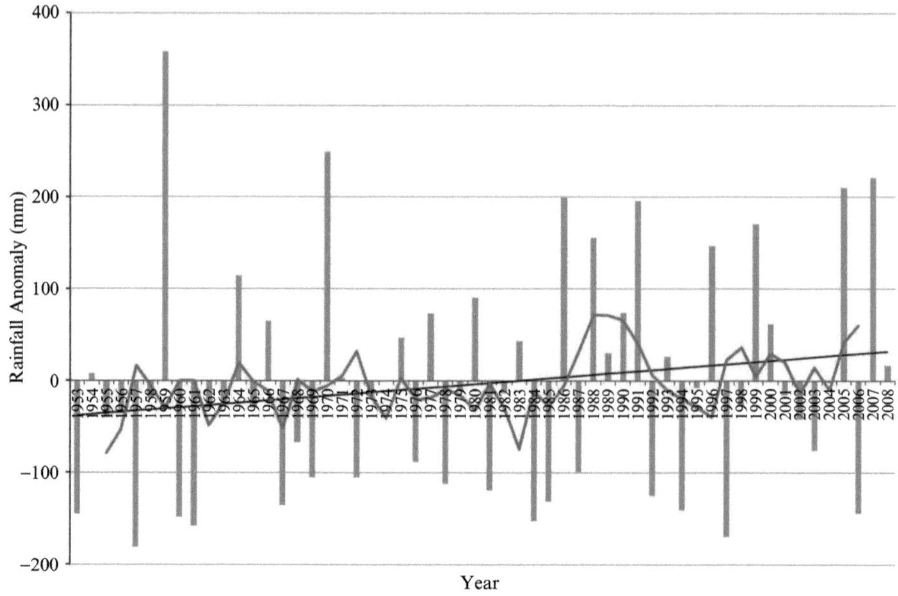

Fig. 4.9 As in Fig. 4.6 for the post-monsoon season (Oct–Nov)

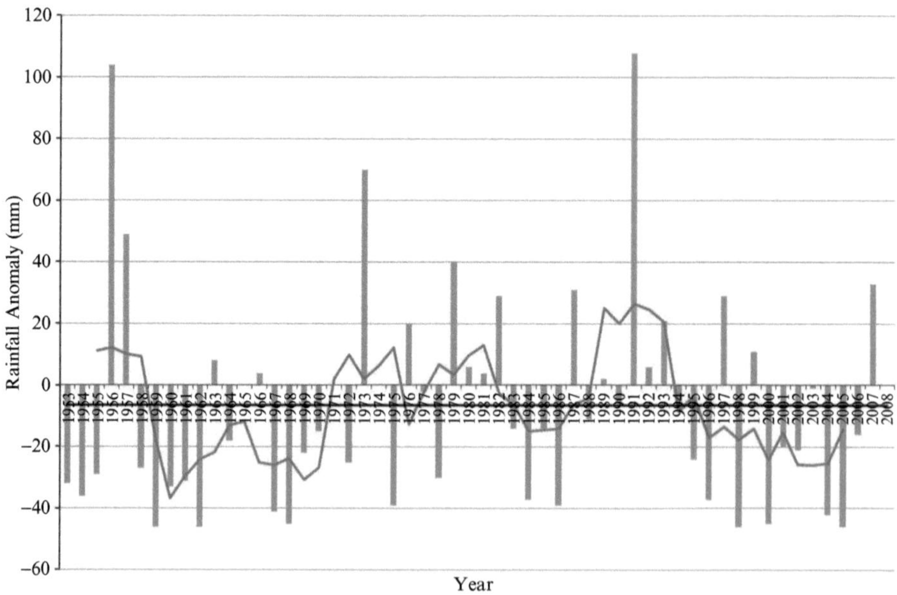

Fig. 4.10 As in Fig. 4.6 for the winter season (Dec–Feb)

relationship between global-scale atmospheric and oceanic variations such as ENSO and Indian Ocean dipole mode and climate change in Dhaka.

Climate change data accompanied by global warming predictions derived from numerical simulation are available on a worldwide basis, and we can investigate the potential future climate change in Dhaka using that model output data.

In order to provide a concrete climate change adaptation plan, a comprehensive understanding of future climate change, obtained by conducting the studies mentioned above, is necessary.

Acknowledgements The dataset used in this analysis is provided by the Bangladesh Meteorological Department (BMD). This research was supported by the Environment Research and Technology Development Fund (S-10) of the Ministry of the Environment, Japan.

References

IPCC (2007) Climate change 2007: synthesis report. Contribution of Working Groups I, II and III to the fourth assessment report of the Intergovernmental Panel on Climate Change [Core writing team, Pachauri RK, Reisinger A (eds)], IPCC, Geneva, 104pp

Pant GB, Rupa Kumar K (1997) Climates of South Asia. John Wiley, New York, p 320

Chapter 5
Monitoring and Prediction of Land-Use and Land-Cover (LULC) Change

Robert J. Corner, Ashraf M. Dewan, and Salit Chakma

Abstract This chapter looks at the use of a Markov chain–cellular automata method to model and then predict land-use change in Dhaka. Initially land-use/land-cover maps for three separate time periods were derived from satellite images and evaluated against ground truth. The Markov chain method was then used to establish transition probability matrices between land-cover categories for the time periods represented. The use of cellular automata in this work enables neighbourhood interactions to be accounted for. After an initial calibration run, the combined method is then used to predict land use and land cover in 2022 and 2033.

Keywords LULC dynamics • Modelling • Landsat TM • Markov-Cellular automata • Built-up areas • Geospatial techniques

5.1 Introduction

Increasing anthropogenic activities around the world are causing large-scale modification of the Earth's land surface which has profound impact on the functioning of global systems (Lambin et al. 2001). One of the most visible human modifications of terrestrial ecosystem is the alteration of land use and land cover (LULC) which significantly affects the local, regional and global environment (Yu et al. 2011; Mitsuda and Ito 2011; Mahmood et al. 2010; Weng 2001). The effects of this include soil degradation (Islam and Weil 2000; Tolba et al. 1992), loss of biodiversity (Yamamura et al. 2009; de Koning et al. 2007), rampant urban sprawl (Wu and Zhang 2012; Rahman et al. 2011; Yuan 2010; Batta et al. 2010; Batta 2009;

R.J. Corner (✉) • A.M. Dewan • S. Chakma
Department of Spatial Sciences, Curtin University, Kent Street, Bentley,
Perth 6102, WA, Australia
e-mail: R.Corner@curtin.edu.au; A.Dewan@curtin.edu.au; salit.chakma@ymail.com

A. Dewan and R. Corner (eds.), *Dhaka Megacity: Geospatial Perspectives on Urbanisation, Environment and Health*, Springer Geography,
DOI 10.1007/978-94-007-6735-5_5,

Taubenböck et al. 2009; Deng et al. 2009; Jat et al. 2008a, b; Martinuzzi et al. 2007; Yu and Ng 2007; Mundia and Aniya 2006; Sudhira et al. 2004), general land degradation by agricultural development and the tourism industry (Shalaby and Tateishi 2007), marked variation in biogeochemical and hydrological cycles (Foley et al. 2005; Meyer and Turner 1994; Ojima et al. 1994), nonpoint source pollution (Xian et al. 2007) and a gradual decline in ecosystem services (Quetier et al. 2007). Furthermore, human-induced changes in LULC influence the global carbon cycle and contribute to the increase of atmospheric CO_2 (Alves and Skole 1996; Dixon et al. 1994). These changes result in an enormous and sometimes irreversible impact on the global environment (Abdullah and Nakagoshi 2005).

Geospatial techniques such as remote sensing (RS) and Geographic Information Systems (GIS) have long been recognised as important and powerful tools in determining LULC changes at a range of spatial scales. Various image analyses and change detection techniques have been used to extract information from remotely sensed data (Lu et al. 2004, 2011). GIS, on the other hand, allows the integration of information derived from remote sensing – into the explicit understanding and modelling of LULC (Mesev 2007). In order to recognise the human dimensions of global change, the International Geosphere-Biosphere Project (IGBP) and International Human Dimensions Programme (IHDP) launched the "Landuse/Cover Change (LUCC)" plan in 1995 (Guan et al. 2011). Since then, mapping and monitoring of LULC change has become a major focus of research in different parts of the world by integrating multispectral and multi-temporal remotely sensed data with GIS (Ahmed and Ahmed 2012; Wu and Zhang 2012; Yu et al. 2011; Islam and Ahmed 2011; Bakr et al. 2010; Chen and Wang 2010; Yuan 2010; Deng et al. 2009; Kamusoko et al. 2009; Jat et al. 2008a, b; Xiao and Weng 2007; Long et al. 2007; Shalaby and Tateishi 2007; Mundia and Aniya 2006; Yin et al. 2005; Muttitanon and Tripathi 2005; Yuan et al. 2005; Seto and Kaufmann 2003; Alphan 2003; Weng 2001, 2002; Yeh and Li 1996, 1997; Giri and Shrestha 1996; Harris and Ventura 1995; Westmoreland and Stow 1992; Meaille and Wald 1990).

As noted by Seto and Kaufmann (2003), there exist two broad LULC models: spatially explicit models and aspatial models. Spatially explicit models include empirical–statistical models, e.g. regression models (Jat et al. 2008a, b; Braimoh and Onishi 2007; Hu and Lo 2007; Long et al. 2007; Fang et al. 2005; Lo and Yang 2002; Lambin et al. 2000; Hazen and Berry 1997), rule-based models such as cellular automata or the Markov-cellular model (Arsanjani et al. 2013; Ahmed and Ahmed 2012; Islam and Ahmed 2011; Mitsova et al. 2011; Guan et al. 2011; Kamusoko et al. 2009; Torrens 2006; Lo and Yang 2002; Clarke et al. 1997; Xie 1996) and agent-based models (Mena et al. 2011; Evans and Kelley 2004; Barredo and Demicheli 2003; Wu 1998). These spatial models are primarily used to determine the pattern and process of LULC change and to project the locations of future changes. Among spatial models, CA in conjunction with Markov chain analysis has historically been the most favoured (Ward et al. 2000). By contrast, aspatial models are largely used to analyse driving factors along with predicting the amount of LULC in a particular geographic region (Yu et al. 2011; Huang et al. 2009; Seto and Kaufmann 2003).

Although Earth observation satellites provide an excellent opportunity to characterise LULC (Loveland et al. 1999) with their high temporal frequency and wealth of historical detail (Donnay et al. 2001), human-induced changes in LULC has typically been poorly enumerated in developing counties. As a result, serious environmental degradation resulted from dramatic change in LULC is believed to be threatening the sustainability of development (Li et al. 2009; Liu et al. 2007; Xiao et al. 2006; Li and Yeh 2004). It has been recognised that the analysis of LULC patterns, including historical patterns, assists in determining the preferred areas for future development (Mitsuda and Ito 2011), helps in assessing the directions and degree of human-related environmental changes (Xiao and Weng 2007), supports informed decision making (Costanza and Ruth 1998) and promotes land-use planning and the formulation of policy (Muttitanon and Tripathi 2005; Verburg et al. 2004), all of which lead to effective management of resources. Geospatial techniques have proved their efficacy for updating and managing spatial data in developing countries (Dong et al. 1997) and are particularly useful for providing accurate and timely geospatial information, such as that which is essential to the efficient management of a large metropolis (Yang 2002).

Bangladesh is one of the most densely populated countries in the world, with a very low per capita income. Until 1970, the major contribution to the gross domestic product (GDP) was from the agricultural sector; currently the service sector has taken that position and now contributes 62 % to the GDP. In contrast, industrial growth shows a very flat trend and contributes only 18 % to the GDP (BBS 2011; The World Bank 2007). These economic changes have led to considerable loss of arable lands and exert tremendous pressure on limited natural resources, particularly on land and vegetation. It is estimated that every year, more than 800 km^2 of agricultural land is converted to cities, roads and infrastructure in Bangladesh (BBS 1996). To elaborate on this, a 0.3 % per annum decline in cultivated areas has been observed by the agricultural census of 2008 (BBS 2010). In addition, the proportion of the country covered by forest has progressively been decreasing in the areas with higher population density (Giri and Shrestha 1996). Simultaneously there has been conspicuous urban growth, with the urban population of the country rising from 14.1 million in 1981 to 33.6 million in 2011 (BBS 2012, 2003, 2001). One of the most important reasons for this population explosion in the cities of Bangladesh is large-scale rural-urban migration, primarily due to collapse of the rural economy (Islam 1999). This loss of agricultural land, once the powerhouse of the national economy, has a number of social and economic effects. An increase in landlessness causes social upheaval in a traditionally agrarian society and leads to unrest. In addition, the decline in agricultural production poses food security risks with the possibility that in future, Bangladesh will be even less able to meet the food demands of its ever-growing population.

Due to the economic and sociopolitical significance of Dhaka, marginalised rural people are often attracted to the area in search of better employment opportunities and improved lifestyles. Dhaka became one of the world's top ten megacities in 2011, and if the current rate continues, Dhaka will be larger than Beijing by 2025, with a projected population of 22.9 million (UN 2012).

LULC change from 1960 to 2005 for Dhaka was first mapped by Dewan and Yamaguchi (2009a, b; 2008) using geospatial techniques and drawing on historical topographic maps, Landsat and IRS-LISS III data. These studies reported that the rate of LULC conversion is extremely fast, which results in rapid depletion of precious natural resources such as cultivated land and wetlands. In contrast, urban growth is prominent mostly in an unplanned and piecemeal manner causing severe environmental degradation (Dewan et al. 2012). For example, built-up areas grew by 344 % between 1960 and 2005 (Dewan and Yamaguchi 2009b). A more recent study reported a total increase of urban built-up areas between 1990 and 2006 of 232 % (Griffiths et al. 2010). Econometric modelling to identify the underlying causes of LULC changes suggests that population growth is the major factor contributing to the rapid changes in LULC in that area (Dewan and Yamaguchi 2009a).

This chapter describes the results of LULC classification in Dhaka megacity that are derived from multi-temporal remotely sensed data but differ from previous studies in terms of their areal extent and method of analyses. Specifically, the aims of this study are to map and monitor LULC changes from 1990 to 2011 and to simulate future land-use change using a combined Markov chain analysis and cellular automata model.

5.2 Materials and Methods

5.2.1 Data Acquisition and Preparation

Six cloud-free Landsat TM 5 scenes were acquired for the years 1990, 2000, and 2011 all from the period late winter to early summer. The study area is on the boundary between two Landsat Rows, so two scenes were required for each year. Each Landsat TM image was enhanced using histogram equalisation to help identify ground control points for rectification. A minimum of 75 ground control points (GCPs), taken from topographic maps of 1990, were used to register each pair of images to the Bangladesh Transverse Mercator (BTM) system, an area-specific standard UTM projection system for Bangladesh. GCPs were well dispersed throughout the scenes, yielding an RMS error of <0.5 pixels. A first-order polynomial fit was applied and images were resampled to 30-m output pixel size using the nearest neighbour method.

Once rectified, the pairs of Landsat scenes were mosaicked. Since the data are historical and non-anniversary, the three resulting mosaicked images were subject to radiometric correction to correct for varying sun angle and changes in surface reflectance (Jensen 1996). An image-based radiometric correction (Chavez 1996) was used to minimise radiometric differences in the Landsat data. Finally, each Landsat scene was clipped using an area of interest (AOI) file derived from a vector dataset of the study area's boundary.

Apart from the Landsat images, a number of reference datasets were also acquired for this study. Due to its retrospective nature, the study relied on a variety of sources for reference data from which training areas could be selected and also

Table 5.1 Land-use/land-cover classification scheme

LULC types	Description
Built-up	Residential, commercial and services, industrial, transportation, roads, mixed urban and other urban
Bare soil	Exposed soils, landfill sites, brick fields and areas of active excavation
Cultivated land	Agricultural area, crop fields, fallow lands and vegetable lands
Vegetation	Deciduous forest, mixed forest lands, palms, conifer, scrub and others
Waterbodies	River, permanent open water, lakes, ponds and reservoirs
Floodplain	Permanent and seasonal wetlands, low-lying areas, marshy land, rills and gully, swamps
Rural settlements	Villages, isolated and clustered settlements having rural characteristics and rural markets

for accuracy evaluation of the resulting LULC maps. These sources include 1:50,000 scale topographic maps of the study area in 1990, 2000 and 2008 from the Survey of Bangladesh (SOB). Three high-resolution optical images were also available. These were an IKONOS 1-m panchromatic image of 2000, an IRS 5.8 m panchromatic image of 2000 and a GEOEYE-1 pansharpened image of 2010. In addition, a number of field visits were made during 2010 and 2011 with a high-resolution GPS (Trimble Nomad 800GXE) and a spatial database was created to support the analysis of the 2011 Landsat TM image.

5.2.2 *Image Analysis*

A modified version of the Anderson Level I Scheme (Anderson et al. 1976) was used in this study of LULC change. Though the scheme was originally developed for the United States, it is widely used across the world (Mundia and Aniya 2006; Shalaby and Tateishi 2007; Yuan et al. 2005; Weng 2002). It is a multilevel LULC classification with classes at Level I able to be mapped from Landsat data or other satellite imagery, whereas the extraction of information for levels II, III and IV requires the use of high-, medium- and low-altitude aerial photographs, respectively. Seven separable LULC types have been identified in this study as waterbodies, floodplain, built-up, cultivated land, vegetation, bare soil/landfill and rural settlements (Table 5.1).

A hybrid classification technique was adopted to digitally categorise each Landsat image since this technique has been shown to perform better in the case of spectral variability of individual cover types (Mas 1999). Several studies have demonstrated that hybrid classification produces superior results to traditional pixel-based supervised or unsupervised classification alone (Xiao and Weng 2007; Garcia and Alvarez 1994). Initially, each image was categorised using the iterative self-organising data analysis (ISODATA) technique, and for each image, 200 spectral signatures were generated. Each of these clusters was subsequently evaluated using histogram plots and transformed divergence (TD) separability methods (Jensen 1996). Signatures were refined, deleted, renamed and merged to ensure uni-modality of histograms and TD values. Finally, a supervised maximum likelihood classification (MLC) algorithm was applied to each image. MLC has generally been proven to provide superior results from remotely sensed data, under the assumption that the pixels in each category follow a Gaussian distribution (Bolstad and Lillesand 1991).

An initial visual comparison of the results of the hybrid classification and the available higher resolution images revealed some degree of misclassification. For example, some urban surfaces were misclassified as bare soil due to their similar spectral characteristics. Likewise, misclassification was found between the wetland/lowland category and the cultivated land, waterbodies and lowland/wetland categories. It should be noted that, initially, wetlands were identified as a separate category but were eventually merged with the floodplain class as these two classes were not spectrally separable. Similarly, rural settlements surrounded and overshadowed by trees were misclassified as vegetation. All these misclassifications were caused by similarities of spectral reflectance among these LULC categories.

Post-classification refinement was, therefore, used to minimise misclassification (Harris and Ventura 1995) and to improve the accuracy of LULC maps. Using the methods suggested by Lo and Faber (1997) and Mundia and Aniya (2006), misclassified pixels were corrected. For instance, pixels that were misclassified by the hybrid technique were first highlighted with the area of interest (AOI) tool and then reclassified to an appropriate category. This approach was repeated for all incorrectly classified pixels as determined from topographic maps, high-resolution satellite images mentioned above. The classified LULC data was then smoothed by the use of a 3-by-3 majority filter which reduces the *salt-and-pepper* effect (Lillesand and Kiefer 1999).

In order to determine the changes in LULC between different years, a post-classification comparison method of change detection was used. Even though this technique has a few limitations (Singh 1989; Coppin et al. 2004), it is the most common approach (Jensen 1996; Lu et al. 2004, 2011) when data from different dates are being compared. The advantage of post-classification comparison is that it bypasses the difficulties associated with the analysis of images acquired at different times of the year and/or by different sensors (Yuan et al. 2005; Coppin et al. 2004; Lu and Weng 2005; Alphan 2003). Moreover, the post-classification method also readily provides information on the amount, location and nature of change (Howarth and Wickware 1981). A major pitfall, however, is that the accuracy of the change maps depends on the accuracy of individual classifications and is therefore subject to error propagation (Zhang et al. 2002). A comparison between the classified maps was subsequently carried out on a pixel-by-pixel basis as recommended by Jensen and Ramsey (1987).

Accuracy assessment of the satellite-derived LULC maps was performed using a random sampling scheme. Generally, classification accuracy refers to the comparison of two datasets, one being the result of the analysis of remotely sensed data and the other based on reference information, referred to as "ground truth"

(Congalton 1991). For this accuracy assessment, 350 pixels (50 per category) were first generated using a stratified random sampling scheme for each of the three land-use/land-cover maps. Field data from 2011 was used as ground truth for that image, whilst ground truth for the 2000 and 1990 was obtained from either near-contemporary high-resolution images or from SOB topographic maps. A cross-tabulation was performed between the class values and the ground truth with the results being presented as an error matrix. A nonparametric kappa test was used to determine the classification accuracy as it accounts for all the elements in the error matrix rather than just the diagonal elements (Rosenfield and Fitzpatirck-Lins 1986). The overall accuracy for 1990, 2000 and 2011 maps was 87.7, 90 and 94.6 %, respectively (Table 5.2).

5.2.3 Land-Use Modelling

A hybrid Markov chain analysis (MCA) and cellular automata (CA) technique were used to predict land-use change from 2011 to 2033 using the IDRISI Taiga software. One of the basic assumptions of the MCA is that LULC change is regarded as a stochastic process, and the different categories are the states of a chain (Weng 2002). MCA describes land-use changes from one state to another and can be used to simulate future changes. It requires a minimum of two distinct land-use maps of different time stamps as inputs from which the estimates of the probabilities of transition between periods are derived. It operates on a cell-by-cell basis examining the states of cells at t_1 and t_2 (for a two-period system), and the states of neighbouring cells do not influence the calculation (Eastman 2006).

The MCA process produces two transition matrices – the transition probability matrix which contains the probability that each land-cover category will change to every other category and the transition area matrix which records expected number of pixels to be converted from each land-use type to every other land-use type over a specific time period. In this study, the MCA was used to compute two transition matrices from the cross-tabulation of LULC maps of 1990, 2000 and 2011. The first of these represents 10 time steps and the second 11 time steps. A background cell value was set to 0 and a proportional error was assigned to 15 %, meaning that the overall accuracy of LULC maps is 85 %. Whilst this proportional error is higher than the measured error as shown in Table 5.2, it was conservatively set to this value to account for the unquantified effects of error propagation.

MCA does not in itself provide the spatial location of future land-use transformation, merely transformation probabilities, so the CA model was used to overcome this problem (Eastman 2006; White and Engelen 2000). The CA model has three physical properties – the cell, the neighbourhood and the rules (Santé et al. 2010; Ward et al. 2000) and considers each land-use cell as a lattice. The contiguity rule is based on proximity, meaning that a pixel has a tendency to change to a new LULC category when it is located near the existing areas of the same category (Cabral and Zamyatin 2006). Prior to running CA, transition potential maps (also

Table 5.2 Accuracy assessment of LULC classification

	User's accuracy (%)							Producer's accuracy (%)								
Year	WB	FP	CL	VEG	BU	BS	RU	WB	FP	CL	VEG	BU	BS	RU	Overall accuracy (%)	Kappa
1990	86	94	94	90	88	86	76	97.7	88.7	87	90	70.9	87.7	100	87.7	0.86
2000	90	88	90	86	90	96	90	91.8	84.6	86.5	93.5	86.5	88.9	100	90.0	0.88
2011	92	94	94	96	96	92	98	93.9	92.2	94	96	90.6	95.8	100	94.6	0.94

WB waterbodies, *FP* floodplain, *CL* cultivated land, *VEG* vegetation, *BU* built-up, *BS* bare soil, *RU* rural settlements known as suitability maps) were generated for each land-cover category from the 2011 LULC map. First of all, each LULC dataset was standardised using the linear standardisation technique. Then a Euclidian distance tool was used to create a distance surface for each land-use category. Using the fuzzy module of IDRISI, suitability maps were generated so that the larger their pixel value, the more likely it is that a particular cell would be subject to conversion. Since CA is based on neighbourhood rules (He et al. 2008; Clarke et al. 1994; White and Engelen 1993), a 3×3 contiguity filter was used to ensure that land-use change does not occur randomly but occurs next to existing LULC categories (Eastman 2006). To simulate LULC of 2022 and 2033, we first simulate LULC of 2011 so that the performance of the simulation could be compared with the actual data obtained from the Landsat-based classification. The transition probability matrix for 1990–2000, suitability maps, and a 3×3 contiguity filter were used for this purpose. Using the kappa statistic, a comparison was carried out between the actual and simulated maps of 2011. Based on the successful simulation, the 2011 observed map was then set as a base map and the transition probability matrix for 2000–2011 and suitability maps were used to simulate LULC in 2022. Likewise, 2022 was set as the starting year and the transition probability matrix of 2000–2011 was used to forecast LULC in 2033.

5.3 Results and Discussion

The spatial patterns of LULC changes in the study area for 1990, 2000 and 2011 are shown in Fig. 5.1. From 1990 to 2000, low-lying areas (collectively termed floodplain), cultivated land and waterbodies all decreased in area whilst built-up areas, bare soil (as landfill) and rural settlements increased over the same period. Between 2000 and 2011, cultivated land and floodplain continued to shrink although slight increases in both waterbodies and vegetation were observed. Since the 2011 data is an early summer image, the increase in vegetation category may simply be due to seasonal variations. Over the same period, remarkable growth in the bare soil category was observed together with an increase in built-up areas. The pattern of LULC changes to 2011 showed that Dhaka is expanding in all directions mainly by converting cultivated land and floodplain to urban area. It is noticeable that the rate of encroachment of built-up areas on other land uses considerably increased following the preparation of a new master plan in 1995 and the construction of a number of new infrastructure projects (Islam 2005; Siddiqui et al. 2000). Field visits confirmed the construction of three bridges over the River Buriganga, the establishment of an export processing zone (EPZ) in the northwest part and a 32-km multipurpose flood embankment. These developments have accelerated expansion of the urban built-up category in the southerly and north-westerly directions during the 1990s (Chowdhury et al. 1998). The spatial distribution of the bare soil/landfill category, visible on the maps, clearly shows the encroachment of urban areas on wetland and low-lying areas in the outskirts of

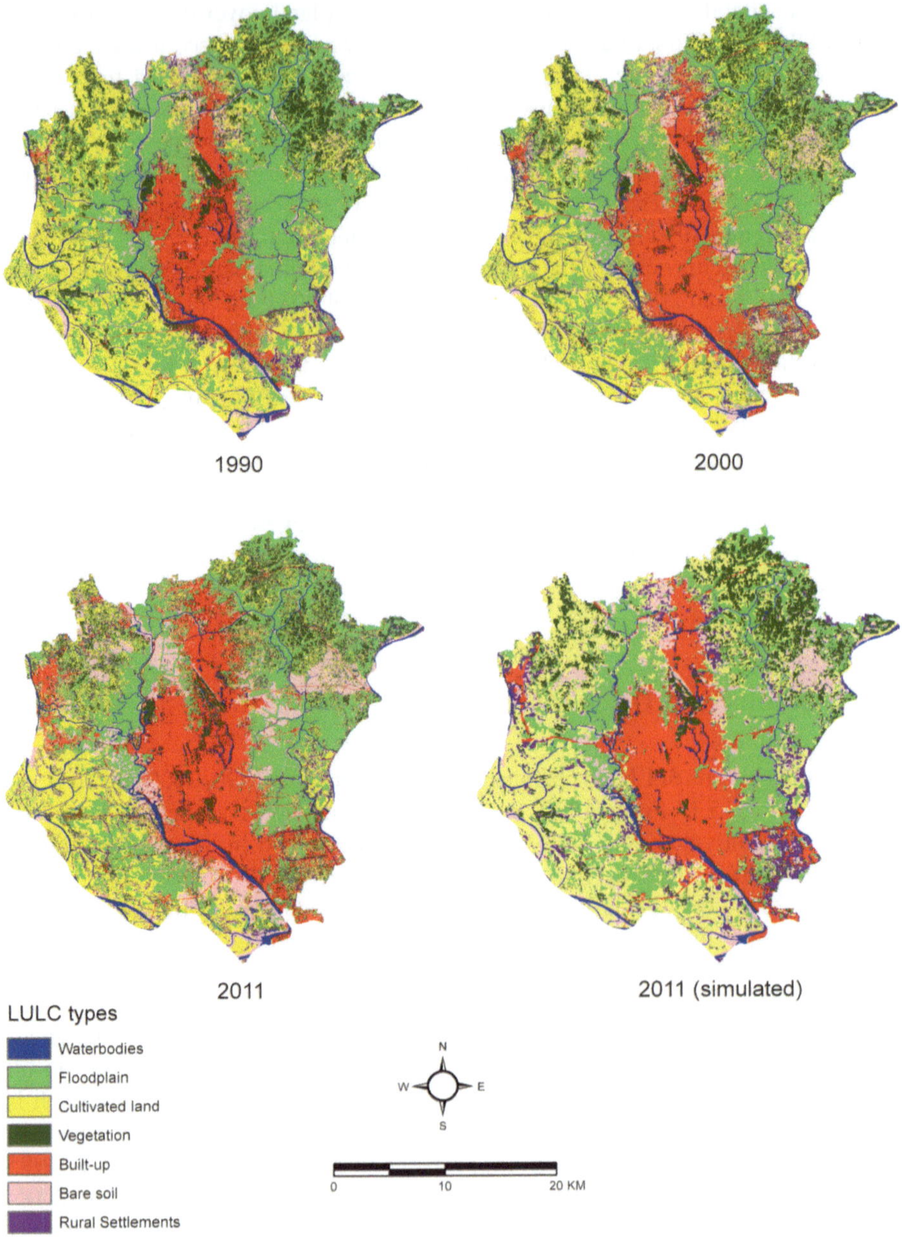

Fig. 5.1 Observed and simulated LULC between 1990 and 2011

Dhaka (Fig. 5.1). Land developments in Dhaka are carried out by three sectors, namely, public, private and individual households. Earlier land developments were mainly the result of ad hoc planning by the public sector primarily on agricultural lands, usually free from inundation (Choudhury 2008). Baridhara, Gulshan Model

Town, Uttara, Dhanmondi etc. are the examples of such developments (Chowdhury 2003). Currently, considerable growth of private real-estate agencies is noticeable in Dhaka. These agencies are illegally developing both wetlands and agricultural lands without considering the environmental cost or with regard to the master plan. Developments in the fringe zone are reported to be largely being carried out by individual households (Islam 1996, 2005). Field observation confirms that development of suburban areas is largely carried out by speculative land conversion by individuals. It should be noted that suburbanisation is a very complex process and that many factors are associated with it.

Speculation and land price are two of them and together may not be adequately explaining the process of suburban development in the study area. Hence, a more detailed study is needed to substantiate factors influencing suburban development in the megacity of Dhaka. Furthermore, poor coordination among state executive agencies is also to blame for the reduction of natural resources in the study area. For example, approximately 6,000 ha of the Dhaka-Narayanganj-Demra (DND) project were originally retained for agricultural production purposes in the 1960s. Since the 1990s, this part of the project area has been used for residential purposes by both local people and internal migrants, without any approval from the relevant authority (Dewan and Yamaguchi 2009a, b). This illustrates the poor level of coordination among organisations involved in the planning and expansion of Dhaka.

The estimates of LULC changes show that the built-up and bare soil categories have both increased significantly (Table 5.3). In 1990, built-up areas covered 11,696 ha (hectares) which increased to 14,641 ha by 2000, indicating a 24 % growth in the 10 years interval. Table 5.3 also shows that between 1990 and 2011, built-up areas increased approximately by 7,860 ha and bare soil by 7,023 ha. Over the same period, agricultural land decreased by 7,234 ha, vegetation decreased by 1,206 ha and floodplain decreased by about 7,206 ha. This result confirms earlier findings made by other researchers based on ground observation (Islam 1996, 2005; FAP 1991; Chowdhury and Faruqui 1989). To accommodate its increasing population, the city has expanded extensively but this spatial expansion has been severely constrained by several physical factors. Therefore, most development has resulted in the loss of natural resources. It has been observed that the growth of Dhaka is a great deal faster than the megacities of North America and Europe. The basic difference is that megacities in the western world grew gradually which enabled them to effectively develop the necessary services and management facilities for their people, whereas, due to the extreme pressure of population, the situation is just the opposite in Dhaka. Consequently, local government is confronting a wide range of challenges in trying to achieve sustainable development which will become more acute in the coming years, if planning regulations are not enforced.

Tables 5.4 and 5.5 show LULC transition probabilities (TPs) for 1990–2000 and 2000–2011 which are the result of cross-tabulation of two time periods adjusted by the proportional error of 15 %. It shows that during 1990–2000, the floodplain category had the highest probability of conversion to cultivated land (TP: 0.1625). Because the highest rate of land conversion to urban occurred in the agricultural category, the probability of this land cover type changing to built-up and bare

Table 5.3 Results of LULC classification for 1990, 2000 and 2011 images showing area of each category, category percentages and area changed

LULC types	1990 Area (ha)	1990 %	2000 Area (ha)	2000 %	1990–2000 Area changed (ha)	2011 Area (ha)	2011 %	2000–2011 Area changed (ha)	1990–2011 Area changed (ha)
Waterbodies	3,561.6	4.0	3,495.5	4.0	−66.1	3,597.5	4.1	102.0	35.9
Floodplain	28,568.8	32.4	25,630.2	29.1	−2,938.6	21,362.0	24.3	−4,268.2	−7,206.8
Cultivated land	27,519.4	31.3	26,215.9	29.8	−1,303.5	20,284.5	23.0	−5,931.4	−7,234.9
Vegetation	11,241.3	12.8	9,416.2	10.7	−1,825.1	10,035.3	11.4	619.1	−1,206.0
Built-up	11,696.0	13.3	14,641.3	16.6	2,945.3	19,556.9	22.2	4,915.6	7,860.9
Bare soil	2,216.9	2.5	4,571.8	5.2	2,354.9	9,239.9	10.5	4,668.0	7,023.0
Rural settlements	3,245.4	3.7	4,078.4	4.6	833.0	3,973.4	4.5	−105.0	728.0

Table 5.4 LULC transition probabilities, 1990–2000 (Bold figures on diagonal are probability of remaining unchanged)

	2000						
1990	Waterbodies	Floodplain	Cultivated land	Vegetation	Built-up	Bare soil	Rural settlements
Waterbodies	**0.6442**	0.1031	0.1006	0.0192	0.0614	0.0573	0.0142
Floodplain	0.0125	**0.6631**	0.1625	0.0336	0.0476	0.0537	0.0270
Cultivated land	0.0075	0.0652	**0.5270**	0.0672	0.1369	0.1044	0.0919
Vegetation	0.0074	0.0522	0.3668	**0.3429**	0.0951	0.0440	0.0916
Built-up	0.0291	0.0279	0.0667	0.0586	**0.7341**	0.0480	0.0356
Bare soil	0.0369	0.1276	0.1868	0.0464	0.3641	**0.1864**	0.0517
Rural settlements	0.0128	0.0924	0.0927	0.1236	0.2897	0.2805	**0.1084**

Table 5.5 LULC transition probabilities, 2000–2011 (Bold figures on diagonal are probability of remaining unchanged)

	2011						
2000	Waterbodies	Floodplain	Cultivated land	Vegetation	Built-up	Bare soil	Rural settlements
Waterbodies	**0.6944**	0.0605	0.0613	0.0328	0.0959	0.0459	0.0092
Floodplain	0.0186	**0.6375**	0.0754	0.0375	0.0530	0.1582	0.0198
Cultivated land	0.0134	0.0548	**0.4624**	0.097	0.1354	0.1585	0.0785
Vegetation	0.0091	0.0358	0.2899	**0.4033**	0.1021	0.0551	0.1047
Built-up	0.0141	0.013	0.0367	0.1254	**0.7505**	0.0487	0.0114
Bare soil	0.0139	0.0497	0.2443	0.0741	0.2636	**0.3032**	0.0513
Rural settlements	0.0119	0.0325	0.2330	0.1927	0.3151	0.0752	**0.1396**

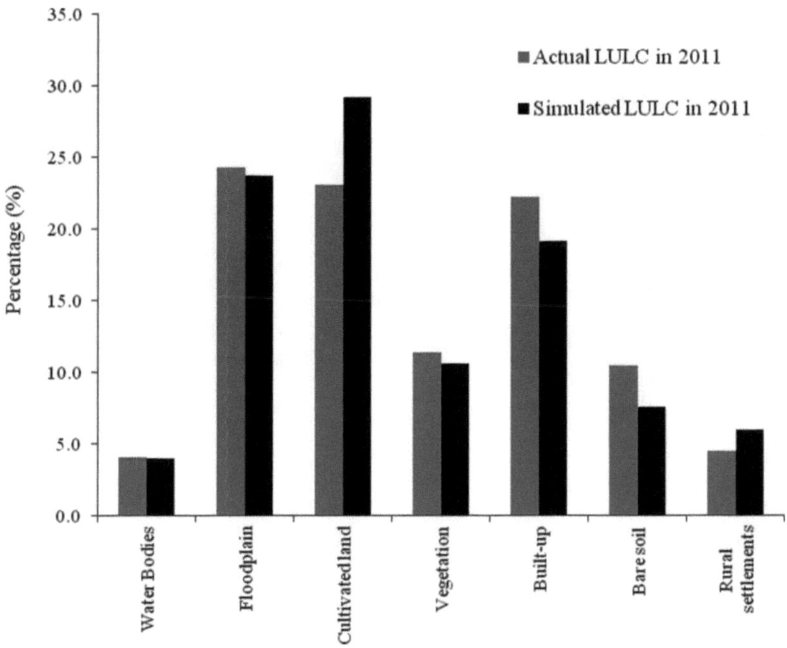

Fig. 5.2 Area percentage of actual versus simulated LULC categories in 2011 (%)

soil categories remains high in the later period as shown in Table 5.5. On the other hand, the TP from vegetation to cultivated category was 0.3668 in 1990–2000 and 0.2899 during 2000–2011.

The MCA-CA model was validated by comparison of the mapped and simulated land-use categories for 2011 as in Fig. 5.1. The agreement of these two maps measured by means of the kappa statistic was 0.745, indicating a reasonable overall performance (74 %) of the model. Further analysis revealed that the cultivated LULC category is most poorly simulated followed by the built-up category. The best agreement was achieved for the waterbodies, vegetation and floodplain categories. For instance, the area of the waterbodies category in the actual map was 4.1 % and the corresponding simulated category was 4 %. Similarly, the observed area in the vegetation category was 11.4 % and the projected value was 10.6 % (Fig. 5.2). Since the Markov-cellular automata technique uses a contiguity rule to simulate LULC growth, the nearness of existing LULC categories may have an impact on the poor simulation of LULC categories (Pontius and Malanson 2005). Furthermore, down-weighting a particular category due to the influence of a few nearby pixels can influence the simulation of LULC growth (Kamusoko et al. 2009).

Based on the actual LULC map of 2011 and the 2001 map, a transition probability matrix and suitability maps, prediction of LULC was carried out for 2022 and 2033 (Fig. 5.3). Since the time-lag was 11 years, we used 11 iterations to predict the future growth in different LULC types. The results showed significant growth in both the built-up and rural settlement categories. For instance, the Markov-CA

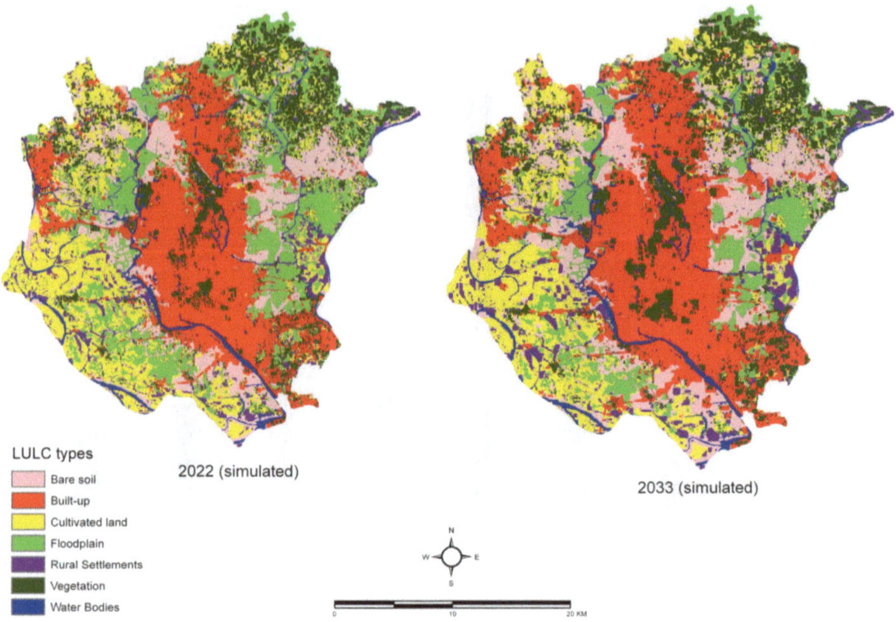

Fig. 5.3 Simulated LULC maps of 2022 and 2033

Table 5.6 Area statistics of Markov-CA projected LULC for 2022 and 2033 (unit: ha)

LULC type	2022 (simulated)	2033 (simulated)	Area change 2011–2022	Area change 2011–2033
Waterbodies	3,827.2	3,883.6	229.7	286.1
Floodplain	16,060.1	10,922.9	−5,302.0	−10,439.1
Cultivated land	18,082.5	15,753.1	−2,202.0	−4,531.4
Vegetation	11,652.7	13,152.5	3,117.2	1,617.4
Built-up	22,723.3	26,038.8	3,166.4	6,481.9
Bare soil	11,336.9	13,530.8	2,097.1	4,290.9
Rural settlements	4,364.5	4,765.4	391.1	792.0

model predicted that the built-up category would increase from 19,556 to 22,723 ha in 2022 and 26,038 ha in 2033, if the current rate of land conversion continues. Likewise, rural settlements would reach 4,765 ha by 2033. In contrast, a significant decrease in the floodplain category was predicted showing a decrease from 21,362 ha in 2011 to 16,060 in 2022 and 10,922 ha in 2033. Similarly, the current area of cultivated land would be reduced by about 4,531 ha between 2011 and 2033 (Table 5.6). Since most of the current land development in Dhaka Megacity is taking place on floodplain and on cultivated land (Ahmed and Ahmed 2012; Islam and Ahmed 2011; Dewan and Yamaguchi 2009a, b), it is no surprise that reduction of these two categories, due to population increase, is predicted. However, the simulation of LULC growth also depends on a variety of factors of which one is the accuracy of the two input LULC maps (Yuan 2010). In addition, various socio-economic and physical factors and government policy may have a vital role in the conversion of LULC of a particular area (Mitsuda and Ito 2011; Guan et al. 2011; Kamusoko et al. 2009). Hence, incorporation of these other factors into the model is warranted.

5.4 Conclusions

Multi-temporal LULC classifications using Landsat data have been described in this chapter. The derived datasets were further used to simulate future LULC changes in Dhaka Megacity using the Markov-CA technique.

The results confirm that Dhaka Megacity is experiencing rapid urban growth leading to the rapid depletion of rural and arable land. Urban area is encroaching rapidly on other forms of land cover causing environmental deterioration. Urban built-up areas significantly increased from 11,696 to 19,556 ha between 1990 and 2011 which is mainly attributed to the rapid increase of population due to large-scale rural–urban migration. Consequently, waterbodies, cultivated lands, vegetation and wetlands/lowlands are all under extreme pressure and rapidly reducing in area. Observations on the ground have found that much of the city's rapid growth in population has occurred in informal settlements with little or no attempt being made to limit the risk of environmental degradation. To alleviate the adverse environmental impacts of urban expansion, planning regulations need to be enforced and effective coordination between government agencies needs to occur to save the fast-declining natural resource base for sustainable development. Urban expansion should be restrained on wetlands, vegetation and expansive floodplains or cultivated lands. This could save productive fertile soils from land degradation and also may contribute to the overall ecological equilibrium. It is increasingly imperative to take a holistic approach to the management of urban land use and environment. Furthermore, regional and local land-use management policy needs to be revised, and integrated multidisciplinary research should be initiated so that a sustainable urban development strategy can be formulated.

Projected changes in LULC to both 2022 and 2033 were explored using the Markov-CA modelling technique. It was found that the cultivated land and floodplain LULC categories would be considerably reduced by 2033 if current activities continue. By contrast, urban built-up surface and rural settlements would increase. Since a number of other issues are associated with LULC change in urban environment, the results obtained here may be instrumental in guiding local authorities to develop pertinent countermeasures to save precious natural resources such as wetlands and arable lands.

This study did not consider various socio-economic factors in the simulation of LULC change due to lack of availability of such data in the study area. A further study could employ biophysical, socio-economic and policy-related factors in a simulation of further LULC changes in Dhaka Megacity which could guide more informed decision making.

References

Abdullah SA, Nakagoshi N (2005) Changes in landscape spatial pattern in the highly developing state of Selangor, Peninsular Malaysia. Landsc Urban Plan 77(3):263–275

Ahmed B, Ahmed R (2012) Modeling urban land cover growth dynamics using multi-temporal satellite images: a case study of Dhaka, Bangladesh. ISPRS Int J Geoinf 1(1):3–31

Alphan H (2003) Land use change and urbanization in Adana, Turkey. Land Degrad Dev 14(6):575–586

Alves DS, Skole DL (1996) Characterizing land cover dynamics using multi-temporal imagery. Int J Remote Sens 17(4):835–839

Anderson R, Hardy EE, Roach JT, Witmer RE (1976) A land use and land cover classification system for use with remote sensor data, Sioux Falls, USA, USGS professional paper 964

Arsanjani JJ, Helbich M, Kainz W, Boloorani AD (2013) Integration of logistic regression, Markov chain and cellular automata models to simulate urban expansion. Int J Appl Earth Obs Geoinf 21:265–275

Bakr N, Weindorf DC, Bahnassy MH, Marei SM, El-Badawi MM (2010) Monitoring land cover changes in a newly reclaimed area of Egypt using multi-temporal Landsat data. Appl Geogr 30 (4):592–605

Bangladesh Bureau of Statistics (BBS) (1996) Agricultural census of Bangladesh. Bangladesh Bureau of Statistics, Ministry of Planning, Dhaka

Bangladesh Bureau of Statistics (BBS) (2001) Population census 2001. Bangladesh Bureau of Statistics, Ministry of Planning, Dhaka

Bangladesh Bureau of Statistics (BBS) (2003) Population census – 2001, Community series. Bangladesh Bureau of Statistics, Ministry of Planning, Dhaka

Bangladesh Bureau of Statistics (BBS) (2010) Census of Agriculture 2008: structure of agricultural holdings and livestock population, vol 1, National series. Bangladesh Bureau of Statistics, Ministry of Planning, Dhaka

Bangladesh Bureau of Statistics (BBS) (2011) GDP of Bangladesh at 2007–08 to 2010–11, BBS, Dhaka. Available at Bangladesh Bureau of Statistics, http://www.bbs.gov.bd/WebTestApplication/userfiles/Image/BBS/GDP_2011.pdf. Accessed 13 Sept 2012

Bangladesh Bureau of Statistics (BBS) (2012) Population and housing census 2011. Bangladesh Bureau of Statistics, Ministry of Planning, Dhaka

Barredo JI, Demicheli L (2003) Urban sustainability in developing countries' megacities: modeling and predicting future urban growth in Lagos. Cities 20(5):297–310

Batta B (2009) Analysis of urban growth pattern using remote sensing and GIS: a case study of Kolkata, India. Int J Remote Sens 30(18):4733–4746

Batta B, Saraswati S, Bandyopadhyay D (2010) Urban sprawl measurement from remote sensing data. Appl Geogr 30(4):731–740

Bolstad PV, Lillesand TD (1991) Rapid maximum likelihood classification. Photogramm Eng Remote Sens 57(1):67–74

Braimoh AK, Onishi T (2007) Spatial determinants of urban land use change in Lagos, Nigeria. Land Use Policy 24(2):502–515

Cabral P, Zamyatin A (2006) Three land change models for urban dynamics analysis in Sintra-Cascais area, First EARSel workshop, Berlin

Chavez PS (1996) Image-base atmospheric corrections – revisited and improved. Photogramm Eng Remote Sens 62(9):1025–1036

Chen Z, Wang J (2010) Land use and land cover change detection using satellite remote sensing techniques in the mountainous Three Georges Area, China. Int J Remote Sens 31 (6):1519–1542

Choudhury AKMK (2008) Land use planning in Bangladesh. A H Development Publishing House, Dhaka

Chowdhury IU (2003) The role of RAJUK in planned urban development. In: Islam N (ed) World habitat day 2003: Souvenir on water and sanitation for cities. Bangladesh Institute of Planners and Center for Urban Studies, Dhaka, pp 88–91

Chowdhury AM, Faruqui S (1989) Physical growth of Dhaka City. In: Ahmed SU (ed) Dhaka: past, present and future. The Asiatic Society of Bangladesh, Dhaka, pp 43–61

Chowdhury JU, Rahman R, Bala Islam SK, Saiful AKM (1998) Impact of 1998 flood on Dhaka City and performance of flood control works. Institute of Flood Control and Drainage Research, Dhaka

Clarke KC, Brass JA, Riggan PJ (1994) A cellular-automaton model of wildfire propagation and extinction. Photogramm Eng Remote Sens 60(11):1355–1367

Clarke KC, Hoppen S, Gaydos L (1997) A self-modifying cellular automation model of historical urbanization in the San Francisco Bay area. Environ Plan B Plan Des 24(2):247–261

Congalton RG (1991) A review of assessing the accuracy of classifications of remotely sensed data. Remote Sens Environ 37(1):35–46

Coppin P, Jonckheere I, Nackaerts K, Muys B, Lambin E (2004) Digital change detection methods in ecosystem monitoring: a review. Int J Remote Sens 25(9):1565–1596

Costanza R, Ruth M (1998) Using dynamic modeling to scope environmental problems and build consensus. Environ Manage 22(2):183–195

de Koning G, Benitez P, Munoz F, Olschewski R (2007) Modelling the impacts of payments for biodiversity conservation on regional land-use patterns. Landsc Urban Plan 83(4):255–267

Deng JS, Wang K, Hong Y, Qi JG (2009) Spatio-temporal dynamics and evolution of land use change and landscape pattern in response to rapid urbanization. Landsc Urban Plan 92(3–4):187–198

Dewan AM, Yamaguchi Y (2008) Effects of land cover changes on flooding: example from Greater Dhaka of Bangladesh. Int J Geoinform 4(1):11–20

Dewan AM, Yamaguchi Y (2009a) Land use and land cover change in Greater Dhaka, Bangladesh: using remote sensing to promote sustainable urbanization. Appl Geogr 29(3):390–401

Dewan AM, Yamaguchi Y (2009b) Using remote sensing and GIS to detect and monitor land use and land cover change in Dhaka Metropolitan of Bangladesh during 1960–2005. Environ Monit Assess 150(1–4):237–249

Dewan AM, Kabir MH, Nahar K, Rahman MZ (2012) Urbanization and environmental degradation in Dhaka Metropolitan Area of Bangladesh. Int J Environ Sustain Dev 11(2):118–146

Dixon RK, Solomon AM, Brown S, Houghton RA, Trexier MC, Wisniewski J (1994) Carbon pools and flux of global forest ecosystems. Science 263(5144):185–190

Dong Y, Forster B, Ticehurst C (1997) Radar backscatter analysis for urban environments. Int J Remote Sens 18(6):1351–1364

Donnay JP, Barnsley MJ, Longley PA (2001) Remote sensing and urban analysis. Taylor & Francis, London/New York

Eastman JR (2006) IDRISI 15: the Andes edition. Clark University, Worcester

Evans TP, Kelley H (2004) Multi-scale analysis of a household level agent-based model of land cover change. J Environ Manag 72(1–2):57–72

Fang S, Gertner GZ, Sun Z, Anderson AA (2005) The impact of interactions in spatial distribution of the dynamics of urban sprawl. Landsc Urban Plan 73(4):294–306

Flood Action Plan (FAP) 8A (1991) Master plan study for greater Dhaka protection project. Japan International Cooperation Agency, Dhaka

Foley JA, DeFries R, Asner GP, Barford C, Bonan G, Carpenter SR, Chapin FS, Coe MT, Daily GC, Gibbs HK, Helkowski JH, Holloway T, Howard EA, Kucharik CJ, Monfreda C, Patz JA, Prentice IC, Ramankutty N, Snyder PK (2005) Global consequences of land use. Science 309 (5734):570–574

Garcia M, Alvarez R (1994) TM digital processing of a tropical forest region in southern Mexico. Int J Remote Sens 15(8):1611–1632

Giri C, Shrestha S (1996) Land cover mapping and monitoring from NOAA AVHRR data in Bangladesh. Int J Remote Sens 17(14):2749–2759
Griffiths P, Hostert P, Gruebner O, van der Linden S (2010) Mapping megacity growth with multi-sensor data. Remote Sens Environ 114(12):426–439
Guan D, HaiFeng L, Inohae T, Su W, Nagaie T, Hokao K (2011) Modeling urban land use change by the integration of cellular automaton and Markov model. Ecol Model 222(20–22):3761–3772
Harris PM, Ventura SJ (1995) The integration of geographic data with remotely sensed imagery to improve classification in an urban area. Photogramm Eng Remote Sens 61(8):993–998
Hazen BC, Berry MW (1997) The simulation of land-cover change using a distributed computing environment. Simul Pract Theory 5(6):489–514
He CY, Okada N, Chang QF, Shi PJ, Li JG (2008) Modelling dynamic urban expansion processes incorporating a potential model with cellular automata. Landsc Urban Plan 86(1):79–91
Howarth PJ, Wickware GM (1981) Procedures for change detection using Landsat digital data. Int J Remote Sens 2(3):277–291
Hu Z, Lo C (2007) Modeling urban growth in Atlanta using logistic regression. Comput Environ Urban Syst 31(6):667–688
Huang B, Zhang L, Wu B (2009) Spatiotemporal analysis of rural-urban land conversion. Int J Geogr Inf Sci 23(3):379–398
Islam N (1996) Dhaka from city to megacity: perspectives on people, places, planning and development issues. Urban Studies Program, Dhaka
Islam N (1999) Urbanization, migration and development in Bangladesh: recent trends and emerging issues, CPD-UNFPA series 1. Center for Policy Dialogue, Dhaka
Islam N (2005) Dhaka now: contemporary development. The Bangladesh Geographical Society, Dhaka
Islam MS, Ahmed R (2011) Land use change prediction in Dhaka city using GIS aided Markov chain modeling. J Life Earth Sci 6:81–89
Islam KR, Weil RR (2000) Land use effects on soil quality in a tropical forest ecosystem of Bangladesh. Agric Ecosyst Environ 79(1):9–16
Jat MK, Garg PK, Khare D (2008a) Monitoring and modeling of urban sprawl using remote sensing and GIS techniques. Int J Appl Earth Obs Geoinf 10(1):26–43
Jat MK, Garg PK, Khare D (2008b) Modelling of urban growth using spatial analysis techniques: a case study of Ajmer city (India). Int J Remote Sens 29(2):543–567
Jensen JR (1996) Introductory digital image processing: a remote sensing perspective. Prentice-Hall, Upper Saddle
Jensen IR, Ramsey EW (1987) Inland wetland change detection using Aircraft MSS data. Photogramm Eng Remote Sens 53(5):521–529
Kamusoko C, Aniya M, Adi B, Manjoro M (2009) Rural sustainability under threat in Zimbabwe – simulation of future land use/cover changes in the Bindura district based on the Markov-cellular automata model. Appl Geogr 29(3):435–447
Lambin EF, Rounseveel MDA, Geist HJ (2000) Are agricultural land-use models able to predict changes in land-use intensity? Agric Ecosyst Environ 82(1–3):321–331
Lambin EF, Turner BL, Geist HJ, Agbola SM, Angelsen A, Bruce JW, Coomes OT, Dirzo R, Fischer G, Folke C, George PS, Homewood K, Imbernon J, Leemans R, Li X, Moran EF, Mortimore M, Ramakrishnan PS, Richards JF, Skanes H, Steffen W, Stone GD, Svedin U, Veldkamp TA, Vogel C, Xu J (2001) The causes of land-use and land-cover change: moving beyond the myths. Glob Environ Chang 11(4):261–269
Li X, Yeh AGO (2004) Analyzing spatial restructuring of land use patterns in a fast growing region using remote sensing and GIS. Landsc Urban Plan 69(4):335–354
Li JJ, Wang XR, Wang XJ, Ma WC, Zhang H (2009) Remote sensing evaluation of urban heat island and its spatial pattern of the Shanghai metropolitan area, China. Ecol Complex 6(4):413–420
Lillesand TM, Kiefer RW (1999) Remote sensing and image interpretation. Wiley, New York

Liu Y, Lv XJ, Qin XS, Gao HC, Yu YJ, Wang JF, Mao G (2007) An integrated GIS-based analysis system for land-use management of lake areas in urban fringe. Landsc Urban Plan 82(4):233–246
Lo CP, Faber BJ (1997) Integration of Landsat thematic mapper and census data for quality of life assessment. Remote Sens Environ 62(2):143–157
Lo CP, Yang X (2002) Drivers of land-use/land-cover changes and dynamic modelling for the Atlanta, Georgia metropolitan area. Photogramm Eng Remote Sens 68(10):1073–1082
Long H, Tang G, Li X, Heilig GK (2007) Socio-economic forces of land-use change in Kunshan, the Yangtze river delta economic area of China. J Environ Manag 83(3):351–364
Loveland TR, Zhu Z, Ohlen DO, Brown JF, Reed BC, Yang LM (1999) An analysis of IGBP global land-cover characterization process. Photogramm Eng Remote Sens 65(9):1021–1032
Lu Q, Weng D (2005) Urban classification using full spectral information of Landsat ETM+ imagery in Marion County, Indiana. Photogramm Eng Remote Sens 71(11):1275–1284
Lu D, Mausel P, Brondiozio E, Moran E (2004) Change detection techniques. Int J Remote Sens 25(12):2365–2407
Lu D, Moran E, Hetrick S, Li G (2011) Land-use and land-cover change detection. In: Weng Q (ed) Advances in environmental remote sensing: sensors, algorithms and applications. CRC Press, Boca Raton, pp 273–291
Mahmood R, Pielke RA Sr, Hubbard KG, Niyogi D, Bonan G, Lawrence P, McNider R, McAlpine C, Etter A, Gameda S, Qian B, Carleton A, Beltran-Przekurat A, Chase T, Quintanar AI, Adegoke JO, Vezhapparambu S, Conner G, Asefi S, Sertel E, Legates DR, Wu Y, Hale R, Frauenfeld OW, Watts A, Shepherd M, Mitra C, Anantharaj VG, Fall S, Lund R, Trevino A, Blanken P, Du J, Chang H-I, Leeper R, Nair US, Dobler S, Deo R, Syktus J (2010) Impacts of land use/land cover change on climate and future research priorities. Bull Am Meteorol Soc 91 (1):37–46
Martinuzzi S, Gould WA, Gonzalez OMR (2007) Land development, land use and urban sprawl in Puerto Rico integrating remote sensing and population census data. Landsc Urban Plan 79(3–4):288–297
Mas J (1999) Monitoring land-cover changes: a comparison of change detection techniques. Int J Remote Sens 20(1):139–152
Meaille R, Wald L (1990) Using geographic information system and satellite imagery within a numerical simulation of regional urban growth. Int J Geogr Inf Syst 4(4):445–456
Mena CF, Walsh SJ, Frizzelle BG, Yao X, Malanson GP (2011) Land use change on household farms in the Ecuadorian Amazon: design and implementation of an agent-based model. Appl Geogr 31(1):210–222
Mesev V (2007) Integration of GIS and remote sensing. Wiley, Chichester
Meyer WB, Turner BL II (eds) (1994) Changes in land use and land cover: a global perspective. Cambridge University Press, Cambridge
Mitsova D, Shuster W, Wang X (2011) A cellular automata model of land cover change to integrate urban growth with open space conservation. Landsc Urban Plan 99(2):141–153
Mitsuda Y, Ito S (2011) A review of spatial-explicit factors determining spatial distribution of land use/land-use change. Landsc Ecol Eng 7(1):117–125
Mundia CN, Aniya M (2006) Dynamics of land use/cover changes and degradation of Nairobi City, Kenya. Land Degrad Dev 17(1):97–108
Muttitanon W, Tripathi NK (2005) Land use/cover changes in the coastal zone of Bay Don Bay, Thailand using Landsat 5 TM data. Int J Remote Sens 26(11):2311–2323
Ojima DS, Galvin KA, Turner BL (1994) The global impact of land use change. Bioscience 44(5):300–304
Pontius RG Jr, Malanson J (2005) Comparison of the structure and accuracy of two land change models. Int J Geogr Inf Sci 19(2):243–265
Quetier F, Lavorel S, Thuiller W, Davies I (2007) Plant-trait-based modeling assessment of ecosystem-service sensitivity to land-use change. Ecol Appl 17(8):2377–2386

Rahman A, Aggarwal SP, Netzband M, Fazal S (2011) Monitoring urban sprawl using remote sensing and GIS techniques of a fast growing urban centre, India. IEEE J Sel Top Appl Earth Obs Remote Sens 4(1):56–64

Rosenfield GH, Fitzpatirck-Lins K (1986) A coefficient of agreement as a measure of thematic classification accuracy. Photogramm Eng Remote Sens 52(2):223–227

Santé I, García AM, Miranda D, Crecente R (2010) Cellular automata models for the simulation of real-world urban processes: a review and analysis. Landsc Urban Plan 96(2):108–122

Seto KC, Kaufmann RK (2003) Modeling the drivers of urban land use change in the Pearl River delta China: integrating remote sensing with socioeconomic data. Land Econ 79(1):106–121

Shalaby A, Tateishi R (2007) Remote sensing and GIS for mapping and monitoring land cover and land use changes in the Northwestern coastal zone of Egypt. Appl Geogr 27(1):28–41

Siddiqui K, Ahmed J, Awal A, Ahmed M (2000) Overcoming the governance crisis in Dhaka City. The University Press Limited, Dhaka

Singh A (1989) Digital change detection techniques using remotely sensed data. Int J Remote Sens 10(6):989–1003

Sudhira HS, Ramachandra TV, Jagadish KS (2004) Urban sprawl: metrics, dynamics and modelling using GIS. Int J Appl Earth Obs Geoinf 5(1):29–39

Taubenböck H, Wengmann M, Roth A, Mehl H, Dech S (2009) Urbanization in India – spatio-temporal analysis using remote sensing data. Comput Environ Urban Syst 33(3):179–188

The World Bank (2007) Dhaka: improving living conditions for the urban poor, Sustainable Development Unit, South Asia Region, Report no. 35824-BD

Tolba MK, El-Kholy OA (eds) (1992) The world environment 1972–1992: two decades of challenge. Chapman & Hall, London

Torrens PM (2006) Geosimulation and its application to urban growth modeling. In: Portugali J (ed) Complex artificial environments: simulation, cognition and VR in the study and planning of cities, Part Three. Springer, London, pp 119–136

United Nations (2012) World population prospects: the 2010 revision, vol 1, United Nations Department of Economic and Social Affairs, New York. Available at: http://eas.un.org/undp/wpp/Documentation/publications.html. Accessed 1 Apr 2012

Verburg PH, Schot PP, Dijst MJ, Veldkamp A (2004) Land use change modeling: current practice and research priorities. GeoJournal 61(4):309–324

Ward DP, Murray AT, Phinn SR (2000) A stochastically constrained cellular model of urban growth. Comput Environ Urban Syst 24(6):539–558

Weng Q (2001) A remote sensing-GIS evaluation of urban expansion and its impact on surface temperature in the Zhujiang Delta, China. Int J Remote Sens 22(10):1999–2014

Weng Q (2002) Land use change analysis in the Zhujiang delta of China using satellite remote sensing, GIS and stochastic modeling. J Environ Manage 64(3):273–284

Westmoreland S, Stow DA (1992) Category identification of changed land-use polygons in an integrated image processing/geographic information system. Photogramm Eng Remote Sens 58(11):1593–1599

White R, Engelen G (1993) Cellular automata and fractal urban form: a cellular modeling approach to the evolution of urban land use patterns. Environ Plan A 25(8):1175–1193

White R, Engelen G (2000) High-resolution integrated modeling of the spatial dynamics of urban and regional systems. Comput Environ Urban Syst 24(5):383–400

Wu F (1998) Simulating urban encroachment on rural land with fuzzy-logic-controlled cellular automata in a geographical information system. J Environ Manag 53(4):293–308

Wu K, Zhang H (2012) Land use dynamics, built-up land expansion patterns, and driving forces analysis of the fast-growing Hangzhou metropolitan area, eastern China (1978–2008). Appl Geogr 34:137–145

Xian G, Crane M, Su J (2007) An analysis of urban development and its environmental impact on the Tampa Bay watershed. J Environ Manag 85(4):965–976

Xiao H, Weng Q (2007) The impact of land use and land cover changes on land surface temperature in a karst area of China. J Environ Manag 85(1):245–257

Xiao JY, Shen YJ, Ge JF, Tateishi R, Tang CY, Liang YQ, Huang Z (2006) Evaluating urban expansion and land use change in Shijiazhuang, China, by using GIS and remote sensing. Landsc Urban Plan 75(1–2):69–80

Xie Y (1996) A generalized model for cellular urban dynamics. Geogr Anal 28(4):350–373

Yamamura Y, Amano T, Koizumi T, Mitsuda Y, Taki H, Okabe K (2009) Does land-use change affect biodiversity dynamics at a macroecological scale? A case study of birds over the past 20 years in Japan. Anim Conserv 12(2):110–119

Yang X (2002) Satellite monitoring of urban spatial growth in the Atlanta metropolitan area. Photogramm Eng Remote Sens 68(7):725–734

Yeh AGO, Li X (1996) Urban growth management in the Pearl River delta – an integrated remote sensing and GIS approach. ITC J 1:77–85

Yeh AGO, Li X (1997) An integrated remote sensing-GIS approach in the monitoring and evaluation of rapid urban growth for sustainable development in the Pearl River delta, China. Int Plan Stud 2(2):193–210

Yin Z-Y, Stewart DJ, Bullard S, MacLachlan JT (2005) Changes in built-up surface and population distribution patterns during 1986–1999: a case study of Cairo, Egypt. Comput Environ Urban Syst 29(5):595–616

Yu XJ, Ng CN (2007) Spatial and temporal dynamics of urban sprawl along two urban-rural transects: a case study of Guangzhou, China. Landsc Urban Plan 79(1):96–109

Yu W, Zang S, Wu C, Liu W, Na X (2011) Analyzing and modeling land use land cover change (LUCC) in the Daqing city, China. Appl Geogr 31(2):600–608

Yuan F (2010) Urban growth monitoring and projection using remote sensing and geographic information systems: a case study in the Twin Cities metropolitan area, Minnesota. Geocarto Int 25(3):213–230

Yuan F, Sawaya KE, Loeffelholz BC, Bauer ME (2005) Land cover classification and change analysis of the Twin Cities (Minnesota) Metropolitan Area by multi-temporal Landsat remote sensing. Remote Sens Environ 98(2&3):317–328

Zhang Q, Wang J, Peng X, Shi P (2002) Urban built-up change detection with road density and spectral information from multi-temporal Landsat TM data. Int J Remote Sens 23(15):3057–3078

Chapter 6
Spatiotemporal Analysis of Urban Growth, Sprawl and Structure

Ashraf M. Dewan and Robert J. Corner

Abstract This chapter demonstrates the use of remote sensing and spatially referenced population data to estimate and model urban sprawl, growth and urban structures. Using spatial analytical tools within a GIS, the typology of urban growth and Dhaka's spatial structure from 2000 to 2011was quantified. The results revealed a 33 % expansion of urban areas during the study period. Analysis of urban growth types showed that the extension growth type being the dominant followed by leapfrogging development. The amount of low-density development is increasing with time, indicating sprawling development. Investigation of changes in the population per unit area of built-up surface indicated that overcrowding and lack of space in the urban core are compelling people to settle in peripheral areas, thereby exerting tremendous pressure on a limited resource base.

Keywords Landsat TM • Growth type • Urban sprawl • Spatiotemporal analysis • Modelling • Population census

6.1 Introduction

More than half of the world's population is now living in urban areas, and this trend is projected to continue. According to a recent estimate by the United Nations (UN), the urban population in the world is expected to rise from 3.6 billion in 2011 to 6.3 billion in 2050. Currently, there are 23 megacities in the world accommodating 9.9 % of the world's urban population, and they are expected to be 37 in 2050 which would house 13.6 % urban population of the world (UN 2012). Urban areas in Asia are expanding more rapidly than in any other region of the world, and by 2030 the 12 Asian

A.M. Dewan (✉) • R.J. Corner
Department of Spatial Sciences, Curtin University,
Kent Street, Bentley, Perth 6102, WA, Australia
e-mail: A.Dewan@curtin.edu.au; R.Corner@curtin.edu.au

A. Dewan and R. Corner (eds.), *Dhaka Megacity: Geospatial Perspectives on Urbanisation, Environment and Health*, Springer Geography,
DOI 10.1007/978-94-007-6735-5_6,

megacities will house 10 % of the Asian population (ADB 2008). Most of the associated urbanisation will take place on agricultural land, forest and other natural land cover. Studies show that the rate of urban physical expansion is faster than urban population growth in most cities (Seto et al. 2010), revealing the increasing pressure on land resources. Since land resources are finite, their increasing scarcity due to human activities – particularly by urban development – is still not given adequate attention, making it difficult to achieve the goals of sustainable development (Haber 2007).

Although urban surfaces currently cover only between 3 and 5 % of the total land surface of the Earth (Schneider et al. 2009), environmental impacts in relation to urban growth have become a major concern around the world. A number of studies have demonstrated that urban expansion is significantly affecting both local and global environments (Seto and Shepherd 2009; Herold et al. 2003; Liu and Lathrop 2002), including influencing climate change (Seto et al. 2010; Kalnay and Cai 2003). For instance, the urban heat island (UHI) phenomenon is found to be spatially correlated with changes in urban land use (Li et al. 2012; Weng 2001a). Dramatic changes in urban land use also enhance regional precipitation variability (Aikawa et al. 2009), consume valuable productive arable lands (Lopez et al. 2001), and cause loss of habitat and biodiversity (McKinney 2002), etc. In order to mitigate the detrimental effects on the environment associated with urban growth and to maintain optimal ecosystem functioning (Fang et al. 2005), timely and accurate information on the spatiotemporal patterns of urban development and the factors influencing this growth are essential to support urban planners, resource managers, economists and environmentalists in solving the problems associated with such growth (Seto et al. 2010; Maktav and Erbek 2005; Stewart et al. 2004; Wu and David 2002).

In the past, various theories such as concentric rings (Burgess 1925), the sector theory (Hoyt 1939) and the multiple nuclei theory (Harris and Ullman 1945) were extensively used to study urban patterns and dynamics. Because of the complexity inherent in urban growth, these theories are unable to characterise the spatiotemporal patterns of urban dynamics (Batty 2002). In response, a variety of techniques such as catastrophe theory (Wilson 1976), cellular automata (Tobler 1979), landscape metrics (O'Neill et al. 1988), fractals (Batty et al. 1989), chaos theory (Wong and Fotheringham 1990), self-organising theory (Portugali 2000), the entropy theory (Yeh and Li 2001) and agent-based modelling (Batty 2005) have emerged since the 1960s to analyse urban structure and dynamics.

Urban sprawl has long been the subject of academic research; however, the term "sprawl" is variously defined by many researchers, and there is no single term that can adequately describe the characteristics of urban sprawl (Tsai 2005; Chin 2002; Torrens and Alberti 2000; Ewing 1994). According to Frenkel and Orenstein (2011), urban sprawl can be defined using three primary ways: (1) urban spatial development, (2) consequences of sprawl and (3) social and/or economic processes that give rise to particular urban spatial development patterns.

As with the definition of sprawl, techniques to characterise the development of sprawl also vary considerably (Jaeger et al. 2010). For example, Galster et al. (2001) used eight indices to calculate urban sprawl. Likewise, Angel et al. (2007) used five metrics whilst Jiang et al. (2007) considered thirteen indices when measuring sprawl. Others have used landscape metrics to study urban composition and structure (Frenkel and Orenstein 2011; Bhatta et al. 2010; Deng et al. 2009; Jat et al. 2008a, b; Huang et al. 2007; Xu et al. 2007; Sudhira et al. 2004). As a result of multiple definitions and techniques, the characterisation and measurement of urban sprawl is a difficult task, and the degree of overlap between some of the indices causes confusion to people trying to decide on appropriate tools (Jaeger et al. 2010; Bhatta et al. 2010; Tsai 2005). Without a universally accepted definition, it is very difficult to quantify and model urban sprawl (Wilson et al. 2003), and the techniques that are used to measure sprawl all have some limitations in capturing the characteristics of sprawling development (Yeh and Li 2001). Therefore, creating an urban growth model instead of an urban sprawl model allows us to quantify the amount of land that has changed to urban areas and lets the user decide what they consider to be urban sprawl (Wilson et al. 2003).

Geospatial techniques such as Geographic Information System (GIS) and remote sensing (RS) are powerful tools that assist in the deeper understanding of urban problems, particularly mapping and monitoring of urban built-up areas through space and time (Yang 2011; Seto and Liu 2003). Since the launch of ERTS-1 (Earth Resources Technology Satellite-1, later renamed to Landsat), numerous studies have employed remote sensing data to map and monitor urban land use/cover (Wu and Zhang 2012; Islam and Ahmed 2011; Yu et al. 2011; Guan et al. 2011; Griffiths et al. 2010; Dewan and Yamaguchi 2009a, b; Deng et al. 2009; Jat et al. 2008a, b; Braimoh and Onishi 2007), urban sprawl (Frenkel and Orenstein 2011; Rahman et al. 2011; Shahraki et al. 2011; Biediger and Mathews 2010; Noda and Yamaguchi 2008), urban spatial structure (Ahmed et al. 2012; Pham and Yamaguchi 2011; Taubenböck et al. 2009), urban growth modelling (Ahmed and Ahmed 2012; Arsanjani et al. 2013; Thapa and Murayama 2011; Mitsova et al. 2011; Yuan 2010) and environmental impacts in relation to urbanisation (Talukder et al. 2012; Suriya and Mudgal 2012; Dewan and Yamaguchi 2008; Grimm et al. 2008; Xian et al. 2007; Weng 2001a, b).

The use of geospatial techniques to study urban expansion in Dhaka is a relatively recent phenomenon despite the relatively long availability of satellite data. Dewan and Yamaguchi (2009a, b) were the first to employ multi-temporal satellite data, with topographic maps to monitor urban expansion between 1960 and 2005. This study was further extended by Ahmed et al. (2012) to analyse urban forms using spatial metrics. Griffiths et al. (2010) used multi-temporal and multisensor data to map megacity growth of Dhaka between 1990 and 2006. Whilst they provide valuable information on the spatiotemporal rate of urban growth, little is known about the degree to which urban expansion is compact or sprawling (Seto and Fragkias 2005).

In addition, the lack of updated spatially referenced statistical land-use data is hindering the effective management of the urban environment which, in turn, has resulted in severe environmental degradation in the recent past (Dewan et al. 2012a). In order to further curb the environmental deterioration associated with rapid and unplanned urbanisation, there is an urgent need to document the changing form of urban areas which will greatly assist in evaluating the efficacy of existing urban planning in directing urban growth or conservation policies to save protected areas or limit urban expansion (Seto and Shepherd 2009).

Primarily, the aim of the work reported in this chapter is to seek the answers to three research questions: (1) What is the extent of urban development in the last decade? (2) To what extent is the urban growth scattered or compact? (3) How are people distributed over urban areas that they occupy? To address these questions, we combined satellite information with population census data to study three issues of Dhaka's landscape: urban growth pattern, sprawl and spatial structure. Firstly, we map the most recent built-up areas. Then, we use indices to quantify urban sprawl; and finally, we integrate spatially referenced population data from two recent censuses with satellite-derived information to investigate the spatial structure of the city.

6.2 Data and Methods

Although the government of Bangladesh conducted censuses in 2001 and 2011, the data with respect to population counts was publically available only in printed tabular form, except at higher administrative level such as *thana*. Therefore, we have created spatial dataset using the smallest possible spatial unit, the census tract, so that population data can be integrated with satellite information for an in-depth analysis.

Geographic data at the lowest level of census geography in Bangladesh, that is, *mauza* and *mahalla* [see Chap. 1 for definitions], were used. Demographic data from the population censuses of 2001 and 2011 were obtained from the BBS community series (BBS 2003, 2012) which were linked with the census tract boundaries with a unique ID. Full details of this procedure can be found in Chap. 3 (Sect. 3.2). Population density was calculated as the total number of people residing in a census tract divided by its total area (km^2).

Land use and land cover (LULC) maps of 2000 and 2011 were derived (see Chap. 5) and reclassified so as to show only the urban built-up area. To characterise the direction of urban expansion, eight directional zones (north, north-west, south, etc.) were established, centred on the central business district (CBD) area of Motijheel and extending for 24 km encompassing all the census tracts (see Fig. 6.1). The built-up area for each zone and for each year was calculated by clipping the built-up surface to the zones and then multiplying the number of pixels in each polygon by the pixel size. Since the satellite information was collected close to the census years, it was not necessary to temporally interpolate the population data (Bhatta 2009, 2010; Yin et al. 2005).

6.2.1 *Analysis of Urban Growth*

There are three primary techniques for the measurement of urban growth; transition matrices, spatial metrics and spatial statistics (Bhatta 2010). Whilst transition matrices are valuable for understanding the dynamics and location of change over

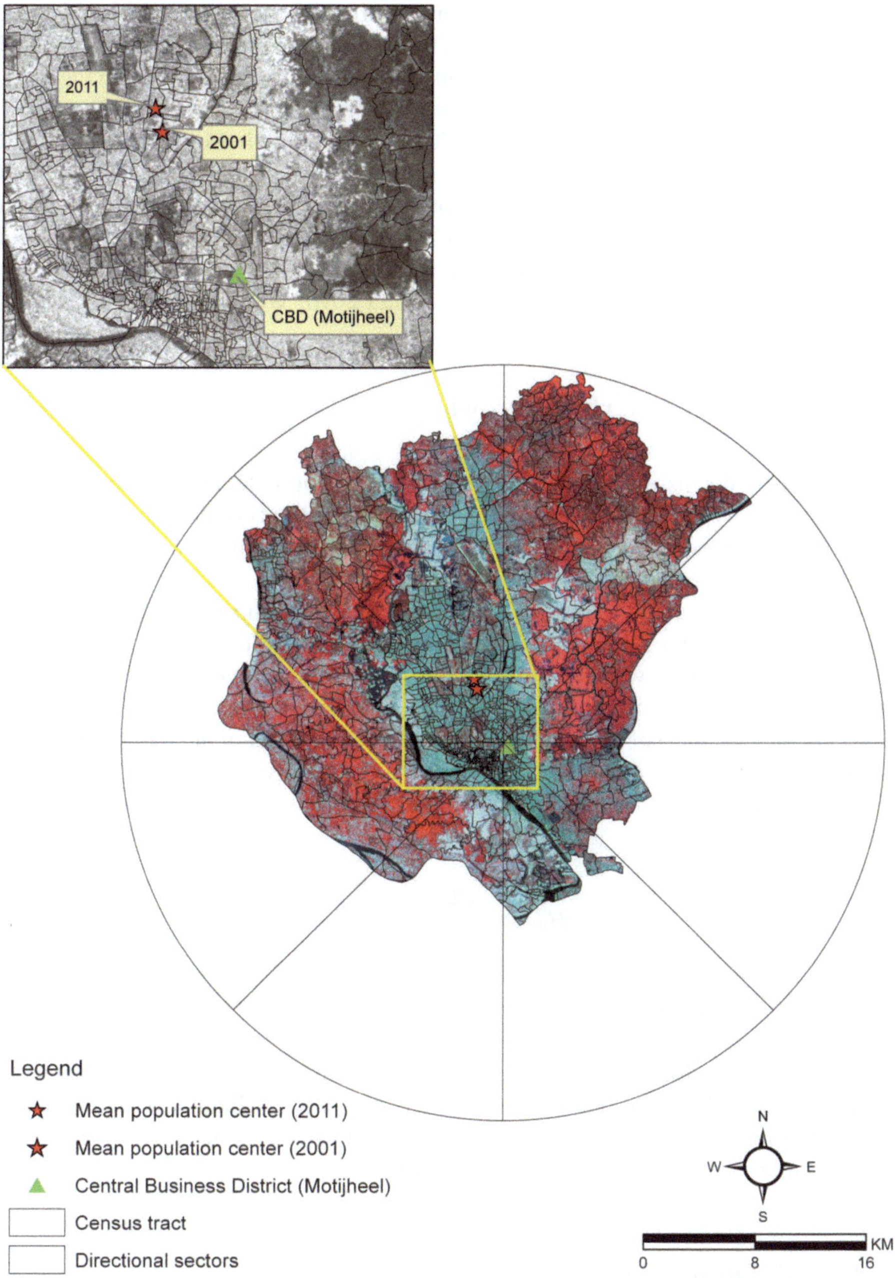

Fig. 6.1 False Colour Composite (*FCC*) of Landsat TM5 of April 2011 of the study area showing CBD and mean population centre in 2001 and 2011

space, they are unable to depict the pattern of urban growth (Ahmed et al. 2012). To overcome this, various spatial metrics have been developed and are being used to elucidate patterns of urban growth through time (Wilson et al. 2003). In the work reported here, urban growth was quantified using the Urban Growth Analysis Tool (UGAT) (Angel et al. 2007). Since UGAT provides a number of metrics, we used those that are adequate to describe forms of new development between 2000 and 2011. Three growth types were defined for built-up areas, namely, infill, extension and leapfrog. The definitions related to these growth types were adopted from Angel et al. (2012). Detailed descriptions of these growth types are widely available in the literature (Angel et al. 2007, 2012; Bhatta 2010; Wilson et al. 2003).

6.2.2 Analysis of Urban Sprawl

As previously discussed, there is considerable variation in techniques for mapping urban sprawl; we therefore considered two popular methods of determining to what extent the urban growth is compact or sprawling. These methods were map density and Shannon's entropy.

Map density is calculated by dividing the number of built-up area pixels by the total number of pixels in a 3 × 3 kernel (Sudhira et al. 2004). This was resulted in nine density classes which were reclassified into three groups using the equal interval method, as low, medium and high density. The area statistics and their relative percentage of each class were then computed to quantify different types of sprawl.

Shannon's entropy is perhaps one of the most widely used indices to analyse urban sprawl phenomenon when integrating remote sensing and GIS (Bhatta 2010). Entropy is calculated by:

$$H_n = -\sum_{i=1}^{n} P_i \log_e(P_i) \tag{6.1}$$

where P_i is the proportion of the variable occurring in the ith zone and n is the total number of zone (here the 1,212 census tracts). Entropy values ranges from 0 to $\log_e(n)$. A value close to 0 indicates the distribution of built-up area is compact whilst larger value signifies the occurrence of sprawl. Shannon's entropy was calculated for each census tract and then summed to determine the distribution of built-up areas over time across the study area.

6.2.3 Relationship Between Built-Up Surface and Population Density

In order to analyse the driving factors for urban growth, regression and econometric models are extensively used (Wu and Zhang 2012; Bhatta 2009; Dewan et al. 2012b; Dewan and Yamaguchi 2009b; Jat et al. 2008a, b; Braimoh and Onishi

2007; Long et al. 2007; Liu et al. 2005; Yin et al. 2005; Tian et al. 2005; Sudhira et al. 2004; Seto and Kaufmann 2003; Lo and Yang 2002). These studies have employed biophysical and/or socioeconomic data to understand the factors influencing urban growth. However, these types of studies are unable to document the spatial location of changes and structure in a city that are driven by population growth. Previous studies in Dhaka have demonstrated that population increase is the major factor contributing to rapid urban growth (Dewan et al. 2012a; Dewan and Yamaguchi 2009b); therefore, in this study we considered only the population data from the 2001 and 2011 censuses to analyse spatiotemporal changes in population density which could more accurately depict the extent and intensity of urbanisation (Stewart et al. 2004; Wang and Zhou 1999; Cowen and Jensen 1998). Since this study is using the smallest unit of census geography, the results will be more useful to determine both population density and the changing extent of urban area. Firstly, population density per unit area of built-up surface was computed for each census unit, and then, a weighted population centre was determined for both 2001 and 2011 using the following formula (ESRI 2012):

$$\bar{x} = \left[\sum_{i=1}^{n} (x_i.pop_i)\right] \Big/ \sum_{i=1}^{n} (pop_i)$$
$$\text{and} \qquad (6.2)$$
$$\bar{y} = \left[\sum_{i=1}^{n} (y_i.pop_i)\right] \Big/ \sum_{i=1}^{n} (pop_i)$$

where $\bar{x}$ and $\bar{y}$ are the coordinates of the weighted population centre; pop_i is the population of each census tract for 2001 and 2011; and x_i and y_i are the coordinates for each census tract. Figure 6.1 shows that the population centre of Dhaka moved towards north and north-west between two census years. Using the 2001 weighted population centre, a multiple ring buffering operation was performed to obtain ten annular buffers with an increment of 2.4 km width from the centre to the furthest census units (24 km). Then, mean population density, proportion of built-up surface and population per unit built-up surface were estimated for each annular buffer for both years. Finally, these ten buffer zones were used to derive a summary overall estimate of population density and built-up surfaces between 2001 and 2011.

6.3 Results and Discussion

6.3.1 Urban Expansion and Typology of Urban Growth

The spatial pattern of urban expansion between 2000 and 2011 is shown in Fig. 6.2. An analysis of the area statistics reveals that urban areas have increased, overall, from 14,641 to 19,556 ha, a 33 % change in the study period. As elevated areas in the city have already been urbanised during the rapid urban development since the

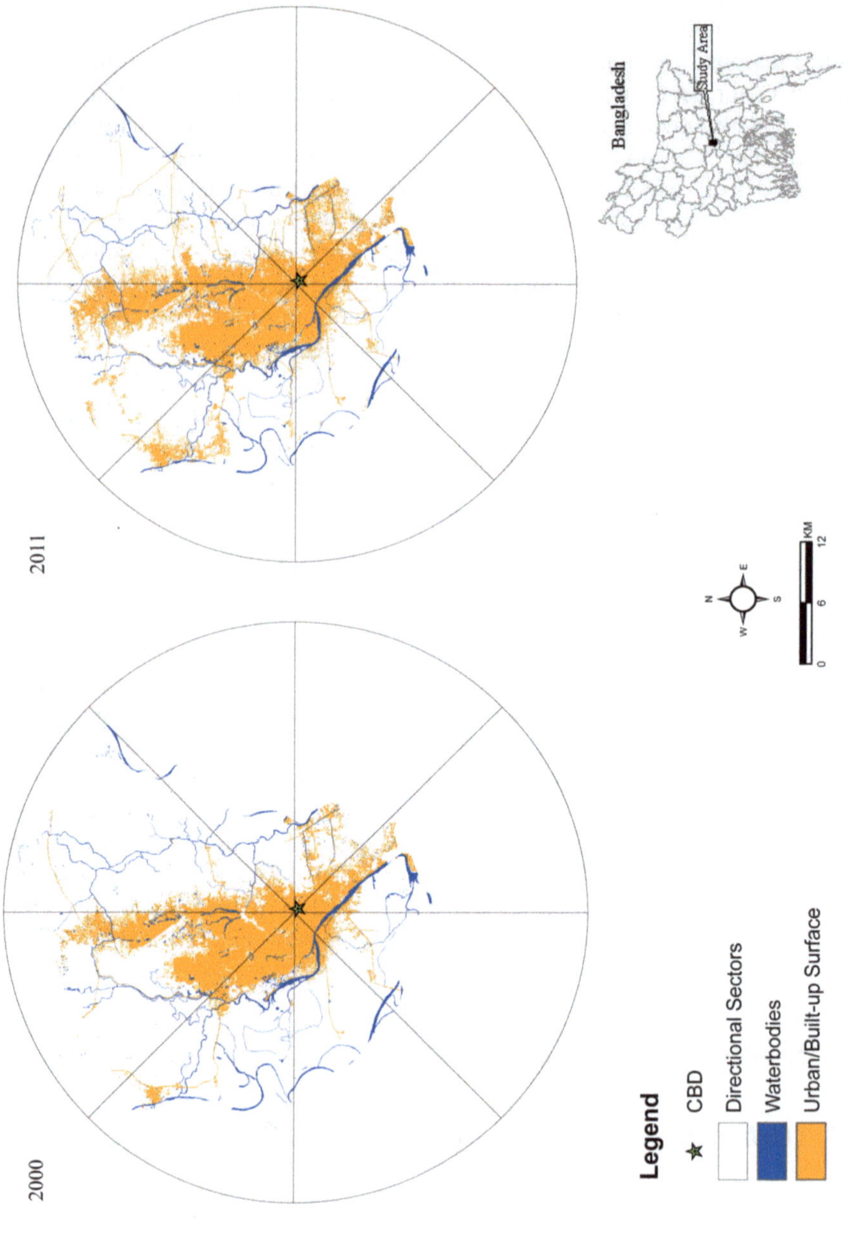

Fig. 6.2 Urban expansion between 2000 and 2011 with directional sectors

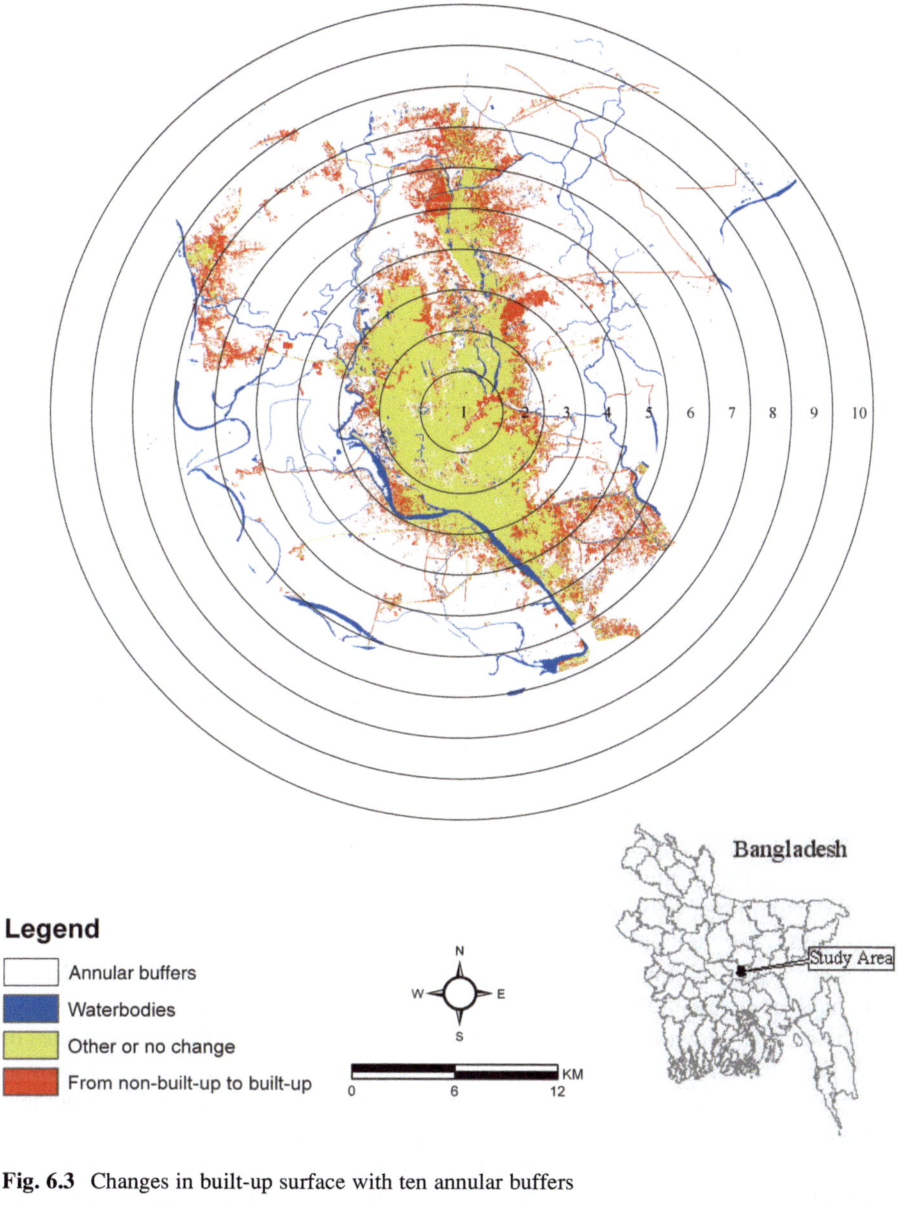

Fig. 6.3 Changes in built-up surface with ten annular buffers

1970s, arable and wetland areas – mainly adjacent to the existing urban edge – have become the target of urban expansion, particularly since the 1990s (Dewan and Corner 2012; Ahmed et al. 2012; Griffiths et al. 2010; Islam et al. 2010; Dewan and Yamaguchi 2009a, b; Begum 2007; Islam 2005b; Gain 2002). The difference between the two images of urban extent (Fig. 6.3) between the two years shows extensive development in peri-urban and suburban areas in recent times. This can be attributed to the construction of extensive transport networks (Islam 2005a). The

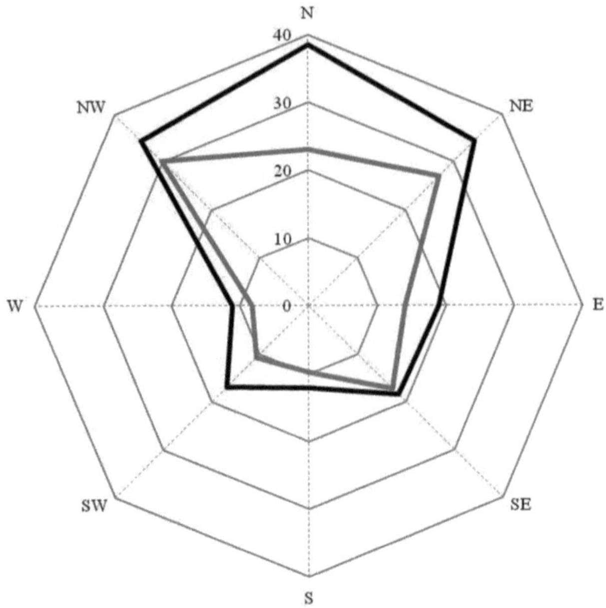

Fig. 6.4 Direction of urban growth, 2000–2011 (*grey* = 2000; *black* = 2011)

Table 6.1 Urban growth types between 2000 and 2011

Growth type	Area (ha)	Percentage (%)
Infill	118.5	1.8
Extension	5,278.9	82.0
Leapfrog	1,042.6	16.2

direction of urban expansion between 2000 and 2011 is shown in Fig. 6.4. This shows that north, north-west and north-east were the major directions experiencing urban development between 2000 and 2011. By contrast, southward and westward expansion is relatively lower except in the south-east zone. Since the south and west zones are primarily dominated by the presence of low-lying and marshy lands, this may have constrained the growth of built-up areas to these directions.

Three urban growth types were identified, using UGAT, and the area statistics of each type are shown in Table 6.1. This reveals that the primary growth type was extension of existing urban areas, accounting for 82 % of total new development between 2000 and 2011. In contrast, the leapfrog growth type used only 16.2 % of urban land, whilst infill growth of the built-up surface accounted for only 1.8 %. This contradicts the earlier findings of Ahmed et al. (2012) who reported that leapfrogging was the dominant growth type in Dhaka. Whilst infill growth is believed to be a remedy for environment degradation caused by sprawling (Wilson et al. 2003), this type of growth is extremely low in the study area. As the city is topographically very flat, filling of land areas for urban expansion is costly both in terms of the provision of fill material and also since additional work is required to offset the recurrent flooding problem that such infill exacerbates. This might be one of the causes of the

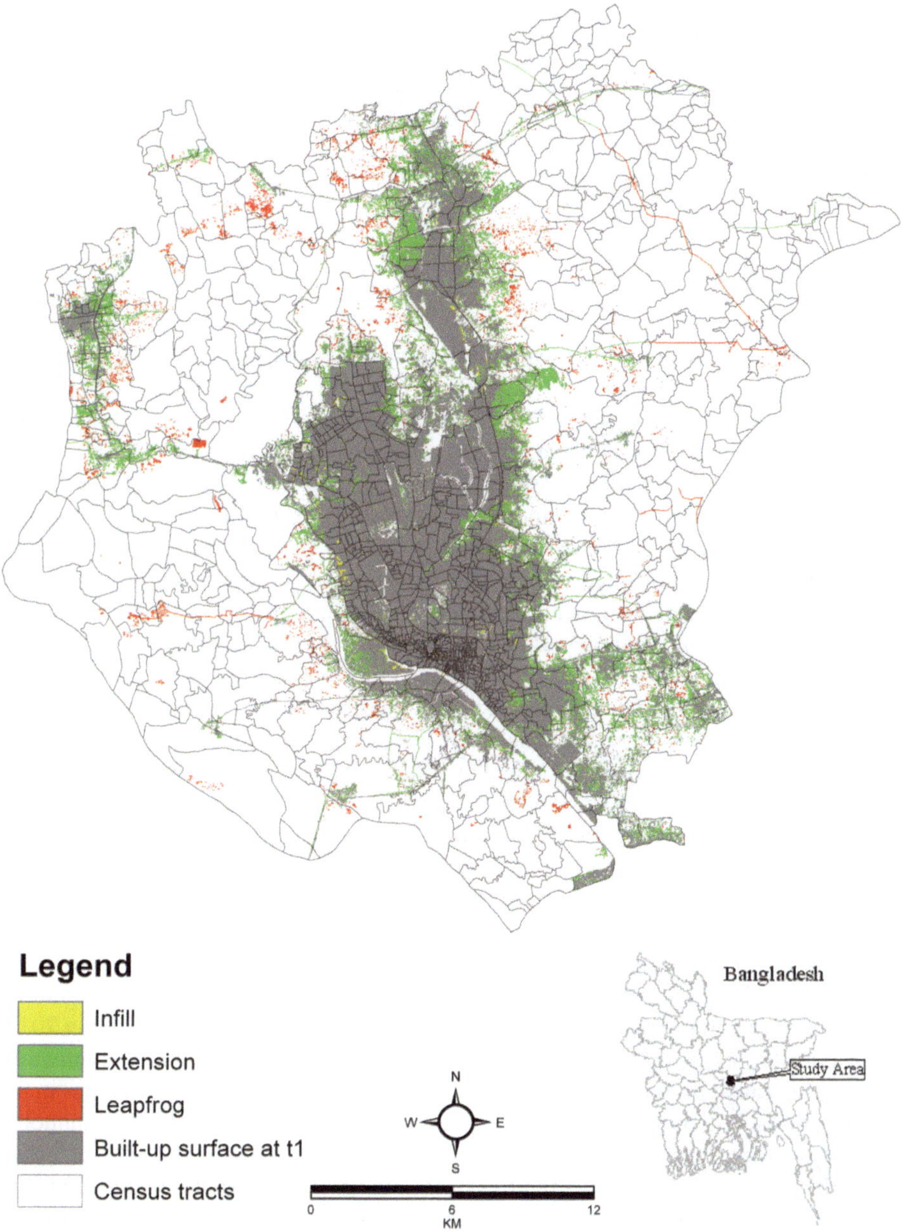

Fig. 6.5 Urban growth types, 2000–2011

considerable extension growth type on the existing urban boundary (Griffiths et al. 2010; Dewan and Yamaguchi 2009b). The results also suggest that leapfrogging is occurring along newly constructed transport networks (Fig. 6.5).

Table 6.2 Built-up area density and Shannon's entropy between 2000 and 2011

	2000		2011		Shannon's entropy	
Category	Area (ha)	Percentage	Area (ha)	Percentage	2000	2011
Low density	736.0	5.0	1,305.5	6.7	2.59	2.66
Medium density	2,160.5	14.8	3,263.4	16.8		
High density	11,694.2	80.1	14,838.8	76.5		

6.3.2 *Urban Sprawl*

As indicated above, two landscape metrics, namely, map density and Shannon's entropy, were used to determine as to what extent the present growth is compact or sprawling. Shannon's entropy was calculated for every census unit and then summed to give an entropy value for the entire urban area for each of the two years (Table 6.2). This shows an increase in the entropy value from 2.59 in 2000 to 2.66 in 2011. Entropy values close to 0 indicates compact distribution, whilst values closer to log *n* (in this case 7.1) show the distribution of a spatial phenomenon is completely dispersed. Since our results give relatively large entropy values for both years, this is a clear indication of dispersed urban growth in Dhaka megacity. Note that because of the large number and heterogeneity of size of the 1,212 census units, it was not possible to present Shannon's entropy value for each unit either as a table or map; hence, only aggregated values are presented.

The result of map density estimation is also shown in Table 6.2 which indicates that low- and medium-density areas have increased since 2000. By contrast, high density has decreased. The percentage of high density of built-up surface was 80 % in 2000 which decreased to 76.5 % in 2011, showing a gradual decline in the compact nature of urban development with time. Low density of built-up areas has increased from 5 to 6.7 %. A similar trend can be observed for the medium-density category (Table 6.2). Figure 6.6 suggests that most of the high-density development is close to established urban areas whilst low- and medium-density development takes place along the road networks and around the city periphery. A substantial increase in medium-density built-up areas is found along peripheral zones in 2011. Low density, on the other hand, can be seen along national and regional highways, suggesting linear branching and ribbon type of development in the outer areas in the later period (Wilson et al. 2003). This result is in agreement with the Shannon's entropy values described above, which substantiate gradual dispersion of urban growth. This also supports the findings of Bhatta (2010) and Mahtab-uz-Zaman and Lau (2000) but contradicts the result of Richardson et al. (2000), who suggest that urban agglomerations in developing countries are less sprawling and denser when compared to western cities (Bhatta 2010).

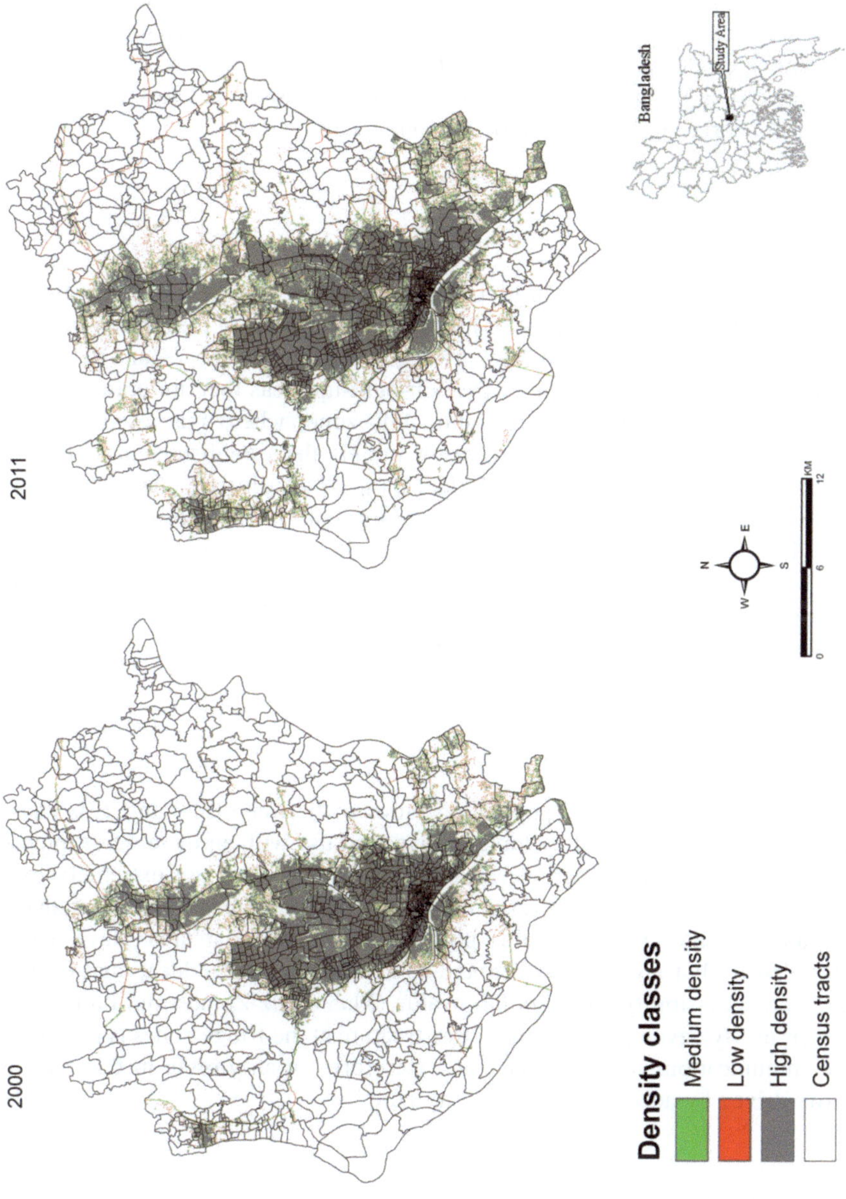

Fig. 6.6 Density of urban areas between 2000 and 2011

6.3.3 Spatial Patterns of Population Density

To determine population density changes from 2001 to 2011 at census tract level, ratios of the 2011 to 2001 values were calculated, and based on this analysis, three types of areas can be identified: areas experiencing little or no change, medium change and high-density change areas. Each of these areas has specific urban characteristics that might be related to changes in government policies, redefinition of urban boundaries and the spread of economic activities.

Stable Areas

These are the areas where little or no population density changes occurred and are represented in Fig. 6.7 as having ratios lower than 0.50. Most of the already built-up areas are within this group, and they are highly saturated with regard to further horizontal expansion. Also, some of the non-built-up areas that are dominated by lowland and agricultural areas are within this zone, where inaccessibility and recurrent flooding in the monsoon season inhibit the establishment of human settlements. However, the central and older parts of the city are characterised by intensive road networks and high building density with intense economic activity. As there is a little room for horizontal expansion in these zones, vertical development is becoming the major feature in this area (Begum 2007). With the exception of some reserve areas, such as the Old Fort, these central areas are also provided with a higher level of urban services and infrastructure.

Medium-Density Areas

These areas have inter-census population ratios of between 0.50 and 1.0. These census units are characterised by mixed urban land use with an increase of intense economic activities. In the peripheral areas, urban amenities are extremely low. However, some of the planned residential areas such as Dhanmondi, Baridhara, Banani and Gulshan are within this group. These residential areas have a low building density but are in great demand as commercial and educational centres due to good accessibility and high land value (Mahtab-uz-Zaman and Lau 2000). Moreover, newly developed transport infrastructure at the fringe zones is providing good communications to these areas, and therefore, urban growth in this zone is primarily driven by accessibility to the city core.

Densifying Areas

Surrounding the medium-density zone (>1.00), this zone includes those census tracts that have exhibited tremendous urban development in the recent past.

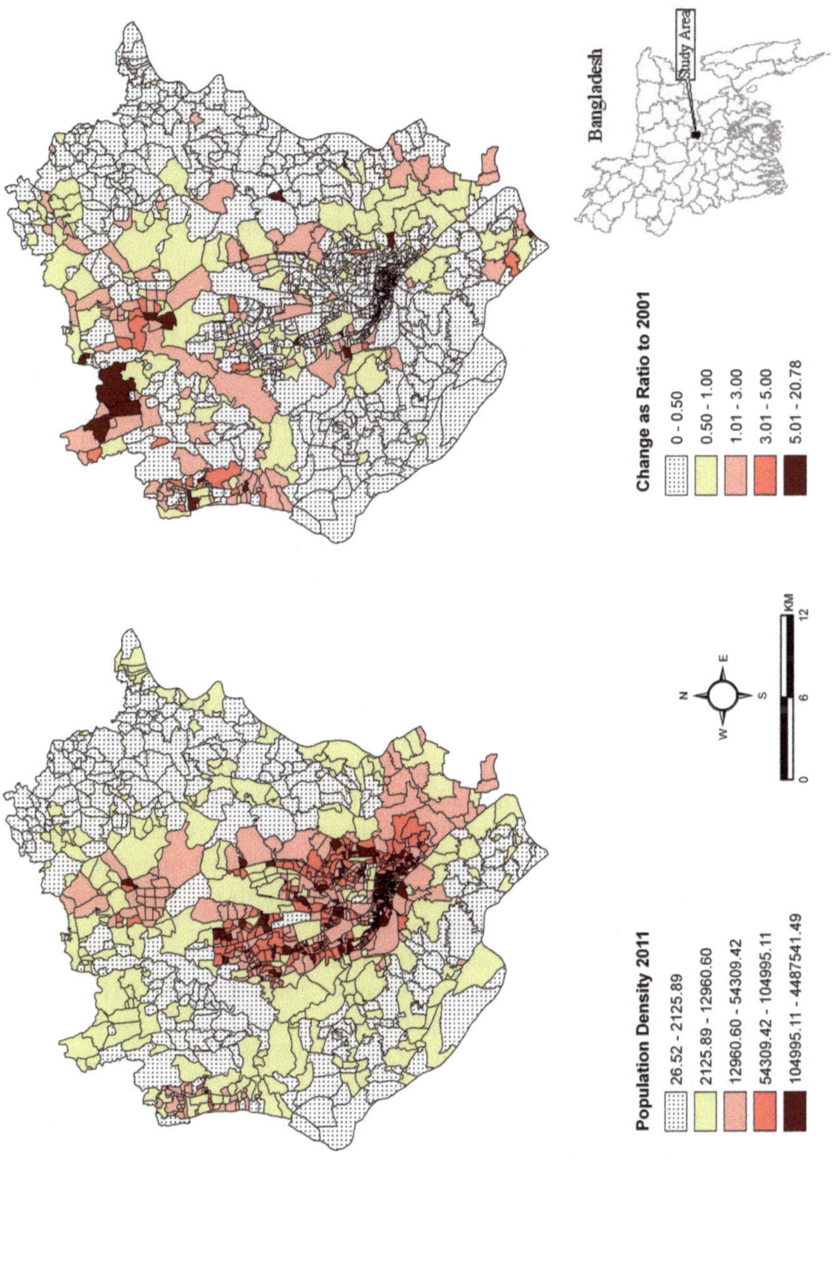

Fig. 6.7 Population density at census tracts in 2011 and change since 2001

Because of newly developed industrial precincts in the north-west corner of the city and the development of new residential plots by public agencies in the peripheral zones, these areas are experiencing high growth as built-up areas encroach on prime agricultural land and floodplains. Figure 6.7 shows clusters of census units in the north-west part of the city that are experiencing incredible growth. Population densification in this zone is strongly influenced by the proximity to regional highways and economic centres such as the export processing zone (EPZ). Ashulia, Savar and Pubail are the examples of such development. Two bridges over the river Buriganga, and a 32-km multipurpose flood control embankment, are stimulating the rapid rise of urban areas in this zone although it has a low level of urban amenities (Chowdhury 2003). Note that the presence of low-lying areas also restrains population growth in many of the census units within the densifying areas.

6.3.4 Analysis of Population Density and Built-Up Area

Using a 2.4-km buffer width, the entire study area was divided into ten zones. Table 6.3 summarises overall population density and built-up surfaces areas between 2001 and 2011. This suggests that the proportion of built-up surfaces is increasing over time in nine out of ten buffer zones, whilst population density has increased in all of the buffer zones. The decline in the proportion of built-up surface in buffer zone 1 may be attributed to consolidation of existing urban areas as this zone is the central and oldest part of Dhaka, indicating compact structure of built-up areas. It may be noted that zone 1 covers most of the urban land and therefore reflects Dhaka's historical development as a dense and compact city (Ahmed et al. 2012). Among other zones, the highest percent of population density changes is found in zone 7 and the lowest being in zone 10. The estimation of proportion of built-up surfaces further revealed that built-up surface in the study area increased rapidly from zone 2 to zone 10, and the highest being found in the buffer zones of 7 and 9. This substantiates the earlier finding of disperse development in these areas. Again, this may be related to the construction in the recent past of transport networks and other infrastructure that may influence the spread of economic activities in the outskirts of Dhaka, thus giving rise to the growth of population and built-up surfaces.

The mean population density in each buffer zone showed an increase from 2001 to 2011. The change is more pronounced in buffer zones 1–4 and 8 (Fig. 6.8a), signifying the decentralising trend of population growth in the recent past. This result disagrees with the result of Wang and Zhou (1999) whose work suggests that the population in Beijing has declined from the city core in the 1980s in response to urban land-use reform, central city renovation and improvement of suburban infrastructure and services. However, in the case of Dhaka, population density is increasing both at the city core and outskirts, revealing extreme pressure on a limited resource base. A remarkable feature can be seen in the estimation of mean built-up surface according to different buffer zones (Fig. 6.8b). For example, zones 1–3 show almost identical patterns of urban development between 2001 and

Table 6.3 Population distribution in annular buffer zones around the 2001 weighted population centre

Buffer zone	Characteristics	2001	2011	Percent change, 2001–2011
Zone 1	Total population	798,381	966,849	
	Population density (persons/km^2)	43,968.5	53,246.4	21.1
	Proportion of built-up surface	0.88	0.88	−0.3
Zone 2	Total population	2,012,984	2,814,811	
	Population density (persons/km^2)	39,637.5	55,426.3	39.8
	Proportion of built-up surface	0.80	0.82	2.2
Zone 3	Total population	2,480,855	3,244,521	
	Population density (persons/km^2)	29,187.3	38,171.8	30.8
	Proportion of built-up surface	0.44	0.53	19.9
Zone 4	Total population	1,100,875	1,629,311	
	Population density (persons/km^2)	8,165.2	12,084.7	48.0
	Proportion of built-up surface	0.17	0.24	39.7
Zone 5	Total population	559,250	872,939	
	Population density (persons/km^2)	3,491.8	5,450.3	56.0
	Proportion of built-up surface	0.07	0.12	68.6
Zone 6	Total population	594,313	891,979	
	Population density (persons/km^2)	3,581.9	5,375.9	50.0
	Proportion of built-up surface	0.05	0.10	107.0
Zone 7	Total population	317,802	649,040	
	Population density (persons/km^2)	2,402.2	4,905.9	104.3
	Proportion of built-up surface	0.04	0.11	172.5
Zone 8	Total population	220,091	417,497	
	Population density (persons/km^2)	2,621.8	4,973.3	89.7
	Proportion of built-up surface	0.04	0.09	121.2
Zone 9	Total population	48,103	80,089	
	Population density (persons/km^2)	1,327.5	2,210.3	66.5
	Proportion of built-up surface	0.007	0.012	180.1
Zone 10	Total population	10,583	13,114	
	Population density (persons/km^2)	1,327.5	2,210.3	23.9
	Proportion of built-up surface	0.004	0.005	43.9

2011; however, considerable growth can be found beyond buffer zone 4, with the greater changes being observed in buffer zones 6–8. These are the areas where intense urban development occurred during the study period that corresponds to the increase of mean population density.

Investigation of changes in population per unit area of built-up surface revealed a considerable increase in all buffer zones with greatest changes being observed in zones 3–4, 6–7 and 9 (Fig. 6.8c). This clearly indicates sprawl development in the study area which is consistent with the overall Shannon's entropy result. The finding of this study contrast with the study of Stewart et al. (2004) in Greater Cairo, Egypt. Their study suggests considerable decrease in population per km^2 of built-up surface, from the city core between 1986 and 1999. But in case of Dhaka, the results are just the opposite, representing a condition of large population with limited liveable space.

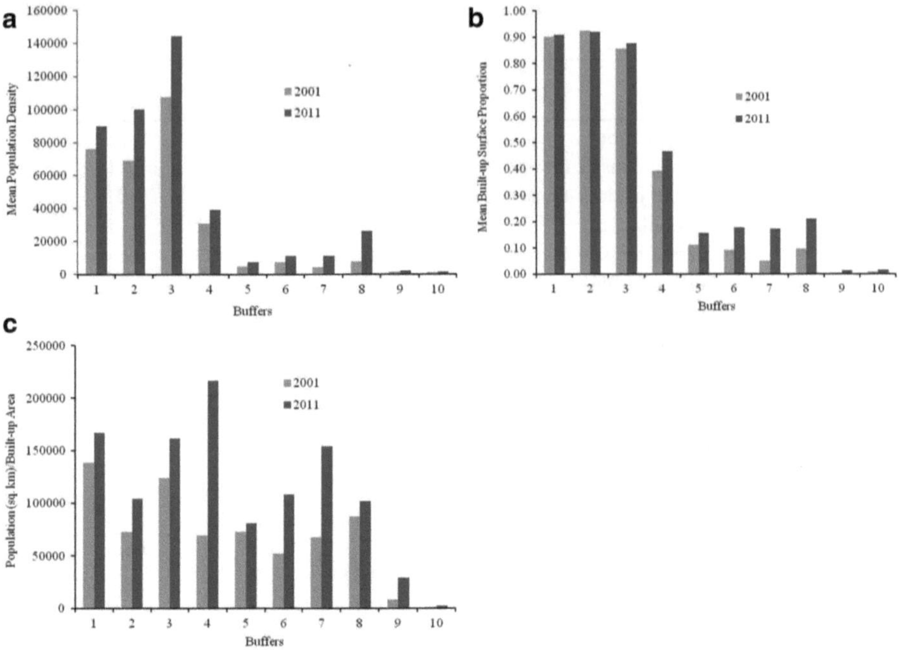

Fig. 6.8 Characteristics of the ten annular buffers in 2001 and 2011: (**a**) mean population density, (**b**) proportion of built-up surface, and (**c**) population per unit area of built-up surface

6.4 Conclusions

The objectives of this chapter were to analyse urban growth, sprawl and spatial structure using remote sensing and census-based population data between 2000 and 2011. The UGAT tools were employed to determine the different types of urban growth whilst two landscape metrics were used to quantify urban sprawl. Finally, spatially referenced population datasets were integrated with urban surface cover, derived from remote sensing to examine the spatial structure of the city. Spatial analysis was carried out in order to investigate the relationships between urban growth and changes in population parameters.

A 33 % expansion in urban areas was found between 2000 and 2011. The direction of growth was mostly confined to north, north-east and north-west zones; however, currently the track of urban expansion extends towards east, south-east and south-west zones. An examination of the types of urban growth showed that the extension growth type is dominant followed by the leapfrogging growth type. The map density metric and Shannon's entropy both indicated that sprawling by means of low-density development has increased from 2000 to 2011. A buffer analysis of three parameters – mean population density, mean built-up surface and population per unit built-up area – revealed that the density of population is increasing towards the outskirts, and that beyond buffer zone 3 (7.2 km) the rate is extremely fast. Likewise, the proportion of mean built-up surface significantly increased beyond buffer zone 4. Analysis of the population per unit area of built-up surface indicated that overcrowding and lack of space in the urban core are pushing people outwards to settle in peripheral areas, thereby exerting tremendous pressure on a limited resource base. Since population increases consistently in Dhaka, this may have resulted in the construction of more residential spaces in the peri- and suburban areas in the recent past.

The results of this study have a number of implications. They highlight the fact that, providing basic services to a large number of people in the expanding city will be a challenging task in the years ahead. Bangladesh is believed to be highly vulnerable to climate change, and increased rainfall is expected in the future. Encroachment of urban areas onto wetlands and floodplains is expected to have considerable impact on the severity of the flooding that this increased rainfall will cause. This and the increase in population will have a negative effect on the currently high level of surface water pollution. Further expansion, particularly into suburban areas, may result in there being more automobiles, contributing to the already bad air pollution of Dhaka. Overcrowding in general could have serious implications for vector- and water-borne diseases which are already prevalent in the megacity of Dhaka leading to increased risk of epidemics. In order to minimise these deleterious effects, it is essential that future urban development is carried out under the proper planning process and that growth only be permitted in a way that is not detrimental either to precious natural resources, such as cultivated land, or to the environment in which the citizens of the megacity must live.

References

Ahmed B, Ahmed R (2012) Modeling urban land cover growth dynamics using multi-temporal satellite images: a case study of Dhaka, Bangladesh. ISPRS Int J Geo-Inf 1(1):3–31

Ahmed SJ, Bramley G, Dewan AM (2012) Exploratory growth analysis of a megacity through different spatial metrics: a case study on Dhaka, Bangladesh (1960–2005). URISA J 24 (1):9–24

Aikawa M, Hiraki T, Eiho J (2009) Changes in atmospheric condition in an urbanized area of Japan from the viewpoint of rainfall intensity. Environ Monit Assess 148(1–4):449–453

Angel S, Parent J, Civco DL (2007) Urban sprawl metrics: an analysis of global urban expansion using GIS. ASPRS annual conference, Florida, 7–11 May 2007

Angel S, Parent J, Civco DL (2012) The fragmentation of urban landscapes: global evidence of a key attribute of the spatial structure of cities, 1990–2000. Environ Urban 24(1):249–283

Arsanjani JJ, Helbich M, Kainz W, Boloorani AD (2013) Integration of logistic regression, Markov chain and cellular automata models to simulate urban expansion. Int J Appl Earth Obs Geoinf 21:265–275

Asian Development Bank (ADB) (2008) Managing Asian cities: sustainable and inclusive urban solutions. ADB, Manila

Bangladesh Bureau of Statistics (BBS) (2003) Population census – 2001, Community series. Bangladesh Bureau of Statistics, Ministry of Planning, Dhaka

Bangladesh Bureau of Statistics (BBS) (2012) Population and housing census – 2011, Community report, Dhaka, Narayanganj and Gazipur Zila. Bangladesh Bureau of Statistics, Ministry of Planning, Dhaka
Batty M (2002) Thinking about cities as spatial events. Environ Plan B 29(1):1–2
Batty M (2005) Agents, cells and cities: new representational models for simulating multiscale urban dynamics. Environ Plan A 37(8):1373–1394
Batty M, Longley P, Fotheringham S (1989) Urban growth and form: scaling, fractal geometry and diffusion-limited aggregation. Environ Plan A 21(11):1447–1472
Begum A (2007) Urban housing as an issue of redistribution through planning? The case of Dhaka city. Soc Policy Adm 41(4):410–418
Bhatta B (2009) Analysis of urban growth pattern using remote sensing and GIS: a case study of Kolkata, India. Int J Remote Sens 30(18):4733–4746
Bhatta B (2010) Analysis of urban growth and sprawl from remote sensing data. Springer, Berlin
Bhatta B, Saraswati S, Bandyopadhyay D (2010) Quantifying the degree-of-freedom, degree-of-sprawl, and degree-of-goodness of urban growth from remote sensing data. Appl Geogr 30(1):96–111
Biediger J, Mathews L (2010) Urban sprawl, Geoworld, January, pp. 24–27
Braimoh AK, Onishi T (2007) Spatial determinants of urban land use change in Lagos, Nigeria. Land Use Policy 24(2):502–515
Burgess EW (1925) The growth of the city: an introduction to a research project. In: Park RE, Burgess EW, McKenzie RD (eds) The city. The Chicago University Press, Chicago, pp 47–62
Chin N (2002) Unearthing the roots of urban sprawl: a critical analysis of form, function and methodology. Working paper series no. 47. Centre for Advanced Spatial Analysis, University College London, UCL, London
Chowdhury MR (2003) The Impact of 'Greater Dhaka Flood Protection Project' (GDFPP) on local living environment–the attitude of the floodplain residents. Nat Hazards 29(3):309–324
Cowen DJ, Jensen JR (1998) Extraction and modelling of urban attributes using remote sensing technology. In: Liverman D (ed) People and pixels: linking remote sensing and social sciences. National Academy Press, Washington, DC, pp 164–188
Deng JS, Wang K, Hong Y, Qi JG (2009) Spatio-temporal dynamics and evolution of land use change and landscape pattern in response to rapid urbanization. Landsc Urban Plan 92 (3–4):187–198
Dewan AM, Corner RJ (2012) The impact of land use and land cover changes on land surface temperature in a rapidly urbanizing megacity. In: IGARSS 2012, Munich
Dewan AM, Yamaguchi Y (2008) Effects of land cover changes on flooding: example from Greater Dhaka of Bangladesh. Int J Geoinf 4(1):11–20
Dewan AM, Yamaguchi Y (2009a) Land use and land cover change in Greater Dhaka, Bangladesh: using remote sensing to promote sustainable urbanization. Appl Geogr 29 (3):390–401
Dewan AM, Yamaguchi Y (2009b) Using remote sensing and GIS to detect and monitor land use and land cover change in Dhaka Metropolitan of Bangladesh during 1960–2005. Environ Monit Assess 150(1–4):237–249
Dewan AM, Kabir MH, Nahar K, Rahman MZ (2012a) Urbanization and environmental degradation in Dhaka metropolitan area of Bangladesh. Int J Environ Sustain Dev 11(2):118–146
Dewan AM, Yamaguchi Y, Rahman MZ (2012b) Dynamics of land use/cover changes and the analysis of landscape fragmentation in Dhaka Metropolitan, Bangladesh. GeoJournal 77 (3):315–330
Environmental System Research Institute (ESRI) (2012) Available at: http://resources.arcgis.com/en/help/main/10.1/index.html#//005p00000018000000. Accessed 15 Sep 2012
Ewing R (1994) Characteristics, causes, and effects of sprawl: a literature review. Environ Urban Stud 21(2):1–15
Fang S, Gertner GZ, Sun Z, Anderson AA (2005) The impact of interactions in spatial distribution of the dynamics of urban sprawl. Landsc Urban Plan 73(4):294–306

Frenkel A, Orenstein D (2011) A pluralistic approach to defining and measuring urban sprawl. In: Yang X (ed) Urban remote sensing: monitoring, synthesis and modeling in the urban environment. Wiley-Blackwell, Chichester, pp 165–181

Gain P (2002) Bangladesh environment facing the 21st century, 2nd edn. Society for Environment and Human Development (SEHD), Dhaka

Galster G, Hanson R, Wolman H, Coleman S, Freihage J (2001) Wrestling sprawl to the ground: defining and measuring an elusive concept. Hous Policy Debate 12(4):681–717

Griffiths P, Hostert P, Gruebner O, van der Linden S (2010) Mapping megacity growth with multi-sensor data. Remote Sens Environ 114(12):426–439

Grimm NB, Faeth SH, Golubiewski NE, Redman CL, Wu J, Bai X, Briggs JM (2008) Global change and the ecology of cities. Science 319:756–760

Guan D, HaiFeng L, Inohae T, Su W, Nagaie T, Hokao K (2011) Modeling urban land use change by the integration of cellular automaton and Markov model. Ecol Model 222(20–22):3761–3772

Haber W (2007) Energy, food, and land – the ecological traps of humankind. Environ Sci Pollut Res 14(6):359–365

Harris C, Ullman E (1945) The nature of cities. Ann Am Acad Pol Soc Sci 242:7–17

Herold M, Goldstein NC, Clarke KC (2003) The spatiotemporal form of urban growth: measurement, analysis and modelling. Remote Sens Environ 86(3):286–302

Hoyt H (1939) The pattern of movement of residential rental neighborhoods. The structure and growth of residential neighborhoods in American cities, Federal Housing Administration. Washington DC, pp 112–122

Huang J, Lu XX, Sellers JF (2007) A global comparative analysis of urban form: applying spatial metrics and remote sensing. Landsc Urban Plan 82(4):184–197

Islam N (2005a) Dhaka now: contemporary urban development. The Bangladesh Geographical Society, Dhaka

Islam N (2005b) Improving the lives of slum dwellers in Bangladesh. CUS bulletin on urbanization and development, no. 48. Center for Urban Studies, Dhaka

Islam MS, Ahmed R (2011) Land use change prediction in Dhaka city using GIS aided Markov chain modeling. J Life Earth Sci 6:81–89

Islam MS, Rahman MR, Shahabuddin AKM, Ahmed R (2010) Changes in wetlands in Dhaka city: trends and physico-environmental consequences. J Life Earth Sci 5:37–42

Jaeger JAG, Bertiller R, Schwick C, Kienast F (2010) Suitability criteria for measures of urban sprawl. Ecol Indic 10(2):397–406

Jat MK, Garg PK, Khare D (2008a) Monitoring and modeling of urban sprawl using remote sensing and GIS techniques. Int J Appl Earth Obs Geoinf 10(1):26–43

Jat MK, Garg PK, Khare D (2008b) Modelling of urban growth using spatial analysis techniques: a case study of Ajmer city (India). Int J Remote Sens 29(2):543–567

Jiang F, Liu S, Yuan H, Zhang Q (2007) Measuring urban sprawl in Beijing with geospatial indices. J Geogr Sci 17(4):469–478

Kalnay E, Cai M (2003) Impact of urbanization and land-use change on climate. Nature 423:528–531

Li Y-Y, Zhang H, Kainz W (2012) Monitoring patterns of urban heat islands of the fast-growing Shanghai metropolis, China: using time-series of Landsat TM/ETM+ data. Int J Earth Obs Geoinf 19:127–138

Liu X, Lathrop RG Jr (2002) Urban change detection based on an artificial neural network. Int J Remote Sens 23(12):2513–2518

Liu J, Zhan J, Deng X (2005) Spatio-temporal patterns and driving forces of urban land expansion in China during the economic reform era. Ambio 34(6):450–455

Lo CP, Yang X (2002) Drivers of land-use/land-cover changes and dynamic modelling for the Atlanta, Georgia metropolitan area. Photogramm Eng Remote Sens 68(10):1073–1082

Long H, Tang G, Li X, Heilig GK (2007) Socio-economic forces of land-use change in Kunshan, the Yangtze river delta economic area of China. J Environ Manage 83(3):351–364

Lopez TdM, Aide TM, Thomlinson JR (2001) Urban expansion and the loss of prime agricultural lands in Puero Rico. Ambio 30(1):49–54
Mahtab-uz-Zaman QM, Lau SSY (2000) City expansion policy versus compact city demand: the case of Dhaka. In: Jenks M, Burgess R (eds) Compact cities: sustainable urban forms for developing countries. Spon Press, London, pp 141–152
Maktav D, Erbek FS (2005) Analysis of urban growth using multi-temporal satellite data in Istanbul, Turkey. Int J Remote Sens 26(4):797–810
McKinney ML (2002) Urbanization, biodiversity, and conservation. Bioscience 52(10):883–890
Mitsova D, Shuster W, Wang X (2011) A cellular automata model of land cover change to integrate urban growth with open space conservation. Landsc Urban Plan 99(2):141–153
Noda A, Yamaguchi Y (2008) Characterising urban sprawl using remote sensing and a simple spatial metric for medium-sized city in Japan. Int J Geoinf 4(1):43–50
O'Neill RV, Krummel JR, Gardner RH, Sugihara G, Jackson B, DeAngelis DL, Milne BT, Turner IMG, Zygmunt B, Christensen SW, Dale VH, Graham RL (1988) Indices of landscape pattern. Landsc Ecol 1(3):153–162
Pham HM, Yamaguchi Y (2011) Urban growth and change analysis using remote sensing and spatial metrics from 1975 to 2003 for Hanoi, Vietnam. Int J Remote Sens 32(7):1901–1915
Portugali J (2000) Self-organizing and the city. Springer, Berlin
Rahman A, Aggarwal SP, Netzband M, Fazal S (2011) Monitoring urban sprawl using remote sensing and GIS techniques of a fast growing urban centre, India. IEEE J Sel Top Appl Earth Obs Remote Sens 4(1):56–64
Richardson HW, Bae CC, Baxamusa MH (2000) In: Jenks M, Burgess R (eds) Compact cities: sustainable urban forms for developing countries. Spon Press, London
Schneider A, Friedl MA, Potere D (2009) A new map of global urban extent from MODIS satellite data. Environ Res Lett 4:044003
Seto KC, Fragkias M (2005) Quantifying spatiotemporal patterns of urban land-use change in four cities of China with time series landscape metrics. Landsc Ecol 20(7):871–888
Seto KC, Kaufmann RK (2003) Modelling the drivers of urban land use change in the Pearl River delta China: integrating remote sensing with socioeconomic data. Land Econ 79(1):106–121
Seto KC, Liu W (2003) Comparing ARTMAP neural network with the maximum-likelihood classifier for detecting urban change. Photogramm Eng Remote Sens 69(9):981–990
Seto KC, Shepherd JM (2009) Global urban land-use trends and climate impacts. Curr Opin Environ Sustain 1(1):89–95
Seto KC, Sánchez-Rodríguez R, Fragkias M (2010) The new geography of contemporary urbanization and the environment. Ann Rev Environ Resour 35:167–194
Shahraki SZ, Sauri D, Serra P, Modugno S, Seifolddini F, Pourahmad A (2011) Urban sprawl patterns and land-use change detection in Yazd, Iran. Habitat Int 35(4):521–528
Stewart DJ, Yin Z-Y, Bullard SM, MacLachlan JT (2004) Assessing the spatial structure of urban and population growth in the Greater Cairo Area, Egypt: a GIS and imagery analysis approach. Urban Stud 41(1):95–116
Sudhira HS, Ramachandra TV, Jagadish KS (2004) Urban sprawl: metrics, dynamics and modelling using GIS. Int J Appl Earth Obs Geoinf 5(1):29–39
Suriya S, Mudgal BV (2012) Impact of urbanization on flooding: the Thirusoolam sub watershed – a case study. J Hydrol 412–413:210–219
Talukder B, Nakagoshi N, Dewan AM (2012) Urbanization and green space dynamics in Greater Dhaka, Bangladesh. Landsc Ecol Eng 8(1):45–58
Taubenböck H, Wengmann M, Roth A, Mehl H, Dech S (2009) Urbanization in India – spatiotemporal analysis using remote sensing data. Comput Environ Urban Syst 33(3):179–188
Thapa RB, Murayama Y (2011) Urban growth modelling of Kathmandu metropolitan region, Nepal. Comput Environ Urban Syst 35(1):25–34
Tian Y, Yue T, Zhu L, Clinton N (2005) Modeling population density using land cover data. Ecol Model 189(1–2):72–88

Tobler WR (1979) Cellular geography. In: Gale S, Olsson GD (eds) Philosophy in geography. Reidel Publishing Company, Dordrecht, pp 3–16

Torrens PM, Alberti M (2000) Measuring sprawl. Working paper series no. 27. Center for Advanced Spatial Analysis (CASA), University College London, London

Tsai Y-H (2005) Quantifying urban form: compactness versus 'sprawl'. Urban Stud 42 (1):141–146

United Nations (2012) World population prospects: the 2010 revision, vol 1. United Nations Department of Economic and Social Affairs, New York. Available at: http://eas.un.org/undp/wpp/Documentation/publications.html. Accessed 1 Apr 2012

Wang F, Zhou Y (1999) Modelling urban population densities in Beijing 1982–90: suburbanization and its causes. Urban Stud 36(2):271–287

Weng Q (2001a) A remote sensing – GIS evaluation of urban expansion and its impact on surface temperature in the Zhujiang delta, China. Int J Remote Sens 22(10):1999–2014

Weng Q (2001b) Modeling urban growth effects on surface runoff with the integration of remote sensing and GIS. Environ Manage 28(6):737–748

Wilson AG (1976) Catastrophe theory and urban modelling: an application to model choice. Environ Plan A 8(3):351–356

Wilson EH, Hurd JD, Civco DL, Prisloe MP, Arnold C (2003) Development of geospatial model to quantify, describe and map urban growth. Remote Sens Environ 86(3):275–285

Wong DSS, Fotheringham AS (1990) Urban systems as examples of bounded chaos: exploring the relationship between fractal dimension, rank-size, and rural-to-urban migration. Geogr Ann B Hum Geogr 72(2/3):89–99

Wu J, David J (2002) A spatially explicit hierarchical approach to modelling complex ecological systems: theory and applications. Ecol Model 153(1–2):7–26

Wu K, Zhang H (2012) Land use dynamics, built-up land expansion patterns, and driving forces analysis of the fast-growing Hangzhou metropolitan area, eastern China (1978–2008). Appl Geogr 34:137–145

Xian G, Crane M, Su J (2007) An analysis of urban development and its environmental impact on the Tampa Bay watershed. J Environ Manage 85(4):965–976

Xu C, Liu M, Zhang C, An S, Yu W, Chen J (2007) The spatiotemporal dynamics of rapid urban growth in the Nanjing metropolitan region of China. Landsc Ecol 22(6):925–937

Yang X (ed) (2011) Urban remote sensing: monitoring, synthesis and modelling in the urban environment. Wiley-Blackwell, Chichester

Yeh AGO, Li X (2001) Measurement and monitoring of urban sprawl in a rapidly growing region using entropy. Photogramm Eng Remote Sens 67(1):83–90

Yin Z-Y, Stewart DJ, Bullard S, MacLachlan JT (2005) Changes in built-up surface and population distribution patterns during 1986–1999: a case study of Cairo, Egypt. Comput Environ Urban Syst 29(5):595–616

Yu W, Zang S, Wu C, Liu W, Na X (2011) Analysing and modelling land use land cover (LUCC) in the Daqing city, China. Appl Geogr 31(2):600–608

Yuan F (2010) Urban growth monitoring and projection using remote sensing and geographic information systems: a case study in the Twin Cities metropolitan area, Minnesota. Geocarto Int 25(3):213–230

Tobler WR (1979) Cellular geography. In: Gale S, Olsson G (eds) Philosophy in geography. Reidel Publishing Company, Dordrecht, pp [illegible]

[illegible] (2000) Measuring sprawl. Working paper series [illegible]

[illegible]

Chapter 7
Key Driving Factors Influencing Urban Growth: Spatial-Statistical Modelling with CLUE-s

Sohel J. Ahmed, Glen Bramley, and Peter H. Verburg

Abstract In a rapidly urbanising megacity such as Dhaka, identifying the driving factors that influence urban growth at different spatio-temporal scales is of considerable importance. In this study, based on literature survey and data availability, a selection of drivers is chosen and then tested through logistic regression. Using the CLUE-s land use modelling framework, the ability of these drivers to simulate urbanisation for the periods of 1988–1999 and 1999–2005 was examined against observed data. The results indicated that the role of these driving factors, as contributors to explaining change dynamics of urban land in Dhaka, changes with time. The overall performance of the model, when validated against observed data, is similar to that reported for other urban growth models.

Keywords Calibration • CLUE-s • Dhaka • Drivers of change • LUCC • Spatial statistical models • Urban growth • Validation

7.1 Introduction

Research in land use change can be grouped into three broad categories (Rindfuss et al. 2004): (1) observation and monitoring of land use change, (2) identification of the drivers of land use change and (3) computer modelling of land use (change) that

S.J. Ahmed (✉)
The Bartlett Development Planning Unit, University College London (UCL), 34 Tavistock Square, WC1H 9EZ London, UK
e-mail: sohel.ahmed@ucl.ac.uk

G. Bramley
School of the Built Environment, Heriot-Watt University, Edinburgh EH14 4AS, UK
e-mail: G.Bramley@hw.ac.uk

P.H. Verburg
Institute of Environmental Studies, VU University, Amsterdam, The Netherlands
e-mail: peter.verburg@vu.nl

A. Dewan and R. Corner (eds.), *Dhaka Megacity: Geospatial Perspectives on Urbanisation, Environment and Health*, Springer Geography,
DOI 10.1007/978-94-007-6735-5_7, © Springer Science+Business Media Dordrecht 2014

combines categories 1 and 2 in a dynamic and integrative manner. Given Dhaka's history of growth and rapid urbanisation, it is vital to know how the various different locational drivers influence its urban footprint, particularly in the years since independence from Pakistan. The key aim of this chapter is therefore to explore and analyse the key driving factors that influence changes in built-up areas of Dhaka. This research embraces all three concepts mentioned above; initially, it analyses land use patterns by identifying relations between these observed patterns and explanatory factors and then aims to link them with the processes that are accountable for those patterns in a GIS environment by dynamic land use modelling (Overmars et al. 2007; Verburg et al. 2006; Geist et al. 2006; Overmars 2006; Verburg et al. 2004a). To accomplish this, selected driving factors are used in multivariate statistical models to explore their degree of influence on the observed urban growth at two different periods. Calibrated results are subsequently used to generate simulations in the CLUE-s raster modelling framework, which are then compared with observed changes in urban land use.

7.2 Data and Methods

7.2.1 CLUE-s: Introduction and Modelling Framework

The CLUE-s (Conversion of Land Use and its Effects at small regional extent) modelling framework was developed at Wageningen University (Verburg et al. 2002). Initially, it was applied at regional scale (Veldkamp and Fresco 1996), but the model has now become one of the most widely applied Land Use Land Cover Change (LUCC) models in various settings, including agricultural intensification, deforestation and urbanisation (e.g. Batisani and Yarnal 2009; Verburg et al. 2002, 2008; Verburg and Overmars 2007). It was developed to simulate land use change using statistically quantified relationships between land use and its driving factors in combination with dynamic modelling of competition between land use types where each cell only contains one land use type. It is capable of identifying areas that have high probabilities for future land use changes and potential 'hot spots' of land use change (Batisani and Yarnal 2009; Verburg et al. 2002, 2008; Verburg and Overmars 2007). The model is subdivided into two distinct modules, a nonspatial demand module and a spatially explicit allocation module (Verburg et al. 2002).

The *nonspatial module*, also known as the land use demand module, can take different model specifications, ranging from simple trend extrapolations to complex economic models. Although the choice for a specific model varies, the results from this demand module provide an annual demand for land, for different land use types, which is then distributed by the allocation module. For the work reported here, a simple piecewise regression was plotted between two observed periods to calculate trend demand for different land use/land cover.

The *spatial module* translates these demands and allocates them spatially (at cell level) which results in a representation of land use changes at different locations within the study region. For every year in the specified time frame, this module creates a land use prediction map taking into account a number of location specific preferences or combinations of them, derived mainly from a multivariate statistical model (e.g. logistic regression) (see Verburg et al. 2002, 2004b; Verburg and Overmars 2007).

To evaluate the probability of a location becoming urban, different driving factors are included as independent variables in the multivariate regression procedure (Verburg et al. 2002, 2004b). The CLUE-s framework assumes that the relationships based on the current land use pattern remain valid during the simulation period (Verburg et al. 2002, 2004b). The dependent variable for the logistic regression is a presence or absence event where 1 denotes that a given cell is certain to convert to urban land, and where 0 shows no change (Lesschen et al. 2005). The logistic coefficients (known as unstandardised logistic regression coefficients or logit coefficients) or the β coefficients vary between plus and minus infinity, with 0 indicating that the given explanatory variable does not affect the logit; positive or negative β coefficients indicate the explanatory variable increases or decreases the logit of the dependent (Lesschen et al. 2005).

Figure 7.1 provides an overview of the CLUE-s model. As seen in the diagram, the allocation is based upon a combination of modelling blocks. In addition, a set of decision rules is specified to restrict the conversions that take place based on the actual land use pattern. These are termed within the model as *Land use specific conversion settings* and can be defined as conversion resistance within a conversion matrix.

Current land use, in many cases, is an important determinator of location preference. Land use conversions are often costly and some conversions are almost irreversible (e.g. residential area is not likely to be converted back into agricultural area). Therefore, a *conversion resistance* factor is assigned to each land use type so that differences in behaviour towards conversion are approximated by conversion costs. For each land use type, a value is specified that represents the relative resistance to change, ranging from 0 (easy conversion) to 1 (irreversible change). The factor can be based on either expert knowledge or observed behaviour in the recent past, or alternatively subjective judgemental parameters can be used to calibrate the model.

The final step before running the model is to set up the conversion settings within a *conversion matrix* where each land use category is allocated values to indicate which other land use categories it can convert to in the next time step. The matrix shows the probability of each current category being converted to another. Rows represent present land use while columns show potential future land use. A value of 1 indicates that the conversion is allowed, and a value of 0 means that the conversion is not allowed.

Thus, this process of spatial analysis and inductive dynamic modelling comprises four categories (i.e. driving factors to change, spatial planning policies and restrictions, neighbourhood effects and land use specific conversion settings) that

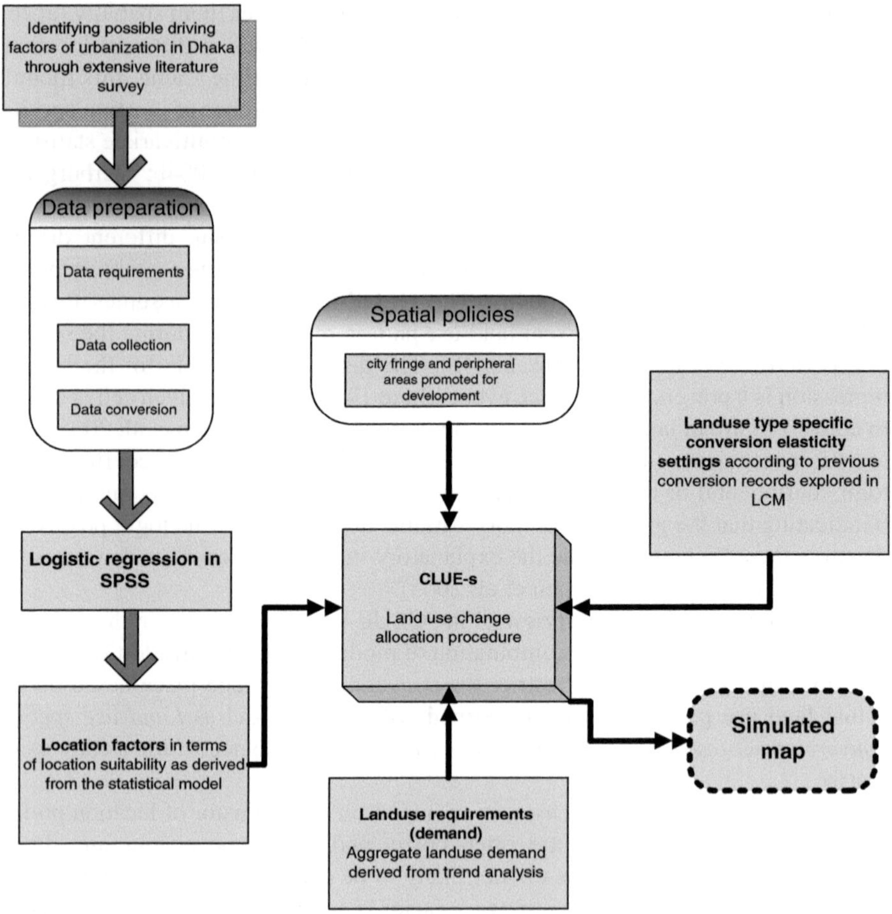

Fig. 7.1 Detailed modelling framework of the CLUE-s model (Modified after Verburg et al. 2002)

together create a set of conditions and possibilities for which the model calculates the best solution in an iterative procedure. Neighbourhood characteristics and spatial policies will be discussed in Sect. 7.3.

7.2.2 Type of Key Variables, Data Access and Management Issues

Two major types of variables are used within empirical land use change models: non-neighbourhood- and neighbourhood-based variables. While the former category is included in all land use studies, only a limited number of studies consider neighbourhood effects and spatial autocorrelation in the modelling of land use change (Braimoha and Onishi 2007; Verburg et al. 2004a; Cheng and Masser 2003). A common set of variables encompassing biophysical and socioeconomic parameters are taken into account in most of the growth modelling studies such as in Clark's SLUETH or Metronomica from the Research Institute for Knowledge Systems (RIKS) or other spatial statistical models (Dietzel and Clarke 2006; White et al. 2004; Barredo et al. 2003, 2004; Engelen et al. 2002; White and Engelen 2000; Clarke and Gaydos 1998). Additionally, the selection of variables depends on both the adopted model and the characteristics of the area being modelled. For instance, Dhaka is surrounded by five major rivers which cause both fluvial and pluvial floods in the monsoon season, and therefore, urban land encroachment generally occurs onto relatively elevated agricultural lands and vegetated areas. In addition, some classical urban economic models like the Bid Rent model, Central Place theory and Tobler's first law of geography hold true in many contexts and provide sound justification in selecting the driving factors influencing urban expansion in Dhaka.

Research indicates that there are four primary factors that drive land use change in an area (Braimoha and Onishi 2007; Hu and Lo 2007; Verburg et al. 2004a, b). These are:

1. Socioeconomic factors
2. Biophysical constraints and potentials
3. Neighbourhood characteristics
4. Institutional factors in the form of spatial policies

The influence of these factors is often determined by an inductive approach, i.e. using multivariate statistical technique (e.g. linear or logistic regression) (Lesschen et al. 2005). One of the major pitfalls of this technique is that it does not account for spatial dependency; hence, autocorrelation may be present in the residuals. To overcome this problem, a stratified random or systematic spatial sampling scheme can be considered for spatial modelling with logistic regression (Cheng and Masser 2003). The major reason for doing this is to expand the distance interval between the sampling sites. Systematic sampling is effective in reducing spatial dependency but may cause loss of important information in isolated areas. In other words, its ability to accurately represent the population/study area may decrease with the increase of distance interval. Conversely, random sampling is efficient in representing population but has low efficiency in reducing spatial dependence, especially local spatial dependence.

In this study, none of the independent variables were found to be strongly correlated when tested using the Moran's I autocorrelation statistic (<0.4 on average). But they are moderately autocorrected, particularly the distance variables, which is to be expected (Braimoha and Onishi 2007; Hu and Lo 2007; Cheng and Masser 2003). Therefore, systematic sampling was initially performed which left one cell distance between sampled cells. This was later modified to spatial random sampling in order to further minimise spatial autocorrelation. Hence, this study uses a randomly distributed sample (30 % of all observations) instead of the full dataset. This does not cause conflict between sample size and spatial autocorrelation as the sample size is still large. This approach is said to be effective at minimising spatial dependency to a level which will not impact the modelling outcome (Verburg et al. 2004a; Cheng and Masser 2003). In addition, the inclusion of neighbourhood characteristics in the model specification explicitly addresses the issue of autocorrelation in land use patterns. The sampling and consequent analysis is run at a 30 × 30 m resolution, which is the original resolution of the land use/cover maps (Dewan and Yamaguchi 2009a, b). Dependent and independent variables are, therefore, kept in or converted to raster format and are resampled where necessary to maintain this resolution. Consequently, datasets for all variables were in raster format with 30 m grid cell size on the Bangladesh Transverse Mercator projection system. In the CLUE-s model only one land use type is allowed for each grid cell. The raster data was then converted to ASCII format for input to the CLUE-s model. Regression modelling was performed in SPSS. Afterwards, the results from the statistical analysis were produced as probability maps using CLUE-s modelling framework and ArcGIS Desktop.

The study area includes Dhaka Metropolitan Area (DMA) (see Chap. 1) and beyond and covers an area of 420 km^2. A detailed description of the preparation of the land use/cover data can be found elsewhere (Dewan and Yamaguchi 2009a, b). This study uses land use/land cover maps for 1988 (t1), 1999 (t2) and 2005 (t2). The original land use/cover data contained six categories which are aggregated into urban and nonurban classes for the purposes of this study. Urban land use change, between two periods (1988–1999 referred to as t1–t2 and 1999–2005 as t2–t3), was considered as the dependent variable. The selection of dependent and independent variables is made between t1 and t2 as most of the independent variables are from t1 to t2, and also changes in urban surface started to be more pronounced during the late 1990s (Islam 1996, 2005). Besides, the lack of data in Dhaka precluded the use of more variables in this study. For instance, variables that are intrinsically dynamic in nature and that change over time play a crucial role in obtaining accurate result from the analysis and modelling. However, spatial data for roads and other facilities were only available for a single year despite the fact that urban growth was continuous. This can also be a crucial barrier to optimum results if the models are used to generate scenarios for the near future. Although it is true that demographic factors are important determinants of urbanisation in the study area, unavailability of finer resolution population data (at neighbourhood level) at the time of study prevented their inclusion in the simulation of urban growth. Similarly, other socioeconomic factors such as economic status, land value, type of water supply and sanitation could not be taken into account due to their unavailability.

7.2.3 Calibration and Validation Techniques

Until recently, efforts to calibrate and validate LUCC models have received scant attention. This lack of attention can be attributed not only to the absence of universal techniques but also to lack of adequate data, particularly in developing countries (Veldkamp and Lambin 2001). However, with recent advances in spatial analysis and modelling techniques, significant progress has been made with calibrating and validating modelling results (Dietzel and Clarke 2006; Gardner and Urban 2005; Hagen-Zanker 2002a, b, 2003; Hagen-Zanker et al. 2005; Pontius and Schneider 2001; Pontius et al. 2004, 2008; Silva and Clarke 2005; Straatman et al. 2004; Visser 2004; Visser and de Nijs 2006).

There is no universally accepted procedure, nor is there an accepted set of guidelines for validating spatio-temporal models. Nevertheless, in principle, this procedure requires land use data at the start and end time of the period being modelled (Vliet 2009). It is also argued that model validation should use different time periods than those for calibration (Vliet 2009; Hagen-Zanker and Lajoie 2008). This is necessary since model results, which are consistent and accurate enough during calibration process, do not guarantee good capacity to predict at different time periods or during the validation process (Hagen-Zanker and Lajoie 2008).

Several map comparison tools were used for both calibration and validation, namely, the kappa statistic and its variants (particularly, fuzzy kappa) (Vliet 2009; Hagen-Zanker and Lajoie 2008; Pontius et al. 2004, 2008). Null or no-change models are also usually used for examining the model accuracy (Pontius et al. 2004, 2008). However, calibration and validation comparisons against null models can be misleading, since land uses with higher conversion cost possess high inertia to change and thereby results in higher false accuracy. This is why a variant of the neutral model was used. This model was created from land use maps that simulated the correct amount of land use change, randomly allocated over the map with minimal adjustments to the original land use map (Vliet 2009; Hagen-Zanker and Lajoie 2008). This neutral model provided additional benchmark maps for calibration and validation. Again as a rule of thumb, a simulation must perform better than the neutral model to indicate that the model possesses predictive power to make scenarios for the future.

The goodness of fit of logistic regression models may be assessed using the Relative/Receiver Operating Characteristic (ROC) value. The ROC method is used in LUCC modelling to measure the relationship between simulated change and actual change (Pontius and Schneider 2001). In this study, the ROC method offers an answer to one important question: How well is urban growth concentrated at the locations of relatively high suitability for urban growth? The ROC statistic ranges from 0 to 1. An ROC value of 1 indicates that there is a perfect spatial agreement between the actual urban growth map and the predicted probability (perfect discrimination). An ROC value of 0.5 is the agreement that would be expected due to chance (Hu and Lo 2007; Lesschen et al. 2005; Verburg et al. 2002, 2004a; Pontius and Schneider 2001).

7.3 Description of Dataset

Based on extensive literature review, three categories of non-neighbourhood variables were used, apart from neighbourhood variables. These were biophysical, access to opportunities and planning variables. Access to opportunity variables were used as a proxy for socioeconomic variables. A brief description of each category is given below, followed by neighbourhood variables.

7.3.1 Biophysical Variables

Previous studies have demonstrated that in a highly flood prone study area such as Dhaka, land elevation plays a crucial role in determining urban expansion (Dewan 2013; Dewan and Yamaguchi 2008; Alam and Rabbani 2007; Dewan et al. 2007; Islam 2005; Huq 1999; RAJUK 1997). Because of recurrent floods and water logging in the monsoon season, flood-free land is usually the target of both public and private organisations for urban development. Therefore, elevation data has been considered here, and land with an elevation above five metres has been regarded as being relatively less vulnerable to flood (Dewan 2013; Dewan and Yamaguchi 2008; Dewan et al. 2007; Dewan et al. 2006a, b).

7.3.2 Access to Opportunities as Proxies to Socioeconomic Variables

Given the paucity of spatially encoded data at the local level in developing countries, information such as the distribution of employment opportunities at different spatial scales can be used as a socioeconomic indicator. The use of approximations like this can be further strengthened by relating this variable to classic utility maximising models like the Bid Rent or Lowry Models (Alonso 1964). A common variable that is used in economic models of land use change is the distance between residential location and employment which is in itself a surrogate for costs associated with travel. A number of studies, in various settings, have employed proxy variables in the absence of actual data (Jat et al. 2008; Braimoha and Onishi 2007; Hu and Lo 2007; Wu et al. 2006; Xiao et al. 2006; Verburg et al. 2006, 2004a, b; Henriquez et al. 2006; Liu et al. 2005; Sui and Zeng 2001; Veldkamp and Lambin 2001; Yeh and Li 1998). Therefore, accessibility representing attractiveness of any given land in terms of employment opportunities, services and facilities was included in this study. Although Euclidian distance has been used as an accessibility indicator in modelling urban land use in some studies (e.g. Cheng and Masser 2003), this may be unrealistic to use given that Euclidian distance is unable to account for the presence of infrastructure and natural features such as rivers and lakes. Apart from economic motives, travel time can be an important factor that influences the decision to live or work at a certain location.

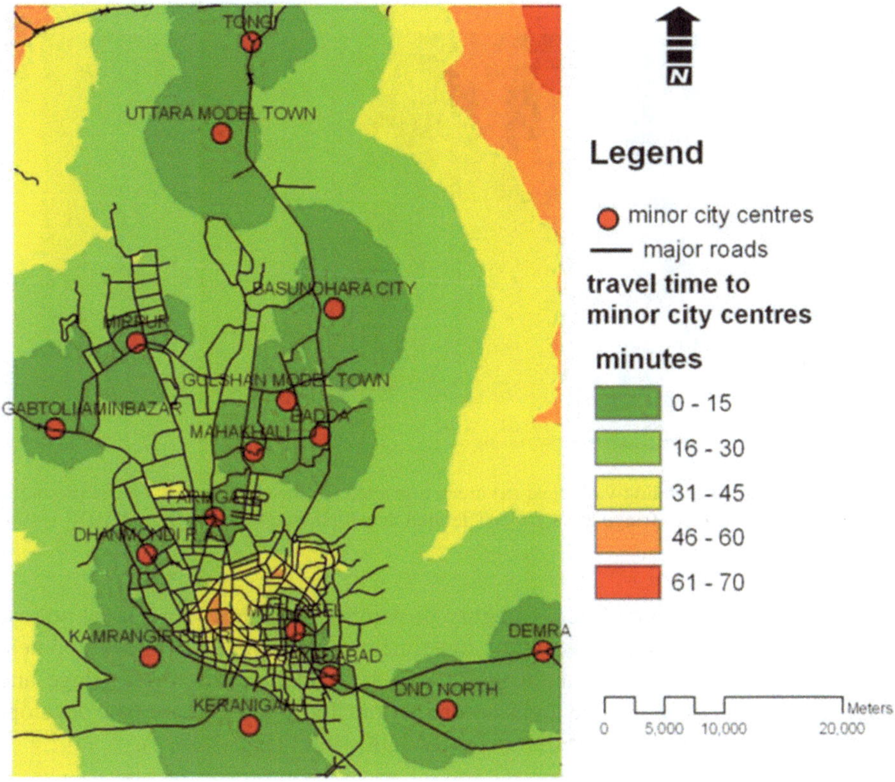

Fig. 7.2 An example of a network-based accessibility surface

Therefore, for this study, network-based accessibility measures were developed using average travel time along major road networks (NPS-ROMN 2008) (see Fig. 7.2 as an example).

7.3.3 Spatial Policy/Planning Variables

Spatial policies that provide guidance for development control or promotion can influence the patterns of land use change. Spatial policies and restrictions mostly indicate areas where land use changes are either restricted or allowed by the relevant authorities (RAJUK 1993). These can take the form of Boolean maps with cells set to 0 (areas not allowed to change) or 1 (areas allowed to change). In case of Dhaka, areas such as land of high agricultural value, areas designated as flood retention ponds and the city's existing natural drainage systems have been selected as restricted planning variables, and the appropriate spatial data were extracted from the Dhaka Metropolitan Development Plan (1995–2015) (RAJUK

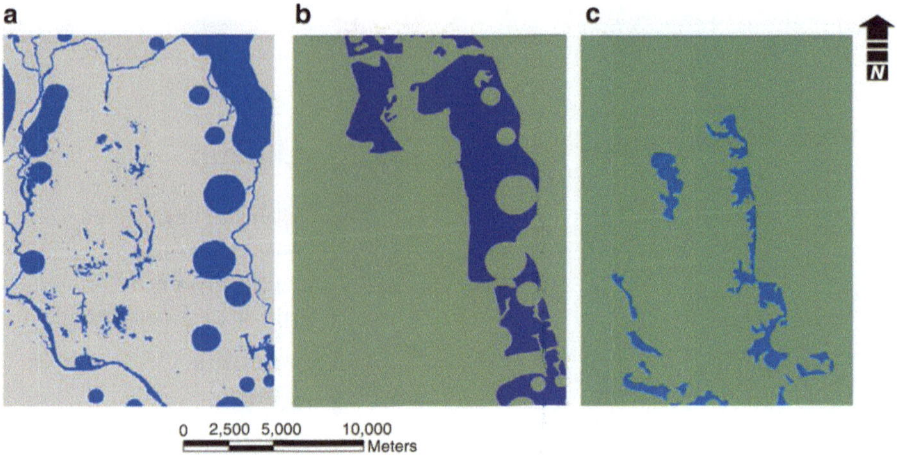

Fig. 7.3 Restricted planning variables: (**a**) areas restricted for development, (**b**) areas promoted for development, and (**c**) fringe areas for accelerated development (Adapted from DMDP 1997)

1997) (Fig. 7.3a). In addition, a category of permitted planning (or spatial policy) variables have been used in this study to delineate areas where urban development is permitted by the relevant authority. These areas include flood-free fringe and edge areas (e.g. northern edge areas around Uttora and Tongi) where new development is encouraged by the DMDP Plan (RAJUK 1997) (Fig. 7.3b, c).

7.3.4 Neighbourhood Variables

The incorporation of neighbourhood effects into a model acknowledges the importance of spatial interaction, proximity and spillover effects. These are known to have crucial roles in the spatial structuring of land use and can be expected in models in which relative location matters (Verburg et al. 2004c; White and Engelen 1993). The inclusion of this variable in empirical statistical models is believed to produce result similar to a constrained cellular automata model (Verburg and Overmars 2007). Spatial interactions can be represented in the model by incorporating a neighbourhood function as one of the drivers of the location preference. The relationship between location preference and neighbourhood composition may be determined either by (calibrated) decision rules, as in most cellular automata models, or can be based on statistical analysis of empirical results (Verburg et al. 2004c). Consequently, Verburg et al. (2004c) introduced the mean enrichment factor, a spatial metric of over- or under-representation of different land use types in the neighbourhood of a specific grid cell. The enrichment factor is presented at a logarithmic scale to obtain an equal scale for land use types that occur above the total

0.25	0.25	0.25	0.25	0.25	0.25	0.25	0.25	0.25	0.25	0.25	0.25	0.25	0.25	0.25	0.25	0.25	0.25	0.25	0.25	0.25
0.25	0.25	0.25	0.25	0.25	0.25	0.25	0.25	0.25	0.25	0.25	0.25	0.25	0.25	0.25	0.25	0.25	0.25	0.25	0.25	0.25
0.25	0.25	0.5	0.5	0.5	0.5	0.5	0.5	0.5	0.5	0.5	0.5	0.5	0.5	0.5	0.5	0.5	0.5	0.5	0.25	0.25
0.25	0.25	0.5	0.5	0.5	0.5	0.5	0.5	0.5	0.5	0.5	0.5	0.5	0.5	0.5	0.5	0.5	0.5	0.5	0.25	0.25
0.25	0.25	0.5	0.5	0.75	0.75	0.75	0.75	0.75	0.75	0.75	0.75	0.75	0.75	0.75	0.75	0.75	0.5	0.5	0.25	0.25
0.25	0.25	0.5	0.5	0.75	1	1	1	1	1	1	1	1	1	1	1	0.75	0.5	0.5	0.25	0.25
0.25	0.25	0.5	0.5	0.75	1	1	1	1	1	1	1	1	1	1	1	0.75	0.5	0.5	0.25	0.25
0.25	0.25	0.5	0.5	0.75	1	1	1	1	1	1	1	1	1	1	1	0.75	0.5	0.5	0.25	0.25
0.25	0.25	0.5	0.5	0.75	1	1	1	1	1	1	1	1	1	1	1	0.75	0.5	0.5	0.25	0.25
0.25	0.25	0.5	0.5	0.75	1	1	1	1	1	1	1	1	1	1	1	0.75	0.5	0.5	0.25	0.25
0.25	0.25	0.5	0.5	0.75	1	1	1	1	1	0	1	1	1	1	1	0.75	0.5	0.5	0.25	0.25
0.25	0.25	0.5	0.5	0.75	1	1	1	1	1	1	1	1	1	1	1	0.75	0.5	0.5	0.25	0.25
0.25	0.25	0.5	0.5	0.75	1	1	1	1	1	1	1	1	1	1	1	0.75	0.5	0.5	0.25	0.25
0.25	0.25	0.5	0.5	0.75	1	1	1	1	1	1	1	1	1	1	1	0.75	0.5	0.5	0.25	0.25
0.25	0.25	0.5	0.5	0.75	1	1	1	1	1	1	1	1	1	1	1	0.75	0.5	0.5	0.25	0.25
0.25	0.25	0.5	0.5	0.75	1	1	1	1	1	1	1	1	1	1	1	0.75	0.5	0.5	0.25	0.25
0.25	0.25	0.5	0.5	0.75	0.75	0.75	0.75	0.75	0.75	0.75	0.75	0.75	0.75	0.75	0.75	0.75	0.5	0.5	0.25	0.25
0.25	0.25	0.5	0.5	0.5	0.5	0.5	0.5	0.5	0.5	0.5	0.5	0.5	0.5	0.5	0.5	0.5	0.5	0.5	0.25	0.25
0.25	0.25	0.5	0.5	0.5	0.5	0.5	0.5	0.5	0.5	0.5	0.5	0.5	0.5	0.5	0.5	0.5	0.5	0.5	0.25	0.25
0.25	0.25	0.25	0.25	0.25	0.25	0.25	0.25	0.25	0.25	0.25	0.25	0.25	0.25	0.25	0.25	0.25	0.25	0.25	0.25	0.25
0.25	0.25	0.25	0.25	0.25	0.25	0.25	0.25	0.25	0.25	0.25	0.25	0.25	0.25	0.25	0.25	0.25	0.25	0.25	0.25	0.25

Fig. 7.4 Size and weights for 21 × 21 cell neighbourhood

study area average in the neighbourhood (enrichment factor >1) and land use types that occur below that average in the neighbourhood (enrichment factor <1).

Based on the analysis of enrichment factors for built-up areas on the 1988 and 1999 land use map, a 21 × 21 cell window was used (i.e. 10 cells from the central cell on all sides of the central cell) for the simulation (Fig. 7.4). Cells close to the central cells are given higher weight, and cells further away are given less weight to reflect the distance decay effect found for built-up areas by the mean enrichment factors.

7.4 Results

7.4.1 Statistical Models of Urban Land Change

Binary logistic regression was used to model the probability of conversion to the urban category in two periods (t1–t2 and t2–t3) for any given cell using a number of aforementioned driving factors (Sect. 7.3) as the independent variables.

From Table 7.1, based on the Wald statistics,[1] it can be deduced that the major determinators of urban development change over time. Distance from the CBD is important in the first period but is replaced by closeness to other minor urban centres in the second; proximity to major roads becomes less important with time,

[1] The Wald statistic is the squared ratio of the unstandardised logistic coefficient to its standard error.

Table 7.1 Logistic regression results for driving factors

	Urban t1–t2			Urban t2–t3		
Driving factors	β^{a}	exp(β)	Wald statistic	β	exp(β)	Wald statistic
Travel time to commercial centres	−.03	.97	1,787.26	−.02	98	444.62
Travel time to historical CBD	−.01	.99	2,256.08	−.01	.99	821.67
Travel time to industrial hubs	−.02	.98	1,117.15	−.02	.98	1,012.45
Travel time to minor city centres	−.02	.97	904.19	−.04	.96	2,586.42
Travel time to major roads	−.17	.84	8,274.91	−.04	.96	1,411.04
elevation	.24	1.28	3,887.09	.13	1.13	1,650.64
Planning variable 1 (urban fringe acceleration)	*	*	*	−.26	.77	104.52
Planning variable 2 (promoted areas for development)	−.10	1.07	4.31	−.10	.90	55.06
Constant	1.37	3.93	2,201.91	.96	5.77	1,300.03
ROC	0.86			0.70		

*Not significant at 0.05 level, rest of the results are significant at 0.05 level
[a]β = unstandardised coefficients and exp(β) = odds ratio

and the physical configuration of sites also becomes less significant. To some extent, this is an example of spatial pattern modelling at various temporal scales. It shows that various factors are changing their roles in the process of land development. It should be noted that this statistical analysis only takes the probability of development into account and does not consider detailed spatial patterns, such as changes in density or vertical expansions of urban growth.

A decentralised, polycentric suburbanizing trend in the metropolitan Dhaka area is also evident from the resultant coefficients. Probability (expressed as log-odds) of converting into urban areas decreases when cells are further away from the CBD, commercial centres, industrial hubs, minor city centres and major roads for both t1–t2 and t2–t3 periods. Nevertheless, the role of CBD becomes less important for the t2–t3 period as evidenced by Wald statistics. This is due to non-availability of land for development in the proximity of CBD since the late 1990s. Dhaka becomes more condensed and consolidated near and around CBD since the 1990s (see Chap. 3). Vibrant activity centres at the east and west fringe areas started to become more urban at or after the period t2–t3. This is reflected in the coefficients and Wald statistics for t2–t3. This can be further exemplified by travel time to minor centres in the later period, which shows the greatest effect in t2–t3. Travel time to industrial areas was found to be negatively related to the urban growth probability (i.e. increases in travel time to industrial areas decreased the logit of urban growth), and it seems to have kept the same importance between the observed periods considered.

Even though the probability of urban growth decreased with increasing travel time to commercial hubs/markets, this also becomes less significant at the period of t2–t3. This may be the result of fragmented growth away from these minor city centres. In other words, people are becoming more vehicle dependent.

Accessibility to major roads is a major determinator of urban land change, and this variable contributed significantly to the model in the period t1–t2; i.e. areas near to major road are 0.171 times more likely to convert to urban land than those further away. But this variable does not predict much growth during t2–t3 since updated data on roads was not available at the time of analysis.

As found in the logistic regression results (Table 7.1), most of the independent variables have significant influence in shaping urban growth in Dhaka despite variations in the degree of influence in different time periods. However, not all the results enable us to draw a solid conclusion. For example, planning variables, particularly urban fringe areas for accelerated development, are not significant during t1–t2. This is due to the fact that during that time, there was room for consolidation within the inner city areas, and this is reflected by the significant influence of factors like travel time to inner city facilities, namely, CBD, commercial and industrial areas.

According to the ROC values in Table 7.1, the driving factors explain the variations in urban growth in t1–t2 better than in t2–t3. While the lower level of fitting for the later period can be attributed to the absence of updated data for the later period, there may be other reasons for overall low performance of the statistical model. This shows that it may be important to include many other possible drivers, particularly those related to population, in the process of spatial allocation of new urban areas. Possibly, the location of urban land can be better explained by self-reinforcement of new urban locations through economies of scale and the emergence of new centres. When the observed pattern of urban locations cannot be explained by determinism alone, chance events become important determinators of city systems. Chance events, self-organisation, self-emergence and coincidences are all characteristics of complex systems (Batty and Xie 1994). Such elements can never be adequately included in an empirical analysis of LUCC. Reinforcement of patterns and path dependence can be studied over shorter time scales which require the application of complex theories (Cheng and Masser 2004; Batty et al. 1999). Previous case studies, therefore, have attempted to include neighbourhood variables in empirical analysis to mimic some of these complexity theories applicable to cities (Hansen 2008; Verburg et al. 2004c).

At this point, inclusion of neighbourhood variables, as enrichment factors, contributes significantly to the model. From Fig. 7.5, where the mean enrichment factor for urban land use 1999 is illustrated, it can be seen that neighbouring cells of existing urban ones are more likely to become urban than those further away. Outward expansion of central Dhaka and continuous infill and consolidation may be the major reason for this characteristic of urban land. Therefore, the enrichment factors developed with urban t1 and t2 as neighbourhood variables are included within the statistical models for t1–t2 and t2–t3, respectively. This has resulted in slightly better prediction overall by the models (with an ROC value of 87 and 73 % for t1–t2 and t2–t3 correspondingly). Both these ROC values are high enough to proceed with urban growth simulation.

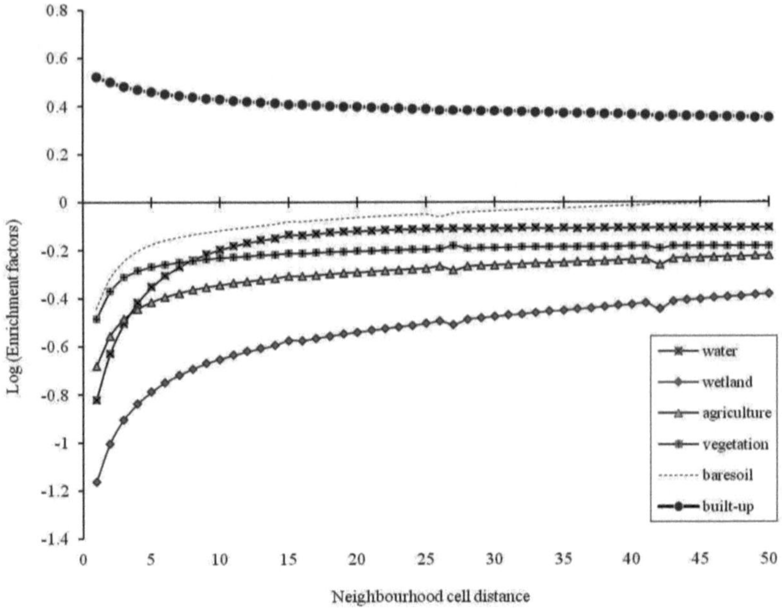

Fig. 7.5 Mean enrichment factors for built-up areas in 1999

Table 7.2 Conversion resistance factor values

	Water	Wetland	Agriculture	Vegetation	Built-up	Bare soil
Initial value	1	0.2	0.2	0.4	0.7	0.4
Final value	0.7	0.1	0.1	0.1	0.8	0.3

Table 7.3 Conversion matrix for CLUE-s simulation

Land use type	Water	Wetland	Agriculture	Vegetation	Built-up	Bare soil
Water	1	0	0	0	1	1
Wetland	0	1	0	0	1	1
Agriculture	0	0	1	0	1	1
Vegetation	0	0	0	1	1	1
Built-up	0	0	0	0	1	0
Bare soil	0	0	0	0	1	1

7.4.2 Conversion Resistance and Conversion Matrix

Based on recent land use change studies (Ahmed et al. 2012; Dewan and Yamaguchi 2009a, b), a tentative conversion resistance table and matrix were specified, and after calibration the values shown in Tables 7.2 and 7.3 were derived for each land use type. As seen in Table 7.2, the water and built-up categories were

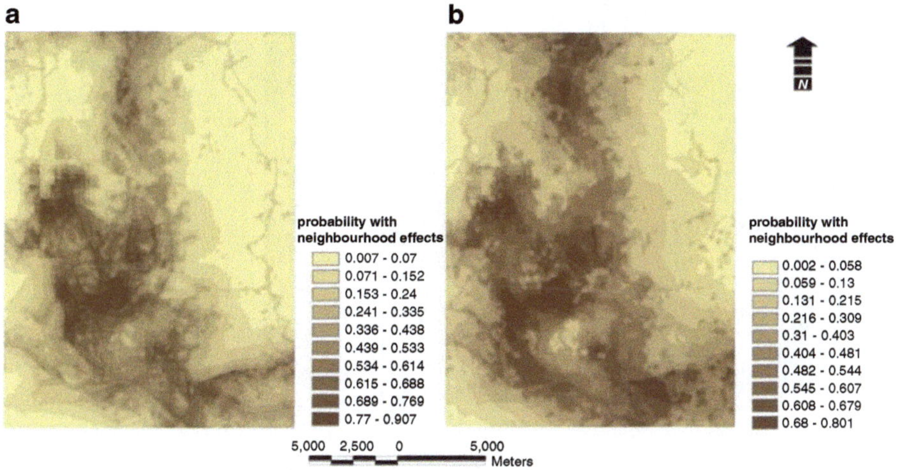

Fig. 7.6 Probability maps with neighbourhood effects: (**a**) t1–t2 and (**b**) t2–t3

given higher values as they are more difficult to convert and more likely to stay in the initial land use type. On the other hand, bare soil and agricultural land are easier to convert and less likely to resist change to built-up.

7.4.3 Simulation Runs in CLUE-s

Probability maps were generated in the CLUE-s model to examine the performance of the driving factors. Probability maps have been generated for t1–t2 and t2–t3, using both the driving factors and neighbourhood effects. For both time periods, areas at mid-north, west, central areas and areas at south-east have higher probabilities to change to urban land. The probability map for the period t2–t3 shows that the eastern part of the study area has the greatest probability of change to urban land. Figure 7.6 shows final simulation for both periods. This shows that the CLUE-S modelling framework produces better simulation maps when neighbourhood effects of the spatial pattern of urban development are included. Although the neighbourhood relationships only capture part of the processes of urban expansion, the empirical approach alone based on locational conditions is not able to capture the complexity of the processes involved. Although the model has been able to capture self-organisation process in urban growth, it lacks elements that can portray spontaneous development that the city is experiencing. Hence, there are many areas at the northern and eastern extensions of the city that are becoming urban but are not at all reflected within the simulation results (Fig. 7.7).

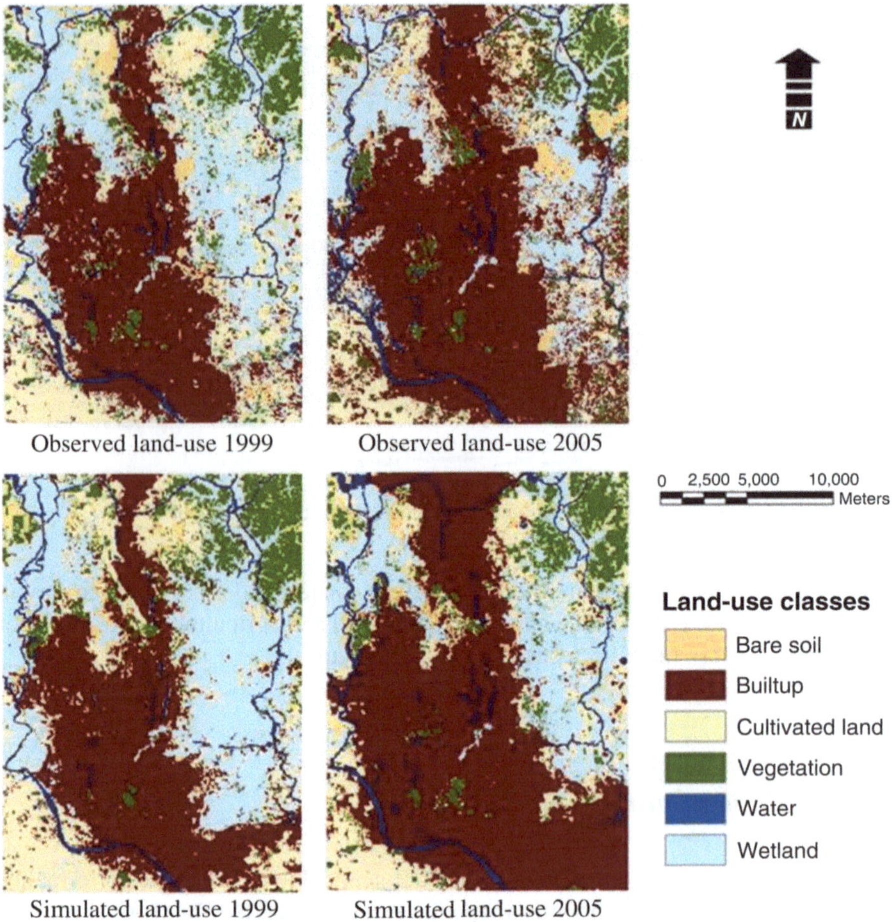

Fig. 7.7 Observed and simulated land-use for 1999 and 2005

7.5 Calibration and Validation

In this study, the model simulates observed change in urban land for t2 using t1 as the baseline during the calibration phase and simulates change in t3 using t2 as the baseline during the validation phase. The number of cells that are converted to urban land (between t1–t2 and t2–t3) is predefined, so errors in the calibration and validation are the result of misallocations in terms of changed cells. To check errors in terms of misallocations of cells, and assess the predictive power of the model, the neutral simulations for t2 are compared to the observed urban land of t2. The comparison is conducted using variants of the kappa statistic (i.e. traditional and fuzzy kappa), and also the percentage accuracy at different scales is derived.

As shown in Table 7.4, the model simulation outperformed the neutral models in terms of kappa statistics, and the percentage accuracy for fuzzy kappa is better than

Table 7.4 Calibration and validation results compared to urban 1999 and urban 2005 (values in parenthesis)

	Methods									
	Overall kappa	KHisto	Kappa Loc	Fraction correct (model scale)	Overall fuzzy kappa	Fuzzy fraction correct (model scale)	Fraction correct at 05 cells aggregation	Fraction correct at 10 cells aggregation	Fraction correct at 20 cells aggregation	Fraction correct at 30 cells aggregation
Urban 1999	1	1	1	1	1	1	1	1	1	1
Urban simulation 1999	0.650 (0.565)	0.993 (0.999)	0.655 (0.566)	0.841 (0.783)	0.167 (−.084)	0.881 (0.835)	0.857 (.808)	0.872 (.827)	0.888	0.900
Random model1	0.476 (.418)	1	0.476 (.418)	0.762 (.709)	−.148 (−.044)	0.836 (0.841)	0.804 (.805)	0.829 (.843)	0.852	0.865
Random model2	0.477 (.420)	1	0.477 (.420)	0.762 (.710)	−.149 (−.041)	0.836 (.841)	0.803 (.806)	0.829 (.847)	0.852	0.865

that of the traditional kappa statistic. In addition, the model results did not improve dramatically when the percentage accuracy for a larger aggregation area is considered (Table 7.4). Nonetheless, the fuzzy kappa variants for the simulation results are lower than for the neutral. The simulation results here show a similar model accuracy to that of other LUCC models reported in literature (Santé et al. 2010).

There may be several reasons why the model has not performed as expected at this stage. One possible reason could be other underlying factors not used here are equally important. These include water and sanitation networks, land value/price for different locations, population density etc. Additionally, according to Wu (2002), urban growth can be influenced by both spontaneous and self-organisational processes. Spontaneous growth results in a spatially homogeneous and sparse pattern, which contains more random components, whereas self-organisational growth results in spatial agglomeration, which is impacted by more self-organised socioeconomic activities. To understand spatial processes and patterns, we must take both types into account. Geographical phenomena can engage in a process of self-organisation in which locations with seemingly identical potential end up playing very different roles. As independent variables, the drivers in this study are able to capture the effect of self-organisation to a considerable extent. Nevertheless, while checking simulation results in the calibration and validation phases, the model results which include neighbourhood effects appear to cover the infill mechanism well but seem to fail to include the chance events of other distant areas becoming urban. This becomes particularly important during the later period, as there was much fragmented urban growth taking place especially in the eastern and north-eastern parts of the study area which are not well positioned in terms of land use preference (mostly low-lying areas that are not well linked to employment attractions of the city). Thus, as far this research is concerned, the results are successful in accommodating self-organisational process but have failed to simulate spontaneous random development.

7.6 Conclusions

Given the fact that significant urban changes have been observed between 1988 and 2005, the role of different spatial drivers, namely, distance to economic opportunities, elevation and planning, has been examined to simulate urbanisation for the periods of 1988–1999 and 1999–2005, using the CLUE-s land use modelling framework. As the city changed its growth pattern quite significantly and away from core areas, the contribution of drivers in explaining urban footprint dynamics also differed substantially. Nevertheless, local knowledge and observations in the study area lead to the deduction that the proposed eastern bypass may have significant influence on city's growth despite the fact that these areas are not free from monsoonal floods. For the same reason, the elevation variable has considerable influence to urban growth at first period of modelling (1988–1999) but became less significant during 1999–2005. This might be due to artificial heightening of land levels from surrounding flood levels. The inclusion of neighbourhood effects improved model performance for both periods. Performance during calibration was better than that during validation, although both calibration and validation results are similar to some other reported urban growth models.

Overall, the methodology is deemed robust enough to be used for exploring and identifying factors responsible for urban growth dynamics that relate to observed patterns in a data poor country. Although the research tries to cover all the available and relevant drivers, the absence of information regarding infrastructural development (e.g. expansion, connection of existing major roads and newly built ones), land value, plans and expansions of utilities, services etc. obviously affects the ability to predict urban change accurately. It was found to be very difficult to simulate spontaneous processes, particularly during t2–t3 as core and adjoining areas have become more consolidated and have left very little room for further development. Consequently, areas with a higher probability of change, as predicted by significant drivers during this period, do not coincide well with areas that actually experienced higher scatter development through leapfrogging. Such mismatches can be attributed partly to the absence of some of the variables mentioned earlier. In addition, the emerging land development pattern in Dhaka can arguably be an outcome of dominant market forces benefiting from largely weak and uncoordinated relevant institutions.

On the other hand, the land-based approach as adopted in CLUE-s model (like many other contemporary LUCC models) where the unit of analysis centres around an area of land, grid cell or census track cannot entirely mimic the decision-making or behavioural process made by individual plot owners or other agents that are involved for land use transition between different time periods. Many examples of agent-based models in land use modelling are being considered (Parker et al. 2003; Parker 2005) to mimic the exact features of urban expansion at local scale by modelling human-environment interactions and their emergent behaviour. However, these models are very data demanding which is a problem in an environment such as Dhaka. Nevertheless, prototype development of such models and their subsequent implementation may be able to complement results from simpler models like CLUE-s and lead to a better and more complete understanding of the behavioural complexity that exists in urban land development process in the megacity of Dhaka.

Acknowledgement This research was conducted as part of a Doctoral project in Urban Studies at the Heriot-Watt University, Edinburgh in the UK. The authors would like to thank the ORS and James Watt Scholarship for funding the study. Also, the authors like to thank Dr Ashraf M. Dewan for providing access to the land cover datasets. The authors are also grateful to Professor Mark Birkin (University of Leeds) and Professor Michael White (Nottingham Trent University) for constructive suggestions on the structuring of the simulation framework used in this study.

References

Ahmed SJ, Bramley G, Dewan AM (2012) Exploratory growth analysis of a megacity through different spatial metrics: a case study on Dhaka, Bangladesh (1960–2005). URISA Journal 24(1):5–20

Alam M, Rabbani MDG (2007) Vulnerabilities and responses to climate change for Dhaka. Environ Urban 19(1):81–97

Alonso W (1964) Location and land use: toward a general theory of land rent. Harvard University Press, Cambridge, MA, 204 p

Barredo JI, Kasanko M, McCormick N, Lavalle C (2003) Modelling dynamic spatial processes: simulation of urban future scenarios through cellular automata. Landsc Urban Plan 64(3):145–160

Barredo JI, Demicheli L, Lavalle C, Kasanko M, McCormick N (2004) Modelling future urban scenarios in developing countries: an application case study in Lagos, Nigeria. Environ Plan B Plan Des 32(1):65–84

Batisani N, Yarnal B (2009) Urban expansion in Centre County, Pennsylvania: spatial dynamics and landscape transformations. Appl Geogr 29(2):235–249

Batty M, Xie Y (1994) From cells to cities. Environ Plan B 21(7):41–48

Batty M, Xie Y, Sun Z (1999) Modelling urban dynamics through GIS-based cellular automata. Comput Environ Urban Syst 23(3):205–233

Braimoha AK, Onishi T (2007) Spatial determinants of urban land use change in Lagos, Nigeria. Land Use Policy 24(2):502–515

Cheng J, Masser I (2003) Modelling urban growth patterns: a multiscale perspective. Environ Plan A 35(4):679–704

Cheng J, Masser I (2004) Understanding spatial and temporal processes of urban growth: cellular automata modelling. Environ Plan B Plan Des 31(2):167–194

Clarke KC, Gaydos LJ (1998) Loose-coupling a cellular automaton model and GIS: long-term urban growth prediction for San Francisco and Washington/Baltimore. Int J Geogr Inf Sci 12(7):699–714

Dewan AM (2013) Floods in a megacity: geospatial techniques in assessing hazards, risk and vulnerability. Springer, Dordrecht

Dewan AM, Yamaguchi Y (2008) Effect of land cover changes on flooding: example from Greater Dhaka of Bangladesh. Int J Geoinf 4(1):11–20

Dewan AM, Yamaguchi Y (2009a) Using remote sensing and GIS to detect and monitor land use and land cover change in Dhaka Metropolitan of Bangladesh during 1960–2005. Environ Monit Assess 150(1–4):237–249

Dewan AM, Yamaguchi Y (2009b) Land use and land cover change in Greater Dhaka, Bangladesh: using remote sensing to promote sustainable urbanization. Appl Geogr 29(3):390–401

Dewan AM, Yeboah KK, Nishigaki M (2006a) Flood hazard delineation in Greater Dhaka, Bangladesh using integrated GIS and remote sensing approach. Geocarto Int 21(2):33–38

Dewan AM, Yeboah KK, Nishigaki M (2006b) Using synthetic aperture radar (SAR) data for mapping river water flooding in an urban landscape: a case study of Greater Dhaka. Bangladesh Jpn J Hydrol Water Resour 19(1):44–54

Dewan AM, Islam MM, Kumamoto T, Nishigaki M (2007) Evaluating Flood hazard for land-use planning in Greater Dhaka of Bangladesh using remote sensing and GIS techniques. Water Resour Manag 21(9):1601–1612

Dietzel C, Clarke K (2006) The effect of disaggregating land use categories in cellular automata during model calibration and forecasting. Comput Environ Urban Syst 30(1):78–101

Engelen G, White R, Uljee I (2002) The MURBANDY and MOLAND models for Dublin. Research Institute for Knowledge Systems (RIKS) BV, Maastricht, p 172

Gardner RH, Urban DL (2005) Model validation and testing: past lessons, present concerns and future prospects. In: Canham CD, Cole JC, Lauenroth WK (eds) The role of models in ecosystem science. Princeton University Press, Princeton, pp 186–205

Geist H, Mcconnell W, Lambin EF, Moran E, Alves D, Rudel T (2006) Causes and trajectories of land-use/cover change. In: Lambin EF, Geist H (eds) Land-use and land-cover change: local processes and global impacts. Springer, Berlin/Heidelberg, pp 41–70

Hagen-Zanker A (2002a) Approaching human judgement in the automated comparison of categorical maps. In: 4th international conference on recent advances in soft computing. Nottingham Trent University, Nottingham

Hagen-Zanker A (2002b) Multi-method assessment of map similarity. In: 5th AGILE conference on geographic information science, Palma

Hagen-Zanker A (2003) Fuzzy set approach to assessing similarity of categorical maps. Int J Geogr Inf Sci 17(3):235–249

Hagen-Zanker A, Lajoie G (2008) Neutral models of landscape change as benchmarks in the assessment of model performance. Landsc Urban Plan 86(3–4):284–296

Hagen-Zanker A, Van Loon J, Maas A, Straatman B, De Nijs T, Engelen G (2005) Measuring performance of land use models: an evaluation framework for the calibration and validation of integrated land use models featuring cellular automata. In: 14th European colloquium on theoretical and quantitative geography, Tomar

Hansen HS (2008) Quantifying and analysing neighbourhood characteristics supporting urban land-use modelling. In: Bernard L, Friis-Christensen A, Pundt H (eds) The European information society: lecture notes in geoinformation and cartography. Springer, Berlin/Heidelberg, pp 283–299

Henriquez C, Azocar G, Romero H (2006) Monitoring and modeling the urban growth of two mid-sized Chilean cities. Habitat Int 30(4):945–964

Hu Z, Lo CP (2007) Modeling urban growth in Atlanta using logistic regression. Comput Environ Urban Syst 31(6):667–688

Huq S (1999) Environmental problems of Dhaka City. In Mitchell JK (eds) Crucibles of Hazard: mega-cities and disasters in transition. United Nations University Press, Tokyo/New York/Paris. http://www.greenstone.org/greenstone3/nzdl?a=d&d=HASH9b6b117419a9f8e9a6e899.7&c=aedl&sib=1&dt=&ec=&et=&p.a=b&p.s=ClassifierBrowse&p.sa=. Accessed 12 Dec 2008

Islam N (1996) Dhaka: from city to megacity: perspectives on people, places, planning, and development issues. Urban Studies Programme, Dhaka

Islam N (2005) Dhaka now: contemporary urban development. The Bangladesh Geographical Society, Dhaka

Jat MK, Garg PK, Khare D (2008) Monitoring and modelling of urban sprawl using remote sensing and GIS techniques. Int J Appl Earth Obs Geoinf 10(1):26–43

Lesschen JP, Verburg PH, Staal SJ (2005) Statistical methods for analysing the spatial dimension of changes in land use and farming systems. LUCC report series no.7. The International Livestock Research Institute & LUCC Focus 3 Office, Wageningen University, Wageningen, 81p

Liu J, Zhan J, Deng X (2005) Spatio-temporal patterns and driving forces of urban land expansion in China during the economic reform era. Ambio 34(6):450–456

NPS-ROMN (2008) Travel time cost surface model, version 1.7. Rocky Mountain Network-National Park Service, Fort Collins

Overmars KP (2006) Linking process and pattern of land use change: illustrated with a case study in San Mariano, Isabela, Philippines. Institute of Environmental Sciences (CML), Leiden University, Leiden, 178 p

Overmars KP, Groot WTD, Huigen MGA (2007) Comparing inductive and deductive modeling of land use decisions: principles, a model and an illustration from the Philippines. Hum Ecol 35:439–452

Parker DC (2005) Integration of geographic information systems and agent-based models of land use: challenges and prospects. In: Maguire DJ, Goodchild MF, Batty M (eds) GIS, spatial analysis and modeling. ESRI Press, Redlands, pp 403–422
Parker DC, Manson SM, Janssen MA, Hoffmann M, Deadman P (2003) Multi-agent systems for the simulation of land-use and land-cover change: a review. Ann Assoc Am Geogr 93(2):314–337
Pontius GR Jr, Schneider CL (2001) Land-cover change model validation by an ROC method for the Ipswich watershed, Massachusetts, USA. Agric Ecosyst Environ 85(1–3):239–248
Pontius RG Jr, Huffaker D, Denman K (2004) Useful techniques of validation spatially explicit land-change models. Ecol Model 179(4):445–461
Pontius RG Jr, Boersma W, Castella JS, Clarke K et al (2008) Comparing the input, output, and validation maps for several models of land change. Ann Reg Sci 42(1):11–37
Rajdhani Unnayan Katripakko (RAJUK) (1993) Strategic growth options – Dhaka 2016. UNDP, UN-HABITAT, RAJUK, Dhaka
Rajdhani Unnayan Katripakko (RAJUK) (1997) Dhaka Metropolitan Development Plan for Dhaka City (1995–2015), Vol I & II. RAJUK, Dhaka
Rindfuss RR, Walsh SJ, Turner Ii BL, Fox J, Mishra V (2004) Developing a science of land change: challenges and methodological issues. Proc Natl Acad Sci U S A 101(39):13976–13981
Santé I, García AM, Miranda D, Crecente R (2010) Cellular automata models for the simulation of real-world urban processes: a review and analysis. Landsc Urban Plan 96(2):108–122
Silva EA, Clarke KC (2005) Complexity, emergence and cellular urban models: lessons learned from applying SLEUTH to two Portuguese metropolitan areas. Eur Plan Stud 13(1):93–115
Straatman B, White R, Engelen G (2004) Towards an automatic calibration procedure for constrained cellular automata. Comput Environ Urban Syst 28(1–2):149–170
Sui DZ, Zeng H (2001) Modeling the dynamics of landscape structure in Asia's emerging desakota regions: a case study in Shenzhen. Landsc Urban Plan 53(1–4):37–52
Veldkamp A, Fresco LO (1996) CLUE-CR: an integrated multi-scale model to simulate land use change scenarios in Costa Rica. Ecol Model 91(1–3):231–248
Veldkamp A, Lambin EF (2001) Predicting land-use change. Editor Agric Ecos Environ 85:1–6
Verburg PH, Overmars KP (2007) Dynamic simulation of land-use change trajectories with the CLUE-s model. In: Koomen E, Stillwell J, Bakema A, Scholten HJ (eds) Modelling land-use change: progress and applications, 40. Springer, Berlin, pp 321–337
Verburg PH, Soepboer W, Veldkamp A, Limpiada R, Espaldon V, Mastura SS (2002) Modeling the spatial dynamics of regional land use: the CLUE-S model. Environ Manage 30(3):391–405
Verburg PH, Schot P, Dijst MJ, Veldkamp A (2004a) Land use change modelling: current practice and research priorities. GeoJournal 61(4):309–324
Verburg PH, Ritsema Van Eck JR, Nijs TCMD, Dijst MJ, Schot P (2004b) Determinants of land-use change patterns in the Netherlands. Environ Plan B Plan Des 31(1):125–150
Verburg PH, Nijs TCMD, Ritsema Van Eck JR, Visser HJ, de Jong K (2004c) A method to analyse neighbourhood characteristics of land use patterns. Comput Environ Urban Syst 28(6):667–690
Verburg PH, Kok K, Pontius RG, Veldkamp A (2006) Modeling land-use and land-cover change. In: Lambin EF, Geist H (eds) Land-use and land-cover change: local processes and global impacts. Springer, Berlin/Heidelberg, pp 117–135
Verburg PH, Eickhout B, Meul HV (2008) A multi-scale, multi-model approach for analyzing the future dynamics of European land use. Ann Reg Sci 42(1):57–77
Visser H (2004) The map comparison kit: methods, software and applications. National Institute for Public Health and the Environment (RIVM), Bilthoven, 124 p
Visser H, De Nijs T (2006) The map comparison kit. Environ Model Software 21(3):346–358
Vliet JV (2009) Assessing the accuracy of changes in spatial explicit land use change models. In: 12th AGILE international conference on geographic information science Leibniz, Universität Hannover, Hanover

White R, Engelen G (1993) Cellular automata and fractal urban form: a cellular modelling approach to the evolution of urban land-use patterns. Environ Plan A 25(8):1175–1199
White R, Engelen G (2000) High resolution integrated modelling of the spatial dynamics of urban and regional systems. Comput Environ Urban Syst 24(5):383–400
White R, Straatman B, Engelen G (2004) Planning scenario visualization and assessment: a cellular automata based integrated spatial decision support system. In: Goodchild MF, Janelle D (eds) Spatially integrated social science. Oxford University Press, New York, pp 420–442
Wu F (2002) Calibration of stochastic cellular automata: the application to rural-urban land conversions. Int J Geogr Inf Sci 16(8):795–818
Wu Q, Li H-Q, Wang R-S, Paulussen J, Hec Y, Wang M, Wang BH, Wang Z (2006) Monitoring and predicting land use change in Beijing using remote sensing and GIS. Landsc Urban Plan 78 (4):322–333
Xiao J, Shen Y, Ge J, Tateishi R, Tang C, Liang Y, Huang Z (2006) Evaluating urban expansion and land use change in Shijiazhuang, China, by using GIS and remote sensing. Landsc Urban Plan 75(1–2):69–80
Yeh AGO, Li X (1998) Sustainable land development model for rapid growth areas using GIS. Int J Geogr Inf Sci 12(2):169–189

Chapter 8
Analysis of Urban Development Suitability

Salit Chakma

Abstract This chapter applies multi-criteria evaluation (MCE) within a GIS to the problem of land suitability assessment in Dhaka. A total of eight factors extracted from a spatial database were used as predictors with the analytic hierarchy process (AHP) followed by weighted linear combination (WLC) being used to identify areas appropriate for future urban development. The results revealed that approximately 21 % of land of the study area is highly suitable for urban development, whereas 29 % of land is moderately suitable. About 26 % of land, comprised of existing urban areas and waterbodies, is shown to be either unsuitable or of very low suitability for urban development although some redevelopment or vertical expansion may be possible. The analysis shows that GIS-based urban development suitability has major implications for a rapidly growing city such as Dhaka which can help urban planners and decision-makers to make better land-use planning decisions.

Keywords Analytical hierarchy process (AHP) • Dhaka • Multi-Criteria evaluation • Suitability analysis • Urban development

8.1 Introduction

Land-use suitability analysis is one of the most useful applications of GI Science (Malczewski 2004, 2006; Collins et al. 2001) and has gained renewed interest in urban, regional and environmental planning (Joerin et al. 2001; Han and Kim 1989) due to its ability to guide various stakeholders about the spatial patterns of future land use that are appropriate to the requirements of society (Malczewski 2004;

S. Chakma (✉)
Department of Spatial Sciences, Curtin University, Kent Street,
Bentley, Perth 6102, WA, Australia
e-mail: salit.chakma@ymail.com

A. Dewan and R. Corner (eds.), *Dhaka Megacity: Geospatial Perspectives on Urbanisation, Environment and Health*, Springer Geography,
DOI 10.1007/978-94-007-6735-5_8, © Springer Science+Business Media Dordrecht 2014

Hopkins 1977). The primary purpose of GIS-based suitability analysis is to define areas in which good sites for a particular activity may exist by considering diverse biophysical and socioeconomic factors using GIS tools (Malczewski 2004). Generally, suitability analysis considers three main factors of an area: location, development activities and biophysical/environmental processes (environmental effects) (Miller et al. 1998; Lyle and Stutz 1983). Although manual techniques such as sieve mapping were popular in the early days of land suitability mapping (McHarg 1969), advances in computing technologies, particularly GIS with its capability of storing and manipulating large volume of data, have given a new thrust to the inclusion of more diverse data in land suitability analysis (Collins et al. 2001). Geospatial techniques have therefore become important tools for analysing and assessing land suitability in a wide range of situations, including habitat suitability (Store and Jokimäki 2003; Store and Kangas 2001; Pereira and Duckstein 1993), urban development (Youssef et al. 2011; Lotfi et al. 2009; Doygun et al. 2008; Dong et al. 2008; Aly et al. 2005), land use and planning (Yu et al. 2011; Pourebrahim et al. 2011; Cengiz and Akbulak 2009; Duc 2006; Malczewski 2006; Joerin et al. 2001; Dai et al. 2001; Steiner et al. 2000; Van der Merwe 1997), agricultural suitability (Wang 1994; Cambell et al. 1992), optimum site selection (Al-Hanbali et al. 2011; Zucca et al. 2008; Irigaray et al. 1994), aquaculture development (Hossain et al. 2009), corridor siting (Aissi et al. 2012), landscape evaluation and planning (Uy and Nakagoshi 2008; Miller et al. 1998), route selection (Jankowski and Richard 1994) and regional planning (Janssen and Rietveld 1990).

In order to relieve population pressure on the environment and natural resources, new guidelines for urban development for Bangladesh were devised as part of a 20-year Structure Plan (1995–2015) by the Capital Development Authority (RAJUK). This new plan suggested an expansion of current growth on suburban and agricultural land to meet the growing demand for urban land (DMDP 1997). However, development has not proceeded as recommended by the planning authority (Mahtab-uz-Zaman and Lau 2000). Apart from developments on agricultural land, urban expansion onto floodplains and wetlands has become pervasive which has resulted in piecemeal and unauthorised development throughout the megacity (Dewan et al. 2012a; Alam and Ahmad 2010; Islam 1996). This kind of urban development introduces a number of socioeconomic and environmental problems (Rana 2011; BBS 2010; Islam 2005). Improper use of land leads to the destruction of land resources, and ill-planned urban development gives rise to urban poverty together with other social problems such as food insecurity (Hossain et al. 2009; Duc 2006). A recent study (CUS et al. 2006), for instance, shows that increased levels of poverty in the city can be attributed to unplanned urbanisation. In addition to this increase in poverty, widespread pluvial and fluvial flooding during the monsoon has increased (Barua and van Ast 2011; Stalenberg and Vrijling 2009). This can be linked with the rapid conversion of wetlands and floodplains to urban lands (Griffiths et al. 2010; Dewan and Yamaguchi 2009a, b, 2008; Begum 2007; Onishi et al. 2013; see Chap. 9) and severely affects the proper functioning of urban systems (Ghosh 2007; Jahan and Maniruzzaman 2007; Alam and Rabbani 2007). It is probable that climatic change will also trigger more frequent flooding in the event of intensified precipitation which will cause further suffering and damage to urban infrastructures since the development of drainage facilities in the city has not kept pace with the rapid urban development (Khan 2006). In addition, non-climatic factors such as haphazard and piecemeal urban growth could limit the ability of communities and authorities to plan climate change adaptation strategies (Dewan 2013; Haque et al. 2012; GoB 2005).

The efficient and thoughtful use of land is an important consideration in managing and developing any area and should guide the direction of future development (Steiner et al. 2000). Since Dhaka is expected to be one of the world's largest megacities by 2025 (UN 2012), consideration of physical factors that influence the expansion of city may be of considerable help in determining the fitness of a particular area for a specified use such as urban development (Steiner 1983). Despite the fact that GIS-based land suitability evaluation is a technique that has been available for decades, there is no such work for Dhaka evident in the literature except for some landfill demand and allocation analysis carried out by Hasan et al. (2009). Although planning agencies have adopted GIS as a tool for urban planning in Dhaka, most of the work in these organisations is still at the database development stage; hence, traditional tools remain in use and lead to a process of mismanaged urban expansion (Mahtab-uz-Zaman and Lau 2000). The use of geospatial techniques with advanced analytical tools for urban growth suitability could help ameliorate the continual degradation of environment associated with the current urban growth (Dewan et al. 2012b). With this in mind, this chapter makes use of geospatial data and spatial analytical tools to examine urban development suitability. This study includes a subset of Dhaka Metropolitan Development Plan (DMDP) area (see Chap. 1).

8.2 Materials and Methods

8.2.1 Data Sources

A number of data sources were considered in this study as a basis for analysing the suitability of land for urban use. These are following: a digital elevation model (DEM) obtained from the Survey of Bangladesh (SOB); a recent land-use/land-cover map classified from a Landsat Thematic Mapper Image of April 2011 (Dewan and Corner 2012); river network and geology vector files, obtained from the Centre for Environmental and Geographic Information Services (CEGIS); administrative boundaries at the ward and union level [UN level 4 administrative areas, see notes in Chap. 1] gathered from Bangladesh Space Research and Remote Sensing Organisation (SPARRSO); and road network data obtained from the Detailed Area Plan (DAP) of RAJUK. Since the DAP data was created during 2004–2005, the road network data were updated using a high-resolution GEOEYE image of 2010 and Google Earth images. Population data were from Bangladesh Bureau of Statistics 2011 census and were aggregated according to administrative boundary (e.g. union/ward) (BBS 2012). The population density was calculated as the population/km^2 and slope information in degrees was derived from the DEM.

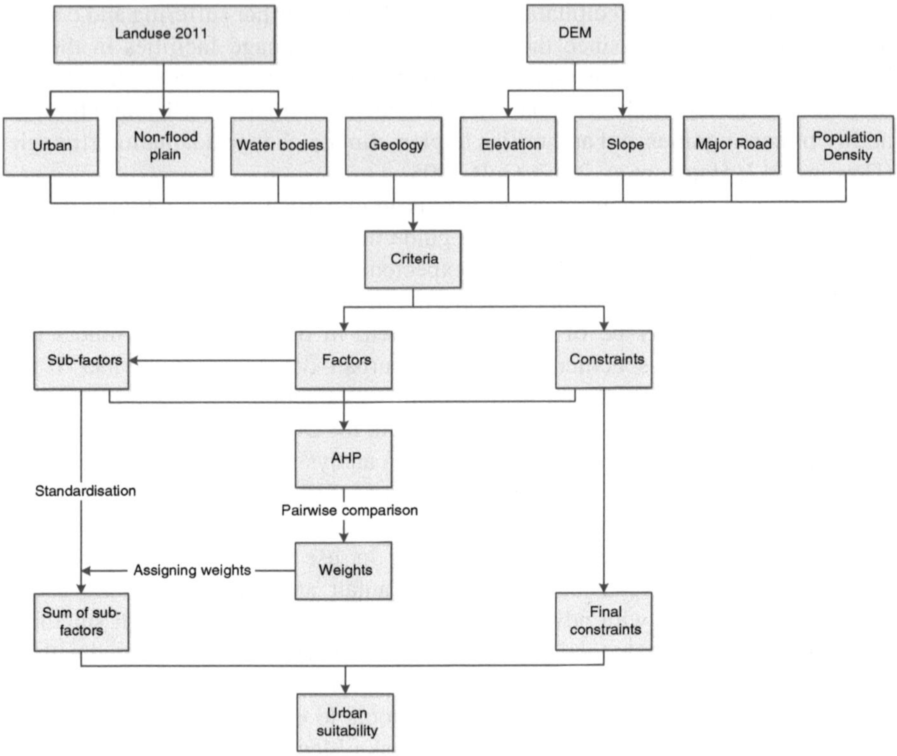

Fig. 8.1 Flowchart of urban suitability analysis

8.2.2 Analytical Method

A range of techniques have been used to evaluate land suitability which can be broadly categorised as computer-assisted overlay mapping, multi-criteria evaluation and artificial intelligence methods (Malczewski 2004). In this study, multi-criteria evaluation (MCE) methods comprising of Analytic Hierarchy Process (AHP) and Weighted Linear Combination (WLC) were used to assess urban development suitability. An overview of the process used in this study is shown in Fig. 8.1.

The idea of AHP was first introduced by Saaty (1977) as a robust and flexible technique for supporting priority setting and improved decision-making when the phenomenon that is being studied involves both the quantitative and qualitative aspects of a decision. It is based on the concept of prioritising multiple criteria identified by different groups of people (stakeholders and experts) involved in the decision-making process, in order to arrive at the best overall decision. It also assists in justifying the optimality of that decision (Saaty 1980). AHP allows a problem to be organised into primary and secondary objectives, hence the term hierarchy, and a matrix is subsequently used to weigh each factor against every other factor within each level of the hierarchy. Every level in the process is tied with its top and bottom levels, resulting in a clear priority statement for a group of criteria or for individual criteria (Ramanathan 2001). AHP generally involves three steps: (1) structuring a hierarchical decision model, (2) development of a pairwise comparison matrix and (3) determining priorities (Vaidya and Kumar 2006).

Model Design

The hierarchical decision model is the design phase of AHP in which the top level exhibits the overall goal of the decision. In this phase, each criterion is divided into several sub-criteria to establish the relationships between the elements.

A total of eight factors were extracted from the dataset described above (Table 8.1). The 2011 land-use/cover map was used to derive the boundaries of the existing urban areas, waterbodies and floodplain, whilst the Euclidean distance function was used to establish subcategories relative to these. Distance to major roads and rivers were calculated using the same tool. Elevation was categorised into four subcategories as 0–5, 5–8, 8–12 and 12>. Population density was classified as very low, low, medium, high and very high density using the natural breaks classification, whilst the geology layer was used with its original classification. Once these factors and subfactors were established, they were standardised so that all inputs would have uniform ratings for the pairwise comparison. Factors were standardised using the fuzzy membership tool with linear scale membership which is a commonly used approach in standardising factors for suitability analysis (Malczewski 2004).

Constraint Maps

Based on the Town Improvement Act and the Building Construction Act (Hossain 2012), three constraint maps were generated for waterbodies, road networks and current urban area maps. These maps were Boolean images showing areas of zero suitability. These areas include existing urban areas, assuming that further horizontal development is impossible, although vertical expansion or redevelopment may be possible. Within 4.5 m of a major road, no development is permitted (Hossain 2012); hence, a constraint image created from the road data showing that urban development potential within this distance is zero. Similarly, a Boolean image was constructed to account for the fact that urban development is prohibited within 250 m of waterbodies (Hossain 2012; GoB 2000).

Weighting

Following standardisation and the creation of constraint maps, factors were weighted to determine their relative influence on suitability of urban growth. It is

Table 8.1 Database specifications

Map layer	Data type	Units	Subcategories	Range
Distance to waterbodies	Continuous	Metres	Reserved	<250
			Near	250–2,469
			Moderately near	2,469–4,425
			Far	4,425–6,662
			Very far	>6,662
Distance to existing urban area	Continuous	Metres	Near	<1,087
			Moderately near	1,087–2,980
			Far	2,980–5,477
			Very far	>5,477
Distance to major roads	Continuous	Metres	Very close	4.5–649
			Close	649–643
			Moderately close	1,643–2,905
			Far	2,905–5,046
			Very far	>5,046
Distance to non-floodplain	Continuous	Metres	Near	<65
			Moderately near	65–255
			Far	> 255
Slope	Continuous	Degree	≤1	<1
			1–3	1–3
			≥3	>3
Elevation	Continuous	Metres	Very low	<5
			Low	5–8
			Medium	8–12
			High	>12
Geology	Categorical	–	Madhupur residuum	–
			Alluvial silt	–
			Alluvial silt and clay	–
			Marsh clay and pit	–
Population density	Categorical	(Km^2)	Very low	<6,883
			Low	6,883–23,396
			Medium	23,396–47,711
			High	47,711–77,107
			Very high	>77,101

important to note that the weights for main factors and their subfactors were based on literature review and a priori knowledge of local environmental conditions of the study area.

The process involves the construction of a pairwise comparison matrix which enables the relative contribution of each factor to the objective to be assessed. Each pair of factors or subfactors are compared in terms of their relative importance using a 9-point system from 1 (if two indicators equally contribute to the objective) to 9 (when one indicator is strongly favoured over another to meet the objective). A score of 1 denotes equal importance, a score of 3 refers to weak preference, whilst scores 5 and 7 represent obvious and strong preferences. The even numbers (i.e. 2, 4, 6 and 8) are used when a compromise is needed between the odd numbers (Saaty 1980).

Table 8.2 Relative weighting of principal factors

Criteria	(1)	(2)	(3)	(4)	(5)	(6)	(7)	(8)	Weights
(1) Distance to waterbodies	1								0.1538
(2) Distance to existing urban area	2	1							0.2160
(3) Distance to major roads	1/2	1/2	1						0.1049
(4) Distance to non-floodplain	1/3	1/4	1/2	1					0.0595
(5) Slope	1/4	1/5	1/3	1/2	1				0.0498
(6) Elevation	3	2	4	6	7	1			0.3503
(7) Geology	1/6	1/8	1/4	1/3	1/3	1/9	1		0.0245
(8) Population density	1/4	1/5	1/3	1	1/2	1/8	2	1	0.0411
Consistency Ratio (CR)	0.0229								

The local priority (weight) for each factor is then computed from the pairwise comparison matrix by normalising the scores in the columns (divide a cell value by the sum of a column) and averaging the normalised scores across the rows of the criterion. The consistency of the comparisons is evaluated by calculating a consistency ratio (CR) which is a measure of the logical consistency of judgements and enables the identification of possible errors of judgement in matrix. If the CR is equal to or less than 0.1, the comparisons are considered consistent, otherwise it should be re-evaluated (Saaty and Vargas 1991). CR is calculated as:

$$\mathrm{CR} = \frac{\mathrm{CI}}{\mathrm{RI}} \tag{8.1}$$

where the random index (RI) refers to a randomly generated reciprocal matrix from the 9-point scale, and it can be obtained by referring to a published RI table. Saaty (1980) provides a function of n in relationships. The consistency index (CI) is defined as

$$\mathrm{CI} = \frac{\lambda_{\max} - n}{n - 1} \tag{8.2}$$

where $\lambda_{\max}$ is the largest eigenvalue derived from the comparison matrix and n is the number of criteria.

Once the consistency is validated, the final priority of the factor at the upper level of the hierarchy model is obtained by aggregating the local priorities of the factors at its lower level. Relative weightings of the eight factors that influence urban growth with their corresponding CR are presented in Table 8.2. This shows a CR was 0.0229, well below than the specified value of 0.10, meaning that the results of the consistency test were acceptable.

Regarding the importance of the criteria at the upper level (level 1), elevation was considered the most important criterion and given the highest weight since ground elevation is an important determinant of urban development suitability in Dhaka. The second highest weight was given to distance to existing urban areas, and subsequent weights were assigned to distance to waterbodies and roads. Relative weights of sub-criteria are shown in Table 8.3.

Table 8.3 Relative weighting of sub-criteria

Factors	(1)	(2)	(3)	(4)	(5)	Weight	CR
Distance to waterbodies (m)							
(1) Near	1					0.4494	0.0231
(2) Moderately near	1/2	1				0.3001	
(3) Far	1/4	1/3	1			0.1327	
(4) Very far	1/6	1/5	1/2	1		0.0722	
(5) Reserved area	1/7	1/6	1/4	1/2	1	0.0455	
Distance to existing urban areas (m)							
(1) Near	1					0.4967	0.0126
(2) Moderately distant	1/2	1				0.3135	
(3) Distant	1/4	1/3	1			0.1213	
(4) Far	1/6	1/5	1/2	1		0.0685	
Distance to major roads (m)							
(1) Very close to road	1					0.4377	0.0092
(2) Close to road	1/2	1				0.2674	
(3) Moderately close	1/3	1/2	1			0.1600	
(4) Far	1/5	1/3	1/2	1		0.0873	
(5) Very far	1/7	1/6	1/4	1/2	1	0.0476	
Distance to non-floodplain (m)							
(1) Near	1					0.0882	0.0061
(2) Moderately near	3	1				0.2431	
(3) Far	7	3	1			0.6687	
Slope (degrees)							
(1) ≤ 1	1					0.7146	0.0158
(2) 1–3	1/4	1				0.2064	
(3) ≥ 3	1/8	1/3	1			0.0789	
Elevation (m)							
(1) High	1					0.4203	0.0075
(2) Medium	1/2	1				0.2407	
(3) Low	1/5	1/3	1			0.0883	
(4) Very Low	1/7	1/5	1/2	1		0.0507	
Geology							
(1) Madhupur Clay residuum	1					0.5737	0.0252
(2) Alluvial Silt	1/3	1				0.2290	
(3) Alluvial Silt and Clay	1/5	1/2	1			0.1361	
(4) Marsh Clay and Peat	1/7	1/4	1/3	1		0.0611	
Population density (km^2)							
(1) Very low	1					0.4491	0.0089
(2) Low	1/2	1				0.2707	
(3) Medium	1/3	1/2	1			0.1487	
(4) High	1/6	1/4	1/2	1		0.0799	
(5) Very high	1/7	1/5	1/3	1/2	1	0.0517	

Combination of Datasets

Following the AHP, a Weighted Linear Combination (WLC) technique was applied to combine factors after weighing each of them. Each factor is multiplied by the corresponding weight and then summed to derive an urban suitability map (Eastman et al. 1995). The following equation was used:

$$S = \sum_{1}^{n} w_i f_i \tag{8.3}$$

where S is the suitability, w_i is the final weight of factor i and f_i is the cell value/attribute of factor i.

Constraint layers are included in the suitability analysis, by modifying Eq. 8.3 to include a Boolean image which is the mathematical product of the various constraint layers (Eastman et al. 1995). Hence, the following equation was applied to derive the final suitability map by considering both the factors and the constraints:

$$S = \sum_{1}^{n} (w_i f_i) \prod_{1}^{m} C_j \tag{8.4}$$

where C_j is the value of constraint j at cell m and Π is the product operator.

The final suitability map obtained from Eq. 8.4 was reclassified into four categories using an equal interval classification as unsuitable, low suitable, moderate and highly suitable.

8.3 Results

The spatial pattern of urban development suitability is shown in Fig. 8.2. It clearly indicates locations suitable for further urban expansion. Since existing urban areas and waterbodies were considered as constraints, a large proportion of the area is exhibited unsuitable for horizontal expansion. However, existing urbanised areas may be used for redevelopment or vertical expansion. Considering the monsoonal flooding of the study area, ground elevations less than 5 m were deemed to have low suitability. These areas are generally wetlands, marshy lands that act as water retention ponds during floods, and therefore should be preserved to alleviate the effects future flooding in Dhaka. Peripheral areas where ground elevation is relatively high and in the proximity of agricultural lands are moderately suitable areas. By contrast, the most suitable areas for urban expansion are primarily located in zones of the higher elevation and closer than 4.5 m to the major road networks. These are also close to the existing urban areas, which adds to their suitability. Use of these areas for urban growth would greatly help to ameliorate the recurrent flood

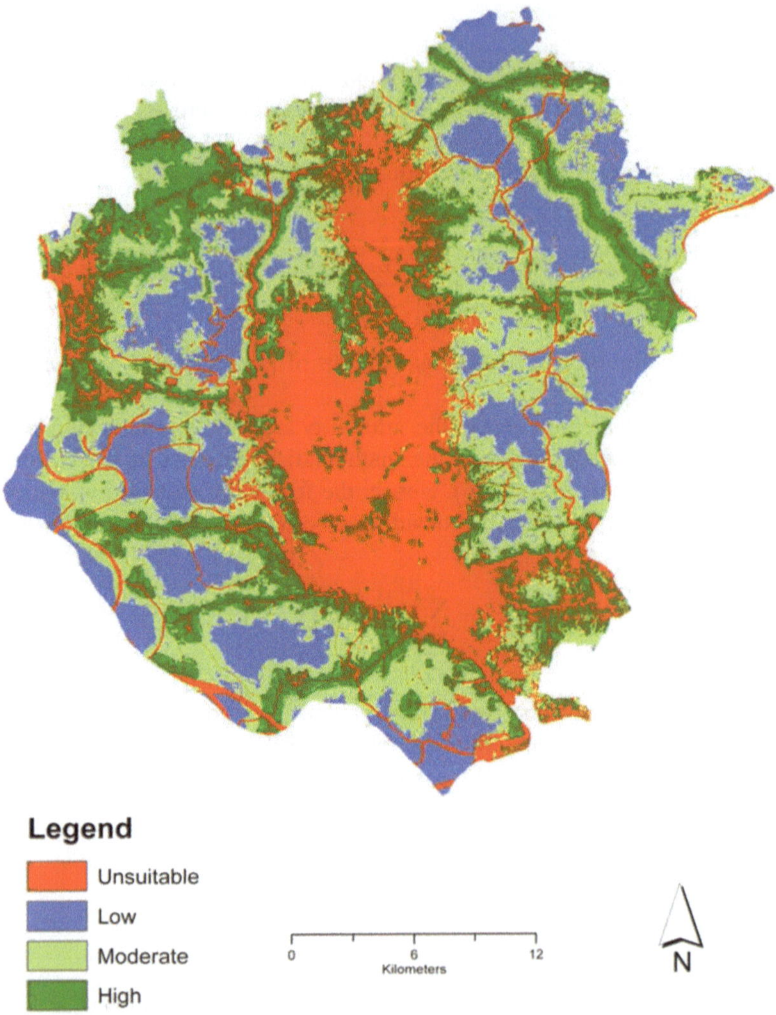

Fig. 8.2 Suitability potential for urban development in the study area

damage as well as saving dwindling precious natural resources (Islam et al. 2010; Alam and Ahmad 2010). In addition, adoption of the suitability map would ensure the optimum use of land resources, an essential component of achieving sustainability in urban land-use management (Cengiz and Akbulak 2009).

Table 8.4 shows area statistics for the four categories of urban development suitability. Around 30 % of the land in the study area is unsuitable for further development. This constitutes current urban land along with major river networks. Approximately 20 % of the land is highly suitable; however, when the moderate suitability category is added to this, the total area suitable for urban development increases to about 50 %. The north-west and further north parts of the city are highly

Table 8.4 Area and percentage suitability classes

Suitability	Area (ha)	Percentage (%)
Unsuitable	25,930.0	29.6
Low	19,202.2	21.9
Moderate	25,295.3	28.8
High	17,285.1	19.7

suitable since these areas are on Pleistocene era terraces. They are generally higher land, usually free from monsoonal flooding. In contrast, about 23 % of the land area was identified as being least suitable, of which low-lying land is the major feature. Unfortunately, these areas are currently undergoing massive unplanned development through illegal landfilling. For the sake of the sustainability of the city, these should instead be prioritised for conservation to save the fast depleting natural resources.

8.4 Discussion

Although AHP and WLC have been extensively used in developing lands suitability maps in a number of diverse situations, there is no commonly accepted method for assigning the relative weights to the criterion maps (Malczewski 2004). As a result, different weighting schemes can produce different outcomes (Chen et al. 2010; Van der Merwe 1997). For instance, Heywood et al. (1995) demonstrated that different multi-criteria evaluation rules results in different suitability patterns. Jankowski (1995) advocated that the approach is more appropriate for vector-based GIS. Although the present study sourced factors from geospatial data to develop urban development suitability, further incorporation of experts' opinion and governmental policies could be of help in identifying the most suitable lands for urban expansion. In addition, a sensitivity analysis, carried out by considering different weighting schemes, would assist in assessing the appropriateness of the relative weights used in this study (Pourebrahim et al. 2011). Therefore, a further study for understanding spatial patterns of urban land suitability may follow which will include more factors and different weighting schemes.

8.5 Conclusions

The objective of this study was to determine appropriate land for urban development in Dhaka. A subset of Dhaka Metropolitan Development Plan (DMDP) was used as a case area. Using analytic hierarchy process (AHP) with weighted linear combination (WLC) in a GIS, spatial pattern of future land uses for urban development was mapped out, and a total of eight factors extracted from spatial database were used for this purpose.

The study demonstrated that nearly 20 % of land is highly suitable, whilst 29 % are moderately suitable for urban development in the study area. The study underscores the fact that urban planners should account for the geo-environmental conditions of particular land parcels for optimum development of an area which can support sustainable use of land resources in fast growing metropolis.

As evident in the literature, Dhaka megacity is experiencing uncoordinated and piecemeal urban growth which should have serious implications for its limited natural resources. The result of this study is deemed to be particularly useful for urban development authority (e.g. RAJUK) to enforce existing urban development policies so that sustainable urban environmental management can be achieved. Furthermore, location of waterbodies to be preserved can be identified from this suitability map for saving flood loss potential from recurrent flood disaster in the city.

Acknowledgement The author wishes to express his gratitude to Dr. Ashraf M. Dewan for his support whilst supervising this work at Curtin University, Western Australia. I am also very grateful to Dr. Robert J. Corner for constructive comments on the manuscript. Finally, I am grateful to AusAid for the scholarship which enabled me to study in the Department of Spatial Sciences at Curtin University.

References

Aissi H, Chakhar S, Mousseau V (2012) GIS-based multicriteria evaluation approach for corridor siting. Environ Plan B 39(2):287–307

Alam MJ, Ahmad MM (2010) Analysing the lacunae in planning and implementation: spatial development of Dhaka city and its impact upon the built environment. Int J Urban Sustain Dev 2(1–2):85–106

Alam M, Rabbani MG (2007) Vulnerabilities and responses to climate change for Dhaka. Environ Urban 19(1):81–97

Al-Hanbali A, Alsaaideh B, Kondoh A (2011) Using GIS-based weighted linear combination analysis and remote sensing techniques to select optimum solid waste disposal sites within Mafraq City, Jordan. J Geogr Inf Syst 3(4):267–278

Aly MH, Giardino JR, Klein AG (2005) Suitability assessment for New Minia City, Egypt: a GIS approach to engineering geology. Environ Eng Geosci 11(3):259–269

Bangladesh Bureau of Statistics (BBS) (2010) Compendium of environment statistics of Bangladesh 2009. Bangladesh Bureau of Statistics, Ministry of Planning, Dhaka

Bangladesh Bureau of Statistics (BBS) (2012) Population and housing census – 2011, Community report, Dhaka, Narayanganj and Gazipur Zila . Bangladesh Bureau of Statistics, Ministry of Planning, Dhaka

Barua S, van Ast JA (2011) Towards interactive flood management in Dhaka, Bangladesh. Water Policy 13(5):693–716

Begum A (2007) Urban housing as an issue of redistribution through planning? The case of Dhaka city. Soc Policy Adm 41(4):410–418

Cambell JC, Radke J, Gless JT, Whirtshafter RM (1992) An application of linear programming and geographic information systems: cropland allocation in antigua. Environ Plan A 24(4):535–549

Cengiz T, Akbulak C (2009) Application of analytical hierarchy process and geographic information systems in land-use suitability evaluation: a case study of Dümrek village (Çanakkale, Turkey). Int J Sustain Dev World Ecol 16(4):286–294
Center for Urban Studies (CUS), National Institute of Population, Research and Training (NIPORT), Measure Evaluation (2006) Slums in urban Bangladesh: mapping and census 2005. CUS, Dhaka
Chen Y, Yu J, Khan S (2010) Spatial sensitivity analysis of multi-criteria weights in GIS-based land suitability evaluation. Environ Model Software 25(12):1582–1591
Collins MG, Steiner FR, Rushman MJ (2001) Land-use suitability analysis in the United States: historical development and promising technological achievements. Environ Manage 28(5):611–621
Dai FC, Lee CF, Zhang XH (2001) GIS-based geo-environmental evaluation for urban land-use planning: a case study. Environ Geol 61(4):257–271
Dewan AM (2013) Floods in a megacity: geospatial techniques in assessing hazards, risk and vulnerability. Springer, Dordrecht
Dewan AM, Corner RJ (2012) The impact of land use and land cover changes on land surface temperature in a rapidly urbanizing megacity. In: IGARSS 2012, Munich
Dewan AM, Yamaguchi Y (2008) Effects of land cover changes on flooding: example from Greater Dhaka of Bangladesh. Int J Geoinform 4(1):11–20
Dewan AM, Yamaguchi Y (2009a) Land use and land cover change in Greater Dhaka, Bangladesh: using remote sensing to promote sustainable urbanization. Appl Geogr 29(3):390–401
Dewan AM, Yamaguchi Y (2009b) Using remote sensing and GIS to detect and monitor land use and land cover change in Dhaka Metropolitan of Bangladesh during 1960–2005. Environ Monit Assess 150(1–4):237–249
Dewan AM, Yamaguchi Y, Rahman MZ (2012a) Dynamics of land use/cover changes and the analysis of landscape fragmentation in Dhaka Metropolitan, Bangladesh. GeoJournal 77(3):315–330
Dewan AM, Kabir MH, Nahar K, Rahman MZ (2012b) Urbanization and environmental degradation in Dhaka metropolitan area of Bangladesh. Int J Environ Sustain Dev 11(2):118–146
Dhaka Metropolitan Development Plan (DMDP) (1997) Dhaka Metropolitan Development Plan (1995–2015): Dhaka Structure Plan I & II, Rajdhani Unnayan Katripakha (RAJUK). Government of Bangladesh, Dhaka
Dong J, Zhuang D, Xu X, Ying L (2008) Integrated evaluation of urban development suitability based on remote sensing and GIS techniques - case study in Jingjinji area, China. Sensors 8(9):5975–5986
Doygun H, Alphan H, Gurun DK (2008) Analysing urban expansion and land use suitability for the city of Kahramanmaraş, Turkey, and its surrounding region. Environ Monit Assess 145(1–3):387–395
Duc T (2006) Using GIS and AHP technique for land-use suitability analysis. In: International symposium on geoinformatics for spatial infrastructure development in earth and allied sciences, organised by Japan-Vietnam Geoinformatics Consortium (JVGC), Institute for Environment and Resources, Vietnam
Eastman JR, Jin W, Kyem PAK, Toledano J (1995) Raster procedures for multi-criteria/multi-objective decisions. Photogramm Eng Remote Sens 61(5):539–547
Ghosh S (2007) A farewell to urban planning: sequel to urbanization in South Asia. In: Jahan S, Maniruzzaman KM (eds) Urbanization in Bangladesh: patterns, issues and approaches to planning. Bangladesh Institute of Planners, Dhaka, pp 163–170
Government of Bangladesh (GoB) (2000) Natural water reservoir conservation Act, 2000, Dhaka
Government of Bangladesh (GoB) (2005) National adaptation programme of action. Ministry of Environment and Forest, Government of Bangladesh, Dhaka
Griffiths P, Hostert P, Gruebner O, van der Linden S (2010) Mapping megacity growth with multi-sensor data. Remote Sens Environ 114(12):426–439
Han S-Y, Kim TJ (1989) An application of expert systems in urban planning: site selection and analysis. Comput Environ Urban Syst 13(4):243–254

Haque AN, Grafakos S, Huijsman M (2012) Participatory integrated assessment of flood protection measures for climate adaptation in Dhaka. Environ Urban 24(1):197–213

Hasan MR, Tetsuo K, Islam SA (2009) Landfill demand and allocation for municipal solid waste disposal in Dhaka city – an assessment in a GIS environment. J Civil Eng 37(2):133–149

Heywood I, Oliver J, Tomlinson S (1995) Building an exploratory multi-criteria modelling environment for spatial decision support. In: Fisher P (ed) Innovations in GIS, vol 2. Taylor & Francis, London, pp 127–136

Hopkins LD (1977) Methods for generating land suitability maps: a comparative evaluation. J Am Inst Plann 43(4):386–400

Hossain MK (2012) Building construction related acts (including latest corrections), 2nd edn. Muhit Publications, Dhaka

Hossain MS, Chowdhury SR, Das NG, Sharifuzzaman SM, Sultana A (2009) Integration of GIS and multicriteria decision analysis for urban aquaculture development in Bangladesh. Landsc Urban Plan 90(3–4):119–133

Irigaray C, Fernandez T, Chacon J, El Amrani-Paaza N, Boussouf S (1994) GIS geotechnical and environmental assessment of site selection for urban waste disposal in the Granada district (Spain). In: Proceedings of the 7th international IAEG congress, Balkema, Rotterdam, pp 46–79

Islam N (1996) Dhaka from city to megacity: perspectives on people, places, planning and development issues. Urban Studies Programme, Dhaka

Islam N (2005) Dhaka now: contemporary urban development. The Bangladesh Geographical Society, Dhaka

Islam MS, Rahman MR, Shahabuddin AKM, Ahmed R (2010) Changes in wetlands in Dhaka city: trends and physico-environmental consequences. J Life Earth Sci 5:37–42

Jahan S, Maniruzzaman KM (eds) (2007) Urbanization in Bangladesh: patterns, issues and approaches to planning. Bangladesh Institute of Planners, Dhaka

Jankowski P (1995) Integrating geographical information systems and multiple criteria decision making methods. Int J Geogr Inf Syst 9(3):251–273

Jankowski P, Richard L (1994) Integration of GIS-based suitability analysis and multicriteria evaluation in a spatial decision support system for route selection. Environ Plan B 21(3):326–339

Janssen R, Rietveld P (1990) Multicriteria analysis and geographical information systems: an application to agricultural land use in the Netherlands. In: Scholten HJ, Stillwell JCH (eds) Geographical information systems for urban and regional planning. Kluwer Academic Publishers, Dordrecht, pp 129–139

Joerin F, Thériault M, Musy A (2001) Using GIS and outranking multicriteria analysis for land-use suitability assessment. Int J Geogr Inf Sci 15(2):153–174

Khan MSA (2006) Stormwater flooding in Dhaka city: causes and consequences. J Hydrol Meteorol 3(1):77–85

Lotfi S, Habibi K, Koosari MJ (2009) An analysis of urban land development using multi-criteria decision model and geographical information system (a case study of Babolsar city). Am J Environ Sci 5(1):87–93

Lyle J, Stutz FP (1983) Computerised land use suitability mapping. Cartogr J 20(1):39–49

Mahtab-uz-Zaman QM, Lau SSY (2000) City expansion policy versus compact city demand: the case of Dhaka. In: Jenks M, Burgess R (eds) Compact cities: sustainable urban forms for developing countries. Spon Press, London, pp 141–152

Malczewski J (2004) GIS-based land-use suitability analysis: a critical overview. Prog Plan 62(1):3–65

Malczewski J (2006) Ordered weighted averaging with fuzzy quantifiers: GIS-based multicriteria evaluation for land-use suitability analysis. Int J Appl Earth Obs Geoinf 8(4):270–277

McHarg IL (1969) Design with nature. Wiley, New York

Miller W, Collins M, Steiner F, Cook E (1998) An approach for greenway suitability analysis. Landsc Urban Plan 42(2–4):91–105

Onishi T, Khan T, Hiramatsu K (2013) Impact of land use change on flooding patterns. In: Dewan A, Corner R (eds) Dhaka megacity: Geospatial perspectives on urbanisation, environment and health. Springer, Dordrecht, pp 163–176

Pereira JMC, Duckstein L (1993) A multiple criteria decision-making approach to GIS-based land suitability evaluation. Int J Geogr Inf Syst 7(5):407–424

Pourebrahim S, Hadipour M, Mokhtar MB (2011) Integration of spatial suitability analysis for land use planning in coastal areas; case of Kuala Langat District, Selangor, Malaysia. Landsc Urban Plan 101(1):84–97

Ramanathan R (2001) A note on the use of the analytic hierarchy process for environmental impact assessment. J Environ Impact Assess 63(1):27–35

Rana MMP (2011) Urbanization and sustainability: challenges and strategies for sustainable urban development in Bangladesh. Environ Dev Sustain 13(1):237–256

Saaty TL (1977) A scaling method for priorities in hierarchical structures. J Math Psychol 15(3):234–281

Saaty TL (1980) The analytical hierarchy process. McGraw-Hill, New York

Saaty TL, Vargas LG (1991) Prediction, projection, and forecasting: applications of the analytic hierarchy process in economics, politics, games and sports. Kluwer Academic Publishers, Boston

Stalenberg B, Vrijling H (2009) The battle of Tokyo and Dhaka against floods. Built Environ 35(4):471–491

Steiner F (1983) Resource suitability: methods for analysis. Environ Manage 7(5):401–420

Steiner F, McSherry L, Cohen J (2000) Land suitability analysis for the upper Gila river watershed. Landsc Urban Plan 50(4):199–214

Store R, Jokimäki J (2003) A GIS-based multi-scale approach to habitat suitability modeling. Ecol Model 169(1):1–15

Store R, Kangas J (2001) Integrating spatial multi-criteria evaluation and expert knowledge for GIS-based habitat suitability modelling. Landsc Urban Plan 55(2):79–93

United Nations (2012) World population prospects: the 2010 revision, vol 1. United Nations Department of Economic and Social Affairs, New York. Available at: http://eas.un.org/undp/wpp/Documentation/publications.html. Accessed 1 Oct 2012

Uy PD, Nakagoshi N (2008) Application of land suitability analysis and landscape ecology to urban greenspace planning in Hanoi, Vietnam. Urban For Urban Green 7(1):25–40

Vaidya OS, Kumar S (2006) Analytic hierarchy process: an overview of applications. Eur J Oper Res 169(1):1–29

Van der Merwe JH (1997) GIS-aided land evaluation and decision-making for regulating urban expansion: a South African case study. GeoJournal 43(2):135–151

Wang F (1994) The use of artificial neural networks in a geographical information system for agricultural land-suitability assessment. Environ Plan A 26(2):265–284

Youssef AM, Pradhan B, Tarabees E (2011) Integrated evaluation of urban development suitability on remote sensing and GIS techniques: contribution from the analytic hierarchy process. Arab J Geosci 4(3–4):463–473

Yu J, Chen Y, Wu J, Khan S (2011) Cellular automata-based spatial multi-criteria land suitability simulation for irrigated agriculture. Int J Geogr Inf Sci 25(1):131–148

Zucca A, Sharifi AM, Fabri AG (2008) Application of spatial multi-criteria analysis to site selection for a local park: a case study in the Bergamo Province, Italy. J Environ Manage 88(4):752–769

Chapter 9
Impact of Land-Use Change on Flooding Patterns

Takeo Onishi, Tahmina Khan, and Ken Hiramatsu

Abstract Recent unplanned and rapid urbanisation of Dhaka has a possibility to induce serious urban flooding in the near future. To evaluate the impacts of land-use changes on flood propagation patterns, we conducted flood inundation simulations using a two-dimensional finite element method with simplified Saint Venant equations. We used as a study site the mid-eastern part of the city of Dhaka, popularly known as "mid-eastern Dhaka". Two different land-cover datasets were prepared—one showing land use in 1990 and the other for 2011. In addition, complete land-cover change scenarios were also considered. As for the boundary conditions for flood simulations, we first estimated river discharge by constructing a kind of conceptual hydrologic model called a tank model, since we only have water-level data with daily time resolution at the Balu River mouth. Changes in inundation areas, related to these different land-cover patterns, were then evaluated. The study shows that although no significant difference was detected between the results for land use in 1990 and in 2011, under the complete land-use conversion scenarios, with all wetlands converted to other uses, both flood propagation time and flooding area will significantly change. Thus, the simulation results prove that the presence of wetland in land cover reduces flood risk, as compared with other land use. While further validations of flood simulation results are required, our results may provide data useful for proper flood management in achieving urban sustainability.

Keywords Flood • Land-use change • Wetlands • Urbanization

T. Onishi (✉) • K. Hiramatsu
Faculty of Applied Biological Sciences, Gifu University, 1-1, Yanagido, Gifu 501-1193, Japan
e-mail: takeon@gifu-u.ac.jp; hira@gifu-u.ac.jp

T. Khan
United Graduate School of Agricultural Science, Gifu University, 1-1, Yanagido, Gifu 501-1193, Japan
e-mail: khan_tahmina@hotmail.com

A. Dewan and R. Corner (eds.), *Dhaka Megacity: Geospatial Perspectives on Urbanisation, Environment and Health*, Springer Geography,
DOI 10.1007/978-94-007-6735-5_9, © Springer Science+Business Media Dordrecht 2014

9.1 Introduction

Bangladesh is one of the most vulnerable countries in the world to natural disasters. It is extremely low lying with most parts of the country being less than 12 m above sea level, and it is believed that about 10 % of the land would be flooded if sea level were to rise by 1 m (Ali 1996). The country is under serious threat from climate change, and the impact will be especially felt in lowland and heavily congested cities like the capital and economic and administrative centre of the country, Dhaka. The city has experienced several disastrous floods, including recent ones in 1987, 1988, 1998 and 2004 that caused serious damage (Dewan 2013; Dewan and Yamaguchi 2008; Dewan et al. 2007; Dewan et al. 2006a, b).

To evaluate the impacts of land-use changes on flood vulnerability, we conducted flood inundation simulations using a two-dimensional finite element method (FEM) with simplified Saint Venant equations. Volume and depth of floodwater in the study area were calculated. We considered several land-cover change scenarios, including land-use patterns of mid-eastern Dhaka in 1990 and 2011 under the same initial and boundary conditions. To evaluate future flooding patterns, we also addressed complete conversions of wetlands to cultivated area.

9.2 Materials and Methods

9.2.1 Study Area

Our study site is to the eastern part of Dhaka city centre in an area popularly known as "mid-eastern Dhaka". The study area is in the watershed of the Balu River (Fig. 9.1), which has an area of 684 km^2. Topographically, this area is the largest flatland in the megacity, occupying the Pleistocene era alluvial terrace, popularly known as the Madhupur terrace (Miah and Bazlee 1968). The study area is part of the Balu River floodplain and contains an area of 104.14 km^2 and surface elevations between 1 and 16 m (Fig. 9.1). Most of the built-up areas have elevations of 6–8 m. Some 60 % of the eastern part of Dhaka is regularly inundated every year between June and October, because of insufficient flood protection.

9.2.2 Materials

To execute numerical simulation of areas inundated by river flooding, observations of river discharge are usually required. However, it is not easy to obtain precise monitored discharge data, since hydrologic monitoring systems are not fully implemented worldwide. In our case, we have water-level data with daily time resolution at the Balu River mouth and at a point just downstream of the junction of the Balu and Lakshya rivers. Locations of both monitoring sites are shown in Fig. 9.1.

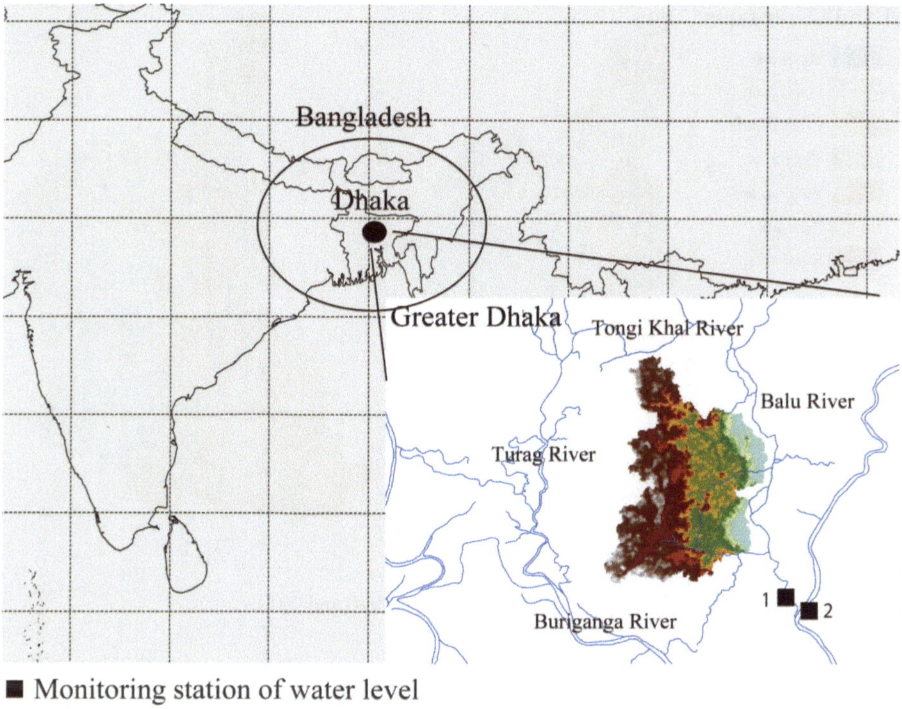

Fig. 9.1 Location of the study site (part of Balu River catchment)

Given an empirical relationship between water level and discharge of a given river, we can convert water-level data into discharge data. We can then use these as input to a flood simulation. Unfortunately, we do not have such data for the study site. Nevertheless, if we accept several assumptions about the relationship between water level and discharge, we can roughly correlate the two. Therefore, our first estimate used a kind of conceptual hydrologic model called a tank model, which was originally developed by Sugawara and Maruyama (1952). To execute the tank model, precipitation and evapotranspiration data are required. For precipitation, we acquired time series data from a metrological station at the centre of Dhaka (see Chap. 1). For evapotranspiration, we used data from the National Centers for Environmental Predictions (NCEP)/Department of Energy (DOE) Atmospheric Model Intercomparison Project (AMIP)-II Reanalysis (Kanamitsu et al. 2002). Although the spatial resolution of these data is coarse ($\approx$ 200 km), each grid value can be considered to provide an average evapotranspiration rate for the region within the grid.

After constructing a tank model and obtaining a discharge estimate, we calculated flood propagation to estimate inundation patterns. To calculate the flood propagation, detailed digital elevation model (DEM) and land-use and land-cover (LULC) data were used. Two different DEM datasets were used—Shuttle Radar Topography Mission 3 (SRTM3) and one compiled by Survey of Bangladesh (SOB). Land-use

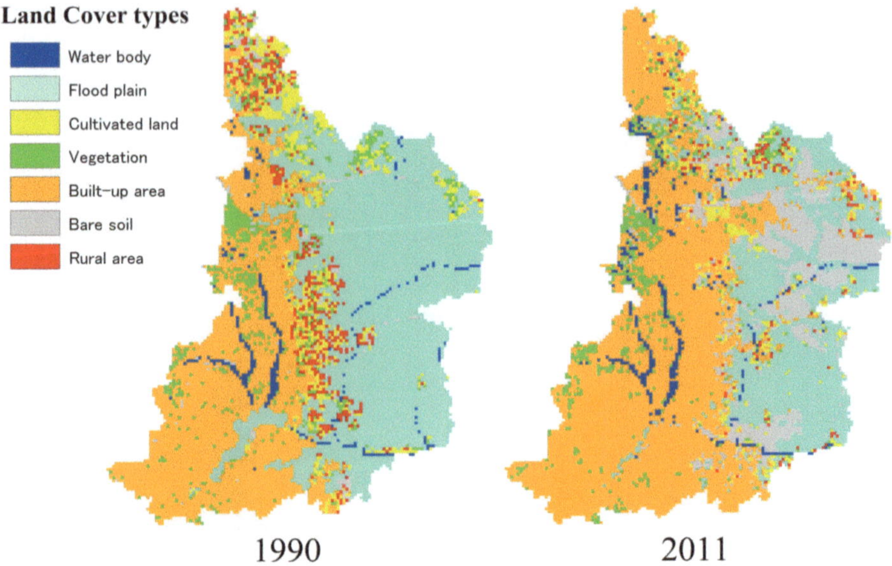

Fig. 9.2 Comparison of land-cover patterns between 1990 and 2011

and land-cover data were obtained from literature (Dewan and Corner 2012). Figure 9.2 shows LULCs at the study site in 1990 and 2011. These were the two LULC datasets that were used in our analysis.

9.3 Methods

9.3.1 Hydrologic Simulation by Tank Model

As indicated above, the tank model is a type of conceptual model. Its typical structure is shown in Fig. 9.3. It is a very simple model, composed of several tanks laid vertically in series (four tanks are commonly used, as shown in the figure). Using a tank model of this structure, we simulated the discharge of the Balu River. The result of tank model calculation is in units of run-off height, which is equivalent to units of discharge. However, we only have observations of river water level. Thus, to compare calculated and observed values, we must convert calculated discharge into water level. To achieve this, we assumed that the Balu River discharge is approximately expressed by the Manning equation under steady uniform flow:

$$V = \frac{1}{n} S_{\mathrm{f}}^{1/2} R_{\mathrm{h}}^{2/3} \tag{9.1}$$

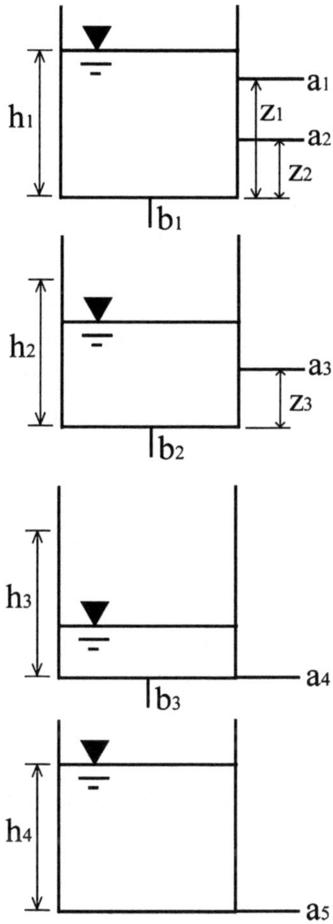

Fig. 9.3 Structure and parameters required to run tank model

where V is the river flow velocity [m s^{-1}], n is the channel roughness coefficient [s m$^{-1/3}$], S_f is the water surface slope [−] and R_h is the hydraulic radius [m].

If we assume that river width and depth are B [m] and H [m], respectively, river discharge Q may be expressed as follows:

$$Q = V \cdot B \cdot H = \frac{1}{n} S_f^{1/2} R_h^{2/3} \cdot B \cdot H \tag{9.2}$$

If B is substantially greater than H and the cross-sectional shape of the river is approximately rectangular, R_h can be expressed as follows:

$$R_h = B \cdot H/(2H + B) \cong H \tag{9.3}$$

Equation (9.2) can then be rearranged as follows:

$$Q = V \cdot B \cdot H = \frac{1}{n} \cdot B \cdot S_f^{1/2} H^{5/3} \tag{9.4}$$

From Eq. (9.4), we see that temporal change of river discharge is correlated with temporal change of S_f and H. Considering the backwater effects that are frequent with flatland flooding, temporal change of S_f is important. In other words, by comparing the difference between calculated discharges without considering the change of S_f with actual discharge, we can roughly assess the extent of backwater effects. With this objective and assuming that $\frac{1}{n} \cdot B \cdot S_f^{1/2}$ is one optimising parameter, we directly estimated H from calculated Q. Here, $\frac{1}{n} \cdot B \cdot S_f^{1/2}$ is defined as C′. However, even if we can estimate H from Q, the estimated H is not the absolute elevation of the river water surface. Thus, when we carry out parameter optimisation, we compare the relative change of H with relative change of observed water levels. Then, comparing calculated and observed water levels, we attained a parameter set which provides the best fit. To select such a set, we randomly generated a parameter set within a reasonable range of each parameter, and sets that minimised errors were evaluated. The total number of parameter sets generated was 100,000. For evaluation, the Nash-Sutcliffe evaluation criteria were used (Nash and Sutcliffe 1970).

9.3.2 *Two-Dimensional Flood Simulation*

The governing equations of flow processes are two-dimensional Saint Venant equations for shallow water flow:

$$\dot{H} + H_{,i}u_i + Hu_{i,i} = 0 \tag{9.5}$$

$$\dot{u} + u_j u_{i,j} - \varepsilon\left(u_{j,i} + u_{i,j}\right)_{,j} + g(H+z)_{,i} + \frac{gn^2\sqrt{u_k u_k}}{H^{4/3}}u_i - \frac{K|W|W_i}{H} + fu_i = 0 \tag{9.6}$$

where H is the water depth [m], u is the depth averaged flow velocity [m s^{-1}], ε is the eddy viscosity coefficient [m^{2}s^{-1}], g is the gravitational acceleration [m s^{-2}], n is Manning's roughness coefficient [s m$^{-1/3}$], K is the wind stress coefficient [−], W is the wind velocity [m s^{-1}] and f is the Coriolis parameter [−]. i (=1, 2) and j (=1, 2) are subscripts for horizontal coordinates (=x, y). Differentiations with respect to x or y and t are denoted by a subscripted comma and superscripted dot, respectively.

The standard Galerkin method was applied to Eqs. (9.5) and (9.6) for spatial discretisation, and the selective lumping two-step explicit method was used for numerical integration in time. This method was originally developed by Kawahara et al. (1978) to simulate tsunami wave propagation. It has been successfully used to solve the shallow water equation of river flow (Kawahara and Yokoyama 1981). Based on those results, Hai et al. (2006) successfully modified the method to evaluate the flood regulation function of paddies during flood season in the Tonle Sap Lake catchment area; we therefore applied their method and simulation code in this study.

9.3.3 LULC Change Scenario

Two different land-cover datasets were prepared—one for the land use in 1990 and the other for 2011. Four land-use types were considered—urban or built-up area, wetland, cultivated land and vegetation. There has been rapid conversion of wetland into cultivated land and other types over the last several decades (Dewan and Yamaguchi 2009). If this conversion continues at the same pace, all wetland will disappear and be transformed into cultivated land or built-up area. Therefore, to evaluate possible flood propagation patterns in the near future, we established a complete scenario in which all wetland has been converted to cultivated land. Thus, two land-cover scenarios were created for both 1990 and 2011, in which such complete conversion was realised.

9.4 Results and Discussion

9.4.1 Hydrologic Modelling of Balu River Discharge

A comparison of the estimated and observed river water levels is shown in Fig. 9.4. The period 1996–1998 is a calibration period, and 2000–2002 and 2004 are validation periods. Optimised parameter values and Nash-Sutcliffe criteria are summarised in Table 9.1. Overall, simulated results agree well with observations, except in summer when flooding occurs. As already explained, the parameter C' depends on parameters n, S_f and B. Although we have no experimental value for n at the study site, values for all rivers are between 0.02 and 0.06 (Brutsaert 2005). S_f is approximately equal to riverbed slope under the condition of steady-state flow. Thus, we estimated a S_f about 1/10,000 by calculating an average slope near the river. Lastly, B is about 80 m. Thus, C' is between 0.025 and 0.075. This rough estimate verifies that an optimised parameter C' of 0.061 is fairly reasonable. Discrepancies between calculated and observed values during summer indicate the possibility of flooding by water backing up the river during that season. Supporting this inference is the fact that river water levels at stations 1 and 2 are sometimes reversed during summer.

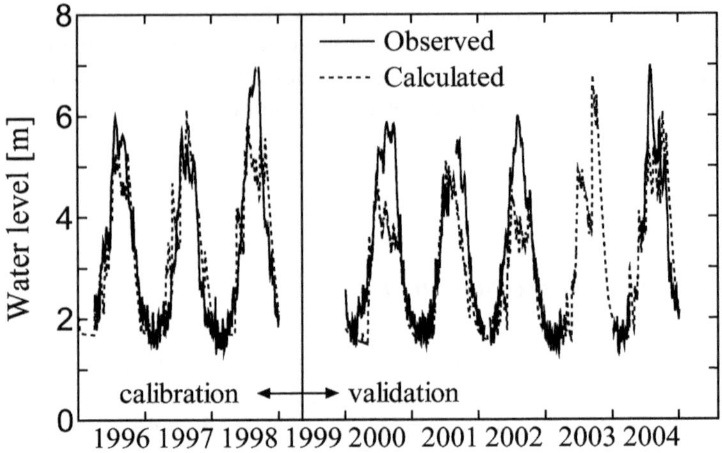

Fig. 9.4 Comparison between calculated and observed river water levels at station 1

Table 9.1 Optimised parameter values and Nash-Sutcliffe criteria

Parameter		Parameter		Parameter			
a1	0.026	z1	44.941	b1	0.001	Nash-Sutcliffe	0.791
a2	0.001	z2	18.840	b2	0.002		
a3	0.003	z3	5.450	b3	0.045		
a4	0.001						
a5	0.002						

9.4.2 Flood Propagation Simulations

Although no significant difference of flooded area was found between the land-cover results of 1990 and 2011, flood propagation speed with the 2011 land-cover condition was faster than that of 1990. The reason for this difference can be attributed to a change of Manning's roughness coefficient, owing to land-cover change from 1990 to 2011. Based on this knowledge, we executed two complete scenarios as follows.

Comparison Between the Results of 1990 Land-Cover and Complete Land-Cover Conversion of 1990

The comparison between the results of 1990 land-cover and complete wetland conversion into cultivated land is shown in Fig. 9.5. This figure shows that all selected nodes had a water depth increase for the land-cover situation of complete wetland conversions. This means that water storage capacity decreased

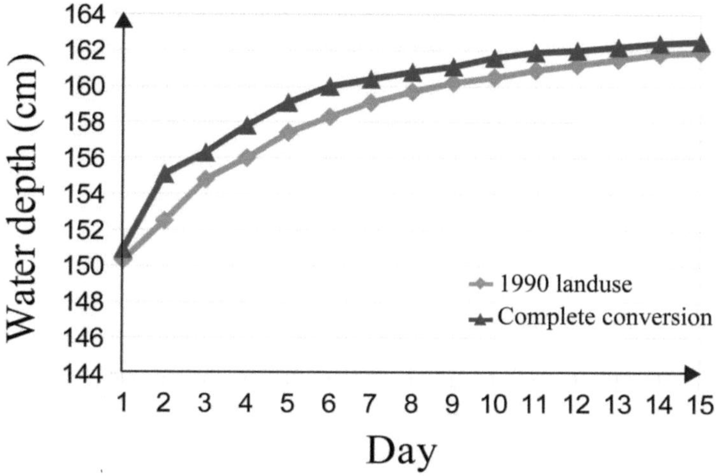

Fig. 9.5 Change of water depth under conditions of 1990 land use and complete land-use conversion

over that of the 1990 land cover, because of the complete conversion of wetland into cultivated land.

Figure 9.6 shows the comparison of inundated area according to land-use change. Total inundated area increased for complete wetland-converted land cover, relative to 1990 land cover. The figure shows the total number of inundated nodes with time, for both 1990 and the wetland-converted land cover. From day 1 through day 15, this number was greater for land cover with all wetland-converted than for 1990 land cover. The difference between node numbers increased daily, and the highest number of inundated nodes was on day 15 for complete wetland-converted land cover. This indicates that the percentage of inundated area increased with time. Removal of wetlands from the land cover increased total inundated area, and water depth in most of this area increased.

Comparing Simulation Results of 2011 and 2011 Complete Land-Cover Conversion

Figure 9.7 shows the total number of inundated nodes with time for 2011 and complete wetland-converted land cover. The total number of inundated nodes from day 1 through to 15 was greater for the complete wetland-converted land cover than for 2011 land cover. The numbers of nodes increased daily, and the highest numbers of inundated nodes were on day 15 for the complete wetland-converted land cover. This means that the percentage of inundated area increased with time. Figure 9.8 shows that the selected node had a higher water depth for complete wetland-

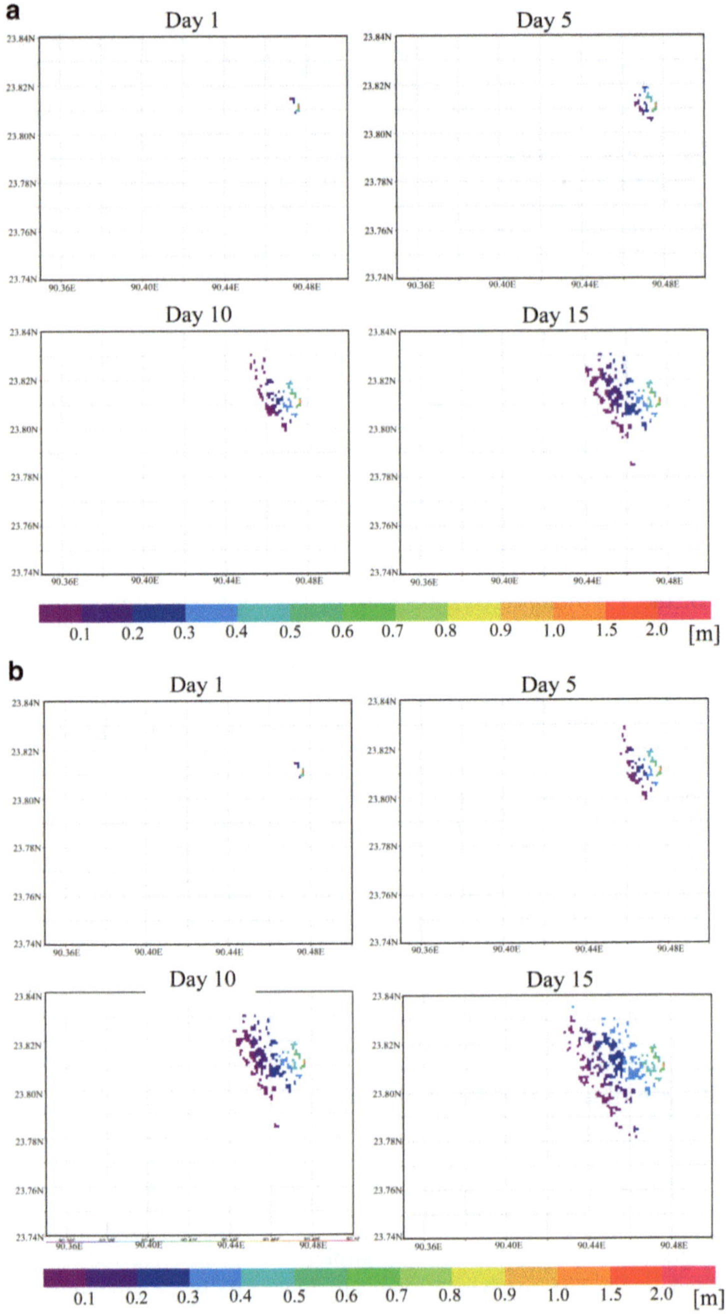

Fig. 9.6 Comparison of inundated area between conditions of: (**a**) 1990 land use and (**b**) complete land-use conversion

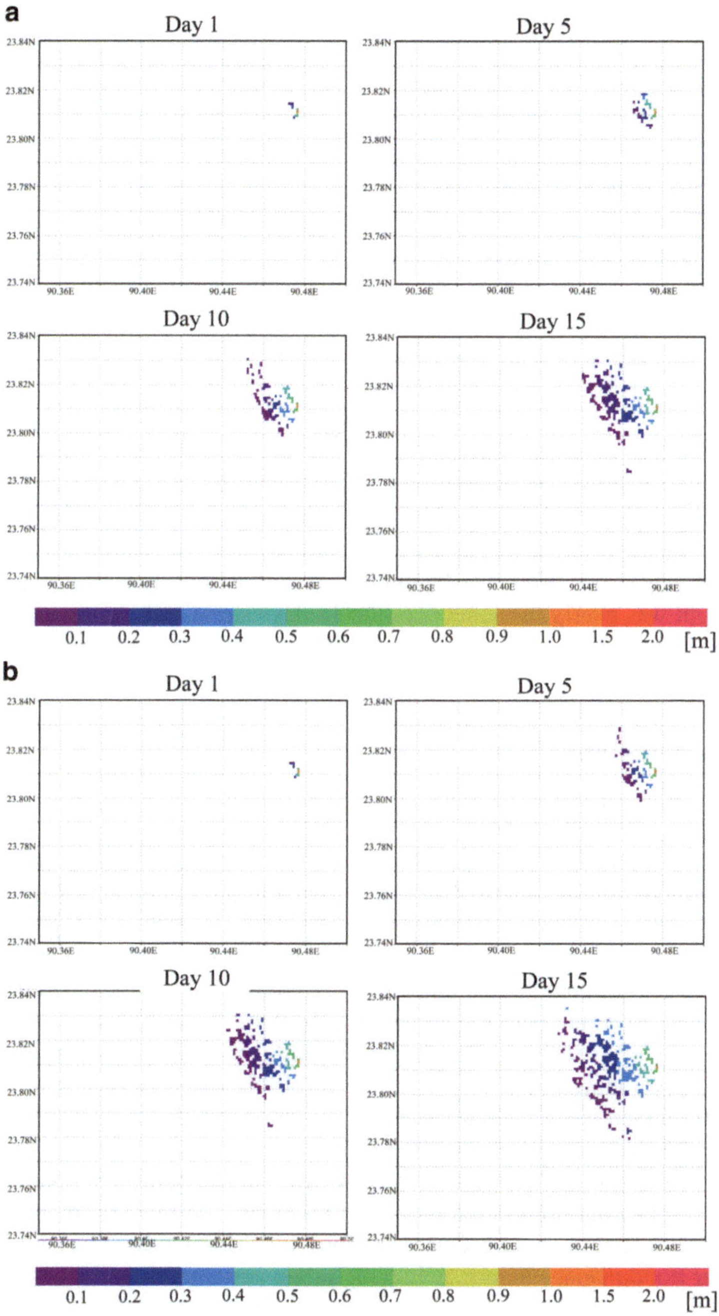

Fig. 9.7 Comparison of inundated nodes between conditions of: (**a**) 2011 land use and (**b**) complete land-use conversion

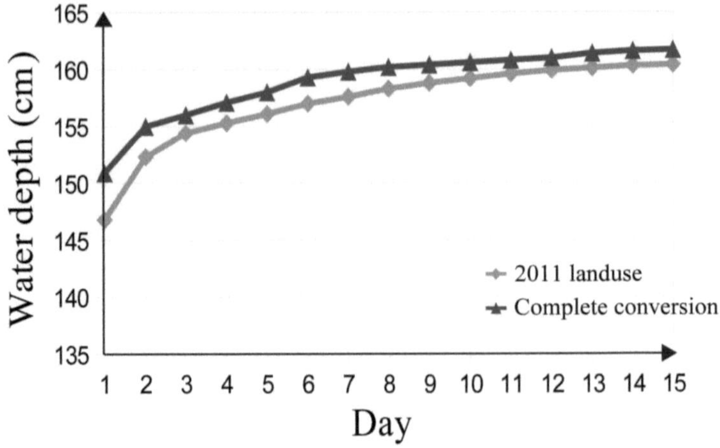

Fig. 9.8 Change of water depth between conditions of 2011 land use and complete land use conversion

converted land cover from the beginning of simulation. The rate of increase in water depth was nearly the same over the 15 days simulated.

9.5 Conclusions

Flooding in urban areas is a perennial problem for many cities in Asia. In Bangladesh, Dhaka has serious problems related to urban flooding, and one of the most important causes of this flooding is very rapid and unplanned urbanisation. The driving force of this urbanisation is economic growth and high population pressure. The environmental and socioeconomic sustainability of Dhaka, which is essential for development planning, has received relatively little attention. The city has experienced a number of devastating floods in recent times, which damaged housing and infrastructure. Flooding represents one of the major obstacles to economic development of the city. The result of this study shows a strong relationship between land use and flooding. The simulation results prove that the presence of wetland in land cover reduces flood risk, as compared with other land use. In actual flood cases in the study area, there were numerous overflow points along the Balu River. We only considered two overflow points, so there is room to improve this work with additional data. Our results provide information that is indispensable for proper flood management in achieving urban sustainability in Dhaka.

Acknowledgements We would like to express special thanks to Dr. Ashraf M. Dewan, who provided valuable hydrological and land-cover data essential to our study. We are also grateful to Dr. Takao Masumoto, who kindly provided the FEM simulation codes for numerical simulations of flood inundation and propagation.

References

Ali A (1996) Vulnerability of Bangladesh to climate change and sea level rise through tropical cyclones and storm surges. Water Air Soil Pollut 92(1–2):171–179

Brutsaert W (2005) Hydrology – an introduction. Cambridge University Press, Cambridge, p 172

Dewan AM (2013) Floods in a megacity: geospatial techniques in assessing hazards, risk and vulnerability. Springer, Dordrecht

Dewan AM, Corner RJ (2012) The impact of land use and land cover changes on land surface temperature in a rapidly urbanizing megacity. In: IGARSS 2012, Munich

Dewan AM, Yamaguchi Y (2008) Effect of land cover change on flooding: example from Greater Dhaka, Bangladesh. Int J Geoinform 4(1):11–20

Dewan AM, Yamaguchi Y (2009) Land use and land cover change in Greater Dhaka, Bangladesh: using remote sensing to promote sustainable urbanization. Appl Geogr 29(3):390–401

Dewan AM, Yeboah KK, Nishigaki M (2006a) Flood hazard delineation in Greater Dhaka, Bangladesh using integrated GIS and remote sensing approach. Geocarto Int 21(2):33–38

Dewan AM, Yeboah KK, Nishigaki M (2006b) Using synthetic aperture radar (SAR) data for mapping river water flooding in an urban landscape: a case study of Greater Dhaka, Bangladesh. Jpn J Hydrol Water Resour 19(1):44–54

Dewan AM, Islam MM, Kumamoto T, Nishigaki M (2007) Evaluating Flood hazard for land-use planning in Greater Dhaka of Bangladesh using remote sensing and GIS techniques. Water Resour Manag 21(9):1601–1612

Hai PT, Masumoto T, Shimizu K (2006) Evaluation of flood regulation role of paddies in the lower Mekong River basin using a 2D flood simulation model. Ann J Hydraul Eng JSCE 50:73–78

Kanamitsu M, Ebisuzaki W, Woollen J, Yang S-K, Hnilo JJ, Fiorino M, Potter GL (2002) NCEP-DEO AMIP-II Reanalysis (R-2). Bull Am Meteorol Soc 83:1631–1643. doi:10.1175/BAMS-83-11-1631

Kawahara M, Yokoyama T (1981) Finite element method for direct runoff flow. Proc Am Soc Civ Eng 106(HY4):519–534

Kawahara M, Takeuchi N, Yoshida T (1978) Two step explicit finite element method for tsunami wave propagation analysis. Int J Numer Methods Eng 12(2):331–351

Miah MM, Bazlee BL (1968) Some aspects of geomorphology of the Madhupur tract. Orient Geogr 12(1):39–48

Nash JE, Sutcliffe JV (1970) River flow forecasting through conceptual models 1. A discussion of principles. J Hydrol 10(3):282–290

Sugawara M, Maruyama F (1952) Statistical method of predicting the runoff from rainfall. In: Proceeding of the 2nd Japanese National Congress for Applied Mechanics, Japan National Committee for Theoretical and Applied Mechanics, Nihon Gakujtsu Kaigi, pp 213–216

Chapter 10
Flood Vulnerability and Risk Assessment with Spatial Multi-criteria Evaluation

Akiko Masuya

Abstract The objective of this chapter is to evaluate flood risk in Dhaka with geospatial techniques. Multi-temporal flood data, derived from digital elevation model and satellite imagery, were used to determine flood hazards. Census and spatial databases were used to evaluate flood vulnerability and risk zoning at a community level. The analytic hierarchy process (AHP) and weighted linear combination (WLC) methods were used to determine flood vulnerability within a geographic information system framework. The results revealed that 45 % of the study area was estimated as highly hazardous, accounting for 7 % of the total study population. Around 40 % of the communities in the study area are highly vulnerable to flood, with 8 % being extremely vulnerable. Further, more than 22 % of the population are in areas that are at high to very high risk of flood. Forty per cent of housing units are located in the high- to very high-risk zone, and around half of these were *katcha* houses, built using fragile construction materials—28 % of the communities in Dhaka were at high risk of flood.

Keywords Flood hazard • Spatial Multi-criteria Evaluation • Analytic Hierarchy Process (AHP) • Weighted Linear Combination (WLC) • Risk assessment • Community • Vulnerability assessment • GIS • Coping capacity • Modelling

10.1 Introduction

Bangladesh is situated in the low-lying Bengal Basin, which embraces one of the world's largest deltas, the Ganges–Brahmaputra–Meghna (GBM) Delta. Along with China and India, Bangladesh is one of the most flood-affected countries in

A. Masuya (✉)
Department of Spatial Sciences, Curtin University, Kent Street,
Bentley, Perth 6102, WA, Australia
e-mail: amasuya@gmail.com

A. Dewan and R. Corner (eds.), *Dhaka Megacity: Geospatial Perspectives on Urbanisation, Environment and Health*, Springer Geography,
DOI 10.1007/978-94-007-6735-5_10,

Asia (PreventionWeb 2011). Floods between 1970 and 2011 affected approximately 295 million people and killed 42,000 (EM-DAT/CRED 2012). It is estimated that every year Bangladesh loses US$175 million due to floods and related hazards, inhibiting its economic progress (Mirza 2011). Despite massive flood control work across the country since the 1960s, flood remains an inordinate threat to the people of Bangladesh, and flood damage continues to rise (Hoque et al. 2011; Rahman et al. 2005; Mahmud 2004; Haque and Zaman 1994). For instance, a moderate flood event in 2007 caused the loss of approximately 1,110 lives and affected nearly 14 million people, and the total economic losses amounted to US $1.4 billion (DMB 2007). Various predictive models indicate that floods continue to be a matter of extreme concern for many parts of the world in the context of climate change, including Bangladesh (Bouwer 2011; Mirza 2003, 2011; Hirabayashi and Kanae 2009). For example, a 23–29 % increase in flooded areas is predicted for Bangladesh if global temperatures rise by 2 °C (Mirza 2003).

Dhaka, the largest city (and capital) of Bangladesh, is the most vulnerable city to flood (Balica et al. 2012). Because of encroachment of wetlands, floodplains and natural waterways linked with rapid and uncontrolled urban development, flooding has become a regular feature during the monsoon, adversely affecting large numbers of urban inhabitants and disrupting the proper functioning of urban systems (Barua and van Ast 2011; Stalenberg and Vrijling 2009). The magnitude of loss due to floods has become massive in the recent past (Alam and Rabbani 2007; Faisal et al. 2003, 1999). At least ten major factors are believed to account for the increasing flood vulnerability of Dhaka, including low elevation, unplanned urbanisation, lack of flood preparedness, high population density and extreme poverty, poor performance of flood control works, sociopolitical structure, governance quality, impact of climate change, dilapidated drainage systems and absence of risk communication tools (Dewan 2013). In addition, flood control work based on an assumption of climatic stationarity (Milly et al. 2008) is of little help in ameliorating flood loss because existing structural measures are no longer applicable (Kundzewicz et al. 2010).

Although various adaptation measures to water-related hazards have been proposed for Dhaka (GoB 2005), the success of achieving efficient flood management depends on various non-climatic factors, such as restricting urban expansion on floodplains and water reservoir areas and introducing efficient drainage facilities (Barua and van Ast 2011; Dewan and Yamaguchi 2008). If adaptation measures are not taken, climate change and widespread ill-structured urban growth may intensify future flooding (Roy 2009), which could overwhelm the majority of urban dwellers, particularly those belonging to marginalised groups (Senga 2004).

In recent years, a paradigm shift in flood policy has been evident across the world, in that flood-risk management has become the focus rather than the traditional concept of flood protection (Meyer et al. 2009; Schanze 2006). The primary aim of risk assessment is to identify risk-related problems and select appropriate measures to manage risks associated with hazards (Smith 2001). Hence, disaster risk reduction has become an integral part of the entire risk management process and could aid in saving lives and property from catastrophic events (Bendimerad 2009). In developed countries, hazard estimation, mapping vulnerability and risk communication are used to reduce risk and vulnerability in relation to natural hazards (Showalter and Lu 2010; Maantay et al. 2010; Cutter et al. 2003; Mitchell et al. 1989; Kasperson 1986); however, structural solutions are the primary means of flood management in many parts of the world, including Bangladesh (Cook 2010; Paul 1997). In addition, flood maps that are available through public agencies such as the Bangladesh Water Development Board (BWDB) are based on subjective judgement (Jakobsen et al. 2005). They are medium to small scale and only used for administrative and forecasting purposes (Jha et al. 2012). Historical hazard and risk maps are lacking, hindering efforts to estimate the likely patterns of risk in relation to flood (Huq 1999).

Geographic Information Systems (GIS) and remote sensing (RS) techniques (collectively termed as geospatial techniques) are increasingly being used to assess vulnerability and risk of natural hazards, including floods. Apart from mapping and monitoring of floods using RS data, flood-risk assessment by integrating geographic information with RS has long been the subject of academic research (Dewan 2013; Gillespie et al. 2007; Sanyal and Lu 2004). Many researchers have used biophysical and socioeconomic parameters to assess flood vulnerability around the world (Suriya and Mudgal 2012; Kienberger 2012; Kazmierczak and Cavan 2011; Wang et al. 2011; Bizimana and Schilling 2010; Pandey et al. 2010; Fekete et al. 2010; Pavri 2010; Meyer et al. 2009; Zheng et al. 2009; Sinha et al. 2008; Azar and Rain 2007; Sanyal and Lu 2006; Tingsanchali and Karim 2005; Yalcin and Akyurek 2004; van der Sande et al. 2003; Wu et al. 2002; Cutter et al. 2000; Islam and Sado 2000a, b). These studies have demonstrated that quantitative analysis of flood hazards using geospatial techniques could assist greatly in the explicit understanding of complex natural hazards in spatiotemporal contexts. This is vital for disaster and emergency management, particularly the management of flood risk (Meyer et al. 2009).

The application of geospatial techniques to flood-risk assessment for Dhaka was first examined by Dewan et al. (2006a, b, 2007a, b). Their study looked at the physical aspects of flood hazards and did not incorporate the human dimensions involved. Moreover, these studies concentrated on only one flood event (1998), whereas effective management of recurrent flood disasters requires the study of multiple events (Hoque et al. 2011). As noted by Cutter et al. (2009) and Rashed and Weeks (2003), the combined use of social and physical (or biophysical) variables could allow rigorous assessment of vulnerability to natural hazards, as they are intrinsically linked. Very recently, Dewan (2013) assessed flood vulnerability in Dhaka using spatial multi-criteria evaluation (MCE) techniques. Although the study used a series of flood maps from multi-temporal RS with a digital elevation model (DEM) from 1988 to 2009, socioeconomic data in the study was from the population census of 2001. Moreover, this study incorporated a built environment indicator as a social component. As suggested by Cutter et al. (2008) in their recent disaster model, separating the built environment factors from the social factors may help explain the degree of influence of urbanisation on overall vulnerability of an area. Since Dhaka has experienced 30 % growth in population between two decennial censuses (2001 and 2011) (BBS 2012), the current study asserts that the spatial patterns of flood vulnerability and risk could have substantially changed during this time, especially considering that the lack of residential space consistently compels a large number of people to live in hazardous places. A new study is therefore needed to identify current spatial patterns of flood vulnerability and the distribution of population at risk.

The objective of this chapter is to assess flood vulnerability at a community level using geospatial techniques. Specifically, this chapter aims to evaluate the flood vulnerability of communities by integrating physical, social, economic and environmental factors through spatial MCE techniques. The lowest-level census tract available, known as a 'community', is used as the spatial unit of study.

10.2 Conceptualising Hazard, Risk and Vulnerability

There have been three different histories of research into the theoretical construct of vulnerability and risk estimation (Lankao and Qin 2011). The first of these stems from the risk and hazard paradigm, which is based on the human–nature interaction (Cutter et al. 2009; Burton et al. 1993). The second emphasises the social dimension of hazards and is rooted in the notion of political economy and political ecology (Wisner et al. 2004; Pelling 2003). The third uses the concept of resilience science to comprehend societal vulnerability to global and climatic change (Eakin and Luers 2006). These three paradigms have resulted in a number of models analysing hazards-related risk and vulnerability (Birkmann 2006). Notable models include the pressure and release model (PAR) (Wisner et al. 2004), hazards-of-place (HOP) model (Cutter 1996), regions of risk model (Hewitt 1997), the disaster resilience of place model (Cutter et al. 2008), the UNISDR framework for disaster risk reduction (UNISDR 2004) and the urban vulnerability framework (Dewan 2013). Despite these efforts, there is considerable disparity in researchers' understanding of the key terms, which is believed to originate from different epistemological orientations and methodological practices (Cutter 1996). The following sections present some of the models in which this study is based on.

10.2.1 Pressure and Release (PAR) Model

One of the best-known models is the PAR model, which defines vulnerability as 'the characteristics of a person or group and their situation that influence their capacity to anticipate, cope with, resist and recover from the impact of a natural hazard'(Wisner et al. 2004, p. 11). This approach sees disaster risk as the product of hazard and vulnerability. This theoretical framework serves to analyse exposure to a natural hazard and vulnerability—focusing on the process and causality of human vulnerability—and to explain how a naturally occurring phenomenon turns into a disaster. Vulnerability is generated by three elements: root causes, dynamic pressures and unsafe conditions (Wisner et al. 2004).

10.2.2 Hazards-of-Place (HOP) Model

Cutter (1996) posits the 'hazards-of-place' model, which integrates biophysical and social characteristics relating to the causal process of hazards. The basic premise of this model is that vulnerability is related to interacting elements, such as location and the people living there, and the interaction and intersection of social and biophysical elements. It defines vulnerability as the combination of social constructs (social vulnerability) and a biophysical condition (potential exposure) in a spatial context. Social and biophysical conditions equally influence the vulnerability of a place and should be examined separately. Risk is defined as the likelihood of the occurrence of a particular hazard (e.g. a 100-year flood) and is seen as the element of 'place vulnerability'.

10.2.3 Disaster Resilience of Place (DROP) Model

An extended version of the HOP model was recently proposed by Cutter et al. (2008), who termed it as DROP (disaster resilience of place). Although the DROP model was originally developed for the study of natural hazards, it is equally applicable to global change research. This model emphasises the relationship between vulnerability and resilience at the community level. It conceives vulnerability and resilience as opposing components with some overlapping features and exists in an interconnection of natural, social and built environment systems. The presence or absence of coping responses in the community can amplify or attenuate the impact of the hazard. In this case, coping responses are related to emergency actions taken during and immediately after the impact, and resilience is related to actions taken before the event (preparedness) or after the event (mitigation).

10.2.4 Urban Vulnerability Framework

Dewan (2013) proposed an urban vulnerability framework, which is primarily based on the HOP model, but also borrowed elements from others. This model incorporates a coping capacity indicator in the estimation of vulnerability to natural hazard. This conceptual framework sees hazard as a pre-existing condition that can lead to devastation depending on the influence of external and internal forces and that is able to overwhelm people, property and the environment. Vulnerability is seen as the interplay between the social and physical aspects of a particular system (such as an urban environment) and is location specific. Coping capacity is a form of adaptive capacity and is perceived as the capability of a community to withstand or recover from disasters. Contrary to some conceptualisations, the model regards overall or total vulnerability as depending on the physical and social attributes, as well as the coping capacity, of a given community. The model stresses that risk is the product of hazard and vulnerability. To lessen the effects of natural hazards, steps should be taken to reduce vulnerability.

10.3 Model Applicability

The PAR, HOP and DROP models have been tested in different environments to investigate societal risk and vulnerability to natural hazards (see Cutter et al. 2009). Although the results of these studies are laudable, data scarcity, particularly in developing countries, may be a serious impediment to the effective use of either of the models (Dewan 2013; Wang et al. 2011; Kiunsi et al. 2006). For example, the PAR model is valuable for descriptive analysis rather than empirical testing, whilst the HOP approach allows the concept to be operationalised through geospatial techniques (Cutter et al. 2009).

On the basis of the models described above, this study conceptualises vulnerability as the combination of exposure, susceptibility to hazards and coping capacity of a community, as determined by social, economic, institutional structure and physical environmental settings. Table 10.1 shows definitions of the key terms used in this study.

10.4 Flooding in Dhaka

The study area described in this chapter encompasses part of the Dhaka Metropolitan Development Plan (DMDP) area and the megacity of Dhaka (see Chap. 1), which includes the historical city core and adjoining areas. A total of 1,212 communities are located in the study area (see Chap. 3). The following section briefly describes recent flooding in Dhaka.

Similar to other parts of Bangladesh, surface water flooding by the spilling of river's water beyond its defined channels is a frequently occurring phenomenon in

Table 10.1 Definitions of the key terminology used in this study

Term	Definition
Hazard	Natural hazards (e.g. flood, earthquake, cyclone) that potentially cause damage to the community or population and property
Vulnerability	The degree of susceptibility to hazards and coping capacity of a community
Coping capacity	The capacity of a community to respond and to withstand the effect of hazards. This encompasses the ability to learn from previous experience of hazards and adapt to adverse changes
Risk	The probability of harmful consequences resulting from natural hazards in a particular area over a specified period of time; the product of hazard and vulnerability

Dhaka. Although this fluvial flooding is the most costly and debilitating natural hazard and remains one of the most serious threats to people and property, pluvial flooding is becoming a grave concern for the inhabitants of Dhaka (Stalenberg and Vrijling 2009; Alam and Rabbani 2007; Tawhid 2004).

Contemporary flood history shows that since 1954, the lowlands of Dhaka have regularly been inundated during monsoons; however, in many instances, the extent of flooding depends both on the amount of precipitation and on the flow from upstream (Faisal et al. 2003). Disastrous floods in 1987 and 1988 inundated areas of 164 and 200 km^2, respectively (FAP 8A 1991). The unprecedented flood of 1998 also severely affected Dhaka and its neighbouring areas (Dewan et al. 2006a, b; DMB 1998), resulting in unusual damage and great suffering to the people. The 1988 flood affected 4.55 million people in total (Hye 2000). These areas were again inundated in 2004 and 2007, however, to a lesser extent and with less destruction than in the 1998 flood. Even though the embankment acted as a buffer that saved lives and property in the 2004 and 2007 floods, pluvial flooding by means of localised ponding and water-logging was abundant, particularly in the embanked part, and resulted in a number of environmental problems, including scarcity of drinking water, disruptions to economic activities and increased prevalence of waterborne diseases (Bala et al. 2009; Alam and Rabbani 2007; Rahman et al. 2005). Details of flood history and damage can be found elsewhere (Dewan 2013; Siddiqui and Hossain 2004; Nishat et al. 2000; Ali et al. 2002).

10.5 Data Acquisition and Preparation

The data used in this study were collected from various sources. Flood maps encompassing the four major flood events of 1988, 1998, 2004 and 2007 were obtained from Landsat Thematic Mapper (TM) and Radarsat Synthetic Aperture Radar (SAR) images (Dewan 2013). Data on land use and cover in 2011 were obtained from a cloud-free Landsat-5 TM image (Dewan and Corner 2012). A DEM was obtained from the Survey of Bangladesh (SOB). River network and geology data were acquired from the Center for Environment and Geographic Information Systems (CEGIS), and road networks and location of buildings data were collected from the Detailed Area Plan (DAP) of the Capital Development Authority (RAJUK).

This study used the lowest-level census tracts, called community, obtained from various sources with a number of field visits (see Chap. 3). The community feature dataset was used to encode demographic and socioeconomic data from the 2011 population and housing census. The census data were in tabular format, and a unique ID was used to aggregate data for each community. The poverty data based on the 2005 Bangladesh Household Income and Expenditure Survey (BBS et al. 2005) was available at a higher administrative unit; hence, areal interpolation was performed to calculate two types of poverty rates, an upper and a lower poverty rate, aggregated to each community.

10.6 Methods of Analysis

10.6.1 Analysis of Flood Hazard

Several hydrological parameters influence the flood hazard of a particular site. These include depth of flooding, rate of water-level rise, flood frequency, water velocity, physical exposure of land and sediment loads (WMO 1999). This study used two parameters to estimate flood hazard in the study area for four major flood events of 1988, 1998, 2004 and 2007: flood-affected frequency and floodwater depth.

The concept of flood-affected frequency was developed by Islam and Sado (2000a) to produce a flood hazard map of Bangladesh on the basis of the 1988 flood. In this study, this concept is borrowed and used to estimate flood-affected frequency by superimposing multi-temporal flood data from four flood events (representing flooded and non-flooded areas). A flood-affected frequency map was prepared by reclassifying the frequency of flooded cells in each location into four classes (high, medium, low and no flooding).

To estimate floodwater depth, the highest water levels (HWL) of five catchments were used, as well as the DEM. Heights of water for each flood event were calculated by subtracting the elevations from the HWL and superimposed with the flood map of the corresponding year to extract the water depths of the flooded zones. The extracted water depths were classified into four depth classes: no water, shallow, medium and deep. The four classified floodwater depth maps were then superimposed to reclassify the depths at each location. Details of the reclassification of the superimposed flood-affected frequency and floodwater depth maps can be found elsewhere (Dewan 2013; Islam and Sado 2000a, b).

The flood hazard map was generated by assigning a hazard rank to each cell in a raster dataset according to the combinations of the classes on the flood-affected frequency and floodwater depth maps (Table 10.2). Each cell classified as 'no hazard' on the flood-affected frequency map is assigned to the 'very low' category on the hazard map. Cells classified as 'shallow' on the floodwater depth map and as 'low', 'medium' or 'high' on the flood-affected frequency map are assigned to the 'low' category on the hazard map. Cells classified as 'medium' or 'deep' on the floodwater depth and as 'medium' or 'high' on the flood-affected frequency map are assigned to the 'high' category on the hazard map (Table 10.2). The ranks of 1, 2, 3 and 4 in the table represent very low, low, medium and high, respectively.

10.6.2 Vulnerability Analysis

The estimation of vulnerability to flood in the study area was carried out using four factors—locational, socioeconomic, built environment and coping capacity—analysed separately to quantify flood vulnerability of the communities. These

Table 10.2 Combinations of classes from maps of flood-affected frequency and floodwater depth, with assigned hazard ranks

	Flood frequency	Flood depth	Rank
0	No hazard	No water	1
1	No hazard	Shallow	1
2	No hazard	Medium	1
3	No hazard	Deep	1
4	Low	No water	2
5	Low	Shallow	2
6	Low	Medium	3
7	Low	Deep	3
8	Medium	No water	2
9	Medium	Shallow	2
10	Medium	Medium	3
11	Medium	Deep	4
12	High	No water	2
13	High	Shallow	3
14	High	Medium	4
15	High	Deep	4

factors included a total of 16 variables. The spatial MCE techniques used for this purpose were analytic hierarchy process (AHP) and weighted linear combination (WLC).

AHP and WLC are techniques that decompose a complex problem into small criteria components and then prioritise and aggregate the criteria to compare or choose between alternatives. AHP is widely applied for environmental impact assessment and draws on the knowledge and opinions of different groups of people (stakeholders and experts) to determine weights for each criterion (Ramanathan 2001). The technique includes structuring a decision hierarchy model with criteria and pairwise comparison, including the evaluation of comparison consistency.

In a hierarchical decision model, criteria are placed immediately after an objective, which can be broken down further into sub-criteria to be placed at the lower levels of the decision hierarchy. Each pair of criteria or sub-criteria in a hierarchical level is compared in terms of relative importance using scale that ranges from 1 (both criteria equally important) to 9 (one criterion relatively extremely important) (Saaty 2000), and a comparison matrix is constructed with these points. The weight of each criterion is calculated by normalising the eigenvector corresponding to the largest eigenvalue of the matrix.

For this study, each indicator was divided into four or five categories. The pairwise comparisons were conducted first within the classes and then among the indicators, in terms of vulnerability to flood, with indicators or classes that increased vulnerability receiving higher points.

The consistency of the comparisons was evaluated by calculating a consistency ratio (CR). If the CR is equal to or less than 0.1, the comparisons are considered to be consistent. CR is defined by the following equation:

$$\mathrm{CR} = \text{Consistency Index}/\text{Random Index} \tag{10.1}$$

The random index (RI) is the index of randomly generated reciprocal matrix from the 9-point scale (Ramanathan 2001; Saaty 1980, 2000). The consistency index (CI) is defined as

$$\mathrm{CI} = (\lambda_{\max} - n)/(n - 1) \tag{10.2}$$

where $\lambda_{\max}$ is the largest eigenvalue derived from the comparison matrix and n is the number of criteria.

WLC is a variant of spatial multi-criteria techniques and is frequently used in GIS to integrate criterion/indicator data layers (Malczewski 2000). It involves the standardisation of criteria/indicator layers, assignment of relative weights to the layers and aggregation of the weighted layers to obtain a single value for each cell or feature. For this study, standardisation was performed by linear scale transformation.

Equation 10.3 was used if there was a greater contribution to the vulnerability of the place or community when an indicator had a higher value. For example, a higher population density makes the place or community more vulnerable to flood than a lower population density. On the other hand, if a low value of an indicator makes a greater contribution to making the place or community vulnerable (e.g. elevation), then Eq. 10.4 was used:

$$s = \frac{x - \min}{\max - \min} \tag{10.3}$$

$$s = \frac{\max - x}{\max - \min} \tag{10.4}$$

where s is the standardised score and min and max represent the minimum value and maximum value of the indicator variable, respectively. x is an indicator value of a cell or community feature. To derive the vulnerability index (VI), the following equation was used:

$$\mathrm{VI} = \sum_{i=1}^{n}\sum_{j=1}^{m} w_i w_{ij} x \tag{10.5}$$

where w_i and w_{ij} are the weight for the ith indicator and the weight for the jth class of the ith indicator, respectively. x represents the value of an indicator.

Analysis of Locational Vulnerability

Three variables were used to assess locational vulnerability of a community to flood: elevation, distance to rivers and geology. A Euclidean distance function was used to calculate distance to water bodies using the river networks data. The study area was divided into four geological types, and all types were included in the analysis.

Table 10.3 Decision hierarchy model and corresponding weights for physical vulnerability criteria

	2		3	
1	Criteria (indicator)	Weight	Criteria (class)	Weight
Locational vulnerability	Distance to rivers (km)	0.145	<2	0.432
			2–3	0.294
			3–4	0.165
			4–5	0.070
			>5	0.038
	Elevation (m)	0.483	<4	0.482
			4–5.5	0.299
			5.5–7	0.122
			7–9	0.064
			>9	0.033
	Geology	0.056	Alluvial silt and clay	0.474
			Alluvial silt	0.320
			Marsh	0.134
			Madhupur	0.072

The intensity of importance in the pairwise comparisons and the class thresholds of distance to river and elevation were determined based on the relationship of these variables with the spatial distribution of the flood hazard. The highly hazardous areas are mostly located within 2 km of the rivers and less than 4 m above the mean sea level; therefore, these areas were considered the most vulnerable. The hazard level gradually decreases as the distance and elevation increase to around 5 km and 7 m, respectively. Locations with alluvial silt and/or clay are highly vulnerable since these areas are usually flood prone. Although areas with marsh clay and peat are also flood prone, the majority of these areas are wetlands; hence, they were considered less vulnerable to flood than alluvial silt and/or clay (Table 10.3).

Analysis of Socioeconomic Vulnerability

Socioeconomic vulnerability was evaluated using demographic characteristics and the proportion of the population living below the poverty line in each community. Gender ratios and percentages of the populations in four vulnerable age groups (0–4, 5–9, 10–14 years and over 60 years) were obtained from the 2011 census dataset. Poverty was represented by the proportion of the population living below the lower level of poverty.[1]

The intensity of importance for the pairwise comparison was determined based on literature (Hossain 2008; DMB 2007; Islam and Sultana 2006; Rahman et al. 2005).

[1] Lower poverty line is defined as those households whose total expenditure on food and nonfood combined is equal to or less than the food poverty line, that is, intake of <2,122 kcal per person/day (BBS et al. 2005).

Table 10.4 Decision hierarchy model and corresponding weights for socioeconomic vulnerability criteria

	2		3	
1	Criteria (indicator)	Weight	Criteria (class)	Weight
Socioeconomic vulnerability	Age	0.240	0–4	0.568
			5–9	0.244
			10–14	0.050
			>60	0.139
	Gender	0.137	Male	0.250
			Female	0.750
	Poverty (%)	0.623	<2.7	0.053
			2.7–5.4	0.112
			5.4–12	0.253
			>12	0.582

In the 2007 floods in Bangladesh (DMB 2007), children were the worst affected; therefore, infants and young children were considered the most vulnerable because of their mobility and lack of resilience. Women in Bangladesh are also highly vulnerable compared to men because they have fewer opportunities to participate in educational and social activities (Islam and Sultana 2006). The poverty rate was accorded the highest importance points because an increase in the number of urban poor may lead to higher vulnerability (Rahman et al. 2005; Hossain 2008). The calculated weights for each indicator and its categories are shown in Table 10.4.

Analysis of Built Environmental Vulnerability

In Dhaka, rapid urbanisation with poor flood-risk management increases vulnerability by exposing more residential or commercial buildings, infrastructures and infrastructure lifelines to flood (Dewan 2013). Indicators used to estimate built environmental vulnerability included land use, housing type, water supply and sanitation, road density and population density at the community level.

As with socioeconomic vulnerability, the intensity of importance in the pairwise comparison was also based on literature (UN-HABITAT 2010; Rabbani 2009; Hossain 2008; Rahman et al. 2002). Considering the impact of flood on infrastructure and humans and economic losses in agriculture, the built-up areas were accorded the highest importance points, followed by rural settlements and cultivated areas. Residential buildings in South Asia, including Bangladesh, may be divided into three types: *katcha, semi-pucca* and *pucca. Katcha* houses are built with fragile materials such as mud, wood and tins (UN-HABITAT 2010; Hossain 2008); therefore, this type of house was considered most vulnerable, whereas the *pucca* houses, which are constructed with solid material such as bricks, were considered the least vulnerable. The highest importance points were assigned to water supplies other than household access to tap and tube-well facilities.

Table 10.5 Decision hierarchy model and corresponding weights for built environmental vulnerability criteria

	2		3	
1	Criteria (indicator)	Weight	Criteria (class)	Weight
Built environmental vulnerability	Housing type	0.183	*Katcha*	0.689
			Pucca	0.067
			Semi-pucca	0.244
	Land use	0.311	Built-up	0.454
			Settlements	0.275
			Cultivated	0.169
			Others	0.050
	Road density (per km^2)	0.033	<10	0.042
			10–25	0.075
			25–50	0.134
			50–100	0.264
			>100	0.485
	Water supply	0.078	Tap	0.067
			Tube well	0.244
			Other supply	0.689
	Sanitation	0.084	Sanitation	0.078
			Others	0.234
			No sanitation	0.688
	Population density (per km^2)	0.311	<2,000	0.034
			2,000–5,000	0.067
			5,000–8,000	0.133
			8,000–10,000	0.245
			>10,000	0.521

Households with no sanitation were given the highest points in the sanitation criteria, as unhygienic practices and unsafe water sources can spread waterborne diseases in the study area during floods (Rabbani 2009; Rahman et al. 2002). Communities presenting higher population density and road density were also assigned higher points (Table 10.5).

Analysis of Coping Capacity

The indicators selected for evaluating the coping capacity of each community included literacy rate, hospitals per capita, flood shelters per capita and flood awareness. The mean flood frequency for each community, based on the four flood events of 1988–2007, served as a surrogate for flood awareness. This was derived by superimposing the flood maps of flooded and non-flooded areas and intersecting the superimposed map with the community boundaries.

In the pairwise comparison, higher importance points were assigned to the classes with higher values or ranges, so that the communities with better coping capacity were emphasised. Of all the coping capacity indicators, the highest

Table 10.6 Decision hierarchy model and corresponding weights for coping capacity criteria

	2		3	
1	Criteria (indicator)	Weight	Criteria (class)	Weight
Coping capacity index	Literacy rate (%)	0.303	<41.7	0.040
			41.7–56.2	0.074
			56.2–68.1	0.142
			68.1–80.0	0.248
			>80.0	0.496
	Number of hospitals (per 1,000 people)	0.074	<1	0.060
			1–3	0.093
			3–7	0.256
			>7	0.591
	Number of shelters (per 1,000 people)	0.068	<3	0.038
			3–8	0.057
			8–13	0.134
			13–20	0.269
			>20	0.502
	Flood awareness (frequency)	0.554	<3	0.062
			3–5	0.108
			5–7	0.267
			>7	0.563

importance was accorded to flood awareness, because this is the most important factor contributing to preparedness against natural hazard (UN/ISDR 2004). Literacy rates were accorded the second highest importance, because literacy helps people acquire the ability to understand warnings and recovery information and to make the right decisions about evacuation (Cutter et al. 2003) (Table 10.6).

Analysis of Total Vulnerability

The total vulnerability index (Dewan 2013) was calculated by performing WLC with equal weights on the indices derived from the four vulnerability analyses described above. Prior to performing the WLC operation, the cell-based locational vulnerability index was intersected with the census tract boundaries to obtain mean vulnerability scores for each community, and the coping capacity index was linearly inverted, so that the lower the coping capacity, the higher the contribution to the increase in flood vulnerability for a particular community.

10.6.3 Flood-Risk Analysis

As in the estimation of the total vulnerability index, the cell-based flood hazard map was intersected with the census tract boundaries, and a mean hazard rank for each

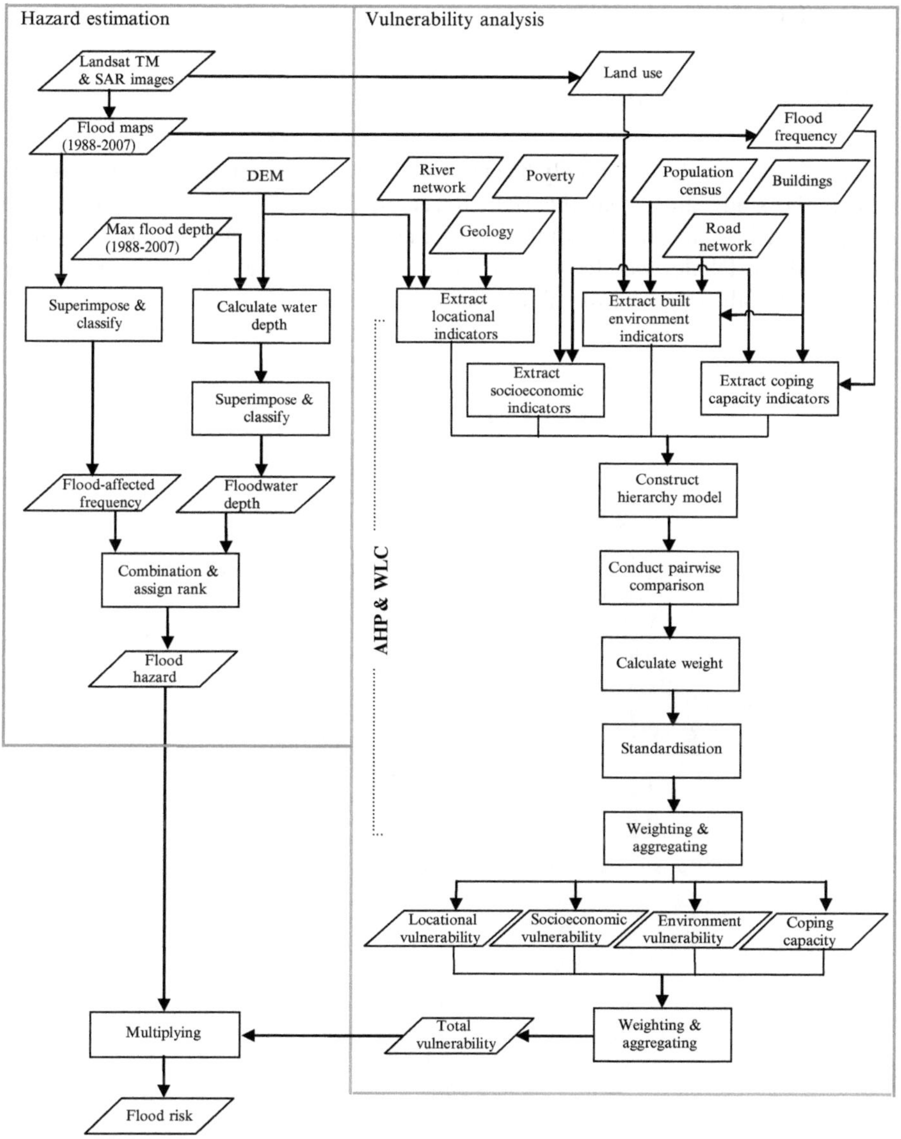

Fig. 10.1 Flood risk and vulnerability analysis model

community was derived. The final risk index was calculated with the following equation:

$$\text{Flood risk} = \text{flood hazard} \times \text{total vulnerability} \tag{10.6}$$

The resulting risk index for the communities was categorised into five classes (very low, low, moderate, high and very high) using the natural breaks algorithm. A flow chart of the overall methodology is shown in Fig. 10.1.

10.7 Results

10.7.1 Flood Hazard

Figure 10.2 shows that the highly hazardous zones are located mainly in the eastern part of the study area, near the Balu River, in the mid- to northern part of the Buriganga River and in the southwest of the study area, between the Buriganga River and the Dhaleswari River. The medium- and low-hazard zones are mostly located along the edges of the highly hazardous zone.

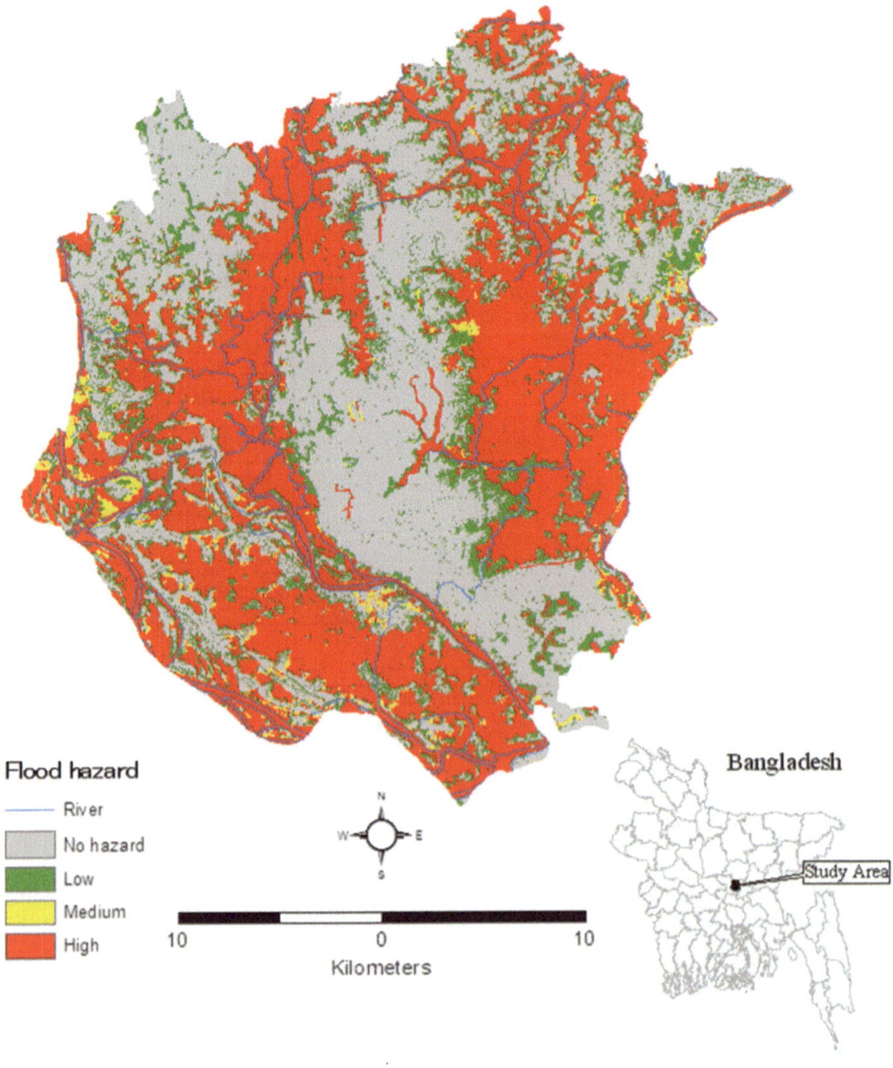

Fig. 10.2 Spatial distribution of flood hazard

Table 10.7 Area and percentage of population in each flood hazard zone

	Area		
Hazard zone	Hectare (ha)	Percentage (%)	Population (%)
Very low	32,252.49	36.75	75.5
Low	13,812.93	15.74	15.3
Medium	2,426.49	2.76	1.9
High	39,275.73	44.75	7.3

More than 60 % of the study area is affected by floods (low to high zones), and approximately 45 % is estimated to be highly hazardous. The medium- and low-hazard zones comprised approximately 3–16 % of the study area (Table 10.7).

It was found that 24.5 % of the total population were exposed to flood hazards. Most of the exposed population were located in the low- to high-hazard zone; however, it is worth noting that over 7 % of the population were located within the highly hazardous flood zone (Table 10.7).

10.7.2 Flood Vulnerability

Figure 10.3 shows the spatial distribution of flood vulnerability in the study area in terms of location, socioeconomics, built environment and coping capacity. Higher levels of locational and socioeconomic vulnerability and lower coping capacity can be found mostly in the outskirts of the city, whereas higher vulnerability of the built environment is concentrated in the central part of the study area. The south to southwestern area and the northeastern fringe between the Balu River and the Lakshya River are particularly vulnerable because of their higher degree of socio-economic vulnerability, lower coping capacity and medium to very high levels of locational and environmental vulnerability (Fig. 10.3).

Another noticeable pattern of vulnerability occurs in the southeastern part along the Balu River and the northern part of the Buriganga River. As these areas are inundated almost every year, they may have higher coping capacity. However, when locational and social vulnerability are combined, they tend to result in high vulnerability to flood (Fig. 10.3).

As indicated in the three vulnerability maps and the coping capacity map, the communities rated as having higher levels of total vulnerability are largely concentrated on the west-to-south and northeastern fringes of the study area. The lower levels of vulnerability can be found mostly in the centre and in the north-western fringe areas (Fig. 10.4).

To investigate underlying causes of these vulnerability patterns, the total vulnerability was intersected with several possible causative variables. It was found that 40 % of the communities were highly vulnerable to floods, whilst 8 % had a very high level of vulnerability, despite their low population density. About half of

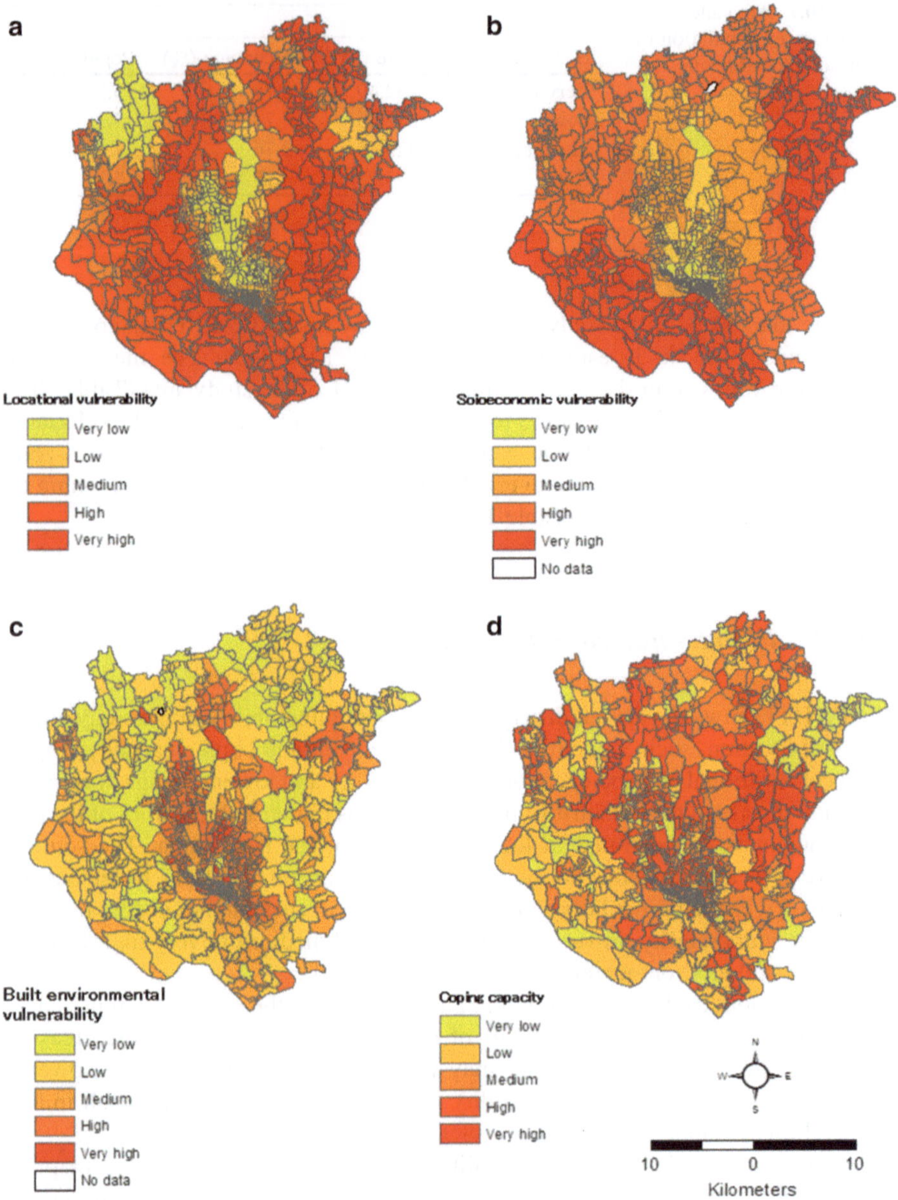

Fig. 10.3 Spatial patterns of vulnerability: (**a**) locational vulnerability, (**b**) socioeconomic vulnerability, (**c**) built environmental vulnerability, and (**d**) coping capacity, by community

the most vulnerable age groups were at high to very high levels of vulnerability. At the very high level of vulnerability, more than 18 % of the population were below the poverty line, and only around half of the population were literate. Housing data revealed that 53 % of houses in this very high-vulnerability zone are *katcha* type,

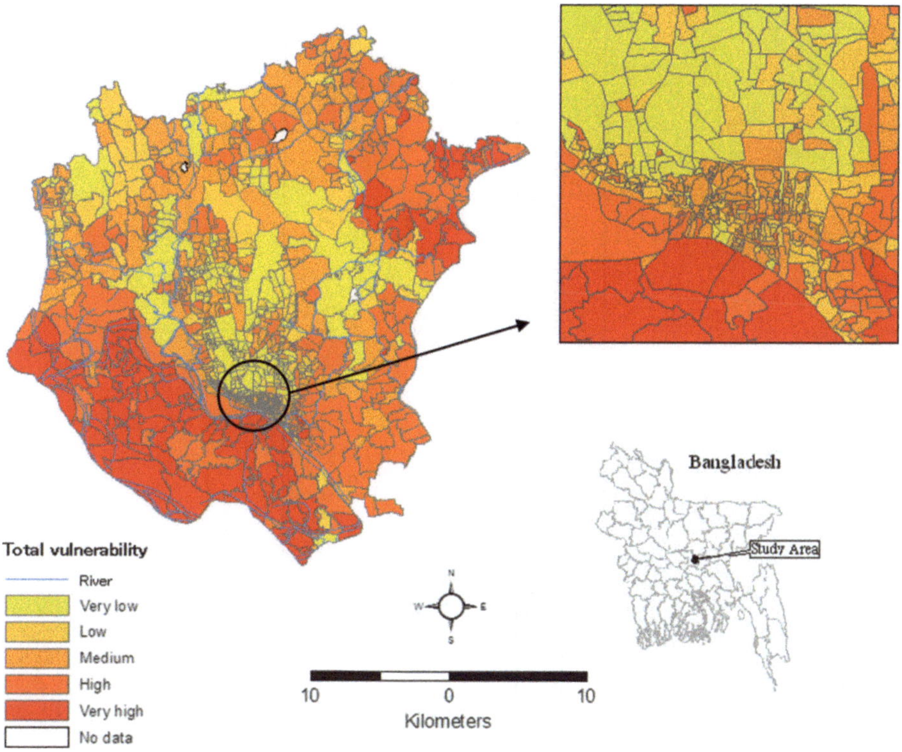

Fig. 10.4 Spatial patterns of total vulnerability, by community

and more than 20 % of households do not have access to a sanitation system. These settlements are located at an elevation of less than 4 m above the mean sea level, and their proximity to major river networks is less than 1 km (Table 10.8).

10.7.3 Risk of Flood

Figure 10.5 shows the spatial distribution of flood risk, derived by multiplying flood hazard with total vulnerability. It shows that high- to very high-flood-risk zones were located mostly in the west to southwest and the northeast through to south-eastern peri-urban areas. Most of the communities at very high risk can be identified in highly hazardous zones on the flood hazard map and in the high level of vulnerability on the total vulnerability map. The central and northern parts of the city are in the low-flood-risk zone, as shown in Fig. 10.5.

Table 10.8 Distribution of different parameters in relation to total vulnerability

Vulnerability level	Population density (km^2)	Community (population) (%)	Dependent population (%)	Poverty rate[a] (%)	Literacy rate[a] (%)	*Katcha* house[a] (%)	Unsafe water supply[a] (%)	No sanitation[a] (%)	Elevation[a] (m)	Distance to river[a] (km)
Very low (<0.43)	3,455	2.4 (1.8)	1.7	5.3	73.6	22.1	2.8	8.3	3.5	0.9
Low (0.43–0.53)	22,081	24.1 (21.4)	19.1	3.4	83.8	6.3	1.1	3.2	7.1	1.5
Medium (0.53–0.63)	17,414	32.4 (31.2)	29.9	4.1	72.8	10.5	2.4	6.4	6.0	1.1
High (0.63–0.76)	12,245	33.1 (40.0)	42.5	8.1	60.2	32.1	3.8	16.6	4.7	0.8
Very high (>0.76)	3,783	8.1 (5.6)	6.8	18.3	53.1	52.7	4.1	21.6	3.8	0.8

[a] Mean value at each level

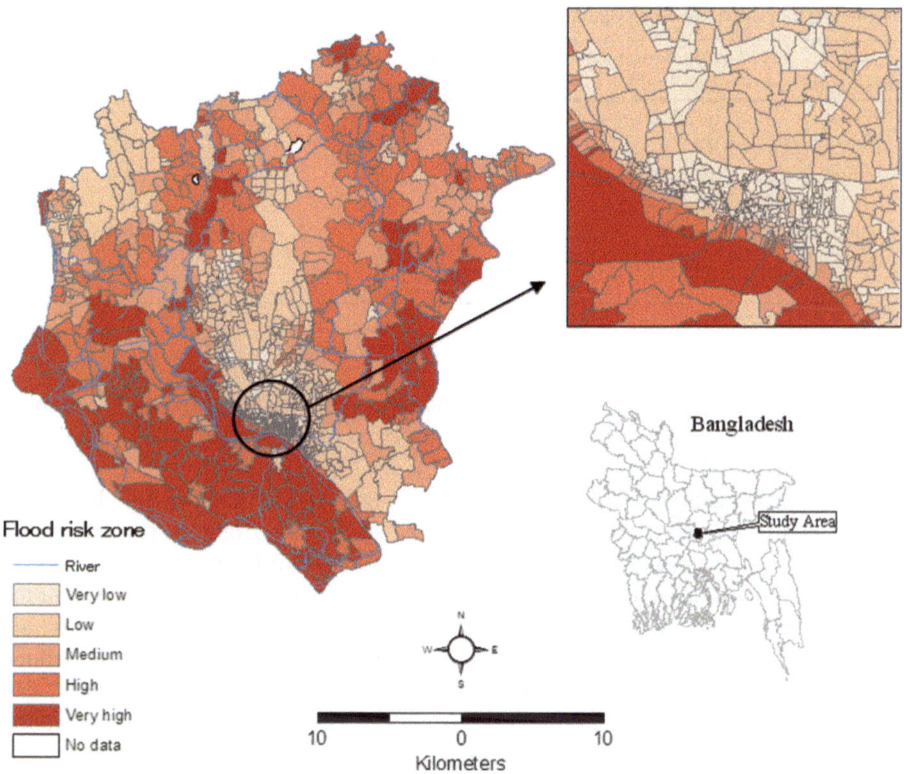

Fig. 10.5 Spatial patterns of flood risk, by community

The flood-risk map was also intersected with population and housing data to estimate the number of people and housing units at risk of floods. About 30 % of the communities, or 22 % of total population, are in high- to very high-risk zones. Approximately 40 % of housing units are in high- to very high-risk zones, with higher proportions of *katcha* houses (Table 10.9).

10.8 Conclusions

This study assessed flood vulnerability and associated flood risk at the community level in Dhaka. Geospatial techniques were used to carry out analyses of flood hazard, vulnerability and flood risk. A ranking matrix was used to estimate flood hazard by taking combinations of the classes from maps of flood-affected frequency and floodwater depth. Spatial multi-criteria evaluation (MCE) techniques were used to evaluate flood vulnerability of the communities, with 16 indicators from

Table 10.9 Distribution of population and other variables in different flood risk zones

Risk zone	Population density (per km^2)	Community (%)	Population (%)	Housing unit (%)	Katcha housing unit (%)
No risk	71,056	24.9	15.5	20.2	1.5
Low (<0.11)	27,701	29.3	42.0	19.8	9.7
Medium (0.11–0.28)	10,619	17.5	20.4	20.0	24.4
High (0.28–0.47)	6,572	15.9	13.7	20.1	40.4
Very high (>0.47)	3,622	12.5	8.3	19.8	47.2

locational, socioeconomic, built environment and coping capacity factors. Finally, a flood-risk map was obtained by multiplying total vulnerability and flood hazard maps.

The highly hazardous zone is concentrated in the areas along the Balu River and Buriganga River and accounts for 45 % of the study area. Eighteen per cent of the communities are estimated to be in low- to medium-hazard zones. More than 24 % of the total population are exposed to flood hazards, with 7 % being located in the highly hazardous zone. Communities with higher levels of vulnerability are distributed on the west to the southern fringes and northeast part of Dhaka. Around 40 % of the communities in the study area are highly vulnerable, and of these, 8 % are extremely vulnerable to flood. This vulnerability is related to community location, socioeconomic conditions such as poverty and rates of literacy, and accessibility to urban services such as sanitation. Similarly, the communities at high risk are mostly concentrated in the outskirts of the city, accounting for 22 % of the total population and 40 % of the housing units in Dhaka.

There are some limitations of the study. They are as follows: results were not validated against data collected from the field due to the difficulty in gathering such data. Although the allocation of weights to the criteria in the vulnerability analysis is based on the literature, it is still more or less subjective. When indices and variable values in the cell-based datasets (30 $\times$ 30 m resolution) are aggregated or averaged for a community, variation among the cell values within the community boundary may be lost since this operation assumes homogeneity in a community. Sensitivity of the allocated weights was not tested. Further study is therefore needed to validate the model, including incorporating major stakeholders' opinions in the model to develop a decision support system for effective management of flood hazards in Dhaka or elsewhere.

Acknowledgement Akiko Masuya acknowledges her gratitude to Dr. Ashraf M. Dewan for basic data preparation and supervision of this work whilst studying for her MSc at Curtin University.

References

Alam M, Rabbani MG (2007) Vulnerability and responses to climate change for Dhaka. Environ Urban 19:81–97

Ali MA, Seraj SM, Ahmed S (eds) (2002) Engineering concerns of flood. Bangladesh University of Engineering and Technology, Dhaka

Azar D, Rain D (2007) Identifying population vulnerable to hydrological hazards in San Juan, Puerto Rico. GeoJournal 69(1–2):23–43

Bala SK, Islam AKMS, Chowdhury JU, Rahman MR, Haque MA, Khan MSA, Salehin M (2009) Performance of flood control works around Dhaka city during major floods in Bangladesh. A paper presented in the 2nd international conference on Water and Flood Management, Dhaka

Balica SF, Wright NG, van der Muelen F (2012) A flood vulnerability index for coastal cities and its use in assessing climate change impacts. Nat Hazards 64(1):73–105

Bangladesh Bureau of Statistics (BBS) (2012) Population and housing census – 2011, Community report, Dhaka, Narayanganj and Gazipur Zila. Bangladesh Bureau of Statistics, Ministry of Planning, Dhaka

Bangladesh Bureau of Statistics (BBS), The World Bank, World Food Programme (WFP) (2005) Bangladesh household income and expenditure survey 2005, Dhaka

Barua S, van Ast JA (2011) Towards interactive flood management in Dhaka, Bangladesh. Water Policy 13:693–716

Bendimerad F (2009) State-of-the-practice report on urban disaster risk management, Earthquakes and Megacities Initiative. Available at www.emi-megacities.org

Birkmann J (ed) (2006) Measuring vulnerability to natural hazards: towards disaster resilient society. United Nations University Press, Tokyo

Bizimana JP, Schilling M (2010) Geo-information technology for infrastructural flood risk analysis in unplanned settlement: a case study of informal settlement flood risk in the Nyabugogo flood plain, Kigali city, Rwanda. In: Showalter PS, Lu Y (eds) Geospatial techniques in urban hazard and disaster analysis. Springer, Dordrecht, pp 99–124

Bouwer LM (2011) Have disaster losses increased due to anthropogenic climate change? Bull Am Meteorol Soc 92(1):39–46

Burton I, Kates R, White GF (1993) The environment as hazard, 2nd edn. Guilford Press, New York

Cook BR (2010) Flood knowledge and management in Bangladesh: increasing diversity, complexity and uncertainty. Geogr Compass 4(7):750–767

Cutter SL (1996) Vulnerability to environmental hazards. Prog Hum Geogr 20(4):529–539

Cutter SL, Mitchell JT, Scott MS (2000) Revealing the vulnerability of people and places: a case study of Georgetown County, South Carolina. Ann Assoc Am Geogr 90(4):713–737

Cutter SL, Bryan J, Boruff W, Lynn S (2003) Social vulnerability to environmental hazards. Soc Sci Quart 84:242–261

Cutter SL, Barnes L, Berry M, Burton C, Evans E, Tate E, Webb J (2008) A place-based model for understanding community resilience to natural disasters. Glob Environ Chang 18(4):598–606

Cutter SL, Emrich CT, Webb JJ, Morath D (2009) Social vulnerability to climate variability hazards: a review of literature, Final report to Oxfam America. Available at http://adapt.oxfamamerica.org/resources/Literature_Review.pdf. Accessed 10 Jan 2012

Dewan AM (2013) Floods in a megacity: geospatial techniques in assessing hazards, risk and vulnerability. Springer, Dordrecht

Dewan AM, Corner RJ (2012) The impact of land use and land cover changes on land surface temperature in a rapidly urbanizing megacity. In: IGARSS 2012, Munich

Dewan AM, Yamaguchi Y (2008) Effects of land cover changes on flooding: example from Greater Dhaka of Bangladesh. Int J Geoinf 4(1):11–20

Dewan AM, Yeboah KK, Nishigaki M (2006a) Flood hazard delineation in Greater Dhaka, Bangladesh using integrated GIS and remote sensing approach. Geocarto Int 21(2):33–38

Dewan AM, Yeboah KK, Nishigaki M (2006b) Using synthetic aperture radar (SAR) data for mapping river water flooding in an urban landscape: a case study of Greater Dhaka, Bangladesh. Jpn J Hydrol Water Resour 19(1):44–54

Dewan AM, Islam MM, Kumamoto T, Nishigaki M (2007a) Evaluating flood hazard for land-use planning in Greater Dhaka of Bangladesh using remote sensing and GIS techniques. Water Resour Manag 21(9):2101–2116
Dewan AM, Kabir MH, Islam MM, Kumamoto T, Nishigaki M (2007b) Delineating flood risk areas in Greater Dhaka of Bangladesh using Geoinformatics. Georisk Assess Manag Risk Eng Syst Geohazards 1(4):190–201
Disaster Management Bureau (DMB) (1998) Report on Bangladesh flood 1998: chronology, damages and responses. DMB, Dhaka
Disaster Management Bureau (DMB) (2007) Consolidated damage and assessment, lessons learnt from the flood 2007 and future action plan. DMB, Dhaka
Eakin H, Luers A (2006) Assessing the vulnerability of social-environmental systems. Annu Rev Environ Resour 31:365–394
EM-DAT (2012) Global flood records, 1950–2011, Centre for Research on the Epidemiology of Disasters (CRED), Universite catholique de Louvian, Belgium. www.emdat.be
Faisal IM, Kabir MR, Nishat A (1999) Non-structural flood mitigation measures for Dhaka city. Urban Water 1(2):145–153
Faisal IM, Kabir MR, Nishat A (2003) The disastrous floods of 1998 and long term mitigation strategies for Dhaka City. Nat Hazards 28(1):85–99
Fekete A, Damm M, Birkmann J (2010) Scales as a challenge for vulnerability assessment. Nat Hazards 55(3):729–747
Flood Action Plan (FAP) 8A (1991) Master plan study for Greater Dhaka protection project. Japan International Cooperation Agency, Dhaka
Gillespie TW, Chu J, Elizabeth FA, Thomas D (2007) Assessment and prediction of natural hazards from satellite imagery. Prog Phys Geogr 31(5):459–470
Government of Bangladesh (GoB) (2005) National adaptation programme of action. Ministry of Environment and Forest, Government of Bangladesh, Dhaka
Haque E, Zaman M (1994) Vulnerability and responses to riverine hazards in Bangladesh: a critique of flood control and mitigation approaches. In: Varley A (ed) Disasters, development and environment. Wiley, London, pp 65–79
Hewitt K (1997) Regions of risk: a geographical introduction to disasters. Longman, Harlow
Hirabayashi Y, Kanae S (2009) First estimate of the future global population at risk of flooding. Hydrol Res Lett 3:6–9
Hoque R, Nakayama D, Matsuyama H, Matsumoto J (2011) Flood monitoring, mapping and assessing capabilities using RADARSAT remote sensing, GIS and ground data for Bangladesh. Nat Hazards 57(2):525–548
Hossain S (2008) Rapid urban growth and poverty in Dhaka City. Bangladesh e-J Sociol 5:1–24
Huq S (1999) Environmental hazards in Dhaka. In: Mitchell JK (ed) Crucibles of hazards: mega-cities and disasters in transition. United Nations University, Tokyo, pp 119–137
Hye A (2000) Effects on livelihood. In: Nishat A, Reazuddin R, Amin R, Khan AR (eds) The 1998 flood: impact on environment of Dhaka City. Department of Environment, Dhaka, pp 201–216
International Strategy for Disaster Reduction (UN/ISDR) (2004) Living with risk: a global review of disaster reduction initiatives. United Nations Publications, Geneva
Islam MM, Sado K (2000a) Development of flood hazard maps of Bangladesh using NOAA-AVHRR images with GIS. Hydrol Sci J 45:337–355
Islam MM, Sado K (2000b) Satellite remote sensing data analysis for flood damage zoning with GIS for flood management. Annu J Hydraul Eng (JSCE) 44:301–306
Islam N, Sultana NY (2006) The status of women in Bangladesh: is the situation really encouraging. Res J Soc Sci 1:56–65
Jakobsen F, Hoque AKMZ, Paudyal GN, Bhuiyan MS (2005) Evaluation of the short-term forcing the monsoon river floods in Bangladesh. Water Int 30(3):389–399
Jha AK, Bloch R, Lamond J (2012) Cities and flooding: a guide to integrated urban flood risk management for the 21st century. The World Bank, Washington, DC
Kasperson RE (1986) Six propositions on public participation and their relevance for risk communication. Risk Anal 6(3):275–281

Kazmierczak A, Cavan G (2011) Surface water flooding risk in urban communities: analysis of vulnerability, hazard and exposure. Landsc Urban Plan 103(2):185–197

Kienberger S (2012) Spatial modelling of social and economic vulnerability to floods at the district level of Buzi, Mozambique. Nat Hazards 64(3):2001–2019

Kiunsi RB, Mechack V, Uhinga G, Mayunga J, Mulenge F, Kabali C, Sigalla N, Bilia M (2006) Disaster vulnerability assessment: the Tanzania experience. In: Birkmann J (ed) Measuring vulnerability to natural hazards: towards disaster resilient society. United Nations University Press, Tokyo, pp 227–245

Kundzewicz ZW, Hirabayashi Y, Kanae S (2010) River floods in the changing climate – observations and projections. Water Resour Manag 24(11):2633–2646

Lankao PR, Qin H (2011) Conceptualizing urban vulnerability to global change and environmental change. Curr Option Environ Sustain 3(3):142–149

Maantay J, Maroko A, Culp G (2010) In: Showalter PS, Lu Y (eds) Geospatial techniques in urban hazard and disaster analysis. Springer, Dordrecht, pp 71–97

Mahmud W (2004) The impact of floods on the economy of Bangladesh. In: Siddiqui KU, Hossain ANH (eds) Options for flood risk and damage reduction in Bangladesh. The University Press Limited, Dhaka, pp 81–92

Malczewski J (2000) On the use of weighted linear combination method in GIS: common and best practice approaches. Trans GIS 4:5–22

Meyer V, Scheuer S, Haase D (2009) A multicriteria approach for flood risk mapping exemplified at the Mulde river, Germany. Nat Hazards 48(1):17–39

Milly PCD, Betancourt J, Falkenmark M, Hirsch RM, Kundzewicz ZW, Lettenmaier DP, Stouffer RJ (2008) Stationarity is dead: whither water management? Science 319:573–574

Mirza MMQ (2003) Three recent extreme floods in Bangladesh: a hydro-meteorological analysis. Nat Hazards 28(1):35–64

Mirza MMQ (2011) Climate change, flooding in South Asia and implications. Reg Environ Chang 1(supp):95–107

Mitchell JK, Devine N, Jagger K (1989) A contextual model of natural hazards. Geogr Rev 79(4):391–409

Nishat A, Reazuddin R, Amin R, Khan AR (eds) (2000) The 1998 flood: impact of environment of Dhaka City. Department of Environment, Dhaka

Pandey AC, Singh SK, Nathawat MS (2010) Waterlogging and flood hazards vulnerability and risk assessment in Indo Gangetic plain. Nat Hazards 55(2):273–289

Paul BK (1997) Flood research in Bangladesh in retrospect and prospect: a review. Geoforum 28:121–131

Pavri F (2010) Urban expansion and sea-level rise related flood vulnerability for Mumbai (Bombay), India using remote sensed data. In: Showalter PS, Lu Y (eds) Geospatial techniques in urban hazard and disaster analysis. Springer, Dordrecht, pp 31–49

Pelling M (2003) The vulnerability of cities. Earthscan, London

PreventionWeb (2011) UN International Strategy for Disaster Reduction, Geneva, Switzerland. Viewed 1 Apr 2012. www.preventionweb.net

Rabbani G (2009) Environmental risks in Dhaka: present initiatives and the future improvements. In: Shaw R, Srinvas H, Sharma A (eds) Urban risk reduction: an Asian perspective, vol 1, Community, environmental and disaster risk management. Emerald Group Publishing, Bingley, pp 319–338

Rahman M, Rahman M, Hassan D (2002) Assessment of water quality in flood affected areas of Dhaka City. In: Ali MA, Seraj SM, Ahmad S (eds) Engineering concerns of flood. Bangladesh University of Engineering and Technology, Dhaka, pp 35–47

Rahman R, Haque A, Khan SA, Salehin M, Bala SK (2005) Investigation of hydrological aspects of flood—2004 with special emphasis on Dhaka City. Final report. Institute of Water and Flood Management, Bangladesh University of Engineering and Technology, Dhaka

Ramanathan R (2001) A note on the use of the analytic hierarchy process for environmental impact assessment. J Environ Manage 63:27–35

Rashed T, Weeks J (2003) Assessing vulnerability to earthquake hazards through spatial multicriteria analysis of urban areas. Int J Geogr Inf Sci 17(6):547–576

Roy M (2009) Planning for sustainable urbanization in fast growing cities: mitigation and adaptation issues addresses in Dhaka, Bangladesh. Habitat Int 33(3):276–286
Saaty TL (1980) The analytical hierarchy process. McGraw-Hill, New York
Saaty TL (2000) Fundamentals of decision making and priority theory with the analytic hierarchy process. RWS Publications, Pittsburgh
Sanyal J, Lu XX (2004) Application of remote sensing in flood management with special reference to monsoon Asia: a review. Nat Hazards 33:283–301
Sanyal J, Lu XX (2006) GIS-based flood hazard mapping at different administrative scales: a case study in Gangetic West Bengal, India. Singapore J Trop Geogr 27:207–220
Schanze J (2006) Flood risk management – a basic framework. In: Schanze J, Zeman E, Marsalek J (eds) Flood risk management – hazards, vulnerability and mitigation measures. Springer, Ostrov, pp 149–167
Senga R (2004) Natural and unnatural disasters: the relative vulnerabilities of Southeast Asian megacities to climate change. World Wide Fund for Nature (WWF), Switzerland. www2.lse.ac.uk/IDEAS/publications/reports/pdf/SR004/WWF.pdf. Accessed 15 Feb 2012
Showalter PS, Lu Y (eds) (2010) Geospatial techniques in urban hazards and disaster analysis, vol 2, Geotechnologies and the environment. Springer, Dordrecht
Siddiqui KU, Hossain ANHA (2004) Options for flood risk and damage reduction in Bangladesh. The University Press Limited, Dhaka
Sinha R, Bapalu GV, Singh LK, Rath B (2008) Flood risk analysis in the Kosi river basin, North Bihar using multi-parametric approach of analytical hierarchy process (AHP). J Indian Soc Remote Sens 36(4):293–307
Smith K (2001) Environmental hazards: assessing risk and reducing disaster, 3rd edn. Routledge, London
Stalenberg B, Vrijling H (2009) The battle of Tokyo and Dhaka against floods. Built Environ 35(4):471–491
Suriya S, Mudgal BV (2012) Impact of urbanization on flooding: the Thirusoolam sub watershed – a case study. J Hydrol 412–413:210–219
Tawhid KG (2004) Causes and effects of water logging in Dhaka city, Bangladesh. TRITA-LWR master thesis, Department of Land and Water Resource Engineering, Royal Institute of Technology, Stockholm
Tingsanchali TA, Karim MF (2005) Flood hazard and risk analysis in the southwest region of Bangladesh. Hydrol Process 19(10):2055–2069
UN-HABITAT (2010) Monsoon flood 2010 Pakistan: rapid technical assessment of damage and needs for reconstruction in housing sector, Islamabad
van der Sande CJ, de Jong SM, de Roo APJ (2003) A segmentation and classification of IKONOS-2 imagery for land cover mapping to assist flood risk and flood damage assessment. Int J Appl Earth Obs Geoinf 4(3):217–229
Wang Y, Li Z, Tang Z, Zeng G (2011) A GIS-based spatial multi-criteria approach for flood risk assessment in Dongting lake region, Hunan, Central China. Water Resour Manag 25(13):3465–3484
Wisner B, Blaikie P, Cannon T, Davis I (2004) At risk: natural hazard, people's vulnerability and disasters, 2nd edn. Routledge, Abingdon
World Meteorological Organization (WMO) (1999) Comprehensive risk assessment for natural hazards, WMO/TD no 955. WMO, Geneva
Wu SY, Yarnal B, Fisher A (2002) Vulnerability of coastal communities to sea-level rise: a case study of Cape May County, New Jersey, USA. Clim Res 22:255–270
Yalcin G, Akyurek Z (2004) Analysing flood vulnerable areas with multicriteria evaluation. Available at: http://www.isprs.org/proceedings/XXXV/congress/comm2/papers/154.pdf
Zheng N, Takara K, Yamashiki Y, Tachikawa Y, (2009) Large scale quantitative vulnerability analysis for regional flood hazard, Annual of Disaster Prevention Research Institute, Kyoto University, B, pp 57–65

Chapter 11
Supplementing Electrical Power Through Solar PV Systems

Md. Humayun Kabir and Wilfried Endlicher

Abstract Dhaka megacity is home to more than 14 million inhabitants. However, the adequate supply of urban utilities, particularly electricity to city dwellers, has been a major challenge mainly due to high population pressure and increasing demand. An insufficient supply of electricity (1,000–1,200 MW) compared to its peak demand (around 2,000 MW) has had a severe negative impact on the city dwellers' lifestyles. Given the shortfall of power supply in Dhaka, electricity generation through rooftop solar PV systems is widely discussed nowadays. Based on secondary data (sunshine duration, GHI values, etc.) and literature (identification of well-illuminated rooftops within the city and calculation of solar PV-based electricity generation potential by the authors), this chapter reveals that the driving forces (geophysical, economic and sociopolitical and environmental factors) are fully supportive of harnessing solar energy for electricity generation. The application of solar PV system on the city's extensive well-illuminated rooftops (>10 km^2) would be very effective and could possibly meet around half of the city's power demand.

Keywords Bangladesh • Dhaka megacity • Driving forces • Power supplement • Solar PV systems

Md.H. Kabir (✉)
Department of Geography and Environment, University of Dhaka, Dhaka 1000, Bangladesh
e-mail: mh_kabir@yahoo.com

W. Endlicher
Department of Geography, Humboldt-Universität zu Berlin, Unter den Linden 6, 10099 Berlin, Germany
e-mail: wilfried.endlicher@geo.hu-berlin.de

A. Dewan and R. Corner (eds.), *Dhaka Megacity: Geospatial Perspectives on Urbanisation, Environment and Health*, Springer Geography,
DOI 10.1007/978-94-007-6735-5_11,

11.1 Introduction

Bangladesh has been experiencing a severe power deficit for the last few decades due to rapid population growth. Both per capita generation and consumption of electrical power are extremely low as compared to the global and neighbouring country rates (Akbar 2004). As at 2011, only around 40 % of the total population is serviced by public utility grids. Although the country's electricity demand has increased enormously over time, it has not been possible to enhance the installed generating capacity for various reasons. A drastic decline in the reserves of indigenous gas, which contributes more than 80 % of the country's electricity generation, has recently created immense interruption in electricity generation and supply. The annual per capita generation and consumption of electricity, both less than 200 kWh in 2011, are among the lowest on the globe, meaning that the country is a land of power crises. Although the installed capacity of electricity generation in Bangladesh accounts for nearly 7,000 MW (as of 2011), only around 4,000–5,000 MW can be regularly generated (BPDB 2009, 2011). Moreover, electricity is currently generated mostly with conventional fuel sources (82 % indigenous natural gas, 9 % imported oil, 5 % coal and 4 % hydropower) (2011 figures), whilst the rapid decline in natural gas reserve has already started hampering electricity generation in the country. In contrast, the megacity of Dhaka, from the viewpoint of social, economic, political, education, health and other services, is of immense importance to the nation, and its central location has further enhanced its significance to the people. But the city struggles to provide adequate urban utilities (particularly electricity supply) to city dwellers which has recently been a major challenge to development authorities mainly due to immense population pressure and increasing demand. Insufficient supply of electricity to the city (1,000–1,200 MW) compared to its peak demand (2,000 MW) has acutely aggravated the lifestyles of the city dwellers. Given the severe deficiency of electrical power, there is a crying need to supplement electricity from renewable sources. As a precondition of being connected with public utility grid in Dhaka, newly constructed buildings in the city must be able to provide at least 3 % of the electricity demand from solar PV systems. Considering the abundance of suitable sites on well-illuminated rooftops and the effectiveness of solar PV systems, their application is shown to be a viable and pragmatic option for providing electricity supplement to the present capacity (Kabir and Endlicher 2012; Kabir et al. 2010). The successful application of solar PV systems requires an in-depth assessment of the driving forces, which include geophysical (absolute location, global horizontal irradiance, temperature, cloud cover, sunshine duration, land availability and so on), economic and sociopolitical (capital investment, technology supports, social acceptability, political commitment, etc.) and environmental factors (climate protection through GHG emission reduction). An assessment of these driving forces reveals that the application of solar PV system can substantially meet the city's power demand. The present chapter, based mainly on secondary data (sunshine duration, GHI values, etc.) and published materials (e.g. Kabir et al. 2010; Kabir and Endlicher 2012), is an attempt to assess the role of the driving forces in harnessing solar energy resource for electricity supplement in Dhaka. The objective of this chapter is to assess the role of the driving forces of solar photovoltaic applications in the megacity of Dhaka. Specifically, this chapter identifies electricity generation potential from solar PV systems through rooftop installations.

11.2 Materials and Methods

This chapter is primarily based on secondary data and published literature. Secondary data on global horizontal irradiance (GHI), sunshine duration, etc. were collected from the Renewable Energy Research Centre (RERC) of the University of Dhaka. GHI data from 2003 to 2007 of various cities of Bangladesh including Dhaka have been prepared by the RERC with the financial support of UNEP and GEF under the Solar and Wind Energy Resource Assessment (SWERA) Project. Kabir et al. (2010) and Kabir (2011) identified well-illuminated rooftops in Dhaka through object-based image analysis (OBIA) from high-resolution (60 cm) Quickbird satellite data of 2006 that covered an area of 431 km^2. Along with solar data, published literature has (mainly by the authors and others) widely been reviewed.

11.3 Power Supply Scenario

With the dramatic growth of the city population, the demand for energy has also been increasing. Whilst the city receives priority supplies of electricity at the expense of other parts of the country, the overall situation is not satisfactory at all. The whole system of electricity distribution is poorly managed and also suffers losses of more than 30 % of the system due to illegal connections (Alam et al. 2004). There have been a number of attempts at reform of the power sector in the country, aimed at reducing system loss, but practically all these reforms have been inconsequential. The first serious attempts at power sector reforms began shortly after the country achieved independence in 1971. In 1972, the Bangladesh Power Development Board (BWDB) was created to ensure satisfactory power supply to feed the newborn country's socio-economic development. Major initiatives include the establishment of the Rural Electrification Board (REB) in 1977, the Dhaka Electricity Supply Authority (DESA) in 1991 and the Dhaka Electric Supply Company (DESCO) in 1997 (Alam et al. 2004). Although DESA initially took over Dhaka District for power supply under the control of BPDB, it has gradually reduced its jurisdiction, handing over the rural areas to REB, and now has been confined to Dhaka megacity (except for a part of the metropolitan area). DESCO, wholly owned by DESA, is responsible for all activities of distribution system in Mirpur, Uttara, Gulshan and its adjoining area under the Company Act. Dhaka

Power Distribution Company (DPDC) was created in 2005 through recent power sector reforms under the Company Act 1994, which is responsible mainly for the southern part of Dhaka City and for the adjacent towns of Narayanganj and Tongi. In Bangladesh, power generation is performed by BPDB, REB and Independent Power Producers (IPPs), whilst distribution is run by BPDB (29 %), REB (12 %) and DESA (59 %). BPDB and Power Grid Company of Bangladesh Limited (PGCB) are the agencies responsible for power transmission in the country.

The electricity demand of Dhaka megacity is increasing at an alarming rate every year due to the rapid growth of population along with the growth of electricity connectivity. Currently, the demand is around 1,500–1,700 MW, but DESA can supply a maximum of 1,000–1,200 MW, which is not adequate to the existing demand. The situation becomes acute in the summer, when the demand of electricity increases to about 2,000 MW. The reasons for this worsening situation in the power supply of the megacity primarily lie with wide-scale corruption of stakeholders, mismanagement or undesirable system loss, ageing of the infrastructure and unpredictable annual increase of demand. The unexpected demand is mainly due to the installation of luxury goods like air conditioner in shopping centres, private universities, commercial office buildings, clinics and hospitals, as well as some public buildings, which lead to the huge increase in electricity demand during the summer. The supply often fails to meet the demand and has led to events such as riots against the electricity providers. In the last few years, the impatience of the city dwellers has risen, and several people were killed in riots over electricity, particularly in 2006. This situation was brought about by load shedding caused by the tremendous shortfall of electricity, particularly in summer when the daily shortfall is around 500–800 MW. In order to minimise the adverse situation, various strategies have adopted by the responsible agencies. The former Caretaker Government (January 2007 to January 2009) prepared a renewable energy policy which is yet to be approved by the Parliament. The sitting government (since 2009) has imposed the obligations that newly constructed buildings must be able to provide at least a 3 % supplement of electricity demand from solar PV systems as a precondition of being connected with public utility grid.

11.4 Solar PV in Bangladesh: Current Installations

Solar PV systems installed in Bangladesh so far are mainly solar home systems (SHS) in off-grid areas where electricity supply through public grids is not feasible and would not be possible in the near future. In contrast, Grameen Shakti (GS) has a target of installing ten million SHS by 2015 which will cover approximately 10 % of the rural people (Barua 2007). At the end of 2011, the installed capacity of solar PV systems (stand-alone) with one million solar home systems all over the country accounted for nearly 50 MW, an increase from 15 MW in 2008. It is assumed that this installation process will continue until the entire country gets connected to public utility grids. The electricity demand in the rural areas is still confined to lighting, using black-and-white television, mobile phone charger and other low-consumption appliances.

In spite of massive power deficit in the megacity of Dhaka, no substantial action has been undertaken to promote large-scale photovoltaic applications except several small-scale rooftop installations (e.g. a pilot project 1.5 kWp grid-connected PV system at University of Dhaka (2007), 10.16 kWp system at the prime minister's office).

11.5 Determining Factors of SPV Application

Considering the abundance and effectiveness of renewable sources, the exploitation of solar energy for electricity generation in Dhaka megacity is the most attractive option (Kabir and Endlicher 2012). However, three major groups of factors (physical, geographical and technical) are essential in order to assess the effectiveness and potential of solar PV systems in a particular area (Izquierdo et al. 2008). In the case of Dhaka, a number of other factors can also be considered as the determining forces. These factors include geophysical, economic and sociopolitical aspects. The geographical location of the site, land availability or available surface area (available roof area of buildings in the built-up areas), global horizontal irradiation (GHI), sunshine hours, temperature, cloud cover, etc. are categorised as geophysical factors. Economic, technical and sociopolitical factors are equally as important as the geophysical viability. Economic and technical aspects include financial arrangements, technology support, performance of the systems and availability of local skilled technicians to maintain the systems. Social factors include consumers' acceptance and willingness to pay for SPV-based electricity technology, whilst political factors include the government's commitment and good governance. Last but not the least, environmental aspects are important and include greenhouse gas emission reduction, ensuring clean environment and protection of climate. The environmental issues can be viewed as the nation's obligation that can be achieved through the exploitation of renewable energy sources.

Geophysical Factors

Geophysical situation of an area is the most important determinant for the consideration of SPV applications. In the present chapter, the geographical location of Dhaka on the globe, space availability (land availability, available roof surface, etc.), sunshine hours, global horizontal irradiance (GHI), etc. have been identified as the main geophysical factors. Based on the suitability of the geophysical situation, other associated factors need to be assessed. If the geophysical state is not supportive to the PV applications, further assessment of other aspects will be futile. An assessment of these factors has been made in the following sections to investigate whether the data on these geophysical factors are supportive of solar PV applications in Dhaka megacity.

Sunshine Hours and Solar Radiation

Sunshine duration, global horizontal irradiance (GHI), cloud cover, temperature, etc. are the most important geophysical factors which influence the energy potential from the applications of solar PV systems. Sunshine duration at the proposed location of an installation is crucially important. The aggregate GHI depends on how long the site receives sunlight. From a solar energy generation perspectives, Bangladesh is located in a global position that offers very satisfactory sunshine duration. Long-term (1988–1997) sunshine data recorded by the Renewable Energy Research Centre of the University of Dhaka reveal that the period from November to May has the maximum sunshine duration and the period from September to October is reasonably satisfactory (SWERA/RERC 2007). The maximum daily mean sunshine hours in Dhaka occur in February (9.1 h), with the minimum daily average being in June (4.9 h). Although the day length remains quite high in June, the sky is overcast most of the time. As a result, direct sunshine hours remain lower for the receipt of solar irradiation. The daily maximum peak sunshine hour also occurs in February (10.7 h). The daily average sunshine recorded in Dhaka is 7.55 h, and nearly 2,800 h of sunshine is received annually. For comparison, the sunshine exposure of the best global locations (in the Sahara desert and a few countries of the Middle East) is around 4,000 h per year. Algeria, located in the northern part of Africa, is a country of high insolation with 3,300 h of bright sunshine annually (Chegaar and Chibani 2000). Therefore, in comparison with the countries of high duration of incoming solar radiation, the sunshine hours of Dhaka seem quite satisfactory.

Due to the availability of sunshine throughout the year, the GHI of Bangladesh is also satisfactory for generating sufficient solar energy. The daily average GHI in Bangladesh is 4.29 kWh/m^2 and the annual estimation is 1,566 kWh/m^2. It has also been determined that on horizontal and tilted surfaces, the average annual availability of solar radiation in Dhaka is 1.73 MWh/m^2 and 1.86 MWh/m^2, respectively (Hussain and Badr 2005). The RERC of the University of Dhaka, with the financial support of United Nations Environment Programme (UNEP) and Global Environmental Facility (GEF), started the 'Solar and Wind Energy Resource Assessment (SWERA)' project in 2003 and produced reliable data on GHI and diffused normal irradiance (DNI) for Dhaka City and other parts of the country. The centre has also acquired GHI measurements for two time periods (1987–1989 and 1992). The maximum amount of radiation is available in the months of April–May (>4.5 kWh/m^2), and the periods from February to March and June to August also provide excellent solar radiation (~4.0 kWh/m^2). The rest of the months in a year still receive a reasonable amount of GHI (just less than 4.0 kWh/m^2/day). During the SWERA project, the RERC recorded hourly basis GHI data from January 2003 to December 2005. The highest hourly GHI (764 Wh/m^2) is received at 12:30 in April and GHI 727 Wh/m^2 at 11:30 in May (SWERA/RERC 2007). The daily aggregate GHI recorded in Dhaka in April and May is 5,286 Wh/m^2 and 5,459 Wh/m^2, respectively. Moreover, the daily total GHI received in March appears also to be excellent (4,884 Wh/m^2) (SWERA/RERC 2007).

Land Availability

Land use of a city provides a good indication of the city planning. In a well-planned city, a suitable proportionate ratio among various uses of land is found. For various reasons this is generally not seen in the cities of developing countries. Dhaka is an example of such city which has experienced very irregular development, leaving very negligible open spaces (below 5 %). Dewan and Yamaguchi (2009a, b) reported in a study of land use in Dhaka metropolitan area (DMA) in 2005 that built-up areas occupied 49.4 % (205 km^2) of total metropolitan area (416 km^2). Around 20 % of the area is cultivated lands, recent fill, and nearly 30 % of the area is covered by wetlands, water bodies and vegetation. Areas filled with earth are generally used for infrastructure. From the perspective of planning with solar PV-based power for Dhaka and the city's existing land cover and land use, it can clearly be seen that ground-mounted PV installations would not be feasible due to low availability of open land. No large areas of land are available for the large-scale PV plants within the urban areas of Dhaka megacity. Even the small existing patch of open space within the city left may not receive sufficient solar radiation. The peripheral areas of the cultivated or fallow spaces towards the north and northeastern parts of the DMA may be the only option to plan for SPV plants. Therefore, lands required for the large-scale (e.g. megawatt size) PV plant for either a newly built housing society or small town or shopping complex could possibly be developed in the northern parts of Dhaka megacity.

Available Roof Area

As a part of the analysis of determining factors, identification and calculation of well-illuminated rooftops in the megacity of Dhaka through object-based image analysis (OBIA) from high-resolution satellite (e.g. Quickbird) data of Dhaka were carried out (Kabir et al. 2010). A Quickbird scene of Dhaka from 2006 shows considerable heterogeneity with buildings of various characteristics, irregular roads, vegetation, water bodies, open spaces, etc. (Fig. 11.1). In order to determine the applicability of solar PV systems for a city, it is necessary to calculate the available roof area (Izquierdo et al. 2008). It is evident that not all rooftops are appropriate for PV applications; some rooftops receive less sunlight due to shading by buildings and vegetation cover. Due to the city's immense heterogeneity (Fig. 11.2), the calculation of well-illuminated roof area is a very complex task. In this study, the Quickbird image was used for this task. Although the image covers an area of 431 km^2, only the area occupied by Dhaka City Corporation (DCC) has been considered for illuminated rooftop identification and calculation (Table 11.1) (Fig. 11.1). In order to reduce complexity and error, the areas with the highest

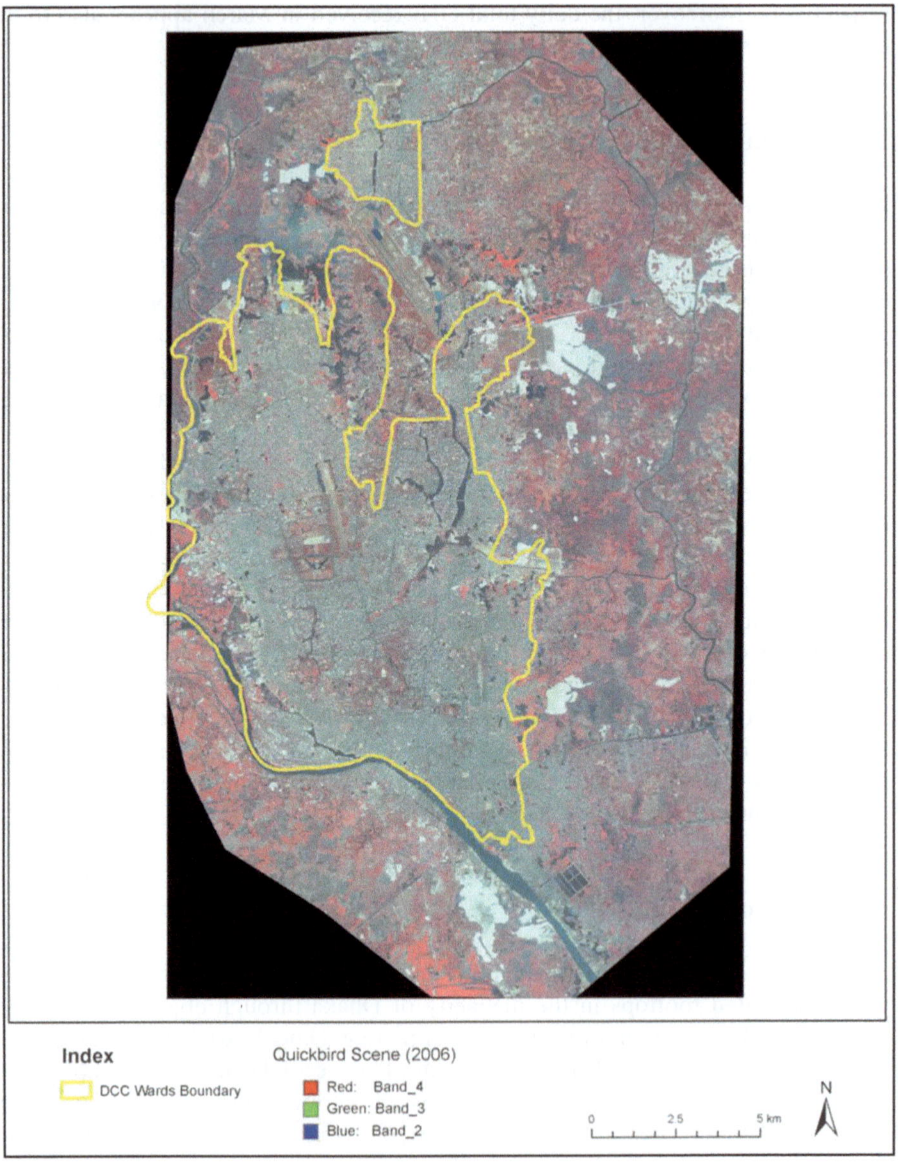

Fig. 11.1 Well-illuminated roof areas (Adapted from Kabir et al. 2010)

building density (DCC area) have been separated from the rest of the Quickbird image based mainly on field experience and visual observation. The other parts of the Quickbird image mostly contain water bodies, agricultural land, vegetation, roads and low-density buildings. It can be considered that in a very heterogeneous city like Dhaka, the calculation of well-illuminated rooftops for PV application

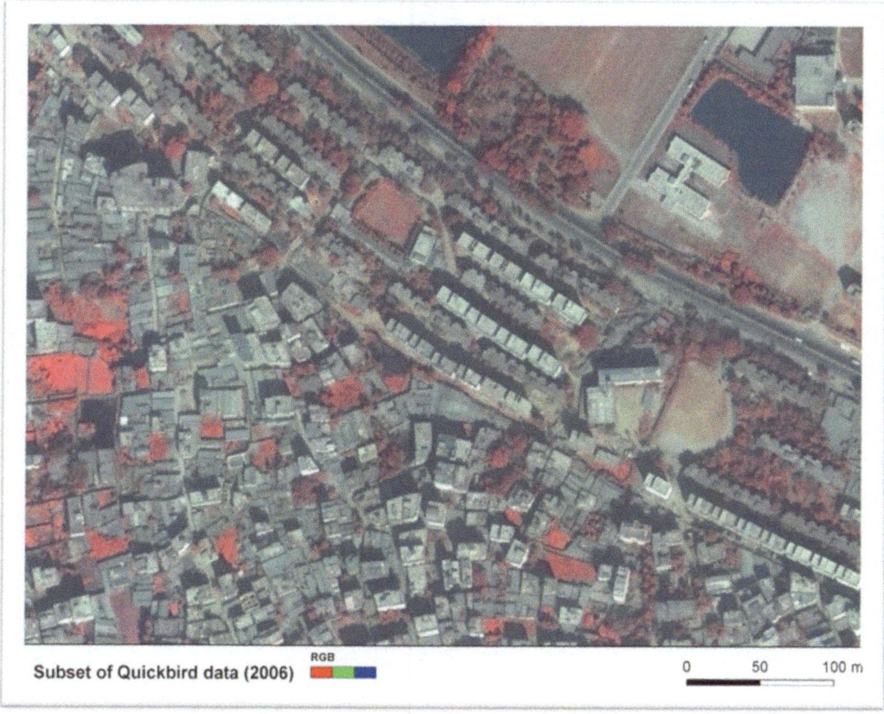

Fig. 11.2 A Zoomed-in well-illuminated roof areas in central Dhaka

Table 11.1 Calculation of well-illuminated roof area from the Quickbird image of 2006

Various objects classified	Area calculated (km^2)	% of total
Bright rooftops	10.554	7.86
Informal settlements	9.646	7.18
Vegetation	53.364	39.74
Water bodies	9.583	7.14
Others (roads, open land, shadow, etc.)	51.135	28.98
Total	134.282	100.0

Adapted from Kabir et al. (2010)

within the DCC areas would substantially contribute as very useful input data. However, from the calculation of bright rooftops within the DCC area, it has been found that the city offers 10.554 km^2 bright roof areas (Kabir et al. 2010). This figure can be considered as a conservative estimate for the entire megacity. The application of solar PV systems on these bright rooftops with 75 Wp solar panels can generate nearly 1,000 MW of electricity preferably through grid-connected PV systems (Kabir and Endlicher 2010). The potential electricity generation can be substantially high (>1,000 MW) with the installation of solar modules with high

Table 11.2 Solar PV-based power generation potential

Bright rooftop (km^2)	Bright rooftop (m^2)	GHI of Dhaka ($kWh/m^2/day$)	Module efficiency (assumed)	Module capacity (Watt)	Area needed/module (m^2)	No. of modules to be installed in given rooftops	Potential power generation (MW)
10.55	10.55×10^6	4.20	10–15 %	75 or more	$1\ m \times 0.8\ m = 0.8$	1.31925×10^7	989.44

(Adapted from Kabir et al. 2010)

capacity (i.e. 100 Wp) and efficiency, which would sufficiently meet up Dhaka City's existing power demand (Table 11.2).

Moreover, the city has remarkable extent of informal settlements, which are widely known as slums. Nearly 40 % of the population of Dhaka megacity are slum dwellers. In 2005, 4,966 slum clusters were identified in the city, contained 3.4 million people out of city's total of nearly 9.1 million (CUS et al. 2006). From this estimation, slum population rose to 5.2 million out of total 14 million in 2009. Within the DCC area, informal settlements occupy nearly 10 km^2 (Table 11.1).

Nearly 96 % of these slum communities are provided with grid electricity but with poor connection facilities. In practice, the slums receive least attention from the authorities with regard to power supply. The rooftops of these informal settlements can be effectively used for stand-alone PV applications, using solar home systems such as those that are used in the rural areas of the country. Electricity demand in informal housing is comparatively low compared to the high- and middle-class residential buildings, and solar home systems can effectively generate enough electric power for these settlements. CUS et al. (2006) also reported that each slum cluster has ten households with at least 25 persons. If solar home systems are installed in each slum cluster commensurate with the demand, between 3 and 5 MW of electricity (600–1,000 W in each slum cluster) can be generated through the application of stand-alone PV systems. This should be adequate to meet the power needs of these communities.

11.5.1 Economic and Sociopolitical Factors

The economy and energy supply of a country are closely linked. If any of these two sectors is weak, then the other will automatically be affected (Ahmad et al. 2005). Strong economic support can ensure a society that is self-dependent in energy and enhances social development through efficient use of energy in every sector. On the other hand, energy sufficiency can help accelerate the country's economic growth. Consideration of the economic situation along with social, political and technological factors is vital for the planning and successful implementation of SPV systems. Electricity generation from solar PV is still an expensive option because the devices themselves are relatively costly. This is one of the major reasons for lower penetration of PV modules in the energy market. The slow development of PV-based power generation in the megacity of Dhaka can be blamed on the country's weak national economy. Economic factors at work include capital investment in the renewable energy sectors, local technology support and trained manpower to maintain the systems.

Financial Arrangements

Solar PV systems have always been capital intensive, but the price of PV components all over the world has been decreasing as the volume of production increases (Muneer 2004). However, it has been reported that in spite of its widespread experience with solar PV applications and their high efficiencies and reliability, the system components still appear to be expensive in developed countries such as in Germany (Erge et al. 2001). In a country like Bangladesh where at least 40 % of the population is living below the poverty line (even worse in the rural areas), it is still unfeasible among the villagers having access to solar home systems. GS has been successful in solar home system installation through its microcredit programme in the rural areas. The initial investment for solar PV installations seems high due to the price of modules, but the operation and maintenance costs are generally low as there are no moving parts in the system (Islam and Islam 2005). However, even the very low amount of payment for solar home systems appears to be impossible to many of the villagers, and this slows down penetration of the technology. The situation with solar PV system installations in Dhaka City is different in the sense that it would be highly cost intensive. Compared to the rural electricity demand, the urban demand would be substantially different, and the city dwellers are likely to be capable of paying for the expenses of the PV-generated electricity.

Local Technology Support

In order to speed up and sustain the installation of SPV systems, support from local technology providers is immensely important. It is always expensive to depend only on the foreign technologic suppliers. In Bangladesh, there are only a few private entrepreneurs involved in the promotion of solar PV system, such as Rahimafrooz which is a private company producing batteries for solar PVs. For the large-scale success of PV technology, devices should be locally available as much as possible. Even in a country where enormous potential exists, it is impossible to achieve satisfactory progress with only a few manufacturers. Along with manufacturing support, local technicians have also to install and maintain the systems. This is well understood by Grameen Shakti that has already developed a workforce of more than 2,500 skilled technicians all over the country with an especial focus on the training of female technicians. In the rural parts of the country, people are relatively conservative in allowing outsiders into their communities and homes, and in this case, female technicians are also effective in providing SPV-related training to the rural women.

Social Acceptance and People's Willingness to Pay

The assessment of social acceptability of a new technology like PV is very important in the sense that the technology has to be sustainable in the country. In the urban areas, it is possible to conduct research into people's attitudes towards adoption of PV systems. Based on the experiences of increasing numbers of solar home systems in the rural areas, it appears that people have widely accepted this technology for electricity supply, and that they are willing to pay for quality power supply. A study was carried out by Bangladesh Centre for Advanced Studies (BCAS) on 'stakeholders' opinion survey on the power sector of Bangladesh and consumers' willingness to pay (WTP) for electricity supply' (BCAS 2003), which identified several technical, financial, management and governance issues. Some of the key issues are frequent load shedding and interruption in power supply; poor power quality; high system loss due to poor technical standards; unscheduled shutdown of power supply; imbalance among generation, transmission and distribution; inadequate load management for efficient power supply; damage of devices due to voltage fluctuations; etc. Given the state of electricity supply and associated issues, the consumers are reluctant to regularly pay the bills. If the consumers are provided with continuous and good quality of power supply, they have much greater willingness to pay. To be practical with the expectation of consumer's quality and uninterrupted power supply, the existing conventional fuel-based power cannot meet their demand. However, extensive installation of solar PV systems both in the urban and rural areas can ensure quality power supply, and people will be willing to pay for the electricity. People's willingness to pay for solar electricity has also been identified by another researcher (Khan et al. 2005). That study was conducted for market assessment of the solar home systems in three subdistricts of Bangladesh that are part of the territory of the Rural Electrification Board (REB). It is found that more than two-thirds of the respondents (80.5 %) expressed their interest in obtaining some kind of solar home system. Higher-income groups showed their willingness to pay for relatively large solar home systems, whilst small-income group intended to pay for small home systems. Moreover, promotion and motivational activities through demonstration and exhibition will need to be continuously carried out for both the city and rural people to be more acquainted with new technology and to improve their willingness to pay for it.

Political Commitment and Good Governance

Political commitment is essential to a project such as the introduction of widespread SPV generation. In most cases, developing nations such as Bangladesh suffer from political conflicts over power structures and from corrupt governance. Shifting the country from a conventional energy base to new and clean energy systems requires national-level decision making, policy formulation and effective policy implementation. In order to adopt, promote and enhance renewable energy technologies (e.g.

solar PV systems) in the country, the government has to play the pivotal role and needs to be the main patron. To do this, massive countrywide PV programmes including research, development and dissemination will have to be conducted with government support. The European Renewable Energy Council (EREC) has set up a binding target for the EU as a whole and each member state, to achieve at least 20 % energy consumption from renewable sources by 2020 (Lins 2009). This is a political target which mainly encourages each member state to work towards this attainment of the goal. This can be a shining example to other developing countries that can also look forward with an aim to reach a certain target in a particular time span. In Bangladesh, the recently devised renewable energy policy is a good reflection of the government's willingness to enhance renewable energy technologies. Moreover, the rapid increase of country's electricity demand and the declining trend of indigenous gas (major ingredient of electricity generation) have also required the government to promptly decide on the massive installations of solar PV systems. In this connection, it has recently been decided that all future multistorey buildings have to have self-supplemented electricity with solar systems. Therefore, the building owners have to plan with sufficient power generation capacity with PV modules before getting the building design approved. It is important to monitor the extent to which this obligation is in fact implemented.

11.5.2 Environmental Considerations

Environmental concerns are of very practical and pragmatic consideration when considering the installation of renewable energy technologies. Given the limited reserve of conventional fuels and tremendous increase of GHG emission due to over-exploitation of fossil fuels, countries around the globe have predicted the impending dangers of inadequate power supply and climate change. On the other hand, the capital city of a developing country like Bangladesh has neither a remarkable amount of industrial development nor a sufficient power supply. The country is also not a big contributor to global greenhouse emission, but the imminent consequences of climate change in the country are likely to be severe due to sea-level rise and other climatic disturbances. Compared to industrialised countries, Bangladesh's per capita CO_2 emission (0.30 metric tonnes) is nominal, when compared to the per capita CO_2 emissions in Germany, Japan and USA which in 2004 were 9.79 tonnes, 9.84 tonnes and 20.4 tonnes, respectively (CDIAC/UN 2009). For two reasons, Bangladesh continues to reduce CO_2 emission, although it is one of the lowest CO_2 emitters on Earth. In one hand, it is an obligation for the country to participate in the efforts of global GHG emission reduction as a signatory of the Kyoto Protocol and United Nations Framework Convention on Climate Change (UNFCCC), and it is an opportunity, on the other hand, to earn money from developed countries by selling the quota of CO_2. Bangladesh has enormous potential in solar energy, and therefore, the application of small- and large-scale PV systems can help reduce its current share of GHG emission. The demand on a typical solar home system (which is normally installed in the rural areas) consists of 4 fluorescent bulbs of 7 W each, 1 black-and-white TV of 15 W and a radio of 5 W. One family using this small system can save yearly 290 l of kerosene by using solar lighting technology and can prevent the emission of 0.76 tonne CO_2 per year (SWERA/RERC 2007).

11.6 Conclusions and Recommendations

In view of the phenomenal growth of population and steady increase of electricity demand in the megacity of Dhaka, the application of PV systems (grid connected or stand-alone on bright rooftops) is essential to supplement to the increasing demand. Among the influential factors, the geophysical characteristics (geographical location, global horizontal irradiance, sunshine duration, available bright rooftops, etc.) have been found to be fully supportive to large-scale PV applications in Dhaka City, whilst sociopolitical and economic factors will not be detrimental. The city offers more than 10 km^2 of well-illuminated rooftops, by the use of which half of the city's electricity demand can be met. The development of solar home systems in the rural areas so far is considered as indicating the people's acceptance of PV technologies. The economic status of the consumers is one of the main factors determining market development of PV technologies, and given the tremendous existing power deficiency, consumers are willing to pay for PV-generated power. However, the government has a leading role to play in the case of initial investment, and there is a scope to manage big funds internationally (e.g. through carbon trading). Moreover, the currently devised renewable energy policy that targets in implementing solar PV application should immediately be in force. From the perspective of climate protection, the application of solar PV systems would be very effective as the country is a signatory to both the Kyoto Protocol and UNFCCC.

Acknowledgements This chapter is based on the first author's doctoral study funded by the Yousef Jameel Scholarship Fund at the Humboldt-Universität zu Berlin, Germany. Therefore, the Yousef Jameel Scholarship Fund is gratefully acknowledged for the financial supports. We greatly acknowledge the support of the German Research Foundation (DFG) for the Quickbird satellite image of Dhaka.

References

Ahmad QK, Azim MA, Mintoo AA (eds) (2005) Energy security in Bangladesh. Academic Press and Publishers Library, Dhaka

Akbar S (2004) Present status of renewable energy and some issues for renewable energy resources development in Bangladesh. In: Proceedings of the regional workshop on renewable energy: progress and prospects, BSES and RERC, University of Dhaka, Dhaka, 20–27 February 2004

Alam MS, Kabir E, Rahman MM, Chowdhury MAK (2004) Power sector reform in Bangladesh: electricity distribution system. Energy 29(11):1773–1783

Bangladesh Centre for Advanced Studies (BCAS) (2003) A report on stakeholder research in power sector of Bangladesh. BCAS, Dhaka

Barua DC (2007) An introduction to Grameen Shakti. Grameen Shakti, Dhaka

BPDB (2009) Annual report. Bangladesh Power Development Board (BPDB), Dhaka

BPDB (2011) Annual report. Bangladesh Power Development Board (BPDB), Dhaka

CDIAC/UN (2009) The list of countries by carbon dioxide emissions per capita 2004, Carbon Dioxide Information Analysis Centre, (US Department of Energy). Retrieved from http://en.wikipedia.org/wiki/Carbon_Dioxide_Information_Analysis_Center. Accessed on 11 April 2009

Centre of Urban Studies (CUS), NIPORT (National Institute of Population Research and Training), Measure Elevation (2006) Slums of urban Bangladesh: mapping and census 2005, Dhaka

Chegaar M, Chibani A (2000) A simple method for computing global solar radiation. Rev Energ Ren: Chemss, 111–115

Dewan AM, Yamaguchi Y (2009a) Using remote sensing and GIS to detect and monitor land use and land cover change in Dhaka Metropolitan of Bangladesh during 1960–2005. Environ Monit Assess 150(1–4):237–249

Dewan AM, Yamaguchi Y (2009b) Land use and land cover change in Greater Dhaka, Bangladesh: using remote sensing to promote sustainable urbanization. Appl Geogr 29(3):390–401

Erge T, Hoffmann VU, Kiefer K (2001) The German experience with grid-connected PV-systems. Sol Energy 70(6):479–487

Hussain AK, Badr O (2005) Prospects of renewable energy utilization for electricity generation in Bangladesh. Renew Sustain Energy Rev 11(8):1617–1649

Islam AKMS, Islam M (2005) Status of renewable energy technologies in Bangladesh. ISESCO Sci Technol Rev 1:21–60

Izquierdo S, Rodrigues M, Fueyo N (2008) A method for estimating the geographical distribution of the available roof surface area for large-scale photovoltaic energy- potential evaluations. Sol Energy 82(10):929–939

Kabir MH (2011) Renewable energy as a perspective for power supply and greenhouse gas emission reduction in Bangladesh. Unpublished PhD thesis, Department of Geography, Humboldt-Universität zu Berlin

Kabir MH, Endlicher W (2010) Analysis of the determining factors of solar PV applications in Dhaka megacity, Bangladesh. In: Proceedings of the international conference on renewable energy, Pacifico Yokohama, Japan, 26 June–2 July 2010

Kabir MH, Endlicher W (2012) Exploitation of renewable energy in Bangladesh: power supply and climate protection perspectives. A H Development Publishing House, Dhaka

Kabir MH, Endlicher W, Jaegermeyr J (2010) Calculation of bright roof-tops for solar PV applications in Dhaka megacity, Bangladesh. Renew Energy 35(8):1760–1764

Khan HJ, Huque AJ, Khatun S, Mannan MA (2005) Commercialization of solar home systems: market assessment survey in Bangladesh. In: Eusuf M (ed) Solar photovoltaic systems in Bangladesh: experiences and opportunities. University Press Limited, Dhaka, pp 95–102

Lins C (2009) The new RES directive and EREC's technology roadmap 20% by 2020. Paper presented in the International RES-Boat conference, Vienna, 23 March 2009

Muneer T (2004) Solar radiation and day light model. Elsevier Butterworth-Heinemann, Oxford

SWERA/RERC (2007) Final report on solar and wind energy resource assessment (SWERA)-Bangladesh, Supported by United Nations Environment Programme (UNEP) and Global Environmental Facility (GEF), prepared by Renewable Energy Research Centre. University of Dhaka, Dhaka

University of Dhaka (2007) Roof-top grid connected solar photovoltaic system for Renewable Energy Research Center, a pilot project implemented by the Department of Applied Physics, Electronics & Communication Engineering, University of Dhaka, Dhaka

Chapter 12
Impact of Land Use and Land Cover Changes on Urban Land Surface Temperature

Ashraf M. Dewan and Robert J. Corner

Abstract The objectives of this chapter are to quantify the impact of land use and land cover (LULC) changes on land surface temperature (LST) and to map the changes in the spatial and temporal distribution of LST. A transition matrix was used to determine LULC changes in a multi-temporal manner. LST was retrieved using Landsat TIR data from 1990 to 2011, and statistical analyses were carried out to determine the relationship between LSTs and biophysical parameters. The results showed that the expansion of urban built-up surface over natural land cover such as floodplain and agricultural land has become conspicuous, significantly affecting the spatial and temporal distribution of surface temperature. Annual mean land surface temperature estimation revealed that urban built-up surface consistently has the highest ambient radiant temperature during the study period. A decrease in vegetation cover and subsequent increase in urban land cover were found to be associated with increased LST, suggesting an amplification of the UHI effect with time.

Keywords Land Surface Temperature • Land use/Land cover • Landsat TIR • Spatial and Temporal Distribution • Biophysical parameters

12.1 Introduction

With the increasing concentration of population in cities of the world, more attention is now being directed to the study of the urban environment, including climate change at local and regional scale (Wilby 2007). More than 50 % of the world total population is currently living in urban area, a figure which is expected to continue to grow in the coming decades (UN 2012; Seto et al. 2010). Urban expansion driven by

A.M. Dewan (✉) • R.J. Corner
Department of Spatial Sciences, Curtin University, Kent Street, Bentley, Perth 6102, WA, Australia
e-mail: A.Dewan@curtin.edu.au; R.Corner@curtin.edu.au

A. Dewan and R. Corner (eds.), *Dhaka Megacity: Geospatial Perspectives on Urbanisation, Environment and Health*, Springer Geography,
DOI 10.1007/978-94-007-6735-5_12,

population growth and economic development has resulted in dramatic changes to the Earth's surface, causing the replacement of soil and vegetation with impervious surface materials such as concrete, asphalt and buildings. One of the most obvious manifestations of such alteration of the Earth's surface is the marked differential in the surface temperature in urban areas compared to its rural counterpart. This is known as the urban heat island (UHI) effect (Voogt and Oke 2003; Landsberg 1981), which has a profound impact on landscape structures, ecosystem processes and local weather and climate (Seto et al. 2010; Yow 2007; Wilby 2007; Kalnay and Cai 2003; Lo and Quattrochi 2003; Carlson and Arthur 2000; Owen et al. 1998; Landsberg 1981). The UHI effect is a direct representation of a worsening urban environment caused by land surface changes (Lu et al. 2009) and has a detrimental impact on human health and on the proper functioning of the ecosystem (Imhoff et al. 2010). Furthermore, higher temperature in urban areas increases the demand for air conditioning, raises pollution levels and may locally modify precipitation patterns (Yuan and Bauer 2007). Therefore, characterising the spatial patterns of UHI driven by rapid land use/cover change enables us to identify human-related environmental changes (Xiao and Weng 2007). This is essential for the development of appropriate measures for urban planning and landscape policies (Nichol 1994).

Based on temperature differences, UHI can be categorised into subsurface, surface and air effects (Oke 1995); however, the processes involved in their formation are different (Yow 2007). Although UHI may be described by multiple ways, two or three subtypes of UHI can be distinguished. These are canopy layer heat island (CLHI), boundary layer heat island (BLHI) and surface urban heat island (SUHI) (Weng 2009, 2010; Lo and Quattrochi 2003). Whilst CLHI and BLHI are atmospheric heat islands, SUHI on the other hand is generally defined as the temperature gradient between urban and rural areas (Yuan and Bauer 2007). As noted by Roth et al. (1989), atmospheric heat island anomalies are higher at night. In contrast, the temperature differential of the surface urban heat island is higher in the day time, making urban areas relatively warmer than fringe zones. A number of factors such as street geometry, sky view factor, generation of anthropogenic heat, reduction in wind speed caused by canyon geometry and decreased evapotranspiration due to construction materials are accountable for increased temperature in urban areas (Oke 1982).

12.1.1 Potential Remotely Sensed Data Sources

A number of different thermal sensor platforms have been used for UHI mapping; these include TIROS, Landsat, ASTER and MODIS. Rao (1972) was the first to illustrate the potential of remote sensing in mapping the thermal footprint of urban areas using infrared data from the TIROS Operational Satellite (ITOS-1). Since then, thermal data from a variety of sensors, both spaceborne and airborne, has been utilised to detect SUHI in a diverse range of environments. Satellite-based remotely sensed coarse-resolution data from NOAA–AVHRR (advanced very high resolution radiometer) was extensively used to study urban thermal landscapes in Los Angeles (Carlson et al. 1977); the Midwest and Northeast United States (Matson et al. 1978);

the New York–New England city area (Price 1979); St. Louis, Missouri (Kidder and Wu 1987); Phoenix in Arizona (Balling and Brazell 1988); Dallas–Ft. Worth (Gallo and Owen 1998); Pennsylvania (Owen et al. 1998); Houston, Texas (Streutker 2002, 2003); Korea (Lee 1993); and multiple cities in North America (Gallo et al. 1993; Roth et al. 1989). Although these studies showed immense potential in mapping urban temperature over large areas, the correlation between image-derived values and ground observation was found to be poor (Weng 2009). By contrast, data from the Landsat TM and ETM+ thermal channels have proved particularly promising for retrieving land surface temperature (LST) and studying UHIs despite some uncertainties in the calculation of ambient radiant temperature (Weng 2003, 2009). Studies that have used the Landsat thermal channel to estimate LSTs and quantify the effect of UHI include work in Singapore (Nichol 1994), India (Srivastava et al. 2009; Ramachandra and Kumar 2009; Rose and Devadas 2009; Mallick et al. 2008; Katpatal et al. 2008), China (Li et al. 2012; Jiang and Tian 2010; Ma et al. 2010; Su et al. 2010; Xu et al. 2009; Lu et al. 2009; Xiao et al. 2008; Chen et al. 2006; Qian et al. 2006; Nichol 2005; Xu and Chen 2004; Weng 2001, 2003), Malaysia (Tan et al. 2009), Vietnam (Van et al. 2009; Hung et al. 2006), Iran (Amiri et al. 2009), Thailand (Hung et al. 2006), Japan (Kato and Yamaguchi 2005; Hirano et al. 2004; Suga et al. 2003), Korea (Jo et al. 2000), on the Israel–Egypt border (Qin et al. 2001) and in the USA (Weng and Lu 2008; Yuan and Bauer 2007; Xian and Crane 2006; Weng et al. 2004; Lo and Quattrochi 2003; Streutker 2002; Aniello et al. 1995; Kim 1992; Carnahan and Larson 1990). The ASTER (advanced spaceborne thermal emission and reflection radiometer) and MODIS (moderate resolution image spectroradiometer) thermal sensors have also been used to determine LST around the world (Pongrácz et al. 2010; Imhoff et al. 2010; Rajasekar and Weng 2009; Nichol et al. 2009; Hung et al. 2006; Kato and Yamaguchi 2005; Nichol 2005; Ochi et al. 2002).

12.1.2 Use of LST and Other Measures in UHI Studies

In addition to LST derived from satellite data, a range of biophysical and non-biophysical factors such as vegetation indices and population density sourced from remote sensing and Geographic Information Systems (GIS) have been widely used to describe UHI, particularly the SUHI phenomenon. For instance, the normalised difference vegetation index (NDVI) is typically used due to the relationship between LST and NDVI such that the lower the vegetation index, the higher the temperature (Weng 2009). Similarly, non-biophysical variables such as population density can also be used to understand the impact of land use change on the effect of UHI (Li et al. 2012). In addition, fractional vegetation cover (FVC), an indicator of vegetation abundance, has recently been proposed to overcome the problem associated with NDVI-based relationships of LST and vegetation index (Weng et al. 2004). These studies demonstrated that satellite-based measurement of LST could assist in developing strategies for sustainable development and to improve urban residential environments (Song 2005). Further, planners can potentially use this information to minimise heat accumulation by urban surfaces, thereby improving human comfort and living quality (Nichol 2005; Liang and Weng 2008).

12.1.3 Potential Application in Dhaka

The megacity of Dhaka has experienced rapid urban development since the independence of Bangladesh in 1971 (Islam 1996). Although the level of urbanisation in Bangladesh is frequently defined using population parameters (Islam 1996, 2005; Karl et al. 1988), multi-temporal remote sensing-based studies demonstrate that the intensity of urban expansion in Dhaka is extremely fast (Dewan and Corner 2012; Griffiths et al. 2010; Dewan and Yamaguchi 2009a, b; 2008; also see Chaps. 5 and 6), causing large-scale changes in land surface parameters. Due to the scarcity of suitable residential plots and growing demand for shelter, vertical expansion has become dominant (Mallick and Mourshed 2012; Begum 2007). Both vertical and horizontal developments have made the city a 'city of concrete' (Asad and Ahsan 2012), and vegetation cover has been greatly reduced (Talukder et al. 2012). Consequently, the local precipitation and temperature regimes have been changed (Alam and Rabbani 2007). For instance, a long-term microclimatic study using time-series climatic data indicates that both maximum and minimum temperature have increased in response to rapid urbanisation (Roy 2011).

Traditionally, records from ground-based weather stations are used to study land surface temperature; however, the sparseness of measuring points can hinder the development of a spatial explicit LST model. Due to their large synoptic coverage, satellite-based LST measurements have received considerable interest as a means of examining the effect of land surface change caused by human activities on the pattern of LST. Since Dhaka possesses only one climatic station, analysing spatial patterns of surface temperature is only feasible with remotely sensed data. As shown above, various sensors have been used to study LST in diverse cities. So far there has been no study of the use of satellite measurements of the impact of land use changes on LST in Dhaka. The effect of UHI is poorly understood despite the fact of the noticeable occurrence of UHI due to unplanned and unregulated urban growth in recent years (Mallick and Mourshed 2012; Rana 2011; Islam 2008). This study intends to fill this gap with twofold aims: firstly, to analyse the impact of land use/cover changes on land surface temperature and, secondly, to map the changes in the spatial and temporal distribution of LST.

12.2 Materials and Methods

12.2.1 Study Area

The study area is a subset of Dhaka megacity and is shown as a natural colour composite image in Fig. 12.1. This Landsat TM image from April 2011 shows that it is a densely built-up area, mostly around a northwest to southeast axis. However, as discussed in Chap. 3 current urban growth patterns indicate that the city is expanding in all directions owing to the fact that the terrain of the city is flat.

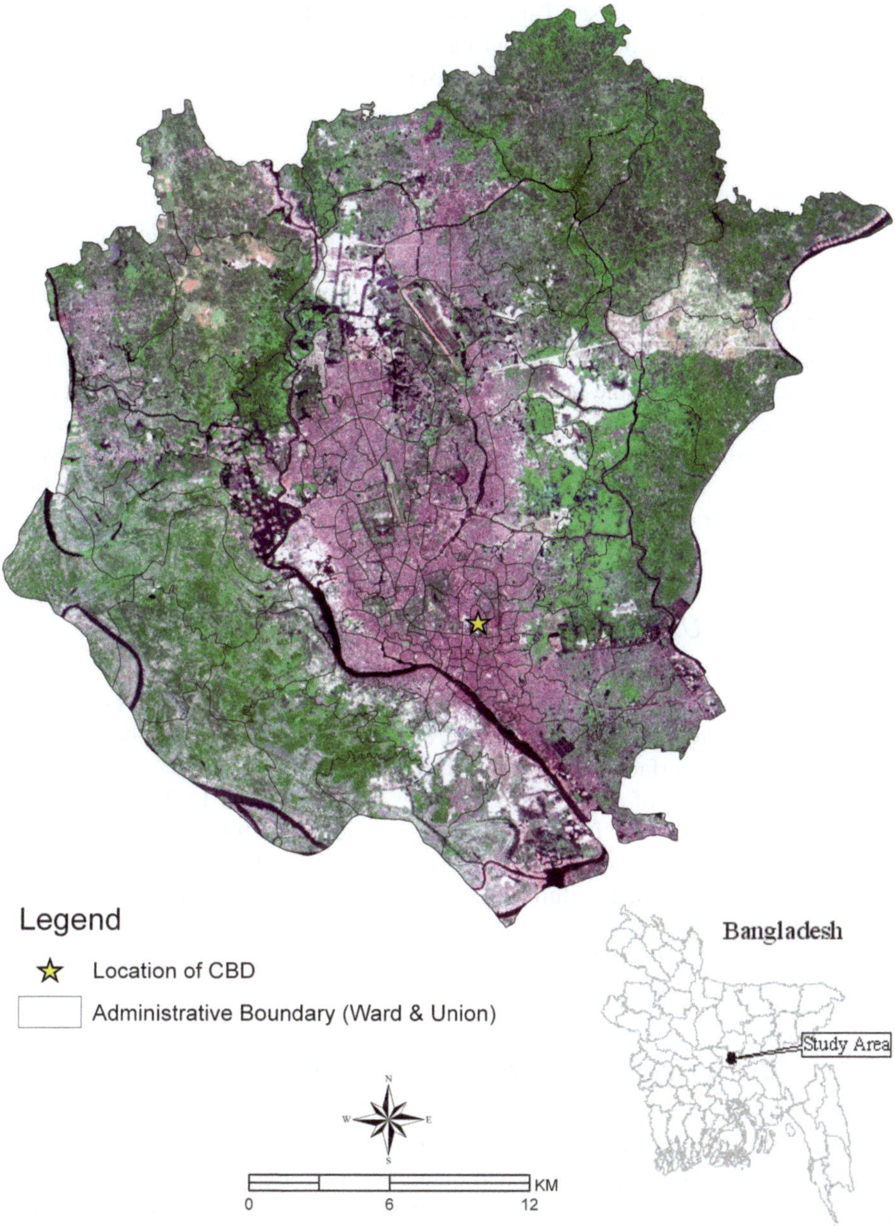

Fig. 12.1 Location of the study area shown in natural colour composite of Landsat TM 2011

Most of the surrounding areas are low-lying and act as water retention basins during the monsoon to alleviate flood damage. Due to ever-increasing population, most of the vegetative and open spaces have been encroached upon, severely impacting urban life and livelihoods (see Talukder et al. 2012). Since rapid urban expansion driven by massive population growth has brought about fundamental changes in land use and land cover patterns, this provides an opportunity for analysing the changing surface temperature patterns.

12.2.2 Data Acquisition and Processing

Three pairs of cloud-free Landsat TM 5 (thematic mapper) images for 07 January 1990, 02 February 2000 and 07 April 2011 were obtained from Geo-Informatics and Space Technology Development Agency (GISTDA), Thailand, and Bangladesh Space Research and Remote Sensing Organization (SPARRSO). A detailed description of image preprocessing is provided in Chap. 5. The 120-m resolution Landsat TM thermal infrared band (TIR) was extracted to derive land surface temperature. In addition, an administrative boundary polygon database was obtained from Bangladesh Space Research and Remote Sensing Organization (SPARRSO).

12.2.3 Land Use/Land Cover (LULC) Transition Matrix

The detailed procedures used to map land use/land cover (LULC) can be found in Chap. 5. The same datasets were utilised in this study to determine the impact of LULC changes on surface temperature. Here, a transition matrix was used to derive statistics and location of major LULC changes during 1990–2011. Specifically, the objective was to generate 'from–to' change maps (Jensen 1996) so that the magnitude and nature of land use/cover changes can be quantified and mapped out. A matrix function embedded within a GIS was used to obtain the transition matrix for two time spans (1990–2000 and 2000–2011), and relative changes between different LULC were calculated based on the derived statistics.

12.2.4 Retrieving Land Surface Temperature (LST)

Several methods exist for retrieving LST from satellite data, including the mono-window routine (Qin et al. 2001) and single-channel algorithm (Jimenez-Munoz and Sobrino 2003). A survey of the available methods of LST retrieval and their merits and demerits can be found elsewhere (Weng 2009; Quattrochi and Geol 1995). However, some of the methods require readily available information such as near real-time atmospheric profiles over the study area during satellite overpasses. In addition, the single-channel method cannot be used without providing atmospheric water vapour content; doing so is believed to increase uncertainties in the calculation of LST (Ma et al. 2010). Therefore, in this study an image-based approach (Lo and Quattrochi 2003; Weng 2001) was adopted to retrieve land surface temperature over different time periods.

The TIR band of Landsat TM 5 was used to retrieve LST for each year. First, the digital number (DN) of the TIR band was converted into spectral radiance using the following formula (Chander and Markham 2003):

$$L_\lambda = L_{\min} + \frac{L_{\max} - L_{\min}}{\mathrm{QCAL}_{\max} - \mathrm{QCAL}_{\min}}(\mathrm{DN} - \mathrm{QCAL}_{\min}) \tag{12.1}$$

where L_λ is the spectral radiance in W/(m^2 sr μm) received by the sensor for the pixel to be analysed. $\mathrm{QCAL}_{\max}$ is the maximum DN (= 255) and $\mathrm{QCAL}_{\min}$ is the minimum DN (= 0), respectively. $L_{\max}$ and $L_{\min}$ are the top-of-atmospheric (TOA) radiances that are scaled to $\mathrm{QCAL}_{\min}$ and $\mathrm{QCAL}_{\max}$ in W/(m^2 sr μm). Next, the spectral radiance was converted to blackbody temperature using Eq. 12.2:

$$T_b = \frac{K_2}{\ln[(K_1/L_\lambda) + 1]} \tag{12.2}$$

where T_b is the effective at-satellite brightness temperature in Kelvin (K), L_λ is spectral radiance in W/(m^2 sr μm) and K_1 and K_2 are prelaunch calibration constants in K. For Landsat TM, $K_2 = 1{,}260.56$ K and $K_1 = 607.76$ W/(m^2 sr μm). The T_b was further corrected for emissivity using the method described by Weng (2001) and Nichol (1994) to retrieve the final LST:

$$\mathrm{LST} = \frac{T_b}{1 + (\lambda T(K)/\rho)\ln\varepsilon} \tag{12.3}$$

where $\lambda = 11.5$ μm (wavelength of emitted radiance in metre) (Markham and Barker 1986), $\rho = hc/\sigma\,(1.438 \times 10^{-2}$ mK), K is the Stefan–Boltzmann's constant (1.38×10^{-23} J K^{-1}), h is Planck's constant (6.26×10^{-34} J s), c is the velocity of light (2.998×10^8 s^{-1}) and ε is the surface emissivity. Generally, the emissivity value for vegetated surface is 0.95, non-vegetated surface is 0.923 and for water areas 0.992 (Nichol 1994, 2005; Artis and Carnahan 1982). The derived LST images were then converted to the most commonly used unit – Celsius – by adding absolute zero (approximately −273.5 °C) (Xu and Chen 2004).

Since the images represent different years and season, derived LST images were normalised using the method suggested by Carlson and Arthur (2000) (Eq. 12.4) so that comparison can be made between years:

$$\mathrm{LST}' = \frac{\mathrm{LST} - \mathrm{LST}_{\min}}{\mathrm{LST}_{\max} - \mathrm{LST}_{\min}} \tag{12.4}$$

Where LST′ is the normalised LSTs and $\mathrm{LST}_{\min}$ and $\mathrm{LST}_{\max}$ are the minimum and maximum values of the each LST image, respectively.

12.2.5 *Extraction of Biophysical Parameters*

To understand the intensity of human activities and the impact of LULC changes on the variation of surface temperature, many researchers have investigated the relationship between LST and biophysical indicators such as normalised difference vegetation index (NDVI), fractional vegetation cover (Gillies et al. 1997) and percent impervious surface (%ISA) (e.g. Yuan and Bauer 2007; Weng 2001; Gallo et al. 1993). Since the megacity of Dhaka is experiencing massive urban development, examining LST in relation to biophysical variables could be useful to see the potential UHI effects. Hence, this study considered both NDVI and vegetation fraction to illustrate the relationship in this rapidly growing city. NDVI was calculated using the standard method (Eq. 12.5):

$$\mathrm{NDVI} = \frac{\mathrm{TM}_4 - \mathrm{TM}_3}{\mathrm{TM}_4 + \mathrm{TM}_3} \tag{12.5}$$

where TM_3 is band 3, the red band (0.63–0.69 μm), and TM_4 is band 4, the near-infrared band (0.76–0.90 μm), of Landsat TM data.

Because NDVI values are subject to error due to seasonal variation (Lo and Quattrochi 2003), a scaled NDVI for each year was computed using the formula (Eq. 12.6) suggested by Gillies et al. (1997):

$$N^* = \frac{\mathrm{NDVI} - \mathrm{NDVI}_{\mathrm{min}}}{\mathrm{NDVI}_{\mathrm{max}} - \mathrm{NDVI}_{\mathrm{min}}} \tag{12.6}$$

where N^* is the scaled NDVI value, $\mathrm{NDVI}_{\mathrm{min}}$ is the minimum NDVI value and $\mathrm{NDVI}_{\mathrm{max}}$ is the maximum NDVI value. The scaled N^* was further used to calculate fractional vegetation cover (F_r) using the equation below:

$$F_r = f_{N*}(N^*) \tag{12.7}$$

where f_{N*} refers to a polynomial fit for F_r as a function of N^* (Gillies et al. 1997).

In order to examine the relationship between LST and F_r, a set of 1,000 randomly distributed sample points was generated. At these points, the values of the mean LST and mean F_r were extracted and used in a simple regression analysis.

12.3 Results and Discussion

To evaluate the results of LULC conversion, matrices of land cover change from 1990 to 2000, 2000 to 2011 and 1990 to 2011 were calculated and the location of LULC changes was determined (Fig. 12.2). Further, relative changes between years were also computed (Fig. 12.3). This reveals that the built-up category has become the dominant LULC type since the 1990s whereas natural cover such as agriculture and wetlands are gradually disappearing, suggesting tremendous pressure on

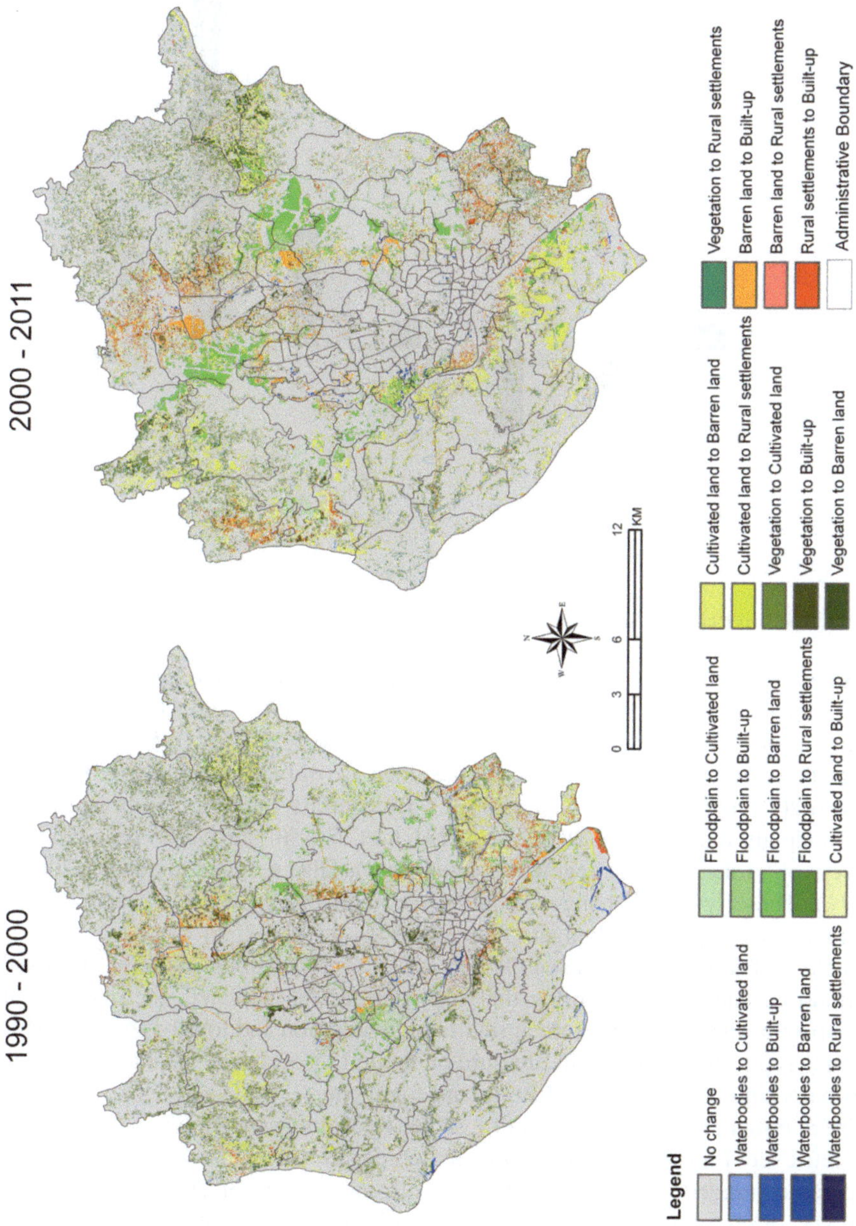

Fig. 12.2 Major LULC conversions in the study area

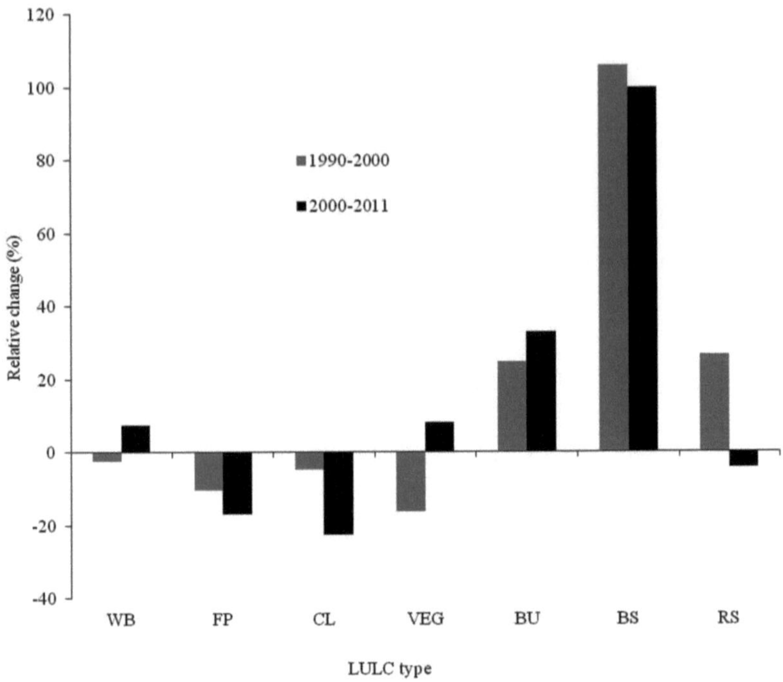

Fig. 12.3 Relative changes in LULC, 1990–2011

a limited resource base. Analysis showed that built-up areas increased from 11,696 ha (13.3 %) to 19,556 ha (22.2 %) between 1990 and 2011 whilst the increment of bare soils was from 2,216 ha (2.5 %) to 9,239 ha (10.5 %), a 311 % change in the bare soil category. The greatest increase of built-up areas was observed between 2001 and 2011 (4,915 ha or 33 %). By contrast, the loss of cultivated land and floodplain during 1990–2011 was 7,234 ha (26.2 %) and 7,206 ha (25.4 %), respectively. Likewise, the vegetation category was also reduced to about 1,206 ha (9.3 %). These types of changes should have noteworthy impact on land surface albedo and surface energy budget (Weng 2009).

Table 12.1 shows a synopsis of the major LULC conversions, namely, 'from–to' information, that occurred during the study period. As indicated, the majority of built-up area was converted from agricultural, vegetation and low-lying areas. This evidence suggests the emergent pressure on natural resources in Dhaka to meet the ever-increasing demand of residential land for shelter.

The summarised characteristics of LST on three dates are shown in Table 12.2. The retrieved LST for the 1990 image represents the winter season, and the temperature ranged from 9.8 to 27.5 °C with an average of 20.7 °C. A similar trend was observed in the year 2000 when the LST ranged from 4.7 to 33.3 °C. In contrast, the 2011 image was acquired during early summer and shows a sharp increase in surface radiant temperature. The minimum and maximum value was found between 25.7 and 38.7 °C with an average of 29.5 °C in 2011 (Table 12.2).

Table 12.1 Major land use/cover conversions from 1990 to 2011

		1990–2000	2000–2011	1990–2011
'From class'	'To class'	Area (ha)	Area (ha)	Area (ha)
Water bodies	Built-up	138.06	201.51	186.93
	Bare soil	121.5	96.39	179.91
	Cultivated land	218.97	128.88	218.97
	Rural settlements	31.14	19.26	34.56
Floodplain	Built-up	837.99	931.41	2,092.14
	Bare soil	2,642.67	1,325.88	2,065.86
	Cultivated land	862.92	2,780.46	3,497.04
	Rural settlements	442.71	347.67	550.17
Cultivated land	Built-up	1,362.6	2,168.55	3,349.62
	Bare soil	1,980.99	3,544.02	3,723.48
	Rural settlements	1,803.33	1,756.44	1,963.44
Vegetation	Built-up	857.25	843.48	1,460.97
	Cultivated land	3,131.46	2,394.63	3,050.73
	Bare soil	347.13	455.13	738.99
	Rural settlements	835.56	864.99	972.81
Bare soil	Built-up	347.13	1113.03	566.1
	Rural settlements	96.48	216.63	97.11
Rural settlements	Built-up	790.56	1,266.84	1,317.87
	Bare soil	287.82	302.22	308.79

Table 12.2 Summary of retrieved LST from 1990 to 2011

Statistics	LST_{1990}	LST_{2000}	LST_{2011}
Minimum	9.8	4.7	25.7
Mean	20.7	23.2	29.5
Maximum	27.5	33.3	38.7
Standard deviation	1.0	1.3	1.9

Examination of the spatial distribution of ambient temperature between 1990 and 2011 reveals a very striking feature (Fig. 12.4) whilst compared with 3-year NDVI maps. Areas of growth in elevated surface temperature tended to increase with concurrent decrease of greenery areas, suggesting increased anthropogenic activities, particularly the conversion of natural surfaces to man-made features (Jo et al. 2000). The image of LST in the winter of 1990 shows that higher temperatures were mostly confined to the city centre, the Tejgaon industrial belt and around the two airports where building density as well as the number of concrete structures are relatively higher. Another pocket of elevated surface temperature appeared in the further north and northwestern part of the study area during the same period. The late winter LST image of 2000 shows that the elevated temperature was distributed over most of the urban areas that could be associated with both season and changes in land surface. Table 12.3 shows the normalised temperature of different land cover classes, as calculated from the three images. The important information here is the relative magnitude of each class in each year, not a comparison of values across years. The temperature of built-up areas and urban

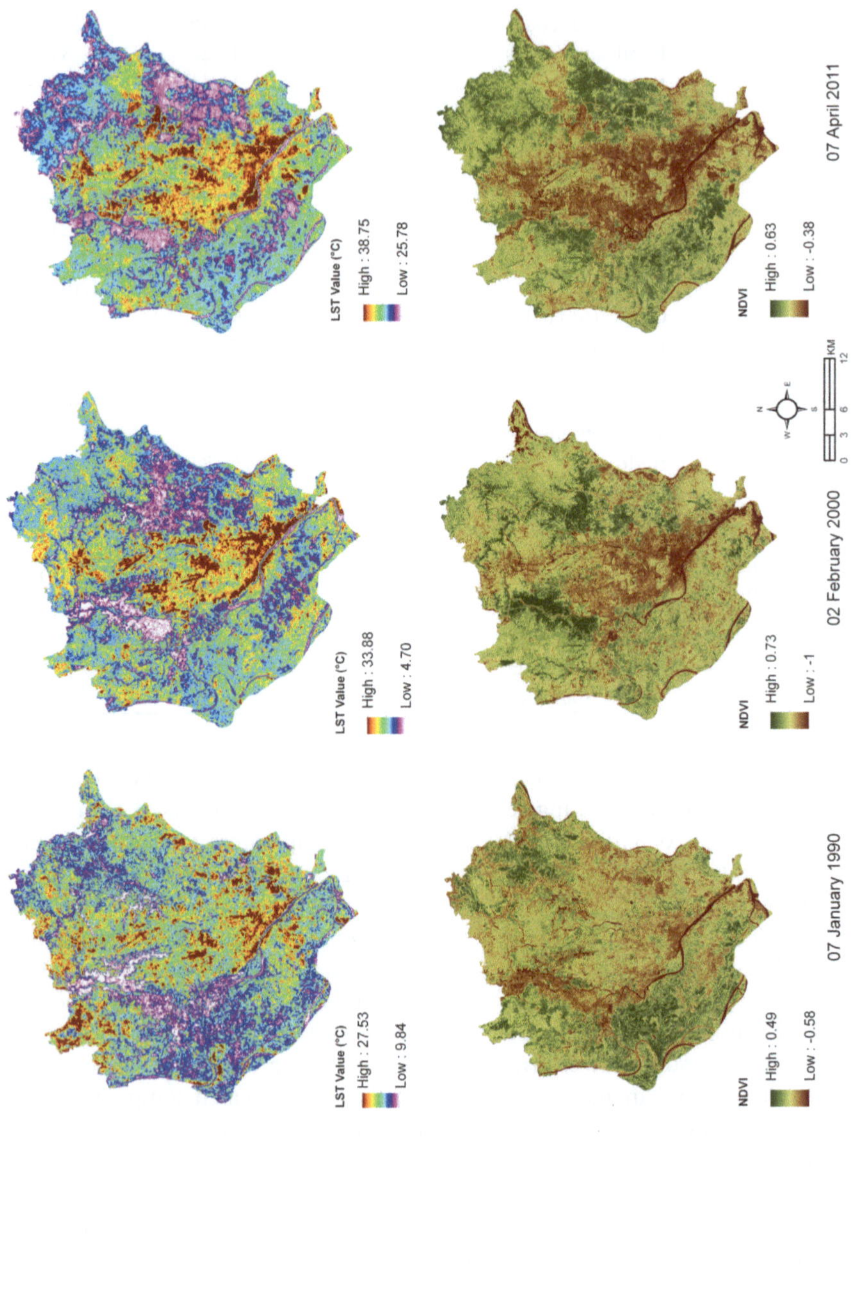

Fig. 12.4 Spatial patterns of LST (°C) and corresponding NDVI, 1990–2011

Table 12.3 Mean normalised surface temperature by land use/cover type

	1990		2000		2011	
Land use/cover type	Mean	Std dev (±)	Mean	Std dev (±)	Mean	Std dev (±)
Built-up	0.66	0.05	0.69	0.04	0.45	0.11
Bare soil	0.64	0.07	0.67	0.04	0.41	0.12
Cultivated	0.63	0.05	0.64	0.03	0.27	0.08
Floodplain	0.59	0.06	0.60	0.03	0.14	0.09
Rural settlements	0.63	0.05	0.65	0.03	0.28	0.08
Vegetation	0.61	0.05	0.64	0.03	0.27	0.08
Waterbodies	0.58	0.05	0.60	0.03	0.18	0.11

areas (collectively termed as built-up) has been consistently higher than that of other land use classes, although in the January and February images (1990 and 2000), there are other categories whose mean temperatures are almost as high. It is the set of values from the April image that is of particular interest. Referring to Table 12.2 shows that the raw temperatures for the study area are considerably higher at this time, yet the mean normalised temperatures for each land use category appear low. The apparent lower values are an artefact of the normalisation method (Eq. 12.4), which is very sensitive to the range and statistical distribution of temperatures within each image. What is however apparent from Table 12.3 is that the mean normalised temperature for the built-up area is considerably higher than that for all other categories, with the exception of bare soil. Since bare soil has similar thermal characteristics to many building materials, this is not surprising. We can conclude from Table 12.3 that in April 2011 the thermal difference between built-up areas and their surroundings was much more pronounced than that in February 2000 or January 1990. From the data at hand, it is not possible to definitively separate seasonal differences from those associated with land use change and further work is required in this area. It is, however, possible to state that the temperature gradients between built-up areas and their surroundings in April 2011 are high. It is these gradients that are diagnostic of the UHI effect.

In order to depict the spatial distribution of temperature changes, a temperature change map was produced by subtracting the normalised LST map of 2000 from that of 2011. This map (Fig. 12.5) helps to visualise the impact of LULC changes and consequent urban expansion on surface radiant temperature in the study area.

In order to quantify the relationship between LST and biophysical parameters, two approaches were used. In the first approach, a correlation analysis was conducted between surface temperature and NDVI for each land use/cover type on pixel-by-pixel basis (Weng 2001). In the second approach, mean LST was regressed against mean F_r, and a prediction model was developed. The correlation analysis revealed that the water bodies category persistently showed positive correlation with NDVI (Table 12.4). This might be attributable to the known serious water pollution problems in Dhaka (see Chap. 16). As seen in Table 12.4, the correlation coefficient for the water bodies category was highest in 1990. The 1990 image was taken in January at a time of low river flow and increasing untreated effluent from surrounding industries which exacerbate pollution loads in the rivers

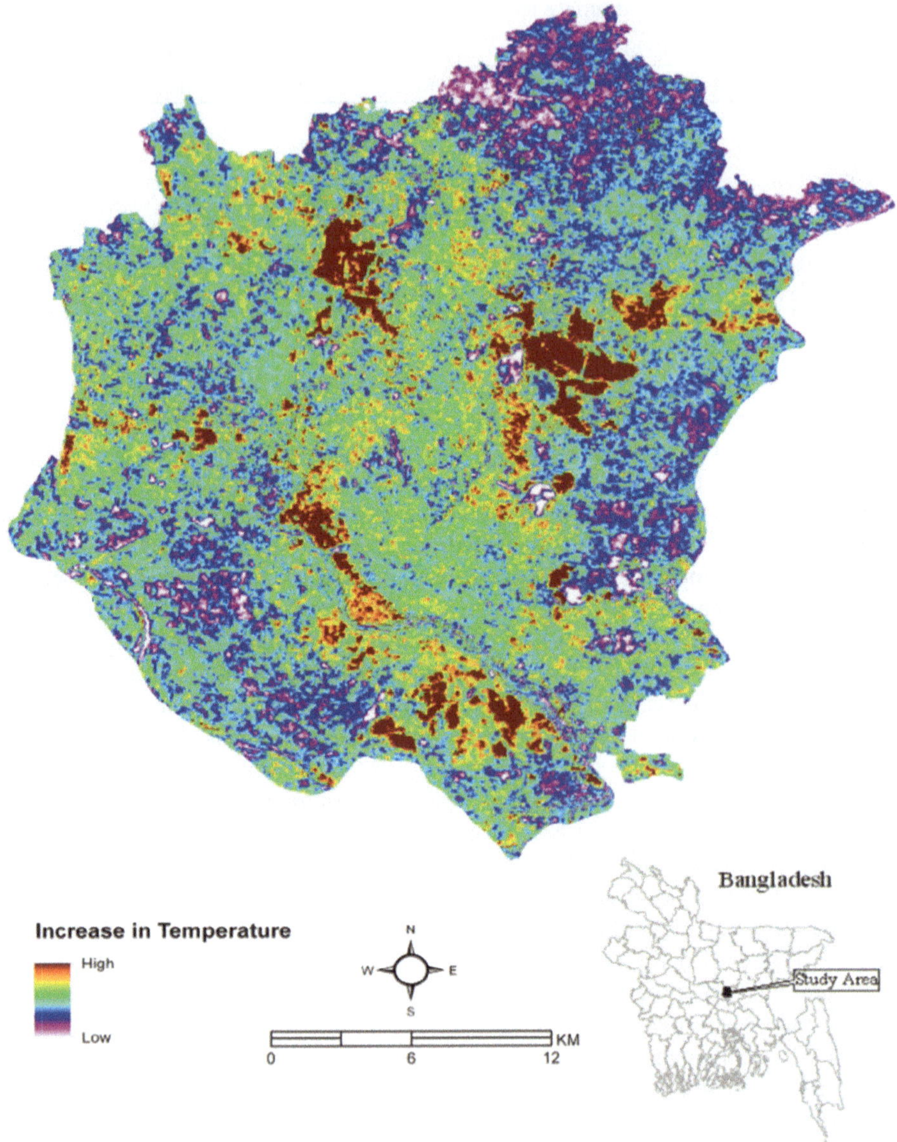

Fig. 12.5 Spatial distribution of temperature increase between 2000 and 2011

(Sohel et al. 2003). The highest negative correlation coefficient was found for the built-up category in 2000 ($r = -0.36$) and 2011 ($r = -0.45$). The vegetation category was also negatively correlated during the study period as were rural settlements. The regression analysis of mean LST and F_r indicated that there exists a significant negative relationship between their annual means of LST and F_r (Fig. 12.6) that mathematically can be expressed as $\mathrm{LST} = 22.873\mathrm{e}^{-0.172Fr}$

Table 12.4 Pearson correlation coefficients between LST and NDVI by LULC type (significant at 0.05)

LULC type	1990	2000	2011
Water bodies	0.30	0.08	0.12
Floodplain	0.20	−0.31	−0.32
Cultivated land	0.34	−0.10	−0.39
Vegetation	−0.33	−0.30	−0.37
Built-up	−0.19	−0.36	−0.45
Bare soil	0.19	−0.21	−0.37
Rural settlements	−0.30	−0.19	−0.32

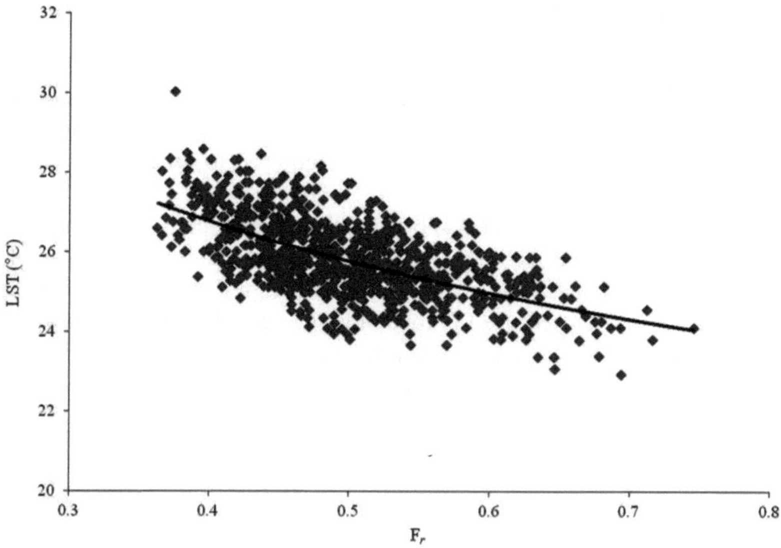

Fig. 12.6 Relationship between annual mean LST and F_r

($r^2 = 0.38$). Though the regression modelling resulted in relatively low correlation coefficient value compared with other studies reported in the literature, this may be attributed to the existence of widespread heterogeneous LULC categories in the study area. Further research is therefore warranted. The result also demonstrates that lack of green space promotes the existence of detectable heat exchange (Xu et al. 2009), which causes the UHI effect. In other words, a decrease in biomass leads to an increase in temperature. This also supports the findings of previous time-series Landsat-based study in Shanghai (Li et al. 2012).

12.4 Conclusions

The objective of this chapter was to determine the impact of LULC changes on surface radiant temperature. Using a transition matrix, land use and cover changes were identified. The TIR band of Landsat TM 5 was used to extract multi-temporal thermal properties of the urban environment. The derived LSTs were statistically analysed, and the relationship between LSTs and biophysical parameters was established through correlation and regression analyses.

The transition matrix of LULC changes revealed that the expansion of urban built-up surfaces has been very rapid, causing depletion of natural resources and concurrently impacting the surface temperature distribution. For instance, 'from class' to 'to class' statistics revealed that 1,362 ha of cultivated land was converted to built-up category during 1990–2000 and the maximum transformation of this category occurred during 2000–2011, amounting a total of 2,168 ha. Likewise, 3,050 ha of vegetated lands were transformed to cultivated land between 1990 and 2011.

Analysis of the normalised surface temperature values suggests that the temperature gradient between the built-up areas and their surroundings is more pronounced during the hottest part of the year. This is commensurate with the existence of UHI effects, but further work is required to quantitatively diagnose its causes.

The relationship between LST and biophysical parameters was explored through correlation and regression techniques which indicated that persistent loss of green space favours an increase in detectable heat exchange that is a driver of the UHI effect. This study could be useful to planners and relevant civic management authority since an increasingly heated urban environment directly affects the comfort and health of urban population.

References

Alam M, Rabbani MG (2007) Vulnerabilities and responses to climate change for Dhaka. Environ Urban 19(1):81–97

Amiri R, Weng Q, Alimohammadi A, Alavipanah SK (2009) Spatial-temporal dynamics of land surface temperature in relation to fractional vegetation cover and land use/cover in the Tabriz urban area, Iran. Remote Sens Environ 113(12):2606–2617

Aniello C, Morgan K, Busbey A, Newland L (1995) Mapping micro-urban heat islands using Landsat TM and a GIS. Comput Geosci 21(8):965–969

Artis DA, Carnahan WH (1982) Survey of emissivity variability in thermography of urban areas. Remote Sens Environ 12(4):313–329

Asad R, Ahsan R (2012) Structuring urban sustainability with water: a case of Kamrangir chor, Dhaka, Bangladesh. J Soc Dev Sci 3(8):293–303

Balling RC, Brazell SW (1988) High resolution surface temperature patterns in a complex urban terrain. Photogramm Eng Remote Sens 54(9):1289–1293

Begum A (2007) Urban housing as an issue of redistribution through planning? The case of Dhaka city. Soc Policy Adm 41(4):410–418

Carlson TN, Arthur ST (2000) The impact of land use – land cover changes due to urbanization on surface microclimate and hydrology: a satellite perspective. Glob Planet Chang 25(1–2):49–65

Carlson TN, Augustine JA, Boland FE (1977) Potential application of satellite temperature measurements in the analysis of land use over urban areas. Bull Am Meteorol Soc 58:1301–1303

Carnahan WH, Larson RC (1990) An analysis of an urban heat sink. Remote Sens Environ 33 (1):65–71

Chander G, Markham B (2003) Revised Landsat-5 TM radiometric calibration procedures and postcalibration dynamic ranges. IEEE Trans Geosci Remote Sens 41(11):2674–2677

Chen X-L, Zhao H-M, Li P-X, Yin Z-Y (2006) Remote sensing image-based analysis of the relationship between urban heat island and land use/cover changes. Remote Sens Environ 104 (2):133–146

Dewan AM, Corner RJ (2012) The impact of land use and land cover changes on land surface temperature in a rapidly urbanizing megacity. In: IGARSS 2012, Munich

Dewan AM, Yamaguchi Y (2008) Effects of land cover changes on flooding: example from Greater Dhaka of Bangladesh. Int J Geoinform 4(1):11–20

Dewan AM, Yamaguchi Y (2009a) Land use and land cover change in Greater Dhaka, Bangladesh: using remote sensing to promote sustainable urbanization. Appl Geogr 29 (3):390–401

Dewan AM, Yamaguchi Y (2009b) Using remote sensing and GIS to detect and monitor land use and land cover change in Dhaka Metropolitan of Bangladesh during 1960–2005. Environ Monit Assess 150(1–4):237–249

Gallo KP, Owen TW (1998) Assessment of urban heat island: a multi-sensor perspective for the Dallas – Ft. Worth, USA region. Geocarto Int 13(4):35–41

Gallo KP, McNab AL, Karl TR, Brown JF, Hood JJ, Tarpley JD (1993) The use of NOAA AVHRR data for assessment of the urban heat island effect. J Appl Meteorol 32(5):899–908

Gillies RR, Kustas WP, Humes KS (1997) A verification of the 'triangle' method for obtaining surface soil water content and energy fluxes from remote measurements of the Normalized Difference Vegetation Index (NDVI) and surface radiant temperature. Int J Remote Sens 18 (15):3145–3166

Griffiths P, Hostert P, Gruebner O, van der Linden S (2010) Mapping megacity growth with multi-sensor data. Remote Sens Environ 114(12):426–439

Hirano Y, Yasuoka Y, Ichinose T (2004) Urban climate simulation by incorporating satellite-derived vegetation cover distribution into a mesoscale meteorological model. Theor Appl Climatol 79(3–4):175–184

Hung T, Uchihama D, Ochi S, Yasuoka Y (2006) Assessment with satellite data of the urban heat island effects in Asian mega cities. Int J Appl Earth Obs Geoinf 8(1):34–48

Imhoff ML, Zhang P, Wolfe RE, Bounoua L (2010) Remote sensing of urban heat island effect across biomes in the continental USA. Remote Sens Environ 114(3):504–513

Islam N (1996) Dhaka from city to megacity: perspectives on people, places, planning and development issues. Urban Studies Program, Department of Geography, University of Dhaka, Dhaka

Islam N (2005) Dhaka now: contemporary urban development. The Bangladesh Geographical Society (BGS), Dhaka

Islam SM (2008) Blazing Dhaka: an urban heat island. The Daily Star (Environment), 25 October 2008

Jensen JR (1996) Introductory digital image processing: a remote sensing perspective. Prentice-Hall, Upper Saddle

Jiang J, Tian G (2010) Analysis of the impact of land use/land cover change on land surface temperature with remote sensing. Proced Environ Sci 2:571–575

Jimenez-Munoz JC, Sobrino JA (2003) A generalised single channel method for retrieving land surface temperature from remote sensing data. J Geophys Res 108:4688

Jo MH, Lee KJ, Shin JH, Suh AS, Oh SN (2000) Surface temperature analysis of urban area using RS and GIS. In: 21st Asian conference of remote sensing (ACRS), Asian Association of Remote Sensing (AARS), Taipei, Taiwan, 4–8 December 2000

Kalnay E, Cai M (2003) Impact of urbanization and land-use change on climate. Nature 423:528–531

Karl TR, Diaz HF, Kukla G (1988) Urbanization: its detection and effect in the United States climate record. J Clim 1(11):1099–1123

Kato S, Yamaguchi Y (2005) Analysis of urban heat-island effect using ASTER and ETM+ data: separation of anthropogenic heat discharge and natural heat radiation from sensible heat flux. Remote Sens Environ 99(1–2):44–54

Katpatal YB, Kute A, Satapathy DJ (2008) Surface- and air-temperature studies in relation to land use/land cover of Nagpur urban area using Landsat 5 TM data. J Urban Plan Dev 134 (4):110–118

Kidder SQ, Wu HT (1987) A multispectral study of the St. Louis area under snow-covered conditions using NOAA-7 AVHRR data. Remote Sens Environ 22(2):159–172

Kim HH (1992) Urban heat island. Int J Remote Sens 13(12):2319–2336

Landsberg HE (1981) The urban climate. Academic Press, New York

Lee H-Y (1993) An application of NOAA AVHRR thermal data to the study of urban heat islands. Atmos Environ 27B(1):1–13

Li Y-Y, Zhang H, Kainz W (2012) Monitoring patterns of urban heat islands of the fast-growing Shanghai metropolis, China: Using time-series of Landsat TM/ETM+ data. Int J Appl Earth Obs Geoinf 19:127–138

Liang B, Weng Q (2008) Multiscale analysis of census-based land surface temperature variations and determinants in Indianapolis, United States. J Urban Plan Dev 134(3):129–139

Lo CP, Quattrochi DA (2003) Land-use and land-cover change, urban heat island phenomenon, and health implications: a remote sensing approach. Photogramm Eng Remote Sens 69 (9):1053–1063

Lu Y, Feng X, Xiao P, Shen C, Sun J (2009) Urban heat island in summer of Nanjing based on TM data. In: Proceedings of Urban Remote Sensing Joint Event (IEEE), Shanghai, China, 20–22 May 2009, pp 1–5

Ma Y, Kuang Y, Huang Y (2010) Coupling urbanization analyses for studying urban thermal environment and its interplay with biophysical parameters based on TM/ETM+ imagery. Int J Appl Earth Observ Geoinf 12(2):110–118

Mallick FH, Mourshed M (2012) Our urban future. The Daily Star, July 7

Mallick J, Kant Y, Bharath BD (2008) Estimation of land surface temperature over Delhi using Landsat – 7 ETM+. J Indian Geophys Union 12(3):131–140

Markham BL, Barker JL (1986) Landsat MSS and TM post-calibration dynamic ranges, exoatmospheric reflectance and at-satellite temperatures. EOSAT Landsat technical notes

Matson M, McClain EP, McGinnis DF Jr, Pritchard JA (1978) Satellite detection of urban heat islands. Mon Weather Rev 106:1725–1734

Nichol JE (1994) A GIS-based approach to microclimate monitoring in Singapore's high-rise housing estates. Photogramm Eng Remote Sens 60(10):1225–1232

Nichol JE (2005) Remote sensing of urban heat islands by day and night. Photogramm Eng Remote Sens 71(5):613–623

Nichol JE, Fung WE, Lam K-s, Wong MS (2009) Urban heat island diagnosis using ASTER images and 'in situ' air temperature. Atmos Res 94(2):276–284

Ochi S, Uchihama D, Takeuchi W, Yasuoka Y (2002) Monitoring urban heat environment using MODIS data for main cities in East Asia. In: Proceedings of the Asian conference on GIS, GPS, Aerial Photography and Remote Sensing, Bangkok, 7–8 August 2002, p 84

Oke TR (1982) The energetic basis of the urban heat island. Q J R Meteorol Soc 108(455):1–24

Oke TR (1995) The heat island of the urban boundary layer: characteristics, causes and effects. In: Cermak JE, Davenport AG, Plate EJ, Viegas DX (eds) Proceedings of the NATO advanced study institute on wind climate in cities. Kluwer Academic Publishers, Waldbronn, pp 81–107

Owen TW, Carlson TN, Gillies RR (1998) An assessment of satellite remotely-sensed land cover parameters in quantitatively describing the climatic effect of urbanization. Int J Remote Sens 19(9):1663–1681

Pongrácz R, Bartholy J, Dezso Z (2010) Application of remotely sensed thermal information to urban climatology of central European cities. Phys Chem Earth 35(1–2):95–99

Price JC (1979) Assessment of the urban heat island effect through the use of satellite data. Mon Weather Rev 107:1554–1557

Qian L-X, Cui H-S, Jie H (2006) Impacts of land use and cover change on land surface temperature in the Zhujiang delta. Pedosphere 16(6):681–689

Qin Z, Karnieli A, Berliner P (2001) A mono-window algorithm for retrieving land surface temperature from Landsat TM data and its application to the Israel-Egypt border region. Int J Remote Sens 22(18):3719–3746

Quattrochi DA, Geol NS (1995) Spatial and temporal scaling of thermal remote sensing data. Remote Sens Rev 12(3–4):255–286

Rajasekar U, Weng Q (2009) Urban heat island monitoring and analysis by data mining of MODIS imageries. ISPRS J Photogramm Remote Sens 64(1):86–96

Ramachandra TV, Kumar U (2009) Land surface temperature with land cover dynamics: multi-resolution, spatio-temporal analysis of greater Bangalore. Int J Geoinf 5(3):43–53

Rana MMP (2011) Urbanization and sustainability: challenges and strategies for sustainable urban development in Bangladesh. Environ Dev Sustain 13(1):237–256

Rao PK (1972) Remote sensing of urban heat islands from an environmental satellite. Bull Am Meteorol Soc 53:647–648

Rose L, Devadas MD (2009) Analysis of land surface temperature and land use/land cover types using remote sensing imagery – a case in Chennai city, India. In: The seventh international conference on urban climate, Yokohama, Japan, 29 June–3 July 2009

Roth M, Oke TR, Emery WJ (1989) Satellite-derived urban heat islands from three coastal cities and the utilization of such data in urban climatology. Int J Remote Sens 10(11):1699–1720

Roy S (2011) Impact of urbanization on microclimatic change of Dhaka City. Lambert Academic Publishing, Germany

Seto KC, Sánchez-Rodríguez R, Fragkias M (2010) The new geography of contemporary urbanization and the environment. Annu Rev Environ Resour 35:167–194

Sohel KMA, Chowdhury MAI, Ahmed MF (2003) Surface water quality in and around Dhaka City. J Water Supply: Res Technol 52(2):141–153

Song C (2005) Spectral mixture analysis for sub pixel vegetation fractions in the urban environment: how to incorporate endmember variability. Remote Sens Environ 95(2):248–263

Srivastava PK, Majumdar TJ, Bhattacharya AK (2009) Surface temperature estimation in Singhbhum Shear Zone of India using Landsat-7 ETM+ thermal infrared data. Adv Space Res 43:1563–1574

Streutker DR (2002) A remote sensing study of the urban heat island of Houston, Texas. Int J Remote Sens 23(13):2595–2608

Streutker DR (2003) Satellite-measured growth of the urban heat island of Houston, Texas. Remote Sens Environ 85(3):282–289

Su W, Gu C, Yang G (2010) Assessing the impact of land use/land cover on urban heat island pattern in Nanjing city, China. J Urb Plan Dev 136(4):365–372

Suga Y, Ogawa H, Ohno K, Yamada K (2003) Detection of surface temperature from Landsat-7/ETM+. Adv Space Res 32(11):2235–2240

Talukder B, Nakagoshi N, Dewan AM (2012) Urbanization and green space dynamics in Greater Dhaka, Bangladesh. Landsc Ecol Eng 8(1):45–58

Tan KC, Lim HS, MatJafri MZ, Abdullah K (2009) Study on land surface temperature based on Landsat image over Penang island, Malaysia., proceedings of the Sixth International Conference on Computer Graphics, Imaging and Visualization (IEEE), Tianjin, China, 11–14 August 2009, pp 525–529

United Nations (2012) World population prospects: the 2010 revision, vol 1. United Nations Department of Economic and Social Affairs, New York. Available at: http://eas.un.org/undp/wpp/Documentation/publications.html. Accessed 1 Apr 2012

Van TT, Trung LV, Lan HT (2009) Application of thermal remote sensing in study on surface temperature distribution of Ho Chi Minh city. In Spatial data serving people: land governance and the environment – building the capacity, 7th FIG regional conference, Hanoi, Vietnam, 19–22 October 2009

Voogt JA, Oke TR (2003) Thermal remote sensing of urban climates. Remote Sens Environ 86 (3):370–384

Weng Q (2001) A remote sensing – GIS evaluation of urban expansion and its impact on surface temperature in the Zhujiang delta, China. Int J Remote Sens 22(10):1999–2014

Weng Q (2003) Fractal analysis of satellite-derived urban heat island effect. Photogramm Eng Remote Sens 69(5):555–566

Weng Q (2009) Thermal, infrared remote sensing for urban climate and environment studies: methods, applications and trends. ISPRS J Photogramm Remote Sens 64(4):335–344

Weng Q (2010) Remote sensing and GIS integration: theories, methods and applications. McGraw-Hill, New York

Weng Q, Lu D (2008) A sub-pixel analysis of urbanization effect on land surface temperature and its interplay with impervious surface and vegetation coverage in Indianapolis, United States. Int J Appl Earth Obs Geoinf 10(1):63–83

Weng Q, Lu D, Schubring J (2004) Estimation of land surface temperature – vegetation abundance relationship for urban heat island studies. Remote Sens Environ 89(4):467–483

Wilby RL (2007) A review of climate change impacts on the built environment. Built Environ 33 (1):31–45

Xian G, Crane M (2006) An analysis of urban thermal characteristics and associated land cover in Tampa Bay and Las Vegas using Landsat satellite data. Remote Sens Environ 104(2):147–156

Xiao H, Weng Q (2007) The impact of land use and land cover changes on land surface temperature in a karst area of China. J Environ Manage 85(1):245–257

Xiao R, Weng Q, Quyang Z, Li W, Schienke EW, Zhang W (2008) Land surface temperature variation and major factors in Beijing, China. Photogramm Eng Remote Sens 74(4):451–461

Xu H-Q, Chen B-Q (2004) Remote sensing of the urban heat island and its changes in Xiamen city of SE China. J Environ Sci 16(2):276–281

Xu H, Ding F, Wen X (2009) Urban expansion and heat island dynamics in the Quanzhou region, China. IEEE J Sel Top Appl Earth Obs Remote Sens 2(2):74–79

Yow DM (2007) Urban heat islands: observations, impacts and adaptation. Geogr Compass 1 (6):1227–1251

Yuan F, Bauer ME (2007) Comparison of impervious surface area and normalized difference vegetation index as indicators of surface urban heat island effects in Landsat imagery. Remote Sens Environ 106(3):375–386

Chapter 13
Illustrating Quality of Life (QOL)

Ashraf M. Dewan, Kamrun Nahar, and Yohei Kawamura

Abstract The objective of this chapter is to develop a quality of life (QOL) index at community level in the Dhaka City Corporation (DCC) area, which could describe the spatial patterns of QOL. Using remote sensing, census and other spatial data, a factor analysis was carried out to develop the different dimensions of QOL. Three principal factors were extracted from the analysis: environmental, economic and demographic. These three factors were then combined in a Geographic Information Systems (GIS) environment to construct a synthetic QOL for the study area. The results were subsequently validated using regression analysis, which revealed a better prediction of QOL based on environmental and socioeconomic variables. Interestingly, only a small portion of the population (1.4 %) in the study area was shown to have good QOL. As higher urban growth driven by rapid rural-urban migration is expected in Dhaka in the coming years, this study will be of substantial help for urban planners and policymakers in formulating related policies to ensure a better living environment for its inhabitants.

Keywords Quality of life • Landsat TM • Urban environmental quality • Dhaka City Corporation • Community • Population census

---

A.M. Dewan (✉)
Department of Spatial Sciences, Curtin University, Kent Street, Bentley, Perth 6102, WA, Australia
e-mail: A.Dewan@curtin.edu.au

K. Nahar
Department of Geography & Environment, University of Dhaka, Dhaka 1000, Bangladesh
e-mail: kamrun73@live.com

Y. Kawamura
Department of Mining Engineering, Curtin University, Locked Bag 30, Kalgoorlie 6430, WA, Australia
e-mail: Yohei.Kawamura@curtin.edu.au

A. Dewan and R. Corner (eds.), *Dhaka Megacity: Geospatial Perspectives on Urbanisation, Environment and Health*, Springer Geography,
DOI 10.1007/978-94-007-6735-5_13,

13.1 Introduction

The world is currently experiencing rapid growth in the urban population. A current estimate of the world urban population is 3.6 billion, but it is predicted that by 2050, this will have risen to 6.3 billion (UN 2012). The pattern of urban expansion is significantly faster in Asia and Africa where large-scale transformation of rural land to urban areas is clearly evident, considerably affecting environmental equilibrium (Seto et al. 2010). Unplanned and unregulated urban growth is leading to widespread environmental pollution and a transfer of urban poor to the cities, both of which result in the deterioration of urban environmental quality, particularly in developing countries (Rahman et al. 2011). As this phenomenon is expected to continue, this could affect the lives of a large number of people, as well as the sustainability of the urban environment. In order to address issues associated with haphazard urban growth, quality of life (QOL) research in cities has received renewed interest around the world (Jensen et al. 2004; Foo 2001; Wong 2001; Mendes and Motizuki 2001; Kirby 1999; Sufian 1993). Assessment of QOL not only allows planners and relevant authorities to identify existing shortcomings with respect to the effects of plans and projects (Massam 2002), it also assists in formulating appropriate policies for sustainable urban planning and management (Li and Weng 2007a, b; Myers 1988).

Although the concept of QOL has long been a subject of academic research, a consensus on its definition is still lacking (Kamp et al. 2003). As a result, there is considerable ambiguity in defining and measuring QOL on any particular scale (Rogerson et al. 1989). It is generally intended to refer either to the conditions of the environment in which people live or to some attribute of the people themselves (Pacione 1982, 2003). QOL is a multidisciplinary construct and used extensively in the fields of behavioural medicine, political science, psychology, policymaking and city planning and management (Foo 2001; Kirby 1999). There are many detailed reviews and application of this concept in different disciplines (e.g. Kamp et al. 2003; Massam 2002; Raphael et al. 1996).

Just as its definition is still the subject of discussion, so indicators and methods to measure QOL also vary largely. Traditionally, questionnaire surveys and census data have been used to assess urban QOL (Mirdha and Moore 2011; Das 2008). Two contrasting approaches, one using subjective and the other using objective indicators, have been taken (Shin et al. 2003; Pacione 1982). In the former approach, the use of surveys is a common method to understand people's perceptions of their lives, and ratings are made of their living environments which depend mainly on each individual's characteristics and cultural background. In contrast, the latter approach considers indicators such as crime, education and pollution to identify the value that people attach to their standard of living (Marans 2003). As these approaches rely heavily on government and survey data, indicators from physical and biophysical environments have rarely been incorporated in the study of QOL in cities (Li and Weng 2007a, b; Lo, and Faber 1997). To overcome this problem, remotely sensed data, primarily from satellite platforms, provide an excellent opportunity for integration with geographic information, including green vegetation and surface temperature, to evaluate QOL in a given area. Green (1957) was the first to employ aerial photography with socioeconomic data to measure 'residential desirability' in Birmingham, Alabama. Inspired by Green's study, various researchers studied QOL by integrating census data with biophysical parameters in different cities. Over the past several decades, macro level studies have become possible using the techniques of Geographic Information Science (GIS). Using various indicators, encompassing both subjective and objective, from census and remote sensing with quantitative techniques, it is now feasible to map out different aspects of QOL (e.g. environmental, economical, demographic) (Nichol and Wong 2007; Nichol et al. 2006). The outcome of these indicators can subsequently be combined to develop a synthetic QOL (Li and Weng 2007a, b), urban QOL (Rao et al. 2012; Jensen et al. 2004; Lo and Faber 1997) and environmental quality of city (Fung and Siu 2000) and to detect urban environmental change (Liang and Weng 2011). In addition, neighbourhood quality, a similar type of concept, can also be developed from multisource databases as has been done to elucidate the factors influencing certain diseases in Saudi Arabia (Khormi and Kumar 2011). A number of techniques have been used to study QOL, namely, principal component analysis (PCA)/factor analysis (see Li and Weng 2007a, b; Lo and Faber 1997), weighted overlay within a GIS environment (Lo 1997), multi-criteria evaluation/queries (Rinner 2007; Nichol and Wong 2005, 2007; Mendes and Motizuki 2001), discriminant function (Sufian 1993) and regression modelling (Weber and Hirsch 1992). While PCA typically applies a correlation matrix to standardised variables, GIS overlay requires subjective determination, such as the ranking of variables (Nichol and Wong 2005). However, PCA-based analysis has been found to be superior for removal of redundant information in data—known as multicollinearity (Li and Weng 2007a). In contrast, multi-criteria assessment suffers from the requirement to set specific thresholds, which may be unique to the study site, and therefore not generally applicable, as well as being dependant on the researcher involved (Liang and Weng 2011).

The majority of QOL research has been conducted in developed countries, and the status of QOL in very few tropical/subtropical cities has been considered. This is due to the fact that cities in developing countries, particularly tropical cities, are growing very fast with little provision of adequate infrastructure and facilities. Environmental degradation caused by unplanned urbanisation could lead to an irreversible condition which could potentially introduce widespread problems, such as the incidence of invasive disease (Khormi and Kumar 2011) and poor quality of living environment (Mirdha and Moore 2011; Chowdhury 2003), which will prove detrimental for a large number of people already on the margins of society (Alamgir and Watanabe 2007). To overcome the problems associated with such urban expansion, timely information on QOL is imperative in order to support the development of appropriate measures. Drawing on information from previous studies in post-industrial cities, this chapter aims to develop a synthetic QOL index for Dhaka City Corporation (DCC) area, a subtropical city which has been experiencing severe environmental degradation in the recent past, primarily caused by unregulated and piecemeal urban growth (Dewan et al. 2012).

13.2 Materials and Methods

13.2.1 Data Preparation

This chapter uses a range of databases, including remote sensing, census and GIS data. The selection of variables for this study was based on previous studies (Li and Weng 2007a, b; Fung and Siu 2000; Lo and Faber 1997). Since the study area is in a developing nation, some of the variables used in developed countries are unavailable. Hence, this study employed related parameters as a surrogate. We have used community-level (see Chap. 3) census boundaries to map QOL indices. A detailed method of creating geographic boundaries of census tracts is described in Chap. 3. Demographic and socioeconomic variables, including total population, literacy rate, access to safe water, percent of people unemployed, housing ownership, total households, population with secondary education and number of households with more than five persons, were derived from the 2011 population and housing census (BBS 2012) and aggregated to community level. Population density was calculated as the total number of people residing in a census tract divided by its total area (km^2). Similarly, the unemployment rate was computed by dividing the total number of unemployed people by the total number of population in each census district.

The mean housing value was computed by considering two spatial datasets, namely, housing types and residential land value data. A spatial database containing various housing types of 2005 was obtained from the Detailed Area Plan (DAP) of the Capital Development Authority (RAJUK). Residential land value for each geographic unit from 2008 to 2012 was collected from the respective sub-registry offices in the DCC area. Note that some of the census tracts do not have updated land value data for 2012. In those situations, the 2008 residential land value data was used. Information for three different types of residential housing were extracted: *katcha*, *semi-pucca* and *pucca* (see notes in Chap. 1). Using the housing and land value datasets, the mean housing price for each census tract was calculated using the following method (Dewan 2013).

The residential building feature dataset was first intersected with the community polygonal database. For each of the intersected buildings, the value was estimated based on its dimensions and the corresponding land value per decimal (one hundredth of an acre or 40.46 m^2). Weights were allocated on the basis of housing types. The highest weight was assigned to the value of *pucca* housing, and the lowest value was given to *katcha* housing because of their contrasting construction materials. The mean housing value was then estimated for each community. It should be noted that the resulting housing value does not indicate the actual real estate or market price of a typical residential house in the DCC area; however, it can be used as a proxy in the absence of an actual median/mean housing price. Due to the highly skewed statistical distribution of population density and residential housing value, they were log-transformed.

Furthermore, housing density was calculated from building footprint information in the RAJUK Detailed Area Plan (DAP) and from the census tract areas. The total building area was divided by the total area of each community, and, finally, housing density was determined per km^2. Poverty data were obtained from the Bangladesh Household Income and Expenditure Survey 2005 (BBS et al. 2005). It should be noted that the poverty data were available at the *thana* level (see notes in Chap. 1). As the spatial unit of study is at the community scale, poverty data were subject to areal interpolation. Using the method suggested by Holt et al. (2004), poverty data were interpolated at the community level to represent the percentage of people living below the poverty line. The percent of people living in *pucca* houses was obtained from 2011 census data and used as a surrogate household income variable, with the assumption that people living in this type of house were more affluent compared with those living in *semi-pucca* and *katcha/jhupri* types (Begum 2007).

13.2.2 Methodology

Previous studies have recognised that surface temperature is an important indicator which is related to human comfort (see Nichol et al. 2006), while green vegetation and low percentage of urban use are also valuable QOL parameters (Weng 2010). Therefore, land surface temperature (LST), normalised difference vegetation index (NDVI) and percent of built-up area according to each community were obtained by analysis of Landsat TM 5 images for April 2011. The procedures for obtaining these variables can be found in Chaps. 5 and 12, respectively. Built-up areas for each zone were calculated by determining the number of the built-up surface pixels in each polygon and then multiplying that number by the pixel size. Subsequently, the proportion of urban areas per census tract was calculated. The mean NDVI and LST values were then derived using the zonal function of a GIS and aggregated with the census boundary file.

A total of 20 variables related to socioeconomics, demographics and the environment were included in the dataset. Since remote sensing, socioeconomic and geographic boundaries had different spatial resolutions and format, they needed to be integrated. A number of approaches such as dasymetric mapping enable the integration of census information with pixel-based remotely sensed data (Chen 2002). Here, we used an approach similar to the one used by Lo and Faber (1997) to combine the diverse database derived from the census, remote sensing and the spatial database.

As mentioned above, various techniques are available to analyse and map QOL in the urban environment. In this study, however, we have adopted a method that includes principal component analysis (PCA)/factor analysis to develop a synthetic QOL for the DCC area. The results were further validated using regression analysis (Weng 2010). PCA was developed to transform a set of original, partially correlated variables into new, uncorrelated variables (axes) known as the principal components (PC). The PCs are weighted linear combinations of the original variables so PCA provides a set of meaningful parameters without losing the original information. It is based on the diagonalisation of the correlation matrix, which in itself is useful to point out associations between variables (Helena et al. 2000). The application of PCA results in a number of factors that are ordered in terms of the percentage of the initial variance that they explain, thus allowing a reduction in the number of variables used in subsequent analysis. Generally, the first factor explains the majority of the variance in the data, and subsequent factors explain less variability (Pallant 2011), thereby containing less information. A Scree plot, a graphical output that shows a change of slope based on eigenvalue, helps to discern which factors explain most of the variance in the data being analysed.

In this analysis, all of the socioeconomic and environmental variables were initially considered, and the suitability of individual variables and variable combinations to be included in the final set of factors is tested using the Kaiser-Meyer-Olkin (KMO) measure and Bartlett's test of sphericity. The KMO measure of sampling adequacy is on a scale of 0–1 and should be greater than 0.50, while the level of statistical significance (*p*-value) for Bartlett's test of sphericity should be less than 0.1 (Norusis 2012). On the basis of these tests, a suite of 13 variables was selected to proceed to the next stage of the factor analysis. This next stage uses Principal Axis Factoring to find a set of new axes in rotated multivariate space which are uncorrelated. From these axes a new set of factors are extracted that together explain the majority of the variance of the input datasets. As a general rule, only those factors in the rotated multivariate space that have eigenvalues greater than 1 (the variance of individual input variables) should be used. The procedure is somewhat iterative in that at this stage the communality of the input variables needs to be examined. The communality of a variable is the proportion of its variance that is explained by the new factors. Only variables exhibiting communalities >.50 should be included.

In our analysis, the final model using 10 variables resulted in a KMO of 0.857 and a Bartlett's sphericity significance of 0.000. Table 13.1 shows the correlation matrix for the 13 variables used. Using the rotated factor loadings, the three principal factors (those whose eigenvalues were greater than 1) were labelled as environmental, economic and crowdedness. A QOL index was then calculated for each census tract using the method of Li and Weng (2007a, b), shown in Eq. 13.1, where n is the number of factors used, F_i is the factor score for the census tract and W_i is the proportion of variance explained by factor. The QOL index was further validated using regression techniques following the example of Li and Weng (2007a, b).

$$\mathrm{QOL} = \sum_{i}^{n} F_i W_i \tag{13.1}$$

13.3 Results

13.3.1 Exploratory Spatial Data Analysis

Raster datasets for the three environmental variables surface temperature, vegetation and proportion of urban area were extracted from remotely sensed images. The mean value for temperature and vegetation for each census tract was calculated using GIS zonal operations. The spatial patterns of normalised difference vegetation index (NDVI) revealed that the central part and popularly known 'well-off' areas, such as

Table 13.1 Correlation matrix

	HH (>5)	TCL	PSS	PD	POV	MHV	TEMP	NDVI	URB	HD	PPP	HO
TCL	.044											
PSS	−.007	.830**										
PD	.963**	−.398**	−.022									
POV	.249**	−.143**	−.114**	.232**								
MHV	.201**	.566**	.012	−.067	−.172**							
TEMP	.129**	−.152**	−.038	.151**	.091*	−.171**						
NDVI	−.279**	.106**	.022	−.148**	.085*	.250**	−.676**					
URB	.290**	−.208**	.219**	.080	.073	−.042	.546**	−.499**				
HD	.110**	−.057	.037	.064	.027	−.337**	.556**	−.542**	.347**			
PPP	−.065	.454**	.524**	−.069	−.213**	.019	.013	−.054	.248**	.052		
HO	−.159**	.086*	.163**	−.191**	−.072*	−.060	.155**	.415**	.155**	.197**	.222**	
UNEMP	.609**	.021	−.004	.654**	.155**	−.012	.115**	−.181**	.028	.059	−.097	.262**

HH (>5) Number of households with >5 persons, *TCL* Total literacy rate, *PSS* percent population with secondary education, *PD* population density, *POV* percentage of family under poverty, *MHV* mean housing value, *TEMP* temperature, *NDVI* vegetation, *URB* proportion of urban area, *HD* housing density, *PPP* percentage of people living in *pucca* houses, *HO* house ownership, *UNEMP* unemployment rate

**Correlation is significant at the 0.01 level (2 tailed); *Correlation is significant at the 0.05 level (2 tailed)

the presidential palace, cantonment and higher-class residential zone, exhibited high vegetation while the areas that are extremely dense in population and economic activity showed low vegetation. Higher temperatures were concentrated in the industrial zones, transport terminals and the older parts of the DCC area where the proportion of built-up areas is comparatively high. These areas also exhibited low vegetation coverage. Among the socioeconomic variables, population density ranged from 121 persons to 44,87,541 persons per km^2. The older parts of the DCC area have the highest population density, followed by the north-west part where elevation compared to surrounding lands is relatively high. The total literacy rate was higher in the central part, while extreme poverty was concentrated in the fringe areas. A similar spatial pattern was observed in the distribution of mean housing value.

Data exploration in terms of the relationships among variables was conducted using a bivariate correlation matrix. The results (Table 13.1) revealed that NDVI was negatively correlated with temperature ($r = -0.676$), housing density ($r = -0.542$) and proportion of urban area ($r = -0.499$). However, it was positively correlated with economic variables such as mean housing value ($r = 0.250$) and with total literacy rate ($r = 0.106$), meaning that well-off and literate people have a tendency to live in greener areas. Likewise, the proportion of the urban area was positively correlated with temperature ($r = 0.546$) and housing density ($r = 0.347$) but had a negative correlation with NDVI ($r = -0.499$), suggesting a contrasting relationship with green areas. Socioeconomic variables, such as total literacy rate, were positively correlated with mean housing value ($r = 0.566$) and the percentage of people living in *pucca* houses ($r = 0.454$) but negatively correlated with population density ($r = -0.398$), indicating that education attainment may be higher in well-off people. On the other hand, population density was weakly positively correlated with temperature ($r = 0.151$) but had a weak negative relationship with NDVI ($r = -0.148$), implying that temperature may be related to crowdedness in the study area. The unemployment rate was positively related with larger households (>5 persons) ($r = 0.609$), demonstrating that a household with a greater number of family members experienced higher rates of unemployment. Unemployment was also weakly negatively correlated with NDVI ($r = -0.181$) and weakly positively correlated with poverty ($r = 0.155$). Further analysis of the correlation matrix showed that population density had a strong positive relationship with larger household size ($r = 0.963$). Similarly, total literacy rate was found to be strongly related to the population with secondary education ($r = 0.830$). These results all indicate a high correlation among variables, revealing the need to reduce the dimensions and redundancy of the data (Li and Weng 2007a).

13.3.2 Factor Analysis

Initially, 13 variables with good correlation coefficients were considered for Principle Factor Analysis to discern their contribution to different aspects of QOL. Communality, an indicator of the suitability of each variable, was used to identify their fitness for use in the factor solution (Table 13.2). It indicated that the

Table 13.2 Communality of 13 and 10 variables

Indicators	Communality 13 variables	Communality 10 variables
Household size (>5 persons)	.901	.969
Total literacy rate	.788	.831
Percent population with secondary education	.789	.808
Population density	.896	.960
Percentage of family under poverty	.532	.571
Mean housing value	.673	.682
Temperature	.738	.778
NDVI	.743	.755
Proportion of urban area	.568	.619
Housing density	.587	.629
Percentage of people living in *pucca* houses	.398	
House owner	.223	
Unemployment rate	.417	

Table 13.3 Rotated factor loading matrix

	Factor 1	Factor 2	Factor 3
Household size (>5 persons)	0.004	– 0.062	0.982
Total literacy rate	−0.174	0.894	0.043
% population with secondary education	−0.053	0.895	0.067
Population density	−0.033	−0.055	0.978
% of family living under poverty line	−0.378	0.637	0.151
Mean housing value	0.289	0.724	−0.273
Temperature	−0.870	−0.053	0.132
NDVI	0.851	−0.043	−0.129
Proportion of urban area	−0.691	0.374	0.040
Housing density	0.156	−0.017	0.750
Initial eigenvalues	3.340	2.252	1.957
% of variance	33.40	22.52	19.57
Cumulative (%)	33.40	55.92	75.49

three variables—house ownership, unemployment rate and percent of people living in *pucca* houses—resulted in low communalities. Since variables with low communality do not fit well into the factor solution (Liang and Weng 2011), they were dropped from subsequent analysis. This resulted in a higher communality for each of the variables to be considered in the final model (Table 13.3).

Table 13.3 shows the three factors which accounted for 75.49 % of the total variance. The first factor explained 33.40 % of the variance, the second factor showed 22.52 %, while the third factor produced 19.57 % of the variation. In Factor 1, NDVI presented a high positive loading (loading (*L*): 0.851), while there was a strong negative loading on temperature (*L*: −0.870) and proportion of urban area (*L*: −0.691), indicating that Factor 1 is descriptive of environmental conditions (score ranged between −2.14 and 2.79) (Fig. 13.1). Factor 2 resulted in strong positive loadings on

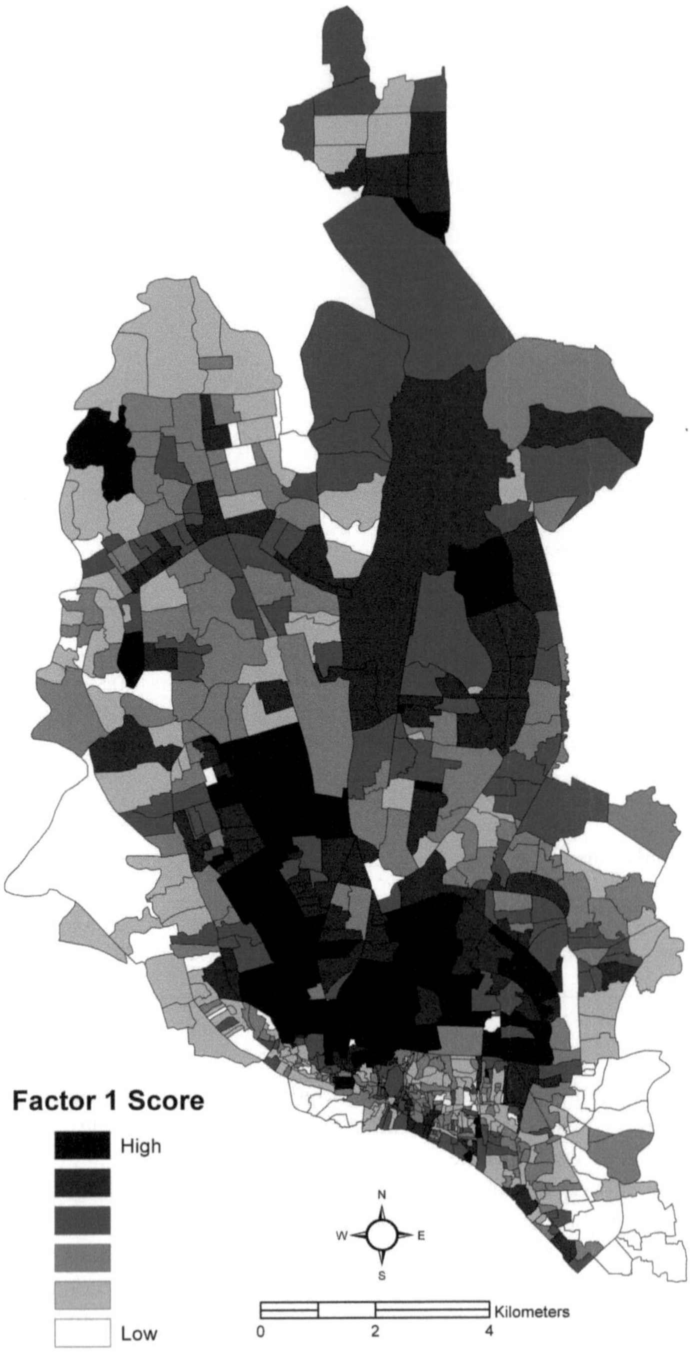

Fig. 13.1 The first factor scores: environmental index

four socioeconomic variables, including mean housing value (*L*: 0.724), percent of families living under the poverty line (*L*: 0.637), percent of the population with secondary education (*L*: 0.895) and total literacy rate (*L*: 0.894). Hence, Factor 2 can be labelled as welfare or economic conditions (Fig. 13.2) (scores ranged from −4.76 to 1.80). Factor 3 showed the strongest loadings on three variables: population density (*L*: 0.978), number of households with size >5 persons (*L*: 0.982) and housing density (*L*: 0.750). Vegetation had negative loadings (*L*: −0.129), but temperature was revealed to have a positive loading (*L*: 0.132), meaning that crowded areas are associated with high temperature. Factor 3 scored between −3.60 and 2.47; higher scores indicated that a large amount of people live in a smaller area. As a result, Factor 3 can be regarded as representing 'crowdedness' (Fig. 13.3).

The three derived factors were mapped out to represent different aspects of QOL. It is important to note that the higher the score, the better the QOL of a particular dimension. For instance, Factor 1, which was labelled 'environmental conditions', revealed the quality of the urban environment in the DCC area. Similarly, Factor 3 was used to discern areas that contained many people in a small area.

13.3.3 Synthetic QOL Index

Synthetic QOL scores for each census district were calculated using the method suggested by Li and Weng (2007a, b), where the weights are the proportions of total variance explained by each factor (Eq. 13.2). A GIS field calculator operation was performed on the attribute table of the census tract polygons to carry this out.

$$\text{QOL} = 0.3340 \times F_1 + 0.2252 \times F_2 - 0.1957 \times F_3 \tag{13.2}$$

The spatial distribution of synthetic QOL scores is presented in Fig. 13.4 and shows a range from −0.74 to 1.70. High scores indicated a better QOL for a particular census district. As expected, census districts with higher vegetation and less population density showed better QOL. The synthetic QOL was divided into six categories using Jenks natural breaks algorithm: 'extremely low', 'very low', 'low', 'medium', 'high' and 'very high' QOL in the DCC area.

In order to investigate the underlying factors influencing QOL in the DCC area, various socioeconomic and environmental indicators were overlaid in GIS with categorised QOL, and a variety of statistics was calculated. Results showed that 7.8 % of the census tracts are within the extremely low QOL zone and about 14.8 % of the total population in the DCC area live in these tracts in which the literacy rate is also low (56 %). These tracts all had low mean NDVI and high mean population density. By contrast, census tracts with very high QOL comprised only 6.5 % of the total DCC area, and only 1.4 % of the population live in these areas, revealing that a very low number of people enjoyed better quality of life in terms of the combined three factors (Table 13.4). However, these statistics are very exploratory in nature and should be used with care since we do not have other instruments to validate these claims.

Fig. 13.2 The second factor scores: economic index

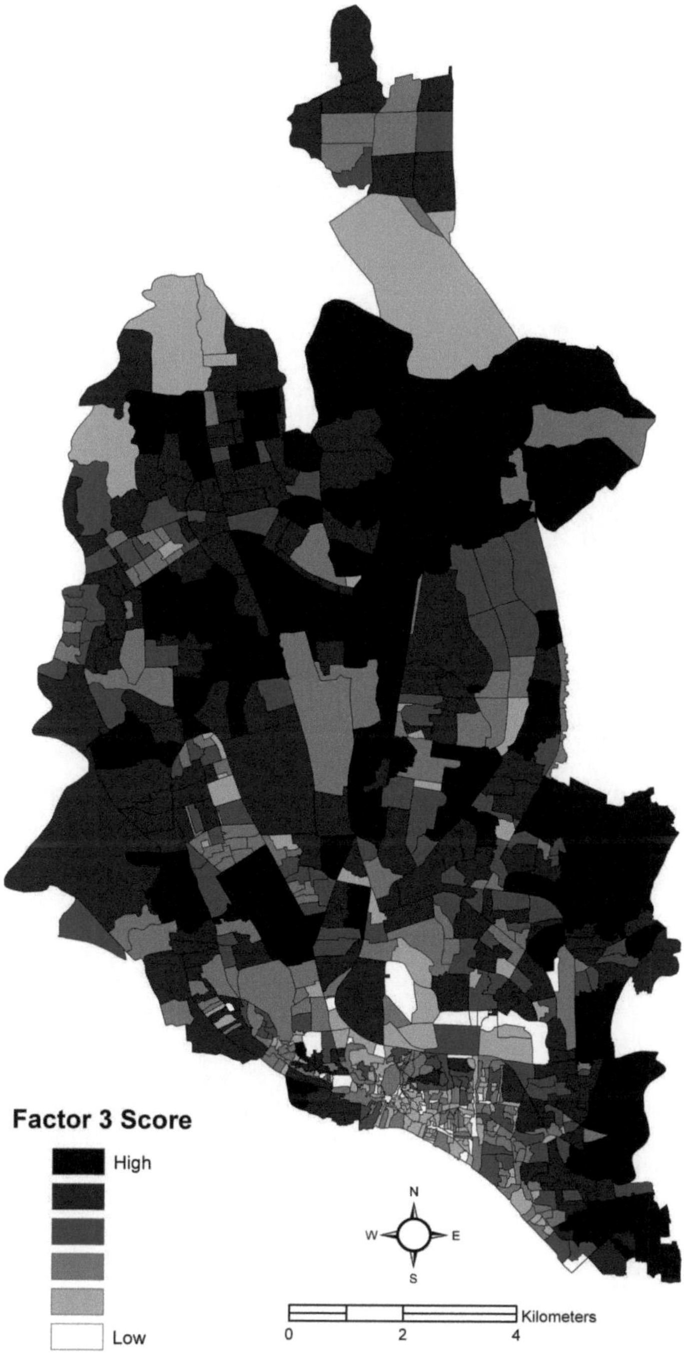

Fig. 13.3 The third factor scores: crowdedness

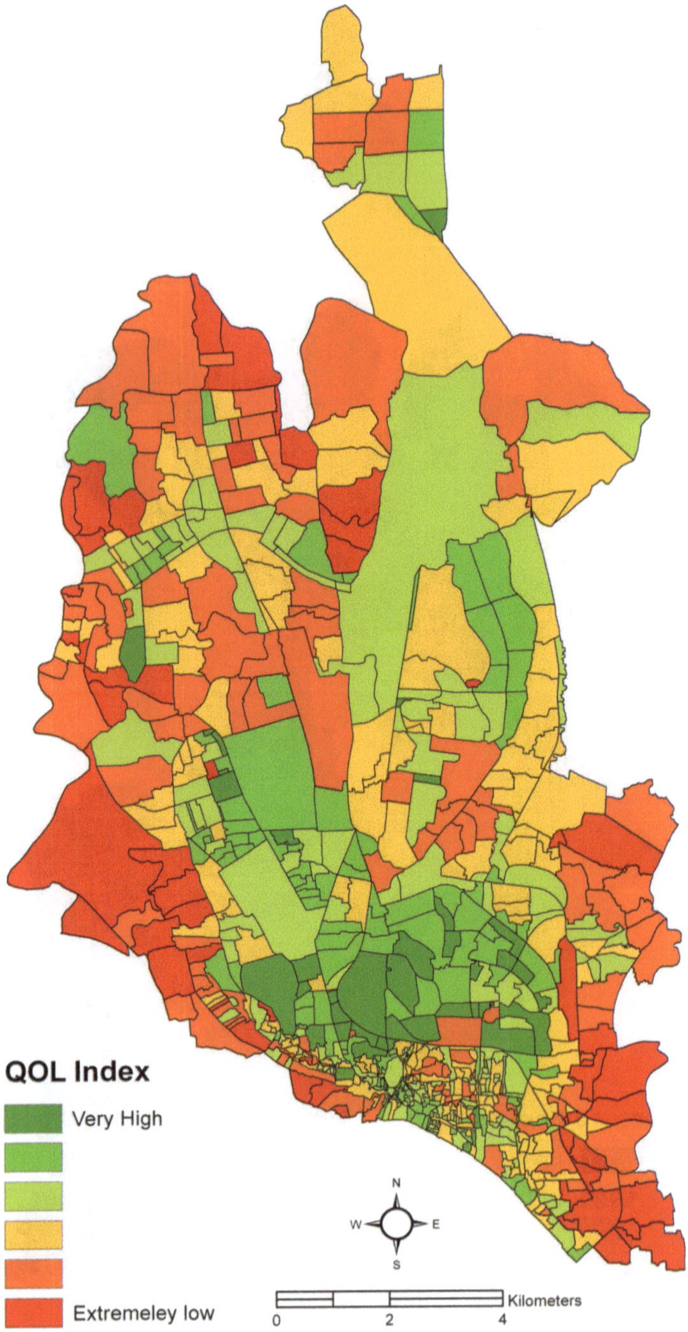

Fig. 13.4 Synthetic QOL index

Table 13.4 QOL categories with different socioeconomic and environmental parameters

QOL category	Census tracts Number	Census tracts %	Population Total	Population %	Mean population density	Mean NDVI	Literacy rate (%)	Mean housing value ('000 BDT[a])
Extremely low	54	7.8	1,052,431	14.9	142,153	0.0468	56.8	825.4
Very low	111	16.1	2,066,982	29.2	144,762	0.0448	66.8	1,889.5
Low	186	26.9	2,112,666	29.9	140,475	0.0228	76.1	1,653.2
Medium	166	24.0	1,183,371	16.7	137,597	0.0250	80.2	2,780.8
High	129	18.7	553,761	7.8	79,059	0.0495	83.9	5,232.5
Very high	45	6.5	100,450	1.4	31,076	0.1134	89.3	10,556.2

[a]*BDT* Bangladeshi currency (TAKA)

Table 13.5 Regression estimation of QOL models

Models	Predictors	Coefficients	R^2
Environmental QOL	Constant	−16.218	.92
	NDVI	1.425	
	Temperature	−6.528	
Economic QOL	Constant	−5.841	
	Total literacy rate	.040	
	Percentage of family with secondary education	8.960×10^{-8}	.87
	Mean housing value	.038	
Crowdedness	Constant	−6.341	.95
	Population density	1.711	
	Housing density	5.915×10^{-6}	
Synthetic QOL	Constant	−5.871	.86
	Vegetation	2.997	
	Literacy rate	.015	
	Population density	−.271	
	Mean housing value	.181	

13.3.4 Regression Analysis

Following the example of Li and Weng (2007a, b), each of the three factors was regressed using the variables with the highest loadings in the rotated factor solution as predictor variables. For example, Factor 1 was regressed against NDVI, temperature and proportion of urban surface. Similarly, synthetic QOL was regressed against four variables: mean housing value, total literacy rate, population density and mean NDVI. Table 13.5 shows the results of the regression analyses. This shows that the crowdedness QOL produced the highest R^2 value (0.95), followed by the environmental condition ($R^2 = 0.92$); however, the synthetic QOL produced the lowest R^2 (0.86). This could be the result of limited data and the use of a few surrogate variables in the absence of actual data. Nevertheless, the results revealed that the regression analysis predicted different dimensions of QOL in the DCC area that were more important to validate.

13.4 Discussion and Conclusions

Using remote sensing, census and GIS data, the work described in this chapter has developed a synthetic QOL index in the DCC area. A correlation matrix was first performed to understand the relationship between different variables. Initially, a total of 20 socioeconomic, demographic and environmental variables were tested. The correlation matrix revealed multicollinearity among variables; therefore, Principal Factor Analysis was carried out to reduce data dimension. Thirteen variables were used as input to the factor analysis, and ten were eventually retained on the basis of communality testing. Three factors were extracted based on the magnitude of their eigenvalues. These factors represented the environmental, economic and demographic aspects of QOL in the DCC area. Using these three factors, a synthetic QOL was developed to determine urban QOL. The derived QOL was further validated using a regression analysis.

The results revealed that overall QOL in the DCC area is extremely poor. For instance, 14 % of people in the study area had extremely low QOL, while only 1.4 % of the total population had good QOL. The standard of living in all QOL zones was mainly influenced by surface temperature, literacy, housing value, population density, etc. As Dhaka is growing rapidly, inadequate provision of urban services results in appalling conditions, thereby affecting a large number of marginal people. Synthetic QOL would be of substantial help for urban planners and city authorities to find necessary solutions to improve people's standard of living.

There are a number of shortcomings of this study. One major issue was data availability. The study could not incorporate economic variables, such as household income, due to lack of such data at Dhaka's community series produced by BBS, which was a great impediment. Instead, we used surrogate variable(s). Census data are collected primarily for socioeconomic and administrative purposes, and, whilst this study looks at some methods to integrate pixel-based information with census data, there is an opportunity for further work in this area. However, GIS provided a good platform to integrate different databases at community level to investigate the spatial patterns of QOL. The study heavily relied on previous work (e.g. Li and Weng 2007a) to determine variables, but it is also equally important to note that indicators related to QOL development depend on the geographical area being analysed. To overcome the problems associated with this work, further studies could utilise survey data from the field and may integrate both subjective and objective indicators to better define and measure QOL. This may be carried out through interdisciplinary work involving government organisations, academics and relevant agencies.

References

Alamgir F, Watanabe S (2007) NGO savings and credit scheme: effectiveness in improving the quality of life of the urban poor in different poverty levels in Dhaka. In: Jahan S, Maniruzzaman KM (eds) Urbanization in Bangladesh: patterns, issues and approaches to planning. Bangladesh Institute of Planners (BIP), Dhaka, pp 105–120

Bangladesh Bureau of Statistics (BBS) (2012) Population and housing census – 2011, community report, Dhaka. Narayanganj and Gazipur Zila, Bangladesh Bureau of Statistics, Ministry of Planning, Dhaka

Bangladesh Bureau of Statistics (BBS), The World Bank, World Food Programme (WFP) (2005) Bangladesh Household Income and Expenditure Survey 2005, Dhaka

Begum A (2007) Urban housing as an issue of redistribution through planning? The case of Dhaka city. Soc Policy Adm 41(4):410–418

Chen K (2002) An approach to linking remotely sensed data and areal census data. Int J Remote Sens 23(1):37–48

Chowdhury MR (2003) The impact of 'Greater Dhaka Flood Protection Project' (GDFPP) on local living environment – the attitude of the floodplain residents. Nat Hazard 29(3):309–324

Das D (2008) Urban quality of life: a case study of Guwahati. Soc Indic Res 88:297–310

Dewan AM (2013) Floods in a megacity: geospatial techniques in assessing hazards, risk and vulnerability. Springer Science + Business Media, Dordrecht

Dewan AM, Kabir MH, Nahar K, Rahman MZ (2012) Urbanization and environmental degradation in Dhaka metropolitan area of Bangladesh. Int J Environ Sustain Dev 11(2):118–146

Foo TS (2001) Quality of life in cities. Cities 18(1–2):1–2

Fung T, Siu W (2000) Environmental quality and its changes, an analysis using NDVI Int J Remote Sens 21(5):1011–1024

Green NE (1957) Aerial photographic interpretation and the social structure of the city. Photogramm Eng 23:89–96

Helena B, Pardo R, Vega M, Barrado E, Fernández JM, Fernández L (2000) Temporal evolution of groundwater composition in an alluvial aquifer (Pisuerga river, Spain) by principal component analysis. Water Res 34(3):807–816

Holt JB, Lo CP, Hodler TW (2004) Dasymetric estimation of population density and areal interpolation of census data. Cartogr Geogr Inf Sci 31(2):103–121

Jensen R, Gatrell J, Boulton J, Harper B (2004) Using remote sensing and geographic information systems to study urban quality of life and urban forest amenities. Ecol Soc 9(5):5

Kamp IV, Leidelmeijer K, Marsman G, Hollander AU (2003) Urban environmental quality and human well-being towards a conceptual framework and demarcation of concepts: a literature study. Landsc Urban Plan 65:5–18

Khormi HS, Kumar L (2011) Modeling dengue fever risk based on socioeconomic parameters, nationality and age groups: GIS and remote sensing based case study. Sci Total Environ 409:4713–4719

Kirby A (1999) Quality of life in cities. Cities 16(4):221–222

Li G, Weng Q (2007a) Integration of remote sensing and census data for assessing urban quality of life: model development and validation. In: Weng Q, Quattrochi D (eds) Urban remote sensing. CRC Press, Boca Raton, pp 311–336

Li G, Weng Q (2007b) Measuring the quality of life in city of Indianapolis by integration of remote sensing and census data. Int J Remote Sens 28(2):249–267

Liang B, Weng Q (2011) Assessing urban environmental quality change of Indianapolis, United States, by the remote sensing and GIS integration. IEEE J Sel Top Appl Earth Obs Remote Sens 4(1):43–55

Lo CP (1997) Application of Landsat TM data for quality of life assessment in an urban environment. Comput Environ Urb Syst 21(3–4):259–276

Lo CP, Faber BJ (1997) Integration of Landsat thematic mapper and census data for quality of life assessment. Remote Sens Environ 62(2):143–157

Marans RW (2003) Understanding environmental quality through quality of life studies: the 2001 DAS and its use of subjective and objective indicators. Landsc Urb Plan 65(1–2):73–83
Massam BH (2002) Quality of life: public planning and private living. Prog Plan 58(3):141–227
Mendes JFG, Motizuki WS (2001) Urban quality of life evaluation scenarios: the case of São Carlos in Brazil. CTBUH Rev 1(2):13–23
Mirdha AMMH, Moore GT (2011) The quality of life in Dhaka, Bangladesh: neighbourhood quality as a major component of residential satisfaction. In: Marans RW, Stimson R (eds) Investigating quality of urban life: theory, methods, and empirical research, vol 45, Social indicators research series. Springer Science + Business Media B.V., Dordrecht, pp 251–272
Myers D (1988) Building knowledge about quality of life for urban planning. J Am Plan Assoc 54 (3):347–358
Nichol JE, Wong MS (2005) Modelling urban environmental quality in a tropical city. Landsc Urb Plan 73(1):49–58
Nichol JE, Wong MS (2007) Assessing urban environmental quality with multiple parameters. In: Weng Q, Quattrochi D (eds) Urban remote sensing. CRC Press, Boca Raton, pp 253–268
Nichol J, Wong MS, Fung C, Leung KKM (2006) Assessment of urban environmental quality in a subtropical city using multispectral satellite images. Environ Plan B Plan Des 33(1):39–58
Norusis MJ (2012) IBM SPSS statistics statistical procedures companion. Prentice Hall, Upper Saddle River
Pacione M (1982) The use of objective and subjective measures of life quality in human geography. Prog Hum Geogr 6(4):493–514
Pacione M (2003) Urban environmental quality and human wellbeing – a social geographical perspective. Landsc Urb Plan 65(1–2):19–30
Pallant J (2011) SPSS survival manual, 4th edn. Open University Press, Berkshire
Rahman A, Kumar Y, Fazal S, Bhaskaran S (2011) Urbanization and quality of urban environment using remote sensing and GIS techniques in east Delhi – India. J Geogr Inf Syst 3:62–84
Rao KRM, Kant Y, Gahlaut N, Roy PS (2012) Assessment of quality of life in Uttarakhand, India using geospatial techniques. Geocarto Int 27(4):315–328
Raphael D, Renwick R, Brown I, Rootman I (1996) Quality of life indicators and health: current status and emerging conceptions. Soc Indic Res 39:65–88
Rinner C (2007) A geographic visualization approach to multi-criteria evaluation of urban quality of life. Int J Geogr Inf Sci 21(8):907–919
Rogerson RJ, Findlay AM, Morris AS, Coombes MG (1989) Indicators of quality of life: some methodological issues. Environ Plan A 21(12):1655–1666
Seto KC, Sánchez-Rodríguez R, Fragkias M (2010) The new geography of contemporary urbanization and the environment. Annu Rev Environ Resour 35:167–194
Shin DC, Rutkowski CP, Park CM (2003) The quality of life in Korea: comparative and dynamic perspectives. Soc Sci Indic Res 62(63):3–36
Sufian AJM (1993) A multivariate analysis of the determinants of urban quality of life in the world's largest metropolitan areas. Urb Stud 30(8):1319–1329
United Nations (2012) World population prospects: the 2010 revision, vol 1. United Nations Department of Economic and Social Affairs, New York. Available at: http://esa.un.org/unpd/wup/index.html. Accessed 1 Apr 2012
Weber C, Hirsch J (1992) Some urban measurements from SPOT data: urban life quality indices. Int J Remote Sens 58(17):439–448
Weng Q (2010) Remote sensing and GIS integration: theories, methods and applications. McGraw-Hill Companies, New York
Wong C (2001) The relationship between quality of life and local economic development: an empirical study of local authority areas in England. Cities 18(1):25–32

Chapter 14
Exploring Crime Statistics

Ashraf M. Dewan, Md. Rafiqul Haider, and Md. Ruhul Amin

Abstract The objective of this chapter is to analyse the spatiotemporal patterns of crime in the Dhaka Metropolitan Area (DMA). Crime data for the period of August 2011 to July 2012 were acquired from Dhaka Metropolitan Police (DMP). Apart from descriptive statistics, spatial pattern analysis was carried out through local indicators of spatial autocorrelation (LISA). Linear regression was conducted using the crime incidence rate as a dependent variable and ten socio-demographic factors as potential explanatory variables. Descriptive analysis showed distinct temporal variation with the pre-monsoon season having the highest crime occurrences. Furthermore, there are three single-centred criminal activity hotspots of crime. Multivariate regression analysis showed that the size of male population and poverty rate were the best predictors of crime incidence in DMA; however, simple regression suggested that total population, number of males and size of unemployment population in the police districts were also suitable predictors. To our knowledge this is the first study that employed spatial techniques to analyse crime in DMA, and the results of this work should be valuable for informed decision-making for the law enforcing agency as well as relevant authorities whose task is to prevent increasing criminal activities in Dhaka.

A.M. Dewan (✉)
Department of Spatial Sciences, Curtin University, Kent Street,
Bentley, Perth 6102, WA, Australia
e-mail: A.Dewan@curtin.edu.au

Md.R. Haider
Centre for Child and Adolescent Health, International Centre for Diarrhoeal Disease Research,
GPO Box No. 128, Dhaka 1212, Bangladesh
e-mail: cslraju@yahoo.com

Md.R. Amin
Traffic North Division, Bangladesh Police, House # 4, Road # 2, Nikunja-1,
Dhaka 1229, Bangladesh
e-mail: ruhul_amin99@yahoo.com

A. Dewan and R. Corner (eds.), *Dhaka Megacity: Geospatial Perspectives on Urbanisation, Environment and Health*, Springer Geography,
DOI 10.1007/978-94-007-6735-5_14,

Keywords Crime • Space-time • Regression analysis • Hot spot • Built environment • Dhaka Metropolitan Area

14.1 Introduction

Crime, particularly violent crime, is increasingly becoming a concern for urban populations both in developing and developed countries (Cozens 2011; Appiahene-Gyamfi 2003; Ceccato et al. 2002). Between the early 1980s and the mid-1990s, for example, the rate of intentional homicides increased by 50 % in Latin America and by more than 100 % in Eastern Europe and Central Asia (Fajnzylber et al. 2002). The situation is particularly alarming in the cities of developing countries, where rapid urbanisation driven by rural to urban migration and increasing transient population are not in pace with economic development (Shafi 2010). The resulting increase in the occurrence of criminal activity and violent behaviour becomes a matter of grave concern, clearly threatening social stability (Fajnzylber et al. 2002) and the quality of urban life (Dempsey 2008). Although a great deal of debate on the causes of crime exists (Fattah 1997), a number of factors have been found to be correlated with the incidence of elevated crimes in cities. These include neighbourhood inequality (Neumayer 2005; Elffers 2003; Morenoff et al. 2001; Krivo and Peterson 1996), unemployment rate and family income (Andersen 2006), demographic composition (Fox and Piquero 2003), poorer housing estates and a mixed society (Wiles and Costello 2000), the presence of large student body in a community (Malczewski and Poetz 2005), ethnic composition (Brimicombe et al. 2001) and an increase in urban poor (Hannon 2005; Stretesky et al. 2004; Bourguignon 1998). A study by Lauritesen and White (2001), for instance, suggests that an increase in poverty leads to an increase in violence in areas that are already disadvantaged socially and that this type of deprivation reduces the ability of a community to function cohesively and causes a loss of collective values (Warner 2003; Kubrin 2003). The size of the young male population has been found to be a good demographic predictor of the level of violence, particularly in deprived areas (Fox and Piquero 2003). Evans (1992) reported that vandalism is correlated with housing ownership and income in the UK. Similarly, the ethnic composition of an area was found to be significantly correlated with racially motivated crimes in the London Borough of Newham (Brimicombe et al. 2001). To create and maintain safer, vibrant and sustainable urban communities (Cozens 2011), it is therefore imperative to investigate spatiotemporal patterns of crime in order to support informed decision-making in crime prevention activities (Ratcliffe 2010).

Although spatial studies of crime have been taking place for nearly 200 years (Chainey and Ratcliffe 2005), the 'place-based' (Anselin et al. 2000) component of crime analysis received a boost when Shaw and McKay (1942) of the Chicago School of Sociology utilised manual techniques to map the residences of juvenile delinquents across Chicago. A number of spatial theories have subsequently evolved to explain crime and space (Chainey and Ratcliffe 2005). Notable amongst them are routine activity theory (Cohen and Felson 1979), the rational choice perspective (Clarke and Felson 1993), social disorganisation theory (Shaw and McKay 1942), crime pattern theory (Brantingham and Brantingham 1993) and multicontextual opportunity theory which is based on both social disorganisation and routine activities (Wilcox et al. 2003).

As can be seen from the literature, geospatial techniques can facilitate the study of crime in a number of ways. Whilst GIS supports relevant database development plus spatiotemporal analysis of crime (Getis et al. 2000), data from satellite remote sensing can be used to retrieve parameters such as land use that allow an integrated assessment of crime patterns (Sparks 2011). Statistical techniques can then be used to determine the relationship of criminal activities with environmental and social parameters. As an example of this approach, land use diversity derived from remotely sensed images was used together with social data to examine the pattern of violent crimes in Texas (Sparks 2011). Likewise, vegetation classification from SPOT XS data was used with socio-demographic variables to investigate the propensity and opportunity for crime in San Diego (Chen et al. 2005).

Apart from studies integrating remotely sensed data, there is an increasing tendency to use crime records together with socioeconomic and demographic data, particularly from census within a geographic information science (GIS) framework, to demonstrate the spatial patterning and the causal processes involved. Although the majority of crime studies have used point data to analyse criminal activities for hotspot detection, areal data is also valuable for identifying spatial clustering of crime in a given environment and was one of the early crime mapping techniques utilised in Europe and the USA (Santos 2013). Despite the fact that area-based analysis is sensitive to the choice of spatial unit (Rengert and Lockwood 2009; Wise et al. 1997), Ceccato et al. (2002) noted that area-based studies of crime provide several benefits. A number of studies have employed GIS as 'enabling technology' in analysing various crimes such as homicide, property crime, burglary, violent crime and racially motivated crime in diverse settings, including in the USA (Ye and Wu 2011; Sparks 2011; Browning et al. 2010; Matthews et al. 2010; Grubesic and Mack 2008; Murray and Roncek 2008; Cahill and Mulligan 2007; Cancino et al. 2007; Harries 2006a, b), Brazil (Ceccato et al. 2007), Australia (Ratcliffe 2005; Murray et al. 2001), India (Kumar and Chandrasekar 2011; Jaishankar et al. 2004), in the UK (Chainey et al. 2008; Law and Haining 2004; Brimicombe et al. 2001; Bowers and Hirschfield 1999), Canada (Kinney et al. 2008; Andersen 2006; Malczewski and Poetz 2005), South Africa (Breetzke 2006), Sweden (Ceccato et al. 2002) and in Germany (Snook et al. 2005). A range of geographic tools such as hotspot analysis and exploratory spatial data analysis (ESDA) techniques embedded within a GIS have been tested to understand the spatial patterning of offences (Ratcliffe 2010). The results are then correlated with criminological theories using a variety of statistical techniques, including spatial statistics, to recognise the causal processes. For instance, using contextual (e.g. cultural) and geographical factors, Ceccato et al. (2002) suggest that socially mixed neighbourhoods with high population turnover experience more crime. A study by Andersen (2006) revealed that areas with higher education rate and family income significantly influence the occurrences of crime as houses in these areas are expected to have expensive goods in them, and this finding is consistent with routine activity theory. Similar results are also reported by Jaishankar et al. (2004), who suggest that upper middle class areas in Chennai city are prone to frequent daytime housebreaking. Sparks (2011) suggests that vacant housing and land use diversity are significantly associated with the incidence of violent crime which is in agreement with social disorganisation theory. Ye and Wu (2011) demonstrated that the level of poverty is correlated with a higher level of homicide which also supports social disorganisation theory. By contrast, built environment variables were significant predictors of property crime in Seattle (Matthews et al. 2010). Kinney et al. (2008) also indicate that assaults and motor vehicle thefts were higher in residential areas, with commercial areas next highest. Since the occurrence of crime varies within cities and regions, there is an increasing interest in the influence of local geography on the incidence of crime and in techniques for modelling it, such as geographically weighted regression (GWR). For instance, Cahill and Mulligan (2007) in a study that highlighted the importance of local processes as drivers of urban violence in Portland, Oregon, found that GWR outperformed global regression. They further used standard structural measures to relate their findings to multicontextual opportunity theory. Other studies have extended their work to identify the diffusion mechanism of crimes amongst urban neighbourhoods (Matthews et al. 2010; Cohen and Tita 1999). The empirical studies discussed above show that the spatial analysis of crime is vital to regional planners, policy makers and law enforcing agencies. Since the incidence of crime is non-random over space, geographic analysis of crime has been found to be very useful in targeting crime reduction measures, particularly in urban areas (Murray et al. 2001).

Bangladesh, one of the poorest countries in the world, is currently experiencing an increase in crime. This is reflected by the fact that it ranks 155th out of 158 countries in the world in terms of lawfulness (Dijk 2006 in Lee and Haider 2012). According to Lambert et al. (2012), the overall crime rate in the country was 78 per 100,000 people in 1996 which increased to 87 per 100,000 people in 2006. A very recent estimate revealed that around 1.4 million criminal offences of various types occurred between 2001 and 2010 of which the miscellaneous category represented the highest (55.9 %) of total crimes, followed by narcotics-related offences (11.1 %) (Table 14.1). Furthermore, violence against women ranks in third position, accounting for 10.5 % of the incidence of total crimes in the form of rape, assault, acid throwing, ransom, etc. followed by theft cases (6.6 %), smuggling (3.5 %) and murder (2.8 %) (Bangladesh Police 2012). As a word of explanation, the heading Speedy Trial Act in Table 14.1 and elsewhere refers to a group of crimes covered by the Speedy Trial Act of 2002. This Act was introduced in an attempt to clear the backlog of cases in the justice system and covers the following crimes: arson, forgery, snatching, hijacking, property damage, fraudulence, breaking and entering, etc.

Although systematic studies similar to those that have been conducted in developed nations are rare, a variety of factors have been found to have contributed to the increase of criminal activities in Bangladesh in the recent past. These include socioeconomic and demographic change (Mas et al. 2011; Ahmed and Islam 2010; Majumder et al. 2008; Ahmad and Baqee 1988), inequality of resources (Sarker 1989, 1991, 1994) and political rivalry (Chowdhury 2003).

Table 14.1 Crime statistics in Bangladesh, 2001–2010

Crime type	2001	2002	2003	2004	2005	2006	2007	2008	2009	2010	Total	%
Robbery	2,023	2,360	2,119	2,092	1,694	1,638	2,345	2,468	2,062	1,715	20,516	1.5
Murder	3,678	3,503	3,471	3,902	3,592	4,166	3,863	4,099	4,219	3,988	38,481	2.8
Speedy Trial Act	2,396	1,693	2,179	2,053	1,814	1,638	1,980	1,700	1,817	1,666	18,936	1.4
Rioting	2,161	1,276	890	754	570	570	263	203	112	130	6,929	0.5
Violence against women	12,958	18,455	20,242	12,815	11,426	11,068	14,250	14,284	12,904	16,210	144,612	10.5
Child abuse	380	512	475	503	555	662	967	962	1,093	1,542	7,651	0.6
Kidnapping	834	1,040	896	898	765	722	774	817	858	870	8,474	0.6
Police assault	344	281	271	280	240	337	278	296	357	473	3,157	0.2
Burglary	3,654	3,959	3,883	3,356	3,270	2,991	4,439	4,552	3,456	3,101	36,661	2.7
Theft	7,432	8,245	8,234	8,605	8,101	8,332	12,015	12,188	9,171	8,529	90,852	6.6
Arms Act	3,151	3,060	2,293	2,370	1,836	1,552	1,746	1,529	1,721	1,575	20,833	1.5
Explosive Act	746	570	499	477	595	308	232	239	227	253	4,146	0.3
Narcotics	5,936	9,018	9,494	9,505	14,195	15,479	15,622	19,263	24,272	29,344	152,128	11.1
Smuggling	3,076	4,746	4,499	4,182	4,334	4,734	5,202	7,962	7,817	6,363	52,915	3.8
Others	65,422	68,898	66,194	67,531	70,046	76,381	93,224	87,417	87,022	87,139	769,274	55.9
Total	114,191	127,616	125,639	119,323	123,033	130,578	157,200	157,979	157,108	162,898	*1,375,565*	100.0

(Adapted from Bangladesh Police 2012)

Despite the fact that all citizens in the country are at risk of becoming victims of crime, women and girls are particularly at risk as evidenced by the high rates reported in Table 14.1. This has largely resulted from the dowry custom (Lee and Haider 2012; Nasrin 2011; Islam 2010; Naved and Persson 2005, 2008; Mannan et al. 2004; Wesson 2002) and family and land disputes (Lee 2010). Interestingly, Islam (2010) compared violence against women during different political regimes from 1996 to 2010. His study revealed that a total of 8,448 crimes were committed against women during 1996–2001, 33,521 during 2001–2006 and 11,184 between 2007 and 2009, and currently the figure stands at 3,818 (2009–2010). This undoubtedly indicates the influence of different political regimes on violence against women in the country.

The police-to-population ratio is low at 1:1,138 which means that it is difficult for the law enforcement agencies to prevent and lower the rate of crimes across the country (Mollah et al. 2012). Furthermore, Bangladesh's police force is governed by an act of the colonial era of 1861 (CHRI 2008) which is a great hurdle for the agency to combat with increasing criminal activities. The CHRI report of 2008 further argues that 'after achieving independence (from Pakistan), Bangladesh continued to struggle with an unprofessional and deeply dysfunctional police force. In addition to tremendous poverty, vulnerability to cyclones and political instability, Bangladesh suffered from an incredibly politicised and unaccountable police cadre' (CHRI 2008, p. 18). This may, at least partly, explain the causes of increased crimes in the country. Nonetheless, Chowdhury (2003) asserts that political violence has become the dominant feature of criminal activities in Bangladesh, overshadowing its economic progress, and the prevention of crime requires that politics be delinked from the activities of criminals and terrorists.

Previous studies have demonstrated that Dhaka is disproportionately affected by crime and related incidences (Khan 2012). Using data from a national daily newspaper from 1973 to 1985, Ahmad and Baqee (1988) showed that nearly 61 % of urban crimes took place in Dhaka and had a distinct seasonal variation. Another study demonstrated that crime situation in Dhaka has been aggravating since 1985 because of the rise of the number of criminals. They estimated that there were about 5,000 criminals in 1984-85 which rose to 100,000 in 2005 (Siddiqui et al. 2010). A total of 47,899 crimes of various types were estimated to have occurred in the capital between 2001 and 2005 of which 8.8 % of total incidences were robbery, 6.8 % murder, 19.1 % violence against women and 15.2 % burglary with various types of theft cases making up 43.7 % (Shafi 2010). A comparison between 2001 and 2005 in the same study shows that the percentage of murder in 2001 was 8.1 % which decreased slightly in 2005 when it amounted to 6.7 % of total crimes. Note, however, that the numerical incidence of murder cases was considerably higher in 2005 (419 cases in 2001 and 1,744 in 2005). Whilst violence against women was 10.5 % in 2001, this figure shot up to 18.1 % in 2005. The occurrence of burglary also increased from 14.1 % in 2001 to 15.3 % in 2005 but the theft cases reduced by 2.5 % (Shafi 2010). Whilst juvenile delinquency was substantially lower during the 1990s (Sarker 1989, 1991), this crime category is perceived to be a matter of serious concern since more than half of crimes are committed by juveniles in the 12–20 age group (Ahmed and Islam 2010; Karzon 2006). Of 80 known crime syndicates across the country, 28 groups are actively operating in Dhaka (Shafi 2010; Karzon 2006), leading to added social insecurity for those going about their daily activities in the capital. Apart from low police-to-people ratio (1:398) for the city (Shafi 2010), an increase in urban poverty and unemployment (Shafi 2011; Majumder et al. 2008; Islam 2005; Sarker 1994), an influx of transient population at certain seasons (Ahmad and Baqee 1988), political rivalry amongst parties involved, ideological indoctrination by religious groups, criminalisation of politics (Chowdhury 2003), increased drug addiction in juveniles (Hossain and Mamun 2006), competition for scarce resources and the 'get-rich-quick' syndrome in society (Siddiqui et al. 2010) are believed to be accountable for the increase in crime in Dhaka. Furthermore, social scientists argue that '[. . .] an additional dimension of the criminal justice system (in particular, the police) was that it had been thoroughly politicised and degraded by the two political parties elected to power in 1991, 1996 and 2001, through brutalising (sic) the recruitment, transfer, posting and promotion process. In addition, the politicians openly resorted to the cash nexus (sic) for favours they showed in these personnel matters. As a result, the quality of the criminal justice system further deteriorated and this has had a telling effect on controlling crime in Dhaka city. The increased of criminals in politics had also increased substantially because of frequent physical confrontation between the two major political parties' (Siddiqui et al. 2010, p. 303).

Even though crime studies in Dhaka seem to show that increased criminal activity is the effect of sociocultural and demographic changes, and economic hardship, there may be another explanation to the correlation between urban neighbourhoods and the development of criminal environment (Knox 1994). According to Knox, the urban neighbourhood life cycle is characterised by five distinct phases (e.g. urbanisation, in-filling, downgrading, thinning out and gentrification or rehabilitation), and each of these stages has a crucial role in the development of criminal activities. At the initial stage (phase 1), existing urban areas expand outwards with young families of middle income or above being attracted to settle in the newly developed areas. Peace reigns for few years as the population is homogenous, social cohesion remains firm and the level of control by older residents is higher in this stage. When the children of these families reach their teenage years, usually at stages 2–3, crimes are likely to increase because social cohesion amongst neighbourhoods and families is reduced significantly with the introduction of different socioeconomic groups. The incidence of crimes continues to mount with the peak population turnover and considerable changes in socio-demographic profile at phases 3–4. However, crime reduces at the last stage due to higher-density development and the reintroduction of more affluent population (Knox 1994). However, this observation, particularly the suggestion that densification of development can help to reduce crime, has been seriously criticised recently (Cozens 2011; Dempsey 2008). Knox's (1994) observations are made in a twentieth-century-developed world context and may not be directly applicable in Bangladesh. As Dhaka progresses from its initial urban development, analysis of urban structure and growth (see Chap. 6) may help relate the development of criminal environment in the city in recent times.

Generally, Dhaka is believed to be experiencing an elevated crime rate because of its capital city status and rapid urban growth; however, no study has yet used geospatial techniques to examine the spatial pattern of crime incidences. The Dhaka Metropolitan Police (DMP) does not have adequate resources (GIS software, people, etc.) to analyse crime from a spatial perspective. As a result, the law enforcement agency still uses the 'pushpin' technique with maps hanging on walls to combat crime with pins representing offences. This technique has elsewhere been found to be largely ineffective in reducing crimes for a given environment (Chainey and Ratcliffe 2005). To address this issue and to explore the potential of GIS in crime mapping in Dhaka Metropolitan Area (DMA), this study applies spatial analysis techniques to the occurrences of crime in the period 2011–2012. We believe that this work will be invaluable for relevant authorities for allocation of resources and, importantly, to prevent spatially targeted crime (Ratcliffe 2010).

14.2 Data and Methods

Unlike countries such as the USA and Canada where crime records are easily accessible either from the Department of Justice or census records, data in relation to crime are only available from the police department in Bangladesh (though some NGOs collect crime records from daily newspaper). The police department collects crime data according to *thana* (police district) with little spatial detail. As a result, the application of spatial techniques to the analysis of crime in Bangladesh and Dhaka is extremely difficult.

The DMP is the authority responsible for recording and managing crime events in DMA. The DMP comprises 41 *thana* which are divided into eight larger units, known as 'police divisions' for administration purpose. There are about 47 crime categories reported by DMP that are broadly divided into 13 headings including robbery (house, bank, navy, industrial, bus/truck, etc.), gang robbery, homicide (political, non-political, murder with riot), riot (political and non-political), burglary, theft of various types (vehicle, cattle, cable and others), violence against women (rape, acid throwing, trafficking, serious injury and others), police assault, child abuse (abduction, trafficking, injury, rape and others), Speedy Trial Act of 2002 (arson, forgery, snatching, hijacking, property damage, fraudulence, breaking and entering, etc.), abduction (ransom, trafficking and others), crimes related to possession of illegal materials (Arms Act, Explosive Act, smuggling, narcotics) and a miscellaneous heading. For this study, we acquired, from DMP a spreadsheet listing (in Bengali) all crimes under these 13 headings for each *thana* for the period of August 2011 to July 2012, constituting a 1-year record across the DMA.

A vector polygon file containing the latest boundaries of the *thana* was acquired from the Bangladesh Space Research and Remote Sensing Organization (SPARRSO), although many studies suggest the use of a smaller spatial unit (e.g. census tracts) for crime mapping (see Rengert and Lockwood 2009). As crime information in DMA is recorded at this scale, we had no choice except using this spatial unit despite the fact that the use of aggregated data at larger spatial unit is sensitive to the well-known modifiable areal unit problem (Openshaw 1984) and to edge effects (Rengert and Lockwood 2009). A vector data file of the road network was obtained from Center for Environmental and Geographic Information Services (CEGIS) and subsequently updated using Google Earth images in 2012. Socioeconomic and demographic data were obtained from the 2011 population and housing census (BBS 2012) as being the best possible match between crime and census data. Social and demographic data included total population, the number of male population, total literacy rate, percent of *pucca* houses, percent of *katcha* houses, percent of house owners, percent of people living in rental housing units, the number of unemployed males and the number of males aged between 18 and 30. It should be noted that housing characteristics will be used with the assumption that affluent people live in *pucca* houses, whilst relatively the lower middle-lower socioeconomic class can afford *katcha* house. The poverty data is based on the 2005 Bangladesh Household Income and Expenditure Survey (BBS et al. 2005). Population density was calculated as the total number of people residing in a *thana* divided by its total area (km^2). Crime and socioeconomic data were encoded in a spreadsheet and subsequently linked with the *thana* boundary feature dataset using a unique ID.

Our intention is not to subscribe to any particular theory rather we are interested to investigate the variables in the available datasets and then relate our findings with criminological theories. We further argue that given the socioeconomic, demographic and cultural conditions, a single theory may not adequately explain the increase in criminal offences in DMA. It should be noted that existing criminological theories are based on empirical findings from the western world and may have little relevance to developing countries, particularly in Dhaka, where acquisition of pertinent variables with crime records is a serious drawback to the application of a particular theory. However, the selection of the sociological and demographic variables used here has been guided by the literature.

14.2.1 Descriptive Statistics

Our emphasis is on the general levels of crime rather than specific crime types; hence, all 47 attributes were used to develop descriptive statistics under the 13 broad categories mentioned previously. Since a preceding study has demonstrated seasonal effect on the occurrence of crime in Dhaka, we categorised the yearly crime data into three distinct season, i.e. pre-monsoon (March to June), monsoon (July to October) and post-monsoon (November to February), in order to discern whether there exist any seasonal variation. In addition, monthly crime statistics according to *thana* were calculated in an attempt to understand the temporal variation of offences.

14.2.2 Cartographic Analysis

To illustrate crime intensity and the number of events in DMA, we used a choropleth mapping technique since the crime information was available in a predefined zoning system. This technique uses the attribute values of zones to shade them with different colours or symbols to display place-to-place variation and respective numbers of events in the corresponding areas. Although choropleth maps cannot illustrate detailed geographic patterns, they avoid concerns about privacy and confidentiality, an issue that has received considerable attention in crime analysis. In addition, when the data refers to rates rather than numbers of occurrences, choropleth mapping is a preferred approach. A crime incident rate for the period being considered was calculated per 100,000 populations using the total crime events per *thana* as numerator divided by the total population in each jurisdiction boundary. Apart from incidence rate mapping, we again mapped crimes using the number of events under six broader categories to illustrate patterns of occurrence of the various types of offence. Note that the appropriate classification technique of attribute values with choropleth mapping depends on the size, shape and number of areas being analysed.

14.2.3 Analysis of Clustering

We postulated the hypothesis that there is no association of crime occurrences amongst neighbouring spatial unit, i.e. that crime distribution obeys complete spatial randomness. The alternative hypothesis is that neighbouring locations have similar crime rates, in other words spatial clustering exists.

We used exploratory spatial data analysis (ESDA) techniques to measure the spatial dependencies of crime events since ESDA techniques have been shown to be important analytical tools for crime analysis (Grubesic and Mack 2008) and capable of revealing complex spatial phenomenon not otherwise identified (Ye and Wu 2011). The GeoDa cluster detection package (Anselin et al. 2006) was used for this purpose on the annualised crime incident data according to each *thana*.

Global Moran's I, an index that is used to measure spatial dependency at global scale, was initially applied to the entire area to explore whether there exists any spatial clustering in the study area. Global scale in this context refers to the fact that the index is derived over the full geographic extent of the available data. A Queen's case contiguity rule was used to conceptualise the spatial neighbourhood structure around each police district. The outcome of global Moran's I is merely an indication of spatial dependencies over the whole area and does not account for the local variability of events. Where spatial autocorrelation was found, we then determined the location of crime clusters through local indicators of spatial autocorrelation (LISA) (commonly referred to as local Moran's I (Anselin 1995)) to locate statistically significant high and low criminal incidents. Despite the fact that LISA is mathematically analogous to global Moran's *I*, local geography is specific in this case (e.g. each *thana*). The inference of significance for both global and local Moran's *I* was based on 199 Monte Carlo randomisations, and an alpha level of 0.05 was set to test the statistical significance. Based on these permutations and threshold, cluster maps were prepared with ArcGIS software for better cartographic visualisation.

14.2.4 Regression Analysis

Two regression techniques were used to determine the relationship between outcome (crime incident rate) and explanatory (independent) variables using SPSS software. Initially, a series of simple regressions were performed considering each individual potentially causative factor (explanatory variable) to confirm its contribution to the occurrence of crime in DMA. Five techniques were utilised in the simple regressions: linear, logarithmic, quadratic, power and exponential. The primary intention was twofold: firstly, to understand the best predictor of crime and, secondly, to compare the results of the different regression analyses. Finally, all the variables used in bivariate analysis were utilised in a multivariate regression. A stepwise regression was conducted with an assumption that the relationship between variables is linear. Based on inspection of the coefficients of determination (R^2) and their associated p-values and F values, the best-fit equation for modelling the driving factors of crime in the study area was derived.

14.3 Results

A total of 23,234 crimes were found to have been committed during a one-year period in DMA. As mentioned previously, these crimes are classified into 13 broad categories. Descriptive statistics for these categories are shown in Table 14.2. This indicates that of the total crimes committed, 49.4 % were related to possession of illegal materials, in which the major proportion was in relation to narcotics. A total of 6,600 (28.4 %) crimes were under the miscellaneous heading which usually covers nuisance crimes such as suspicious behaviour and complaints. Of all the crimes, 8.7 % were theft cases of various types including vehicle and cattle theft and breaking and entering. Consistent with the literature (Lee and Haider 2012; Islam 2010), violence against women in Dhaka is also high, as in other parts of the country, amounting to 5.9 % of total offences. This includes rape, assault, acid throwing and kidnap/ransom. A matter of serious concern is the occurrence of violent crimes such as robbery, murder, stabbing and homicide as shown in Table 14.2. A total of 236 people (1 % of the total) were killed during the study period whilst robbery- and gang robbery (dacoity)-related offences accounted for 1.4 %, with burglary accounting for 2.5 %. A comparison between the occurrence of

Table 14.2 Crime statistics in DMA between August 2011 and July 2012, by crime type

Crime type	Frequency	%	Min	Max	Mean	Std dev.
Gang robbery (dacoity)	41	0.2	1	7	1.0	1.5
Robbery	274	1.2	1	29	6.9	7.0
Murder (political + non-political)	236	1.0	1	23	5.8	4.2
Riot (political + non-political)	8	0.03	3	5	0.2	0.9
Burglary	586	2.5	1	41	14.3	12.0
Theft cases (vehicle, break and enter, etc.)	2,012	8.7	7	125	49.0	36.0
Violence against women	1,371	5.9	2	108	33.4	28.0
Police assault	160	0.7	1	15	3.9	4.2
Child abuse	48	0.2	1	9	1.2	1.8
Speedy Trial Act-2000	307	1.3	1	31	7.5	7.2
Abduction	115	0.5	1	17	2.8	3.2
Rescue-related (e.g. narcotics, Arms Act)	11,476	49.4	46	636	279.9	152.9
Miscellaneous-head	6,600	28.4	38	414	160.9	103.9

Fig. 14.1 Seasonal variations in the occurrences of crime

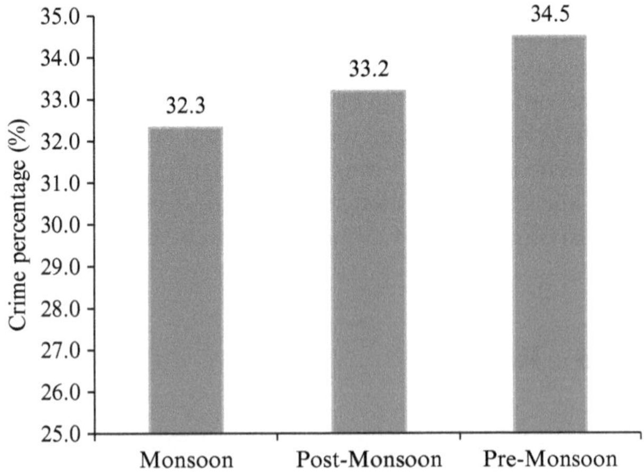

crimes in 1985 and 2012 showed that a total of 10,526 offences, of various types, were committed in 1985 in Dhaka (Siddiqui et al. 2010) which increased to 23,234 in 2012. A serious concern is that violent crimes and crimes against women and children abuse are increasing apace with time, primarily, due to politicisation of every sphere of society, including law enforcing agencies.

An examination of the seasonal variation in the occurrence of crimes revealed that pre-monsoon season from March to June had the highest level of criminal activities (Fig. 14.1). Of total 23,234 crime records, 8,014 crimes (34.5 %) were committed in the pre-monsoon season, 7,510 (32.3 %) were committed during the monsoon season, and the remaining 7,710 (33.2 %) took place in the post-monsoon period. Further inspection of monthly crime occurrences showed that the highest crime rate occurred in March with 2,181 crimes, and February had the lowest incidence during the study period, accounting for 1,716 crimes (Fig. 14.2). This finding is clearly in line with the previous study. The pre-monsoon season is

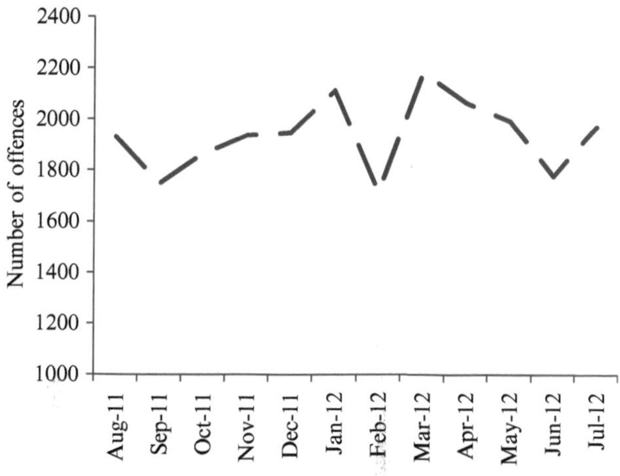

Fig. 14.2 Number of crimes in DMA, by month

characterised by low agricultural activity in the countryside, causing both high food prices and an increased unemployment rate, the resulting increase in transient people may have a considerable impact on the occurrence of crimes in this particular period in Dhaka (Ahmad and Baqee 1988). In addition, political parties in the county often use this time of the year for processions, demonstrations, strikes, etc. which are mainly concentrated on Dhaka as this is the focal point of the country. This may be another factor that contributed to the highest incidences of crime during pre-monsoon season. Another peak can also be seen in the month of January during the post-monsoon period.

The map of crime incidence rate (per 100,000 populations) shows that generally the highest criminal activities are concentrated in the most densely populated places in Dhaka that are comprised of both affluent and socially disadvantaged areas. Figure 14.3 was created using the quantile classification scheme to reflect the spatial variation of various crimes. This cartographic representation illustrates the geographic patterns of crime occurrences in DMA. For example, a total of six police districts were classified as having high number of crime incidences per 100,000 persons of which three (Jatrabari, Sabujbagh and Pallabi) are in the middle-low socioeconomic strata. Further, residential instability is also very high in these areas as indicated by the high percentage of rental housing units (BBS 2012) in these police districts. In addition, these six districts are mainly residential, which could be a factor contributing to high crime rate, a similar finding having been reported in Baltimore, Maryland (Canter 1998). Well-off residential areas such as Gulshan also exhibited high crime rates owing to the fact that the possession of valuable household goods by the residents may attract criminals to commit property crime and burglaries, which supports findings from Canada (Andersen 2006) and also the concept of opportunity theory (Wilcox et al. 2003). By contrast another affluent area, Dhanmondi, had the lowest crime rate (Fig. 14.3). Whilst land use in both of these affluent areas is mainly residential, Gulshan has recently seen an influx of commercial and educational establishments. This may be contributing to higher crimes in that district. In addition, the existence of a few large slums in the

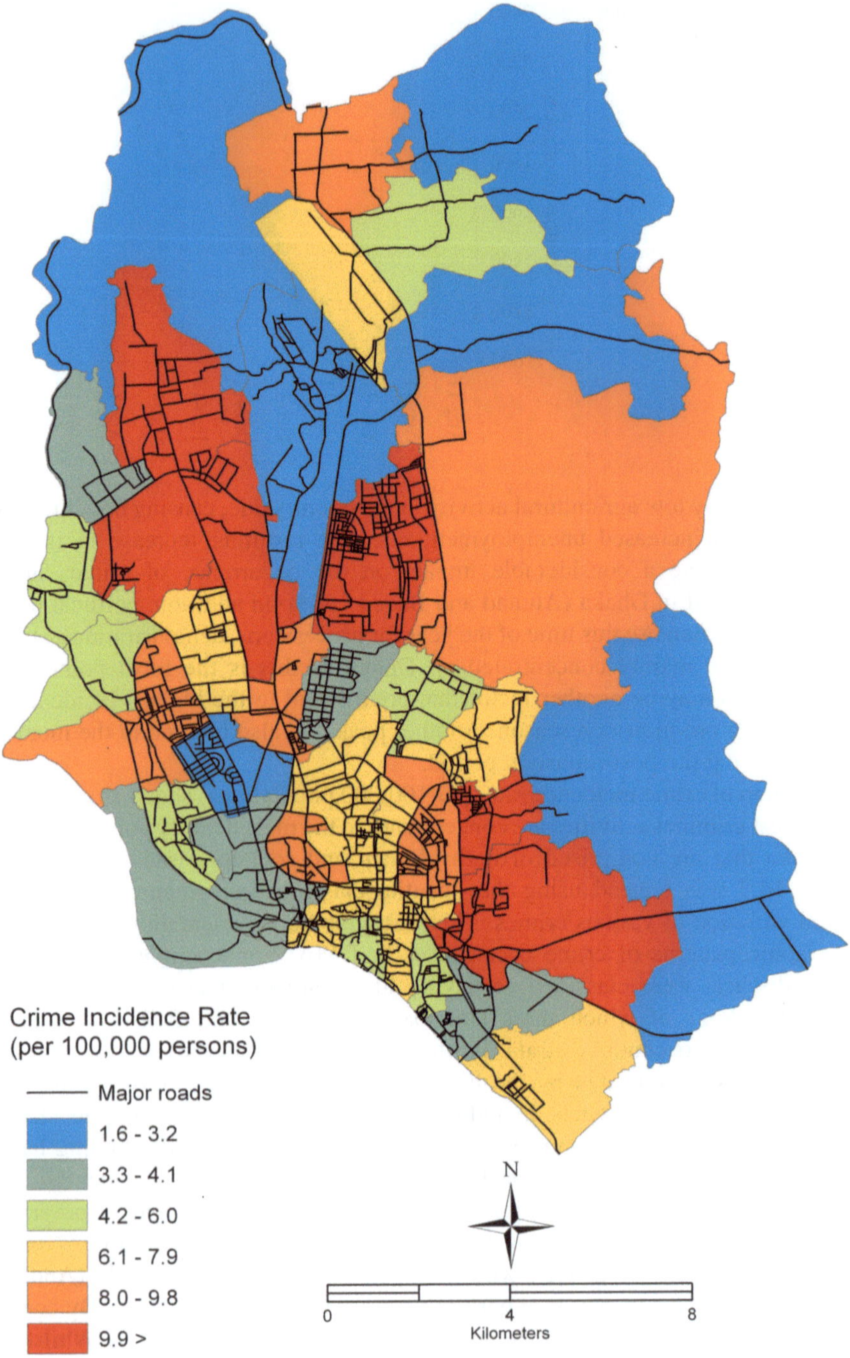

Fig. 14.3 Spatial variation of raw incidence rate of crimes

Gulshan area (CUS et al. 2006) could also account for high rate of offences, an opinion that relates to social disorganisation theory.

The second highest crime rate appears to be in the areas where the proportion of commercial and educational land use is predominant together with middle-low socioeconomic residential units. The presence of students in a community is related to criminal offences (Malczewski and Poetz 2005), and particularly in Bangladesh, most of the student organisations are extremely politicised which contributes to the recent increase in violent crimes (Siddiqui et al. 2010; Chowdhury 2003). This might be linked with the routine activity theory and situational factors that have the propensity to create conditions that foster crime (Ceccato et al. 2007). Another reasonable explanation could be the presence of members of the transient population in some of the districts. A distinct feature of the spatial variation in crime incidence (Fig. 14.3) was that police districts, such as Biman Bandar, with small population appeared highly variable as they contain a disproportionate number of high (or low) parameter estimates which is simply a factor of the relative sizes of the spatial units being considered.

Since the raw crime incidence rate is unable to indicate the spatial distribution of crime by types, we have refined the 13 broad categories of crimes into six classes. They are (i) violent crime (dacoity, robbery, murder and riot), (ii) abuse of women and children, (iii) property crime (burglary and theft cases), (iv) rescue-related offences (abduction, narcotics, Arms and Explosive Act), (v) Speedy Trial plus police assault and (vi) miscellaneous. Using the total number of events in each police jurisdiction, six choropleth maps have been developed with quantile classification scheme and are shown in Fig. 14.4. This may assist in understanding the location of specific offences in DMA. This figure also shows the general patterns of differences in criminal activities over space. For example, the geography of violent crime differs considerably from the geography of abuse of women and children. Although there is some indication of the distribution of various types of crimes, their patterns do not enable us to draw any concrete conclusion. For instance, violent crimes such as homicides are not only confined to areas with lower socioeconomic status, well-off residential areas also experienced violent crime. A similar distribution holds true for property crime. However, abuse of women and children is widespread in the areas of low socioeconomic strata with transient population which may have stemmed from poverty and long-term deprivation of a neighbourhood (Stretesky et al. 2004), and these circumstances can partly be explained by social disorganisation theory.

Examination of spatial patterns through analysis of global spatial autocorrelation revealed that weak spatial clustering existed in DMA as indicated by the Moran's I statistics of 0.1195 ($p = 0.04$). Thus, we can disregard the null hypothesis of spatial independence to the occurrence of crimes. A local indicator of spatial autocorrelation (LISA) was therefore carried out which provides information in the form of significant locations of spatial autocorrelation. Whilst high-high and low-low locations are typically known as hot and cold spots of an event (e.g. crime), high-low and low-high clusters indicate neighbours that have high-low levels of crime but are proximate to high or low levels of offence (Anselin 2005). The LISA map of crime incidences (Fig. 14.5) indicated that there were three single-centred

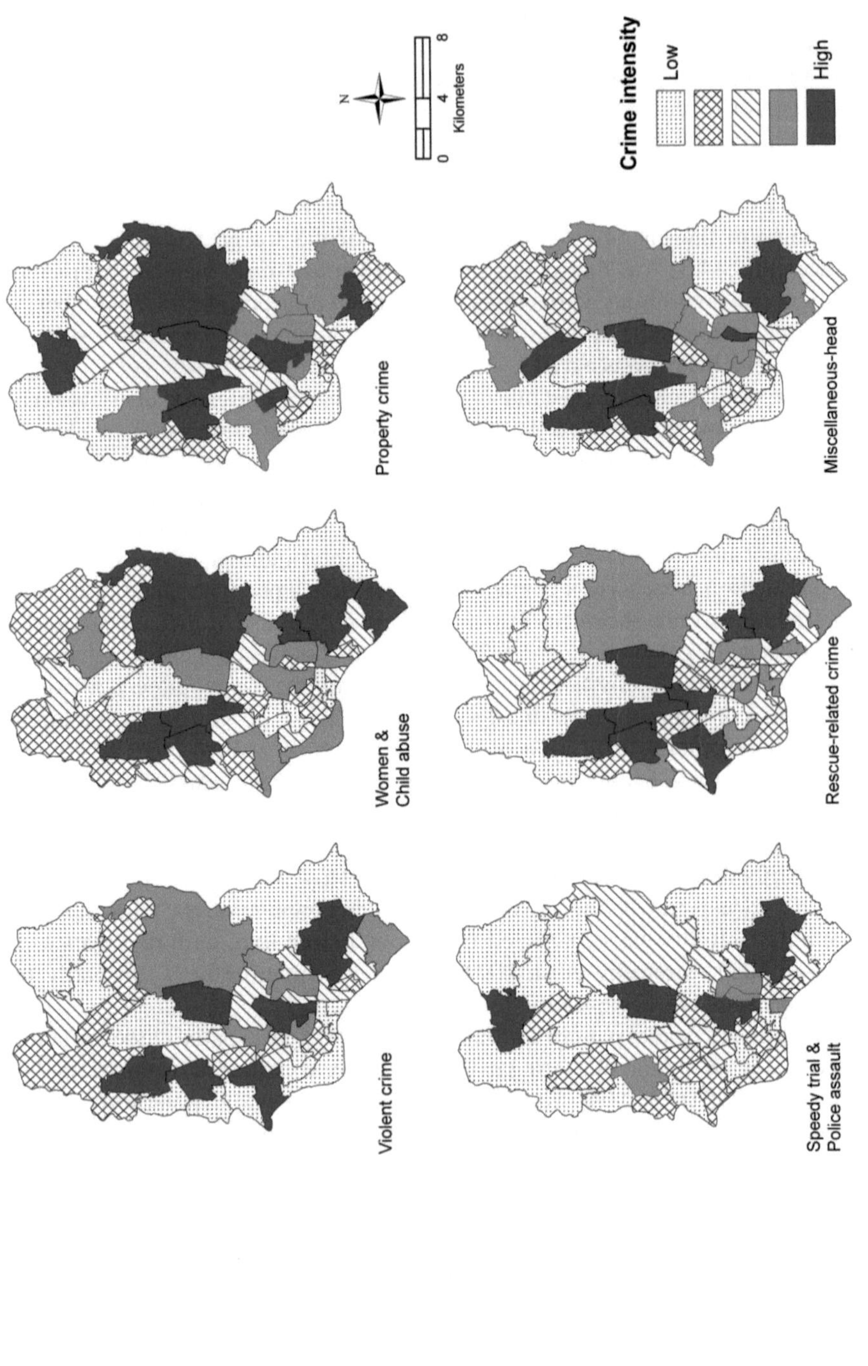

Fig. 14.4 Spatial distribution of the number of different crimes in DMA (see text for explanation)

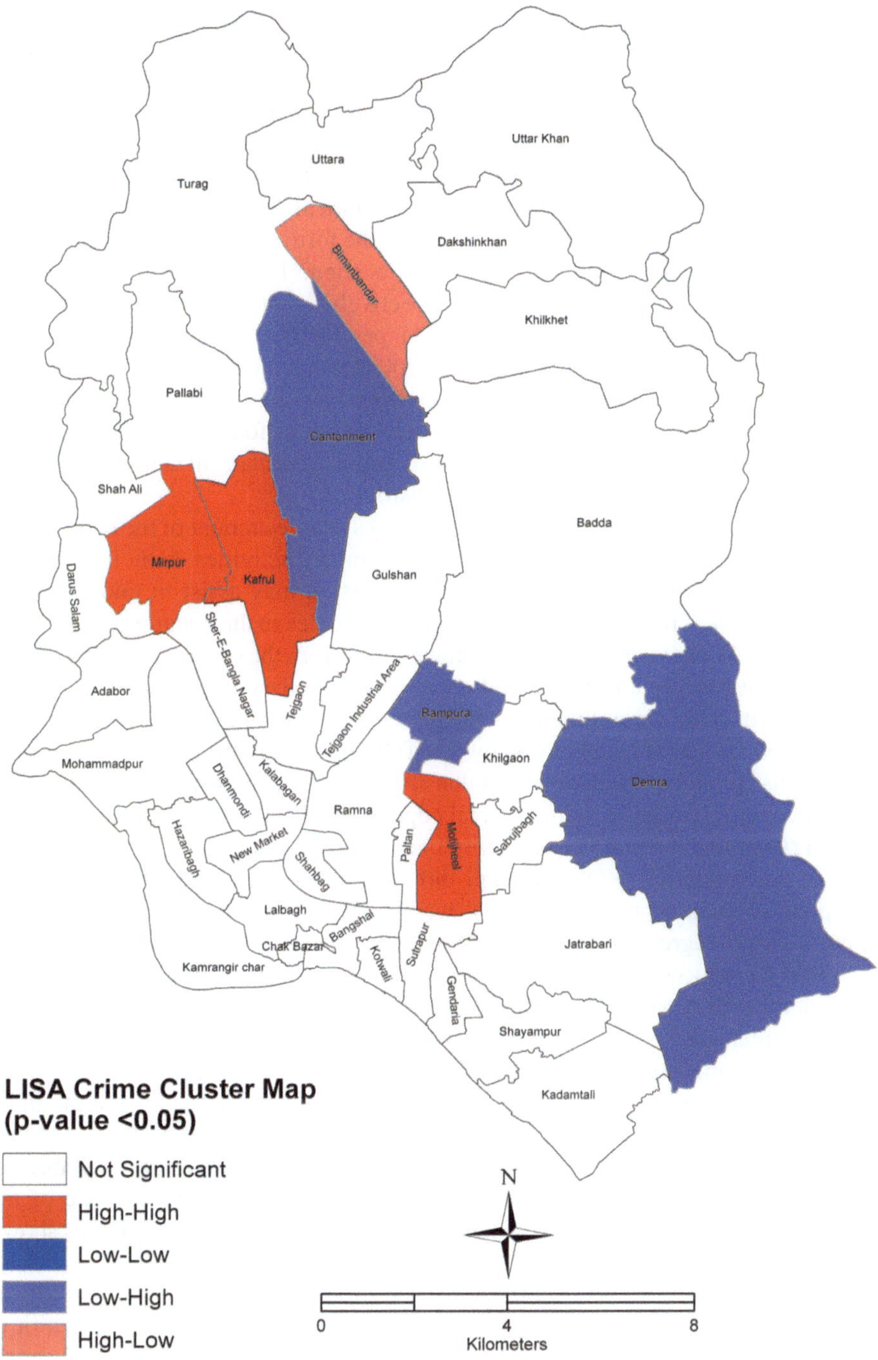

Fig. 14.5 Spatial clusters (hotspots) of crime in DMA during August 2011–July 2012

crime hotspots in the study area. They are the police districts of Kafrul, Mirpur and Motijheel, two of which are highly populated. Interestingly, Kafrul and Mirpur police districts are located next to each other which may be explained by the fact that when two neighbourhoods with differing socioeconomic status are in close proximity, crime rate increases as suggested by Bursik and Grasmick (1993). On the other hand, Motijheel is a commercial district in the middle of the city, and its residential population is relatively low. Since political activities of the country are largely concentred in and around this district in the form of rioting, picketing, police assault, etc., this could result in high–high location. The LISA *I* for these three locations was 0.931, 0.942 and 0.663, respectively, at the 95 % confidence level. The crime incidence rate was more than 11 per 100,000 persons for Mirpur and Kafrul and in Motijheel it was 9.8 per 100,000 persons. Interestingly, there was no cold spot found in the study area, indicating that crimes are omnipresent in the study area. On the other hand, spatial outliers were found in four police districts, they were Cantonment, Demra, Rampura and Biman Bandar, meaning that these districts are adjacent to high (low)-crime districts.

As this study is exploratory in nature and lacks a large amount of the data usually applied, in post-industrial cities, to relate criminal activities with the theories mentioned previously, analysis of simple regression may assist in understanding the contribution of the various socioeconomic variables to the occurrence of crimes (Table 14.3). We have used a total of ten variables from the census and other spatial data to investigate the contribution of each variable to the crime incidences, but the outcome of only six variables has been reported here since the other four parameters did not produce useful results. As shown in Table 14.3, only the linear and quadratic regressions produce creditable results. As far as explanatory variables are concerned, three variables, namely, total population, the number of male population in each *thana* and the number of unemployed males, produced a reasonable coefficient of determination (R^2). The other three variables, population density and the two poverty rates (upper and lower), were extremely poor at predicting crimes through simple regression analysis. The male population and the total population in each *thana* produced the best results through quadratic regression with correlation coefficients of 0.40 and 0.39, respectively. Linear regression also showed a similar type of agreement (Table 14.3). Even though other regression techniques showed poor results, some sort of effect was noticeable for all of them.

Multivariate regression provides the cumulative relationship amongst the independent and dependent variables. We incorporated ten variables in a stepwise regression, and the following relationships were found to be the most suitable for crime prediction. These relationships include only two variables, the size of the male population and percentage of families under the upper poverty line as shown below:

$$\mathrm{CIR} = 274.069 + .002(\mathrm{M})\left(R^2 = 0.389,\ p\text{-value} = .000\right) \tag{14.1}$$

$$\begin{aligned} \mathrm{CIR} &= 408.069 + .003(\mathrm{M}) - 18.455(\mathrm{UP}) \\ &\quad (R^2 = 0.475,\ p\text{-value} = .000) \end{aligned} \tag{14.2}$$

Table 14.3 Coefficients of causal factors and crime incidence using linear regression

Dependent variable (y)	Independent variable (x)	Equation	R^2	p-value
Linear		Equation ($y = mx + c$)		
CIR	TP	CIR = 1.442 E-5 (TP) + 3.230	0.383	.000
	M	CIR = 2.731 E-5 (M) + 3.077	0.389	.000
	PD	CIR = 2.170 E-6 (PD) + 6.239	0.001	.814
	UM	CIR = .018 (UM) + 4.352	0.273	.000
	LP	CIR = −.293 (LP) + 7.325	0.028	.298
	UP	CIR = −.090 (UP) + 7.190	0.017	.413
Logarithmic		Equation ($y = mLnx + c$)		
CIR	TP	CIR = 1.89 Ln (TP) − 16.509	0.197	.004
	M	CIR = 2.010 Ln (M) − 16.729	0.203	.003
	PD	CIR = .798 Ln (PD) − 2.048	0.065	.109
	UM	CIR = 1.795 Ln (UM) − 1.521	0.244	.001
	LP	CIR = −.394 Ln (LP) + 6.771	0.005	.667
	UP	CIR = −.469 Ln (UP) + 7.339	0.006	.617
Quadratic		Equation ($y = ax^2 + bx + c$)		
CIR	TP	CIR = 4.362 E-6 (TP) +1.753 E-11 (TP^2) + 4.259	0.398	.000
	M	CIR = 9.66 E-6 (M) + 5.63 E-11 (M^2) + 4.086	0.401	.000
	PD	CIR = 3.800 E-5 (PD) − 1.219 E-10 (PD^2) + 4.970	0.076	.224
	UM	CIR = .027 (UM) − 2.314 (UM^2) + 3.813	0.302	.001
	LP	CIR = .864 (LP) − .134 (LP^2) + 5.416	0.073	.238
	UP	CIR = .284 (UP) − .016 (UP^2) + 5.496	0.047	.401
Power		Equation ($y = mx^z$)		
CIR	TP	CIR = .284 $\text{Ln(TP)}^{.179}$	0.142	.015
	M	CIR = 304 $\text{Ln(M)}^{.167}$	0.150	.012
	PD	CIR = .196 $(\text{PD})^{.701}$	0.125	.023
	UM	CIR = 289 $\text{Ln(UM)}^{1.553}$	0.203	.003
	LP	CIR = −.108 $(\text{LP})^{6.164}$	0.012	.504
	UP	CIR = −.159 $\text{Ln(UP)}^{7.672}$	0.024	.334
Exponential		Equation ($y = me^x$)		
CIR	TP	2.214 E-6 $(\text{TP})^{3.408}$	0.291	.000
	M	4.217 E-6 $(\text{M})^{3.320}$	0.299	.000
	PD	1.109 E-6 $(\text{PD})^{5.174}$	0.012	.493
	UM	CIR = .003 $(\text{UM})^{4.039}$	0.224	.002
	LP	CIR = −.061 $(\text{LP})^{6.746}$	0.039	.214
	UP	CIR = −.023 $(\text{UP})^{6.795}$	0.036	.237

CIR crime incident rate per 100,000 persons, *TP* total population, *M* number of male population, *PD* population density, *UM* number of unemployed adult male, *LP* percent of family under lower poverty line, *UP* percent of family under upper poverty line

Equation 14.2 was found to be the best fit that produced highest coefficient of determination (R^2) and lowest-significance F values.

Whilst this study found a relationship between total population and the occurrences of crime, no relationship was found between population density parameter and criminal activities. As noted by Harries (2006a, b), the concept of population density is somewhat ambiguous. Whilst some argue that population density is a crime generator for particular type of offences such as property crime (Sampson 1983), Harries argues that socioeconomic status mediates population density and that crime is likely to be relatively high regardless of density (Harries 2006b). As population turnover in an urban neighbourhood, particularly with heterogeneous groups, leads to anonymity and decreasing adult supervision, neighbourhood instability resulting from high population growth is related to the generation of crimes as suggested by routine activity theory. Our result provides evidence to support this argument. Thus, a conclusion can be drawn that population size of *thana* influences the occurrences of crime in Dhaka.

Generally, Bangladesh is a male-dominated society and most of the crimes are committed by males, and this may simply be the reason why the number of males is associated with the incidence of crimes. A recent study suggests that more than 81 % of male, aged between 15 and 35, are addicted to drugs in Dhaka (Hossain and Mamun 2006). The same study found that these drug addicts are involved in committing crimes of various types including property crimes such as burglary and vehicle theft. Thus, our finding is consistent with literature (Matthews et al. 2010; Fox and Piquero 2003), suggesting that the male population has the greatest propensity of committing crimes in Dhaka. Although simple regression using the number of unemployed male population showed a low coefficient of determination, a relationship is nevertheless apparent. Similar findings have been reported in Canada (Andersen 2006; Malczewski and Poetz 2005) and in the USA (Kohfeld and Sprague 1988).

The two poverty variables predicted crime incidence very poorly in our study. As Andersen (2006, p. 274) argues, 'unemployment is a far better predictor of criminal activity than social disorganization theory's claim to poverty as the best predictor'. Notwithstanding that, the percentage of families under the upper poverty line, in conjunction with the number of male population, provided the best fit in multivariate regression. This may provide evidence of the area of affluence/ deprivations hypothesis (Sampson and Wooldredge 1987).

14.4 Discussion

Though this is the first study of its kind in Bangladesh, particularly in Dhaka, it has raised a few issues to be considered whilst carrying out further work. The first concerns the applicability of the various criminological theories which have been developed over the years in post-industrial cities. They appear to be of little relevance in DMA where rapid urbanisation driven by rural-urban migration is causing radical changes in the composition of urban neighbourhood that are now highly heterogeneous in terms of population characteristics. Secondly, as described in Chap. 1, extreme inequality in resource distribution exists in the city, and this may also be responsible for criminal activity. Hence, we argue that a hybrid approach is required to analyse offences in Dhaka megacity. Furthermore, socioeconomic variables that are usually used elsewhere to determine the linkage between crimes and theoretical notions are not available in the study area. Therefore, the use of surrogate variables in lieu of actual data may not provide adequate evidence for crime-related inferences. The use, in this study, of a relatively large spatial unit may be masking some of the short-range variability of predictor variables; thus, using a smaller spatial unit may assist in the deeper understanding of crimes in DMA. Instead of using crime incidence rate, crime density (Harries 2006b) can be employed to identify a suitable predictor variable. Finally, advanced spatial statistics such as GWR may assist in exploring the influence of local geography on crime incidence and to examine the spatial non-stationarity. Nevertheless, a major contribution of this study is that it demonstrates the importance of spatial analysis in understanding spatial variability of patterns of crimes in Dhaka.

14.5 Conclusions

The objective of this chapter was to analyse crime incidence in Dhaka Metropolitan Area. Using 47 measures of crime from Dhaka Metropolitan Police during the period of August 2011–July 2012 together with socioeconomic and demographic data, descriptive, cartographic and spatial pattern analysis techniques were employed to explore the crime statistics. Police jurisdiction areas were used as the spatial analysis units. Finally, linear and multivariate regression techniques were used to determine the variables that are the most suitable predictors of criminal activities.

The study revealed that although rescue-related and miscellaneous offences comprise the majority of crimes in Dhaka, violent crimes such as murder, violence against women, property crimes and burglary are also affecting every sphere of social life in the city. Seasonal variation showed that pre-monsoon season had the highest occurrence of crimes followed by the post-monsoon season. The calculation of crime incidence rate indicates that, on average, there are six crimes committed per 100,000 persons. Choropleth mapping of crime incidence revealed that both affluent and disadvantaged areas are affected by a range of offences. Through the use of local indicators of spatial autocorrelation, LISA indicated that there are three single-centred spatial clusters. Crime incidence rate was subsequently used as an outcome variable with ten potential explanatory variables to explore the factors contributing most to the occurrence of crime. Linear regression analyses showed that the size of total population, number of males, unemployed males and the percent of families below the lower poverty level are better predictors when compared to other independent variables.

Despite the fact that this study was exploratory in nature, the results of this study should have considerable implications for a resource-poor country where law enforcing agencies are constantly struggling to secure the normal social activities in the city from various offences. Using cluster and incidence maps, resources can be mobilised and priority areas can be set to prevent crimes in Dhaka.

References

Ahmad N, Baqee MA (1988) Urban crimes in Bangladesh. Orient Geogr 32(1–2):65–75
Ahmed MB, Islam MR (2010) Treatment of juvenile offenders: Bangladesh perspective. Bangladesh Res Publ J 4(3):274–285
Andersen MA (2006) Crime measures and the spatial analysis of criminal activity. Br J Criminol 46(2):258–285
Anselin L (1995) Local indicators of spatial association – LISA. Geogr Anal 27(2):93–115
Anselin L (2005) Spatial statistical modelling in a GIS environment. In: Maguire DJ, Batty M, Goodchild MF (eds) GIS, spatial analysis and modelling. Environmental System Research Institute (ESRI), Redlands, pp 93–111
Anselin L, Cohen J, Cook D, Goor W, Tita G (2000) Spatial analysis of crime. Crim Justice 4:213–262
Anselin L, Syabri I, Kho Y (2006) GeoDa: an introduction to spatial data analysis. Geogr Anal 38 (1):5–22
Appiahene-Gyamfi J (2003) Urban crime trends and patterns in Ghana: the case of Accra. J Crim Justice 31(1):13–23
Bangladesh Bureau of Statistics (BBS) (2012) Population and housing census 2011, community series – Dhaka, Gazipur, Narayanganj. BBS, Ministry of Planning, Dhaka
Bangladesh Bureau of Statistics (BBS), The World Bank, World Food Programme (WFP) (2005) Bangladesh Household Income and Expenditure Survey 2005, Dhaka
Bangladesh Police (2012) http://www.police.gov.bd/index5.php?category=48. Accessed 25 Dec 2012
Bourguignon F (1998) Crime as a social cost of poverty and inequality: a review focusing on developing countries. The World Bank, Washington, DC
Bowers K, Hirschfield A (1999) Exploring links between crime and disadvantage in north-west England: an analysis using geographical information systems. Int J Geogr Inf Sci 13(2):159–184
Brantingham PL, Brantingham PJ (1993) Environment, routine, and situation: toward a pattern theory of crime. In: Clarke RV, Felson M (eds) Routine activity and rational choice: advances in criminological theory, vol 5. Transaction Publishers, New Brunswick, pp 259–294
Breetzke GD (2006) Geographical information systems (GIS) and policing in South Africa: a review. Polic: Int J Police Strateg Manag 29(4):723–740
Brimicombe AJ, Ralphs MP, Sampson A, Tsui HY (2001) An analysis of the role of neighbourhood ethnic composition in the geographical distribution of racially motivated incidents. Br J Criminol 41(2):293–308
Browning CR, Byron RA, Calder CA, Krivo LJ, Kwan M-P, Lee J-Y, Peterson RD (2010) Commercial density, residential concentration, and crime: land use patterns and violence in neighborhood context. J Res Crim Delinq 47(3):329–357
Bursik RJ Jr, Grasmick HG (1993) Economic deprivation and neighborhood crime rates, 1960–1980. Law Soc Rev 27(2):263–268

Cahill M, Mulligan G (2007) Using geographically weighted regression to explore local crime patterns. Soc Sci Comput Rev 25(2):174–193

Cancino JF, Varano SP, Schafer JA, Enriquez R (2007) An ecological assessment of property and violent crime rates across a Latino urban landscape: the role of social disorganization and institutional anomie theory. West Criminol Rev 8(1):69–87

Canter PR (1998) Geographical information systems and crime analysis in Baltimore County, Maryland. In: Weisburd D, McEwen T (eds) Crime mapping and crime prevention. Criminal Justice Press, Monsey, pp 158–190

Ceccato V, Haining R, Signoretta P (2002) Exploring offence statistics in Stockholm City using spatial analysis tools. Ann Assoc Am Geogr 92(1):29–51

Ceccato V, Haining R, Kahn T (2007) The geography of homicide in São Paulo, Brazil. Environ Plan A 39(7):1632–1653

Center for Urban Studies (CUS), National Institute of Population, Research and Training (NIPORT), Measure Evaluation (2006) Slums in urban Bangladesh: mapping and census 2005. CUS, Dhaka

Chainey S, Ratcliffe JH (2005) GIS and crime mapping. Wiley, London

Chainey S, Tompson L, Uhlig S (2008) The utility of hotspot mapping for predicting spatial patterns of crime. Secur J 21(1–2):4–28

Chen D, Weeks JR, Kaiser JV Jr (2005) Remote sensing and spatial statistics as tools in crime analysis. In: Wang F (ed) Geographic information systems and crime analysis. Idea Group Publishing, Hershey, pp 270–291

Chowdhury MH (2003) Violence, politics and the state of Bangladesh, Conflict. Secur Dev 3(2):265–276

Clarke RV, Felson M (1993) Criminology, routine activity, and rational choice. In: Clarke RV, Felson M (eds) Routine activity and rational choice: advances in criminological theory, vol 5. Transaction Publishers, New Brunswick, pp 1–16

Cohen LE, Felson M (1979) Social change and crime rate trends: a routine activity approach. Am Sociol Rev 44(4):588–608

Cohen J, Tita G (1999) Diffusion in homicide: exploring a general method for detecting spatial diffusion processes. J Quant Criminol 15(4):451–493

Commonwealth Human Rights Initiative (CHRI) (2008) Feudal forces: reform delayed. Moving from force to service in South Asia policing. CHRI, New Delhi. Available at: http://www.humanrightsinitiative.org/publications/police/feudal_forces_reform_delayed_moving_from_force_to_service_in_south_asian_policing.pdf

Cozens PM (2011) Urban planning and environmental criminology: towards a new perspective for safer cities. Plan Pract Res 26(4):481–508

Dempsey N (2008) Quality of the built environment in urban neighbourhoods. Plan Pract Res 23(2):249–264

Elffers H (2003) Analysing neighbourhood influence in criminology. Stat Neerl 57(3):347–367

Evans DJ (1992) Left realism and the spatial study of crime. In: Evans DJ, Fife NR, Herbert D (eds) Crime, policing and place: essays in environmental criminology. Routledge, London, pp 35–59

Fajnzylber P, Lederman D, Loayza N (2002) What causes violent crime? Eur Econ Rev 46(7):1323–1357

Fattah E (1997) Criminology: past, present and future. Macmillan, Basingstoke

Fox JA, Piquero AR (2003) Deadly demographics: population characteristics and forecasting homicide trends. Crim Delinq 49(3):339–359

Getis A, Drummy P, Gartin J, Gorr W, Harries K, Rogerson P, Stoe D, Wright R (2000) Geographic information science and crime analysis. URISA J 12(2):7–14

Grubesic TH, Mack EA (2008) Spatio-temporal interaction of urban crime. J Quant Criminol 24(3):285–306

Hannon LH (2005) Extremely poor neighbourhood and homicide. Soc Sci Q 86(s1):1418–1434

Harries K (2006a) Property crimes and violence in United States: an analysis of the influence of population density. Int J Crim Justice Sci 1(2):24–34
Harries K (2006b) Extreme spatial variations in crime density in Baltimore County, MD. Geoforum 37(3):404–416
Hossain MF, Mamun M (2006) A critical analysis of the impact of drug addiction in urban life of Bangladesh. Soc Sci 1(1):60–64
Islam N (2005) Dhaka now: contemporary urban development, Golden Jubilee series no. 2. The Bangladesh Geographical Society (BGS), Dhaka
Islam T (2010) Violence against women in Bangladesh. Pak J Criminol 2(2):53–76
Jaishankar K, Shanmugapriya S, Balamurugan V (2004) Crime mapping in India, a GIS implementation in Chennai city policing. Ann GIS 10(1):20–34
Karzon SHR (2006) Theorising the crime and violence of Dhaka city. The Daily Star, Dhaka, 1 Apr 2006
Khan MA (2012) Arrest surging crime. The Daily Star, Dhaka, 5 Sept 2012
Kinney JB, Brantingham PL, Wuschke K, Kirk MG, Brantingham PJ (2008) Crime attractors, generators and detractors: land use and urban crime opportunities. Built Environ 34(1):62–74
Knox PL (1994) Urbanization: an introduction to urban geography. Prentice Hall, Upper Saddle River
Kohfeld CW, Sprague J (1988) Urban unemployment drives crime. Urb Aff Quart 24(2):215–241
Krivo LJ, Peterson RD (1996) Extremely disadvantaged neighborhoods and urban crime. Soc Force 75(2):619–650
Kubrin CE (2003) Structural covariates of homicide rates: does type of homicide matter? J Res Crim Delinq 40(2):139–170
Kumar MV, Chandrasekar C (2011) GIS technologies in crime analysis and crime mapping. Int J Soft Comput Eng 1(5):115–121
Lambert EG, Khondaker MI, Elechi OO, Jiang S, Baker DN (2012) Formal and informal crime control views in Bangladesh and the United States. Justice Ethn Crim Justice 10(3):199–222
Lauritesen JL, White NA (2001) Putting violence in its place: the influence of race, ethnicity, gender, and place on the risk for violence. Criminol Public Policy 1(1):37–59
Law J, Haining R (2004) A Bayesian approach to modelling binary data: the case of high-intensity crime areas. Geogr Anal 36(3):197–215
Lee M (2010) Gender violence in Bangladesh. Pak J Criminol 2(2):viii–x
Lee M, Haider SJ (2012) A letter from Bangladesh – developing gender – responsive community policing in Bangladesh. Crim Prev Comm Saf 14(1):69–77
Majumder MMA, Basher A, Faiz MA, Kuch U, Pogoda W, Kauert GF, Tonnes SW (2008) Criminal poisoning of commuters in Bangladesh: prospective and retrospective study. Forensic Sci Int 180(1):10–16
Malczewski J, Poetz A (2005) Residential burglaries and neighborhood socioeconomic context in London, Ontario: global and local regression analysis. Prof Geogr 57(4):516–529
Mannan A, Ghani S, Sen SL, Clarke A, Butler PEM (2004) The problem of acid violence in Bangladesh. J Surg 2(1):39–43
Mas AA, Rahman Z, Ahmad M, Wahab MA, Ali M, Khalil MI (2011) Socio-demographic characteristics of alleged sexual assault (rape) cases in Dhaka city. J Armed Force Med Coll Bangladesh (JAFMC) 7(2):21–24
Matthews SA, Yang T-C, Hayslett KL, Ruback RB (2010) Built environment and property crime in Seattle, 1998–2000: a Bayesian analysis. Environ Plan A 42(6):1403–1420
Mollah MB, Islam SS, Ullah MA (2012) Proposed e-police system for enhancement of e-government services of Bangladesh. In: IEEE/OSA/IAPR proceedings of the conferences on informatics, electronics and vision, pp 881–886
Morenoff JD, Sampson RJ, Raudenbush SW (2001) Neighborhood inequality, collective efficacy, and the spatial dynamics of urban violence. Criminology 39(3):517–560
Murray RK, Roncek DW (2008) Measuring diffusion of assaults around bars through radius and adjacency techniques. Crim Justice Rev 33(2):199–220

Murray AT, McGuffog I, Western JS, Mullins P (2001) Exploratory spatial data analysis techniques for examining urban crime. Br J Criminol 41(2):309–329
Nasrin S (2011) Crime or custom? Motivations behind dowry practice in rural Bangladesh. Indian J Gend Stud 18(1):27–50
Naved RT, Persson LA (2005) Factors associated with spousal physical violence against women in Bangladesh. Stud Fam Plan 36(4):289–300
Naved RT, Persson LA (2008) Factors associated with physical spousal abuse of women during pregnancy in Bangladesh. Int Fam Plan Perspect 34(2):71–78
Neumayer E (2005) Inequality and violent crime: evidence from data on robbery and violent theft. J Peace Res 42(1):101–112
Openshaw S (1984) The modifiable areal unit problem, CATMOG 38. GeoBooks, Norwich
Ratcliffe JH (2005) Detecting spatial movement of intra-region crime patterns over time. J Quant Criminol 21(1):103–123
Ratcliffe JH (2010) Crime mapping: spatial and temporal challenges. In: Piquero AR, Weisburd D (eds) Handbook of quantitative criminology. Springer Science + Business Media, New York, pp 5–24
Rengert GF, Lockwood B (2009) Geographical units of analysis and the analysis of crime. In: Weisburd D, Bernasco W, Bruinsma GJN (eds) Putting crime in its place. Springer Science + Business Media, New York, pp 109–122
Sampson RJ (1983) Structural density and criminal victimization. Criminology 21(2):276–293
Sampson RJ, Wooldredge JD (1987) Linking the micro- and macro-level dimensions of lifestyle-routine activity and opportunity models of predatory victimization. J Quant Criminol 3(4):371–393
Santos RB (2013) Crime analysis with crime mapping. Sage Publications Inc., Thousand Oaks
Sarker AH (1989) Juvenile delinquency in Bangladesh with special reference to Dhaka city. Indian J Criminol 27(96):5–15
Sarker AH (1991) Situation of crime in Bangladesh: an implication for social planning. Soc Sci Rev 8(1&2):163–175
Sarker AH (1994) Family poverty as a delinquency causation in Bangladesh. Indian J Criminol 22(2):88–92
Shafi SA (2010) Urban crime and violence in Dhaka. The University Press Limited, Dhaka
Shafi SA (2011) Improving law and order. The Daily Star, Dhaka, 7 Apr 2011
Shaw CR, McKay HD (1942) Juvenile delinquency and urban areas: a study of rates of delinquents in relation to differential characteristics of local communities in American cities. University of Chicago Press, Chicago
Siddiqui K, Ahmed J, Siddiqui K, Huq S, Hossain A, Nazimud-Doula S, Rezawana N (2010) Social formation in Dhaka, 1985–2005: a longitudinal study of society in a third world megacity. Ashgate Publishing Group, Surrey
Snook B, Cullen RM, Mokros A, Harbort S (2005) Serial murderers' spatial decisions: factors that influence crime location choice. J Investig Psychol Offender Profiling 2(3):147–164
Sparks CS (2011) Violent crime in San Antonio, Texas: an application of spatial epidemiological methods. Spat Spatio-Temp Epidemiol 2(4):301–309
Stretesky PB, Schuk AM, Hogan MJ (2004) Space matters: an analysis of poverty, poverty clustering, and violent crime. Justice Quart 21(4):817–841
Warner BD (2003) The role of attenuated culture in social disorganization theory. Criminology 41(1):73–97
Wesson K (2002) A situational assessment study of acid violence in Bangladesh. Dev Pract 12(1):96–100
Wilcox P, Land KC, Hunt SA (2003) Criminal circumstance: a dynamic multicontextual criminal opportunity theory. Aldine de Gruyter, New York
Wiles P, Costello A (2000) The road to nowhere: the evidence for travelling criminals. Home Office Research Study, London

Wise S, Haining R, Ma J (1997) Regionalization tools for the exploratory spatial analysis of health data. In: Fischer MM, Getis A (eds) Recent developments in spatial analysis: spatial statistics, behavioural modelling, and computational intelligence. Springer, Berlin, pp 83–100

Ye X, Wu L (2011) Analyzing the dynamics of homicide patterns in Chicago, ESDA and spatial panel approaches. Appl Geogr 31(2):800–807

Chapter 15
Environmental Problems and Governance

Salim Momtaz and S.M. Zobaidul Kabir

Abstract The city of Dhaka has been experiencing significant environmental problems due mostly to a massive increase in population and unplanned development in the last few decades. The city administrators have resorted to a number of measures to address these problems, including the enforcement of environmental impact assessment procedures, with limited success. Based on analyses of relevant documents and interviews with officials involved in urban environmental governance, this chapter identifies some of the major environmental problems facing the city dwellers. This study critically analyses the status of environmental impact assessment and reveals that the government has established a good legal and organisational framework of environmental management but that there are deficiencies in enforcement. This chapter makes recommendations for improvements in environmental governance.

Keywords Administrative framework of EIA • Environmental governance • Environmental Impact Assessment • Environmental problems • Legal framework of EIA

15.1 Introduction

During the decades between 1970 and 1990, the Government of Bangladesh, with a view to alleviating poverty and resolving the country's unemployment problem, initiated many industrial and agricultural development projects. Much of these

S. Momtaz (✉)
School of Environmental and Life Sciences, University of Newcastle, Chittaway Drive, Ourimbah, NSW 2258, Australia
e-mail: Salim.Momtaz@newcastle.edu.au

S.M.Z. Kabir
Centre for Environmental Management, Central Queensland University, Bruce Highway, North Rockhampton, QLD 4702, Australia
e-mail: z.kabir@cqu.edu.au

A. Dewan and R. Corner (eds.), *Dhaka Megacity: Geospatial Perspectives on Urbanisation, Environment and Health*, Springer Geography,
DOI 10.1007/978-94-007-6735-5_15,

development activities took place with little consideration of their environmental consequences. As a result, the country suffered from environmental degradation in many areas. Dhaka, being the centre of all major economic activities, experienced these problems more acutely than any other place in the country.

As discussed earlier in the book (see Chap. 3), the population of Dhaka has grown at a very rapid rate in the recent past (Newgeography 2012), as a result of massive rural to urban migration. The efforts by the government and related agencies to respond to this growth have led to many development activities in Dhaka, particularly, infrastructure, housing and land reclamation. However, it seems that most of these development activities have occurred in an unplanned and unsustainable fashion much to the detriment to the natural environment. In addition, illegal development, such as the development of slums and squatter settlements, where most poor rural to urban migrants end up, the establishment of small industries and businesses in areas not designated for these purposes, unplanned growth of the informal sector and vehicular traffic over unfit roads, have added to the problem. The expansion of urban infrastructure, services and utilities could not keep pace with the increase in demand caused by population growth. This large gap between demand and supply has placed extreme pressures on the existing infrastructure and services.

The government and the city administrators have taken many initiatives to address these issues at different levels often with no tangible positive outcomes. However, one important initiative was the promulgation of the Environmental Conservation Act (ECA) 1995 (the country's first environmental impact assessment (EIA) legislation) and the related Environmental Conservation Rules (ECR) 1997 (hereafter referred to as the Act and the Rules, respectively). Under the Act, the Department of Environment (DOE) was given authority to enforce EIA and other provisions of the law to keep the environment of Dhaka from degrading further and to stop unlawful development. The Act and the Rules were intended to formalise the role of environmental impact assessment in development planning and implementation. Under the Act all major development activities were required to conduct an EIA to avoid, minimise and compensate for any negative impacts as appropriate. There was a high expectation on the part of the government that these legal instruments would bring some respite from environmental problems and that some control over development activities in Dhaka could be exercised. Despite the promulgation of the EIA law, environmental degradation in the city of Dhaka has continued to be a major concern for the city dwellers and the administrators.

Based on interviews with relevant officials and document analysis, this chapter highlights some of the major environmental problems of Dhaka and examines the role of the three major organisations responsible for environmental governance. It critically reviews the effectiveness of the EIA system designed to enforce the Act and the Rules in managing the development activities in Dhaka. It also provides an overview of two major initiatives by the government to address the main causes of environmental problems of Dhaka and makes recommendations for improvement in environmental governance.

15.2 Major Environmental Problems

The unplanned development of Dhaka can be regarded as the single most important reason for its environmental deterioration. Much of these activities have happened without either adequate approval from the authorities or regard to the construction rules and regulations and environmental laws. High-rise buildings are being built to meet the ever-increasing demand for residential spaces – either replacing low-rise buildings, taking over urban green spaces by cutting down trees, or being built on filled-in low-lying areas often without adequate provisions for infrastructure, utilities and amenities.

A recent study (Islam et al. 2010), covering the development of the city from 1960 to 2008, notes that about 33 % of the water bodies and 53 % of lowlands around Dhaka have been lost to urbanisation during this period. Additionally, Dewan et al. (2012) show through spatial analyses how land use and land cover has changed in Dhaka as a result of population increase and how the city has become extremely vulnerable to the risks of devastating flood and other natural hazards (Dewan et al. 2012; Rabbani 2009).

Section 12 of the Act stipulates that 'No industrial unit or project shall be established or undertaken without obtaining, in the manner prescribed by rules, an Environmental Clearance Certificate from the Director General' (DOE 1995, p. 9). Stringent measures have been put in place for environmentally significant development to go through environmental impact assessment processes so as to avoid significant environmental impacts, minimise minor impacts and develop mitigation measures for compensation. Development projects are also required to develop management and monitoring plans along with their environmental impact assessment. However, Dhaka megacity continues to experience unplanned development activities.

Section 9 Clause 1 of the Act provides guidance on the protection of water quality by taking appropriate measures to stop discharge of excessive pollutants. It states, 'where, due to an accident or other unforeseen incident, the discharge of any environmental pollutant occurs or is likely to occur in excess of the limit prescribed by the rules, the person responsible and the person in charge of the place of occurrence shall take measures to control or mitigate the environmental pollution' (DOE 1995, p. 7). In spite of these legal mandates for the protection of water quality, the surface water quality in Dhaka has become a major cause of concern for both city dwellers and administrators. The groundwater is also increasingly becoming exposed to pollutants and effluents, including all the untreated sewage and wastewater coming from many sources such as the slums where millions of poor people live, industries, factories and residential areas. None of the slums are provided with proper sewerage services by the Water and Sewerage Authority (WASA), and only 9% of this population receive solid waste management services (UNEP 2006). The rivers Buriganga, Balu, Turag and Lakhya together regularly receive huge amounts of untreated sewage and industrial liquid waste as well as municipal waste through the three major canal systems and direct disposal. Water in the surrounding rivers and lakes has already passed the standard limits of many water quality parameters, for example, DO, BOD, COD and pH (UNEP 2006; Karn and Harada 2001).

There had been significant improvement in the air quality of Dhaka City with the replacement of two-stroke-engined three wheelers in January 2003 (UNEP 2006) by four-stroke-engined vehicles for public transport and introduction of the use of compressed natural gas (CNG). This success was short lived. With the unplanned urbanisation and expansion of Dhaka, the number of motor vehicles has increased significantly in the last few years. The road network and infrastructure, however, did not increase proportionately. The transport sector is also largely responsible for noise pollution. Under the Rules, Ambient Air Quality Standards, Vehicular Exhaust Emission Standards, River Transport (Mechanized) Emission Standard and Gaseous Emission of Industries or Projects Standards have been set. The Act also contains laws in regard to the protection of environmental health and the control of environmental pollution. The air in Dhaka contains lead, in concentrations reportedly almost ten times higher than the government safety standard set by the Department of Environment (DOE 2012). Section 6 (1) of the Act clearly states, 'A vehicle emitting smoke or gas injurious to health or environment shall not be operated nor shall such vehicles be switched on except for the purpose of test-operation for stopping the emission of such smoke or gas' (DOE 1995, p. 6). Furthermore, the Rules, Section 4, No.1, strengthen the provision of the Act by elaborating on the responsibility of motor vehicle owner 'every motor vehicle using petrol, diesel or gas as fuel shall be fitted with catalytic converter or oxidation catalyst or diesel particulate filter, or with such other instrument or device as may be approved by the Director General for ensuring that emission from the vehicle does not exceed the standards specified in schedule 6' (DOE 1997a, p. 4). There are penalties for those found in violation of these clauses of the law. Despite the existence of these provisions of the law and the regulations, vehicles emitting black smoke in clear violation of this law are commonplace in Dhaka (Mahmud et al. 2012).

15.3 Major Organisations in Urban and Environmental Governance

There are three major agencies in Dhaka entrusted with the responsibility of various aspects of urban governance. These are described below.

15.3.1 Rajdhani Unnayan Kartripakkha (RAJUK or the Capital Development Authority)

The 'Dhaka Improvement Trust' (DIT) was established in 1956 under the provision of the 'Town Improvement Act, 1953' (TI Act 1953). The objectives of this Act were to improve the physical and urban condition of Dhaka City. DIT had the supreme planning and development control power within its jurisdiction.

RAJUK was established in April 30, 1987, replacing the Dhaka Improvement Trust (DIT) and assuming most of its roles and responsibilities. Additionally, it was given the sole authority to develop, improve, extend and manage the city and the peripheral areas through a process of proper development planning and development control. RAJUK controls all development activities in Dhaka under the provisions of the Town Improvement Acts, the Building Construction Rules and Land Use Regulations (RAJUK 2012).

15.3.2 Dhaka City Corporation (DCC)

With the introduction of the Paurashava Ordinance in 1977, the city area was divided into 50 wards, and an election of Ward Commissioners was held in 1977. In 1978, Dhaka Municipality was awarded Municipal Corporation status. In 1990, Dhaka Municipal Corporation was renamed as Dhaka City Corporation and was divided into zones to fulfil the objectives of decentralisation. The Corporation received a legal mandate with the introduction of the Dhaka Municipal Corporation Ordinance, 1993. Its major responsibilities include building control, development, social welfare and urban planning (DCC 2012).

15.3.3 Department of Environment (DOE)

The DOE, within the Ministry of Environment and Forests (MoEF), is the authority responsible for implementing and enforcing the EIA (MoEF 1992, p. 327). It has responsibility to formulate EIA guidelines for different sectors and to review and approve EIA reports. It has the authority to institute proceedings against any agency or persons who are polluting the environment and breaching the environmental directions served by the DOE. Thereby DOE has been given extensive powers to deal with environmental pollution across all sectors (Gain 1998). In addition to its EIA-related functions, the mandate of DOE is to broadly ensure conservation of the environment, assessment and improvement of environmental standards and the monitoring and mitigation of pollution control (DOE 2012).

15.4 Legal Frameworks for the EIA System

Legal arrangements for the EIA system in Bangladesh (also one of the most important legal instruments for urban environmental governance of Dhaka City) broadly include constitutional rights; national environmental policy, plan and strategy; and, more importantly, the Environmental Conservation Act and Environmental Conservation Rules (Fig. 15.1). Some of these are described below.

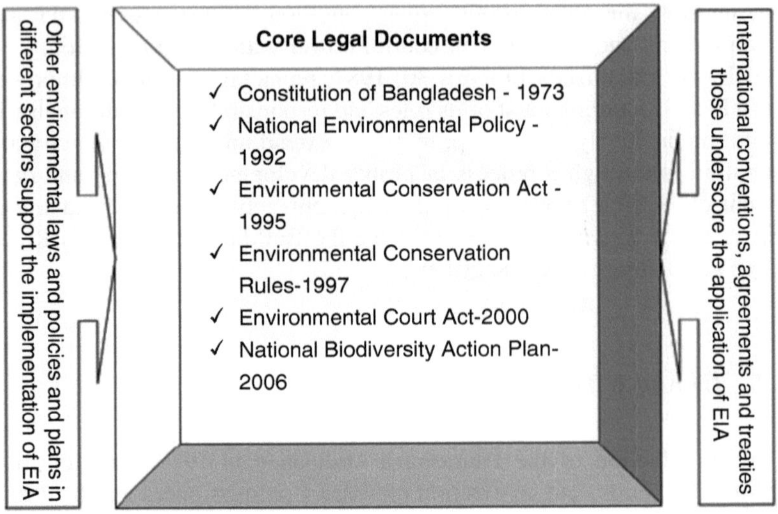

Fig. 15.1 Legal framework of EIA in Bangladesh

15.4.1 The Constitution of Bangladesh

The constitution of Bangladesh protects 'the right to life and personal liberty' (Articles 31 and 32) as a fundamental human right (GOB 1973, p. 43). Although it does not explicitly recognise the right to a clean environment as a fundamental right, in two cases (Gain 1998) related to the environmental impacts of development projects, the Supreme Court of Bangladesh has resolved that the 'right to life' is enshrined as a fundamental right and includes the 'right to a healthy environment' (Gain 1998).

15.4.2 The Environmental Conservation Act

The first major national law for the protection and conservation of the natural environment was the Environmental Conservation Act (ECA) of 1995. The Act provides the legal basis for EIA in Bangladesh (Fig. 15.2). It calls for assessment of the environmental impact of new development projects. The Act stipulates the actions for enforcement of EIA at post-EIS (Environmental Impact Statement) stage. There are penalties specified by the ECA when a project damages the ecosystem or pollutes the environment during its implementation or operation. The DOE can take administrative action including determination of compensation. It can file a compensation case or a criminal case at an environment court in Dhaka (Section 7) against the polluters.

Environmental Conservation Act (1995)

1. Assistance from law enforcing agencies
2. Polluters pay principle
3. Formulation of environmental guidelines
4. Claim for compensation by the affected people
5. Penalties for violation of project approval conditions
6. Offences committed by proponents and action against proponents
7. Power to make rules

Environmental Conservation Rules (1997)

1. Complain by community to DG against non-compliance of project approval conditions
2. Power to collection of sample
3. ECC procedure with specified time limit
4. Screening of both public and private projects
5. Review of EIS
6. No objection Certificate from Local Government
7. Appeal by proponents against order or directions from DOE

Fig. 15.2 Major features of the Act and the Rules (Adapted from DOE 1995, 1997a, b)

15.4.3 The Environmental Conservation Rules

The Environmental Conservation Rules (ECR) of 1997 were promulgated for the implementation of the Act with the Rules elaborating in greater detail on the requirements of EIA procedure and implementation (Fig. 15.2). The Rules explicitly prescribe the EIA requirements with particular focus on the process of environmental clearance. According to the Rules, all types of environmentally significant projects will be subject to EIA, and the EIA reports will need to be submitted to DOE for approval. The ECR also stipulate the responsibilities of DOE to implement, coordinate and enforce the EIA provisions properly during an EIA study and to approve the EIA report.

15.5 Administrative/Organisational Frameworks and Process for EIA

According to the Act, the DOE is the responsible authority to implement and enforce EIA (MoEF 1992, p. 327) within the Ministry of Environment and Forest. It controls and manages the administrative procedure of EIA. A tiered approach is adopted in the assessment of environmental impacts of projects. The EIA procedure passes through the following three tiers – screening, initial environmental examination and environmental impact assessment. This project screening enables DOE to determine the environmental significance of projects and recommend an appropriate environmental clearance process. Accordingly, the proponents have to conduct an initial environmental examination (IEE) or EIA and submit the IEE or EIA report (environmental impact statement or EIS) to the DOE (Fig. 15.3).

15.5.1 Review of EIA Reports and Issuance of Environmental Clearance Certificate (ECC)

EIA reports are reviewed by the review committee in the DOE. The committee consists of DOE officials and is chaired by the Director General of DOE. However, the DOE contracts EIA experts and academics from outside as members of the review committee, for any special cases, on an ad hoc basis. Based on the review reports, the Director General either accepts the EIS, asks for more information or rejects the EIS.

If accepted, the DOE issues an ECC to the proponents outlining terms and conditions. The terms and conditions usually require that the proponents take adequate and appropriate mitigation measures as designated in the Environmental Management Plan (EMP) and implement the measures properly during the implementation and operation of the project. It is the responsibility of the proponents to arrange and monitor the implementation of mitigation activities and report to the DOE regularly. The DOE is responsible for checking regularly that the proponent is properly implementing the EMP. After the completion of the project, a final evaluation (environmental auditing) of the EIA is undertaken jointly by the DOE and the proponent.

15.5.2 Penalties for Non-compliance with ECC

The DOE is empowered to take punitive action against non-compliant proponents. Initially, the DOE sends a warning notice to the proponent. If the proponent does not rectify the problem after the notice, the DOE directs the utility service

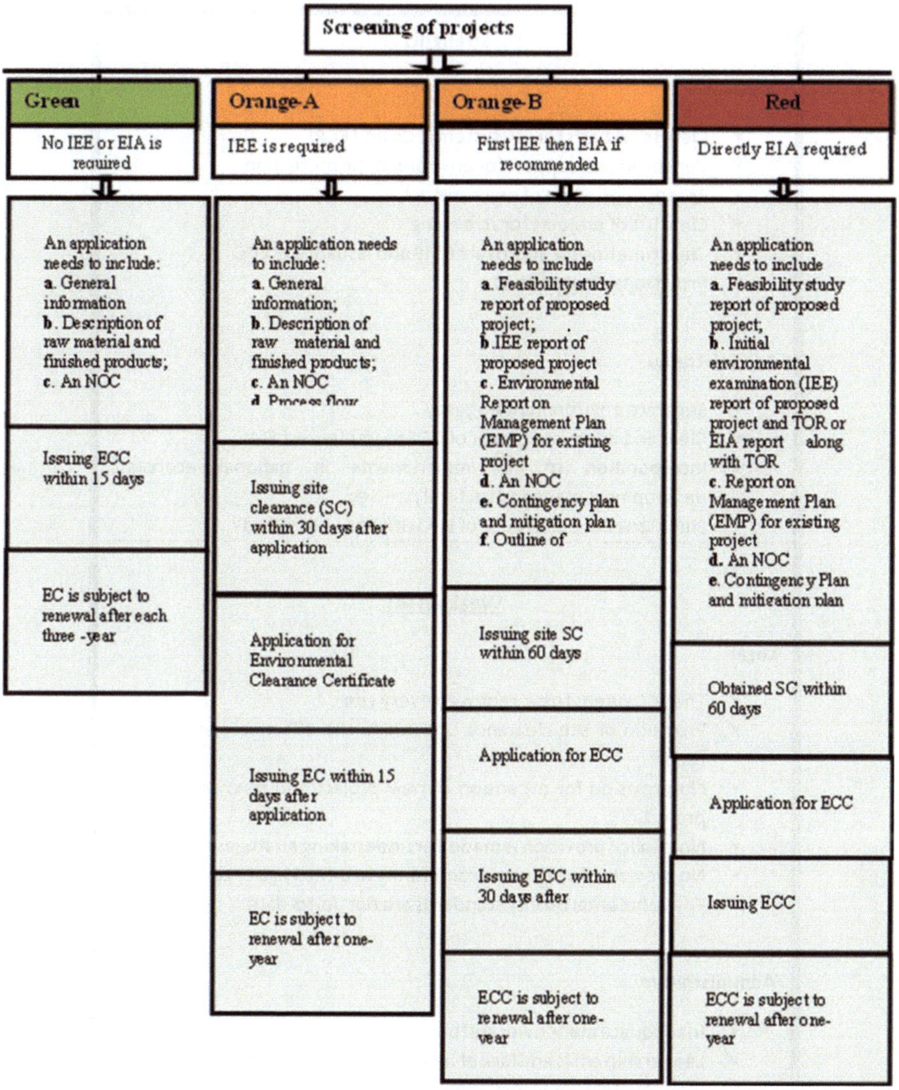

Fig. 15.3 Screening and environmental clearance process of DOE (Adapted from DOE 1997b)

departments such as electricity, gas and telephone service authorities to stop all services. According to the Rules, the DOE is authorised to fine the non-compliant proponent. The organisation also has the authority to file cases in the courts against the proponents. If the offences are proved, the proponents may be jailed for a maximum of 7 years as well as being fined.

Strengths

Legal

- ✓ Clear legal basis for EIA system (ECA and ECR)
- ✓ Constitutional support for environmental protection
- ✓ Broader definition of environment
- ✓ Clear list of projects for screening
- ✓ Clear timeline for approval of EIS and issuance of ECC
- ✓ Provisions for penalties

Administrative

- ✓ Separate environmental agency
- ✓ Clear and adequate power of DOE to implement EIA
- ✓ Incorporation of EIA requirements in national economic development plans and sectoral policies
- ✓ Simultaneous occurrence of EIA with Feasibility Study

Weaknesses

Legal

- ✓ The ECC needs to be renewed every year
- ✓ Provision of site clearance undermine the effective review of EIA report
- ✓ No provision for expansion of new project and renovation of old project.
- ✓ No explicit provision is made for undertaking all stages of EIA
- ✓ No provision for affected community to directly go to court
- ✓ Environmental quality standards are not up-to-date

Administrative

- ✓ Inadequate manpower and budget
- ✓ Leadership crisis and lack of incentives for DOE staff
- ✓ Inadequate interagency cooperation at national and local level

Fig. 15.4 Key strengths and weaknesses of legal and organisational arrangements of EIA in Bangladesh (Adapted from NVivo analysis)

15.6 Strength of the Legal Provisions of EIA

A review of the legal system for EIA shows that well-defined and specific legal provisions for EIA exist in Bangladesh, namely, the ECA of 1995 and the ECR of 1997. These documents clearly specify the responsibilities of the DOE and project proponents to carry out EIA (Fig. 15.4).

The 1995 ECA defines environment in the following terms: 'environment means the inter-relationships existing between water, air, soil and physical property and their relationship with human beings, other animals, plants and micro-organisms' (DOE 1995, p. 155). Thus, the definition recognises not only environmental issues but also social issues.

Although EIA is undertaken by proponents and reviewed centrally by the Department of Environment, there is a provision that the proponent must obtain a No Objection Certificate (NOC) from the relevant local government (DOE 1997a). This shows a strong commitment by the government to democratise the EIA process through the involvement of elected local government representatives.

The effectiveness of the EIA process in protecting the environment also depends on the degree of environmental protection offered by the standards (Lohani et al. 1997). The formulation of environmental quality parameters serves as the basis for planning of development projects (ADB 1986) particularly for designing Environmental Management Plans. The environmental quality standards set by the DOE include both ambient environmental standards and discharge standards (DOE 1997b).

15.7 Strengths of the Administrative Arrangements of EIA

Bangladesh has a separate ministry, the Ministry of Environment and Forests, to deal with environmental issues. This is headed by one full Minister and one State Minister. Within the MoEF, the DOE has the sole responsibility and authority to implement the intents of the Act, to administer EIA and to act as a coordinating 'one stop' facility for applicants. The DOE has the power to declare any area as an Environmentally Critical Area when necessary. Under the terms of the Act, the DOE also has the statutory power to revise and set environmental quality standards.

The establishment of Environmental Courts under the Environmental Court Act of 2000 to enforce the Rules has further strengthened environmental governance. The Environmental Court Act sets out the jurisdiction of Environmental Court (Section 5). It has established the procedures and power of the Environmental Court (Section 8), the right of appeal by aggrieved persons and the arrangement of Environmental Appeal Court (Section 12) (DOE 2000).

15.8 Weaknesses in the Legal Provisions of EIA

Whilst the legislation and policies that underlie EIAs are well founded, the legislation does not clearly and comprehensively spell out EIA procedures. The EIA legislation mentions screening only as one of the stages of the EIA process (DOE

1997a). The legislation does not explicitly define the procedure for other stages, such as scoping, alternative analysis, prediction and evaluation of impacts, monitoring and auditing.

Neither the Act nor the Rules specifically mention social impacts and their mitigation. As inhabitants of a densely populated city in one of the most densely populated countries in the world, the people of Dhaka are frequently affected by development projects as indicated earlier. There is also no explicit legal provision for public participation during the preparation or review of the EIA report or during the implementation of mitigation measures. The Act only includes the provision (in Section 8) that any person affected or likely to be affected as a result of pollution or degradation of the environment may apply to the Director General of the DOE for remedy of the damage (DOE 1995). Although all EIA guidelines require public participation in all stages of EIA (DOE 1997b), it is not sufficient to force the proponent to involve the public in the EIA process since it has no legal backing (Ahammed and Harvey 2004).

In the Rules, under Schedule 7, there is provision for the DOE to issue a site clearance certificate to project proponents before issuing the Environmental Clearance Certificate. Once the proponents get a site clearance certificate, they are allowed to undertake work such as land development in the project site. However, this provides a loophole in the existing EIA rules. Effectively, the site clearance certificate allows the proponents to start working at the project site before the approval of EIS and the issuance of Environmental Clearance Certificate. Since the proponent has already invested in the site, it often becomes morally and technically difficult for the DOE to reject the EIS even if it is of poor quality. Rather, in many cases, the provision of a site clearance certificate forces the DOE to approve the EIA reports hurriedly.

15.9 Weaknesses of the Administrative Arrangements of EIA

The DOE, with its relatively broad mandate that includes the implementation of EIA, is under-budgeted. The budget only covers the salaries of the DOE staff, limited internal travel and certain basic logistics. There is no budget allocation for activities such as environmental awareness programmes, research and regular monitoring and auditing. The DOE therefore depends on donor funds for these activities. Therefore, continuous enforcement and improvement of EIA becomes difficult.

Whilst the Act has burdened the DOE with broad environmental responsibilities, its manpower is very inadequate to effectively discharge its responsibilities in Dhaka. The volume of tasks has been increasing over the last 18 years due to the increase in population and associated development activities, but manpower has not increased since its establishment in 1993. At the time of writing, the DOE has only

244 staff positions of which 101 are managerial or technical. This workforce not only is responsible for the capital city but also has the responsibility of ensuring the enforcement of environmental laws in the rest of the country.

15.10 Brief Overview and Examination of Two Recent Environmental Documents

Recently there have been two major government-funded environmental studies. Their completion enabled two projects designed to address two most important environmental problems of Dhaka – illegal discharge of untreated effluent in the waterways and traffic congestion and related pollution problems. The reports of these studies have been critically examined by the authors for the purpose of this chapter. A brief overview of the findings of this examination has been provided here as examples of recent initiatives by the government to address environmental problems.

15.10.1 Report 1: *Limited Environmental and Social Impact Assessment and Environmental and Social Management Framework, May 2010 (LGED 2010)*

This is an EIA and SIA report of a proposed project designed to address Dhaka's wastewater problem. The project has three components: develop and implement awareness programmes among the worst polluters, strengthen capacity of environmental agencies to better monitor and enforce environmental compliance, and build and operate common effluent treatment plants (CEPTs) for the major polluting zones in Dhaka.

It is proposed that the project be implemented in two phases. The first phase will focus on building the institutional and organisational framework to support and encourage environmental compliance and pollution prevention behaviour in the industries and sectors involved. The second phase will construct the CETP. The study considered various alternatives and conducted water quality tests in 41 major outfalls from the 9 industrial clusters in Dhaka. It recommended three sites that were suitable for construction of the CETP. To compare the relative environmental and social impacts and to identify the development option with the least environmental impacts, the study considered a number of alternatives. The consideration of alternatives was not limited to site selection; rather, it considered individual versus common effluent treatment plant, alternative pretreatment options and alternative treatment options. The report then went on to identify the potential environmental and social impacts of the CETP. The report was conducted in the pre-feasibility phase, hence its title (LGED 2010). It reflects the good intentions of both the proponent (in this case the Bangladeshi government) and the financier of the project (in this case the World Bank). It should be noted here that WB (World Bank) is a major international development partner, and all projects funded by it are conditional on some form of EIA and SIA that examines the potential environmental and social impacts. If implemented, the project has the potential to significantly reduce the water pollution problems of Dhaka.

15.10.2 Report 2: *Environmental Assessment and Review Framework (Greater Dhaka Sustainable Urban Transport Project, February 2012) (GOB 2012)*

This is a report on the mass transit system which has been proposed to address Dhaka's massive traffic congestion problem. This environmental assessment describes the proposed project and its institutional framework and provides guidance on how the project will be screened, how EIA and SIA will be conducted, what environmental and social impacts may be expected, what community consultation processes should be in place and how monitoring and reporting will be performed in the post-EIS phase. The document clearly identifies the relevant legal framework and provides a list of the anticipated environmental, social and biological impacts of the proposed development. One important aspect of the document is that it describes community consultation and grievance processes and outlines the obligations of the proponents towards the affected community. This ambitious project, if completed properly and on time, will bring considerable relief to each and every city dweller.

Although this is a 14-page guidance report for the project, prepared for the funding agency (Asian Development Bank), it is also, in a sense, a summary strategic environmental assessment (SEA) conducted at the policy formulation phase of the project. SEA is a better form (compared to project level site-specific environmental and social impact assessment) of environmental assessment which is conducted at the policy level. As such, like the study above, this study represents good practice in the consideration of project impacts. Like the World Bank, the ADB also has strict guidelines for EIA of its funded projects in developing countries.

These two recent examples of environmental studies indicate that the Government of Bangladesh is moving, either willingly or as a result of pressure from donor agencies, towards the practice of strategic environmental assessment, which is current practice in developed nations, ahead of project level environmental impact assessment. The benefits of doing an SEA are well documented in the literature. The benefits of these documents and initiatives will, however, depend on their proper and timely implementation.

15.11 Conclusions and Recommendations

This study reveals that the government has made a significant effort to legally tackle the environmental problems of Dhaka through the enactment of one of its most important pieces of legislations – the Environmental Conservation Act 1995 and the associated Rules of 1997. There have been other recent initiatives to systematically address problems. It appears that there are strengths in the EIA process but there are weaknesses too. To strengthen the provisions of EIA and for it to be more effective in addressing the environmental governance issues of Dhaka, the Government of Bangladesh needs to take some major steps to rectify the weaknesses of the EIA system.

Coordination is one of the central pillars of effective regulatory EIA process management. As indicated at the beginning of this chapter, there are three main agencies entrusted with the responsibility to enforce urban and environmental governance for Dhaka. There are also a number of other ministries involved in the environmental management of the country. Currently, there is no effective communication, coordination and collaboration between these agencies.

The issue of corruption is a major problem in the country, and it appears in the press on a daily basis. It is regarded as the major inhibitor to the country not achieving its expected economic development goals. Bribery is rampant in financial dealings at all levels of administration. Much of the unplanned development of Dhaka and the illegal activities related to the filling of wetlands around the capital and to tree felling, for example, can be attributed to corruption and the taking of bribes by the relevant officers. Many un-roadworthy vehicles are polluting the air of Dhaka because the owners had the opportunity to renew their licences from the transport authority through under-the-table dealings. Multistorey buildings are being built without full compliance to construction codes, rules and regulations because the developers obtain permission through dishonest means. These are open secrets and everyday facts in the lives of millions of people in Dhaka. Whilst this study did not focus on this particular aspect of urban governance problem, a lot of participants in this study have mentioned corruption as a major problem in law enforcement.

In order to address the environmental problems of Dhaka, the government and the administration will have to address the causes of these problems. One major cause is unchecked population growth in Dhaka due to rural to urban migration. This is a bigger issue and requires comprehensive planning in decentralising the country's economic activities which is beyond the scope of this chapter. The other major causes are related to urban environmental governance including proper implementation and enforcement of rules and regulations by the relevant agencies. The following recommendations are directed to addressing these causes:

1. The government should ensure better accountability and transparency in the agencies in urban environmental governance and all levels of government administration. Corruption is regarded as all pervasive and a major obstacle to progress and advancement of Bangladesh. Better accountability and transparency should help reduce corruption.
2. Inadequate coordination and communication is a major area of deficiency in urban governance. It creates misunderstanding and organisational inertia leading to inefficiency. Better coordination and communication between all the actors in urban governance should help address this problem.

3. Unplanned urban expansion or urban sprawl is a major cause of environmental problem in Dhaka megacity. There is a need for a better planning process in urban development activities. Better planning means that there will be provisions for adequate infrastructure, services and amenities in proportion with the new industrial and residential developments.
4. There is a need to strengthen the legal aspects of EIA by clearly incorporating social impact assessment and community participation in environmental decision-making. The use of SIAs would help to identify the social issues of development activities in more detail and allow for the development of better mitigation measures. In addition, the affected communities would have greater opportunities to influence the outcomes of decision-making through participation.
5. Effective implementation of rules and regulations require better equipped and resourced institutions. Whilst the DOE has a legal mandate, it does not have the adequate staffing or financial resources to properly enforce the provisions of the Act and the Rules. The government needs to ensure that skilled people are recruited in adequate numbers for proper delivery of various technical aspects of the legal provisions and that the DOE is well resourced to properly implement its enforcement activities.

The Government of Bangladesh has established a good legal and administrative framework for effective environmental and urban governance of the megacity of Dhaka. The success of this framework will depend largely on proper enforcement of these rules and regulations in a transparent and accountable manner. In the absence of these, the megacity of Dhaka will continue to remain one of the least desirable places on earth to live in.

References

Asian Development Bank (ADB) (1986) Environmental planning and management-regional symposium on environmental and natural resources planning, Manila, The Philippines

Ahammed AKMR, Harvey N (2004) Evaluation of environmental impact assessment procedures and practice in Bangladesh. Impact Assess Proj Apprais 22(1):63–78

DCC (Dhaka City Corporation) (2012) About Dhaka City Corporation, http://www.dhakacity.org/Page/About/Link_1/2/List_id_1/23/Subid_1/44/Urban_Planning. Accessed 25 Sept 2012

Dewan A, Md Kabir H, Nahar K, Rahman MZ (2012) Urbanisation and environmental degradation in Dhaka Metropolitan Area of Bangladesh. Int J Environ Sustain Dev 11(2):118–147

DOE (Department of Environment) (1995) The environmental conservation act. Ministry of Environment and Forest, Dhaka

DOE (Department of Environment) (1997a) The environmental conservation rules. Ministry of Environment and Forest, Dhaka

DOE (Department of Environment) (1997b) EIA guidelines for industries. Ministry of Environment and Forest, Government of Bangladesh, Dhaka

DOE (Department of Environment) (2000) Environmental court act. Ministry of Environment and Forest, Government of Bangladesh, Dhaka

DOE (Department of Environment) (2012) Clean air and sustainable environment project, http://www.doe-bd.org/projects.html. Accessed 20 June 2012

Gain P (1998) Bangladesh environment: facing the 21st century. SHED, Dhaka

GOB (Government of Bangladesh) (2012) Environmental assessment and review framework: greater Dhaka sustainable urban transport project. GOB, Dhaka

Government of Bangladesh (1973) The constitution of the People's Republic of Bangladesh. Ministry of Law, Justice and Parliamentary affairs, Dhaka

Islam S, Rahman R, Shahabuddin A, Ahmed R (2010) Changes in wetlands in Dhaka city: trends and physico-environmental consequences. J Earth Sci 5:37–42

Karn SK, Harada H (2001) Surface water pollution in three urban territories of Nepal, India and Bangladesh. Environ Manag 28(4):483–496

Lohani BN, Evans JW, Everitt RR, Ludwig H, Carpenter RA, Tu S (1997) Environmental impact assessment for developing countries in Asia, vol 1. Asian Development Bank, Manila, pp 1–356

LGED (Local Government Engineering Department) (2010) Limited environmental and social impact assessment and environmental & social management framework. LGED, Dhaka

Mahmud K, Gope K, Chowdhury M (2012) Possible causes & solutions of traffic jam and their impact on the economy of Dhaka City. J Manag Sustain 2(2):112–135

MoEF (Ministry of Environment and Forest) (1992) National environmental policy. Government of Bangladesh, Dhaka

Newgeography (2012) World Urban Areas Population and Density: A 2012 Update, http://www.newgeography.com/content/002808-world-urban-areas-population-and-density-a-2012-update. Accessed 21 Nov 2012

Rabbani G (2009) Environmental risks in Dhaka: present initiatives and the future improvements. In: Shaw R, Srinivas H, Sharma A (eds) Urban risk reduction: an Asian perspective, vol 1, Community, environment and disaster risk management. Emerald Group Publishing Limited, New Delhi, pp 319–338

RAJUK (Rajdhani Unnayan Kartipakhha) (2012) http://www.rajukdhaka.gov.bd/rajuk/developmentHome. Accessed 28 July 2012

UNEP (United Nation Environment Program) (2006) Dhaka City: state of environment 2005. UNEP, Dhaka

Chapter 16
Assessing Surface Water Quality Using Landsat TM and In Situ Data: An Exploratory Analysis

Razia A. Chowdhury, Towhida Rashid, and Sirajul Hoque

Abstract The objective of this chapter was to explore the potential of Landsat TM data, calibrated by in situ measurements, to map the spatial distribution of water quality in the rivers and lakes of Dhaka. The relationship of satellite brightness values and ground measurement was established through correlation and regression analyses. The results showed that the ratio of TM1 and TM3 was highly correlated with Secchi disk transparency (SDT), a measure of water clarity, while total suspended sediment (TSS) was strongly correlated with brightness values in the near-infrared portion of the electromagnetic spectrum. Regression analysis indicated that TM1 and the ratio of TM1/TM3 was the best predictor for SDT, and TM3 and the ratio of TM1 and TM3 was suitable for the estimation of TSS in waters. Maps of SDT and TSS are presented that illustrate the spatial variation of water quality in the inland water systems of Dhaka.

Keywords Dhaka city • Landsat • Multiple regression • Remote sensing • Secchi disk transparency/depth • TSS • Water quality

R.A. Chowdhury (✉)
Department of Geography, Faujderhat Cadet College, Faujderhat, Chittagong 4316, Bangladesh
e-mail: razuenv_0607@yahoo.com

T. Rashid
Department of Geography and Environment, University of Dhaka, Dhaka 1000, Bangladesh
e-mail: towhida_rashid@yahoo.com

S. Hoque
Department of Soil, Water and Environment, University of Dhaka, Dhaka 1000, Bangladesh
e-mail: sirajswedu@gmail.com

A. Dewan and R. Corner (eds.), *Dhaka Megacity: Geospatial Perspectives on Urbanisation, Environment and Health*, Springer Geography,
DOI 10.1007/978-94-007-6735-5_16,

16.1 Introduction

Water quality monitoring has traditionally been conducted using discrete sample locations which is expensive, time consuming, and often inaccurate (Curran and Novo 1988). In addition, this conventional method is unable to offer synoptic coverage of the entire water body (Curran et al. 1987) or observe the spatial variation of parameters (Brivio et al. 2001; Tassan 1997; Baban 1996, 1995). In overcoming this problem, satellite technology has already shown immense potential as a research tool for providing spatially unbiased water quality parameter estimates across large areas (Kloiber et al. 2002a; Nelson et al. 2003) which is particularly useful for monitoring water pollution (Johnson and Harriss 1980) and for the management of water quality (Hadjimitsis et al. 2010).

Numerous studies have utilized remote sensing data together with in situ measurements to examine water quality around the world. Color aerial photography was initially used during 1960s to examine water quality (Schneider 1968). One of the first contributions of water quality assessment was carried out by Clark et al. (1970) using aircraft remote sensing, and with the launch of ERTS-1 (Earth Resources Technology Satellite-1) (later renamed to Landsat) in July 1972, water quality research using remote sensing gained new momentum (Klemas et al. 1974). Since then, satellite data have been extensively used in a variety of environments (Hellweger et al. 2004; Curran and Novo 1988; Witzig and Whitehurst 1981; Moore 1980; Johnson and Harriss 1980). The basic premise of these studies is that a strong correlation exists between spectral reflectance measured by onboard sensors and in situ measurements. A number of sensors from air- and space-borne platforms have been used to map water quality at a range of scales in what can broadly be divided into two sets of condition, inland and coastal or nearshore ocean water environments. The Thematic Mapper (TM) carried on the Landsat series of satellites has become the most widely used research instrument, as it offers improved spatial, spectral, and temporal resolution (Usali and Ismail 2010).

Primarily, three approaches are used to establish the relationship between remotely sensed spectral properties and ground measurements (Ma and Dai 2005). They are: (i) the establishment of a relationship using formulae derived from theory, (ii) the empirical method in which curve fitting (and sometimes neural networks) is utilized, and (iii) the semiempirical method which is the combination of both theoretical and empirical techniques. The regression method is one of the most widely used techniques for investigating the relationships between satellite-derived brightness values and ground-based measurements. The degree to which a regression equation expressing the relations between variables is measured using the coefficient of determination (r^2) (Wang et al. 2006).

Three generations of sensors on the Landsat series of satellites, the MSS (Multi-spectral Scanner), TM (Thematic Mapper), and ETM+ (Enhanced Thematic Mapper Plus), have been used in assessing water quality with varying degrees of success. For example, the MSS sensor was used to measure suspended sediment, chlorophyll *a*, Secchi disk transparency (SDT), turbidity, total phosphorus, total nitrogen, etc., both in inland and coastal and nearshore ocean environments (Peckham and Lillesand 2006; Kloiber et al. 2002a, b; Baban 1997, 1996; Mertes et al. 1993; Harrington et al.

1992; Ritchie et al. 1990, 1989, 1987, 1986; Lo and Hutchinson 1991; Lyon et al. 1988; Wen-yao and Klemas 1988; Aranuvachapun and Walling 1988; Ritchie and Cooper 1987; Lindell et al. 1985; Almanza and Melack 1985; Verdin 1985; Khorram and Cheshire 1985; Lillesand et al. 1983; Carpenter and Carpenter 1983; Khorram 1981; Aranuvachapun and LeBlond 1981; Scarpace et al. 1979; Klemas et al. 1973). These studies recognized that satellite data can be an excellent and dependable source for determining water quality from reflected solar radiation. Generally, the visible and near-infrared bands of MSS have been found to be appropriate for predicting surface water quality characteristics (Harrington et al. 1992). For instance, MSS data was very reliable for estimating Carlson's (1977) Trophic State Index (TSI), moderately reliable for the prediction of chlorophyll concentration but proved unreliable for phosphorus estimation (Lillesand et al. 1983). A similar observation was made by Ritchie et al. (1987) and Ritchie and Cooper (1988). They reported that MSS band 5 was the best predictor for the estimation of chlorophyll, while suspended sediment estimation was best predicted using MSS band 2 or band 3. Using a semiautomated technique, Scarpace et al. (1979) demonstrated that the MSS band 5 was suitable for the evaluation of Lake trophic levels in Wisconsin, USA. Simple and multiple regression analyses have also shown that the MSS band 3 (700–800 μm) was the best single-band data for explaining variations in suspended sediment (Schiebe and Ritchie 1986).

Despite the fact that Landsat MSS and TM were identical in terms of their geographical coverage, TM data, due to its higher spatial resolution, has been found to be superior as evidenced by many studies for revealing the spatial pattern of water quality. In general, TM data acquired nearly simultaneously with in situ measurements are used to establish the relationships with water quality parameters through simple and multiple regression techniques. Single TM band, ratio of bands, and the combination of TM bands have been extensively tested. For example, TM1, TM2, TM3 (Dekker and Peters 1993) and TM1/TM2 (Dwivedi and Narain 1987) have been found to be suitable for the detection of chlorophyll concentration in water. However, seasonal variation of algal composition could be an important confounding issue in determining chlorophyll through TM data (Allee and Johnson 1999). TM band 3 has provided a significant relationship with suspended sediment concentration (Nas et al. 2010; Bilge et al. 2003; Tyler et al. 2006), while TM1 and TM2 (Östlund et al. 2001), TM2/TM1 (Lathrop 1992), and the average value of TM2 and TM3 (Dekker et al. 2002) have all been used to establish relationships between suspended sediment load and reflectance values. Islam et al. (2003) also showed that TM1/TM2 is useful to study suspended sediment and Secchi disk transparency in Australia, while Svab et al. (2005) reported that TM3 provided the best relationship for mapping suspended sediment concentration in Hungary. A study by Olmanson et al. (2002) found that a strong relationship exists between Secchi disk transparencies and TM1 and TM3. Similarly, Fraser (1998) demonstrated that water turbidity can be effectively measured using the TM bands 1–4. TM band 3 showed good predictive power in identifying turbidity, and band 6 has been used to estimate water temperature in the USA (Lathrop and Lillesand 1986). Pattiaratchi et al. (1994) found that TM bands 1–3 were important in predicting chlorophyll concentration and Secchi disk depth (SDD) in Australia.

Compared with Landsat MSS and TM, relatively few studies have used the ETM+ sensor to assess water quality (Volpe et al. 2011; Hadjimitsis et al. 2010; Peckham and Lillesand 2006; Wang et al. 2007; Tyler et al. 2006; Svab et al. 2005; Stefouli et al. 2005; Ma and Dai 2005; Lambrix and Naugle 2003; Nelson et al. 2003). The ETM+ band 3 was found to be reliable in detecting turbidity in the USA (Lambrix and Naugle 2003). Also, as a single band the ETM+ band 3 showed the best correlation in analyzing chlorophyll concentration in China followed by band 1 and band 2. While ETM+ band 4 provides the strongest correlation with total suspended matter (TSM) as a single band, the correlation coefficient was higher when used with different bands in combination such as the ratio of band 4/band 1 (Ma and Dai 2005). Landsat ETM+ was also tested in Hungary to determine suspended sediment concentration (SSC), and the study revealed that ETM+ band 3 showed significant positive correlation with SSC (Svab et al. 2005). However, in a study by Nelson et al. (2003) which reported a low correlation coefficient, Landsat ETM+ bands 1–3 were found to be sensitive to statistical distribution of the calibration dataset.

Apart from data from the Landsat sensors, data from other spaceborne satellites have also been used in studying inland and coastal water resources with varying degrees of success at various scales. For instance, the Indian Remote Sensing (IRS)-1A band 3 provided the best result for turbidity measurement in India (Choubey 1997). The ASTER bands 1–4 were shown to be highly correlated with chlorophyll distribution in a study in Turkey (Nas et al. 2009). Potes et al. (2011) demonstrated the utility of the MERIS sensor on the ENVISAT platform for evaluating cyanobacteria and chlorophyll. This study shows very high correlation coefficient between brightness values and in situ measurements. A very few studies have employed high-resolution multispectral data such as IKONOS to evaluate water quality in the USA (Hellweger et al. 2007; Sawaya et al. 2003) and in Turkey (Ekercin 2007). These studies indicated promising outcomes showing good correspondence between retrieved and verified results through statistical analyses. In contrast, multispectral Landsat data together with synthetic aperture radar (SAR) data from ERS-2 were also exploited to determine water quality in the Gulf of Finland (Zhang et al. 2003a, b, 2002). All of the studies recognized that satellite data remain a good alternative to the traditional point-based measurement which is not only useful for determining water quality over space but also provides information necessary for scientific or management purposes (Kloiber et al. 2002a, b).

The surface water quality of the major rivers and lakes in Dhaka has been severely degraded due to the discharge of untreated wastewater from municipal, agrochemical, and industrial sources into surface waters (Dewan et al. 2012; BBS 2010; Aktar et al. 2009). A number of point-based studies showed that dissolved oxygen (DO), for instance, reaches zero while biological oxygen demand (BOD) increases concurrently during the dry season (Begum and Tanvir 2010; Rahman and Hossain 2008; Sohel et al. 2003; Karn and Harada 2001), indicating the existence of substantial amounts of inorganic and nonbiodegradable components in the water (BBS 2010; Karn and Harada 2001; Kamal et al. 1999). Likewise, the level of nephelometric turbidity unit (NTU) and suspended sediment of major rivers is considerably higher than the permissible limit (Sohel et al. 2003).

Since 17 % of the water for public use in Dhaka is derived from surface water (Rahman and Hossain 2008), further deterioration of water quality of rivers and lakes resulting from increasing human activities will have serious implications for public health in that area (BBS 2010). A recent study, for example, found that the rate of typhoid infection, a water- and food-borne disease, is higher among residents living closer to waterbodies than those living further away which is potentially linked with increasing water pollution (Dewan et al. 2013). Indeed, changes in land use and consequent urban development triggered by ever-increasing population in Dhaka lead to more impervious surface, with increased runoff and a rise in nonpoint source pollution (Slonecker et al. 2001). Even though water quality in Dhaka has been recognized as being a problem for a long time, particularly in the event of uncontrolled urban development (Aktar et al. 2009), relatively little is known about the spatial pattern of water quality. Additionally, satellite remote sensing which is proven a viable alternative to traditional ground-based measurement has never been tested for assessing the water quality of major waterbodies in the megacity of Dhaka. The objective of this study is to fill this gap and to evaluate water quality of surface waterbodies of Dhaka by using remote sensing with in situ measurements.

Despite the fact that satellite data are capable of predicting a number of water quality parameters, this study concentrated on only two parameters, Secchi disk transparency and total suspended sediment (TSS) due to a number of reasons. Firstly, funding and time constraints; secondly, SDT is an important measure of water quality which can be carried out relatively cheaply with limited resources (Kloiber et al. 2002a, b); finally, the sediment load in waterbodies is an indicator of pollution potential (Julien 1995) and has an impact on aquatic ecosystems (Ritchie and Cooper 1988). Thus, we believe that this study could advance the current knowledge base of water quality in Dhaka by providing information both from in situ and remote sensing measurement.

16.2 Data and Methods

16.2.1 Image Processing

Two cloud-free Landsat TM 5 images covering the study area were obtained from the Center for Earth Observation and Digital Earth (CEODE), China. The images were taken on 02 February 2011. The TM data were first geometrically corrected using a 1997 topographic map. A total of 40 ground control points (GCPs), distributed across the two TM scenes, were used to rectify the TM data, resulting in a root mean square error of <1 pixel. A first-order polynomial fit with nearest-neighbor resampling scheme was used to preserve the original digital numbers (DN) of the image. The Bangladesh Transverse Mercator (BTM) was used as the projection system. After geometric rectification, an image-based approach was used to convert DN to spectral radiance and to minimize the atmospheric effects (Chavez 1996). The next step was to convert radiance value to at-satellite reflectance so that spectral brightness values on TM data can be compared with in situ measurement. The algorithm proposed by Chander and Markham (2003) was used for this purpose.

16.2.2 Sampling Location and Laboratory Analysis

Eleven sample sites were established within the two major rivers and some of the lakes located in Dhaka with an assumption that the samples will be representative of the waterbodies at the time of satellite overpass (Pulliainen et al. 2001). Most of the samples were taken on the same day of Landsat overpass as previous studies suggested that ground observations coinciding with the image yield the best calibrations (Kloiber et al. 2002a, b); however, other studies concluded that a few days difference (±3 to 10 days) between satellite overpass and ground sampling does not pose a problem (Wu et al. 2010). Therefore, some field measurements were performed as late as 7 days after the passage due to weather and transportation problems. Using a boat, samples were collected from the rivers and lakes, and a hand-held global positioning system (GPS) was used to obtain Universal Transverse Mercator coordinates for each station. Surface water samples were collected in polyethylene bottles that were cleaned by soaking and rinsing twice with distilled water. Prior to sampling, each bottle was rinsed again with water to be sampled to avoid potential contamination, and at each sample point a 2,500 ml water sample was collected near the surface (50 cm depth). Secchi disk transparency was measured with a 0.20 diameter, black-and-white quadrated disk. It is worth noting that the water was sampled at the central portion of a river/lake cross section. Each sample was labeled with date and time of collection and brought to the laboratory for analysis and TSS measurement. Firstly, a known volume of water sample from each site was evaporated to dryness for weighing the total solids. Secondly, Whatman filter paper was used to determine the TSS. The estimated dissolved solids ranged from 4.8 to 373 mg/l with distinct locational variation, while the range of SDT was 7–36 cm from the water surface.

16.2.3 Extraction of Water-Only Image

In this study, the measure of water clarity is based on SDT (Olmanson et al. 2001). Therefore, methods in relation to water clarity mapping draw on the work of Brezonik et al. (2005), Kloiber et al. (2002a, b), and Olmanson et al. (2001, 2002). Even though a variety of algorithms such as the normalized difference water index (NDWI) (McFeeters 1996) have been used to extract water body from Landsat TM data, the presence of blue-green algae floating over water surface could hinder obtaining a water-only image (Ma and Dai 2005). Therefore, we used an unsupervised classification method based on a clustering algorithm to extract the water-only image. A total of 10 clusters were extracted that were aggregated to water and land areas. A binary raster image was finally derived by masking out land areas from the aggregated data; hence, only waterbodies are retained in the water-only image.

16.2.4 *Correlating Spectral Values with Water Quality Parameters*

To examine the relationships between TM bands and in situ measurements, Pearson correlation and a stepwise regression model were formulated. Single TM band, the ratio of bands, and band combinations were used to determine if significant correlations occurred between variables. Two approaches were used to investigate the relationships as provided below.

For water clarity mapping using SDT field observation and TM data, a method similar to that of Kloiber et al. (2002a, b) was used. First of all, a vector feature file of the ground sample points was overlaid on the water-only image. Then using each sample location, an area-of-interest (AOI) tool was used to define AOI polygons. The recommendation of Lillesand et al. (1983) was adopted to determine the size of the AOI. An AOI polygon ranging from of at least nine pixels upward was used for each sample point. This allowed for a large enough sample to avoid areas with differing transparencies or affected by vegetation or bottom effects. Once the AOI location has been selected, a spectral signature can be acquired for it and labeled with the appropriate ground identification number. The mean brightness values of TM bands in the AOI were then used to explore the relationship with ground-based SDT. The original SDT and log-transformed SDT values were used to establish the relationship (Islam et al. 2003; Olmanson et al. 2001); however, for regression analysis only log-transformed SDT was used.

In order to map and relate TSS with sample data, the first four TM bands were used as previous studies demonstrated that longer wavelengths provide little information for mapping sediment concentration in waters (Wang et al. 2006; Tan et al. 2003; Dekker and Peters 1993). To determine the window size for TSS estimation, a 3×3 pixel window was used (Reddy 1997). Using the average spectral values with corresponding ground truth data, correlation and regression analyses were then carried out.

16.3 Results and Discussion

Table 16.1 shows the correlation coefficient *r* of SDT, log-transformed SDT, and TSS with the digital data of TM bands in terms of single bands, ratio, and different band combinations. It demonstrated that SDT was negatively related to the individual single bands in the visible and near, mid-infrared while ratios of TM bands, particularly the ratio of TM bands 1/2, 1/3, and 3/1, showed strong positive correlation with SDT value. By contrast, the band combination of TM1 + 2 + 3 showed negative correlation, and the ratio of TM1 with combined bands of TM1 + 2 + 3 produced high positive relationships. The log-transformed SDT showed very little improvement compared with the actual SDT, but it was sufficient to suggest that the log transformation of SDT produces a better prediction. In this respect, these results are in

Table 16.1 Correlation between SDT, logarithmic transformation of SDT, and TSS with TM bands

Parameter	TM1	TM2	TM3	TM4	TM5	TM1/2	TM2/1	TM2/3	TM3/2	TM1/3	TM3/1	TM (1 + 2 + 3)	TM1/(1 + 2 + 3)
SDT	−0.05	−0.46	−0.59	−0.71	−0.45	0.58	−0.56	0.20	−0.22	0.77	0.76	−0.39	0.79
LN(SDT)	−0.10	−0.39	−0.62	−0.70	−0.44	0.46	0.44	0.33	−0.36	0.78	−0.83	−0.40	0.72
TSS	0.00	0.14	0.52	0.60	0.49	−0.19	0.18	−0.49	0.51	0.69	0.70	0.23	−0.52

agreement with previous studies (Islam et al. 2003; Kloiber et al. 2002a, b; Olmanson et al. 2002). In contrast, the strength of relationships between TSS and each TM band increases from visible blue to red band; however, the ratio of bands 1/3 and 3/1 showed the highest correlation with ground truth data. TM bands 3 and 4 showed the strongest relationships with measured TSS of all the bands tested. In this study, near-infrared band (TM4) had the highest correlation, suggesting suspended sediment-dominated surface waters with relatively low phytoplankton algal concentrations (Dekker et al. 2002; Jensen 2000; Lodhi et al. 1998). This study supports the findings of studies in China (Zhou et al. 2006; Ma and Dai 2005) and in the USA (Lodhi et al. 1998) but differs from other studies in Turkey (Nas et al. 2010; Bilge et al. 2003), in the USA (Lathrop 1992), the Netherlands (Dekker et al. 2002), Israel (Mayo et al. 1995), and Hungary (Tyler et al. 2006). Studies from these countries showed that TM3 was most strongly correlated with ground-based measurement of TSS. Further, the additive TM band combination showed disappointing results in the calculation of Pearson correlation statistics.

Since the log-transformed SDT showed slightly better correlation than actual SDT, this variable was used to develop the regression model. The analysis of multiple regression of log-transformed SDT indicated that the ratio of TM bands 1 and 3 produced the highest coefficient of determination (r^2) compared with other models developed in this study (Table 16.2). The best-fitting regression equation for SDT was LN(SDT) = 0.073(TM1) + 4.129 (TM1/TM3) – 13.493 which explained 66 % of the variance in SDT. TM band 1 together with the ratio of TM3/TM1 or TM3 with the ratio of TM1/TM3 or TM2 with the ratio of TM1/(TM1 + 2 + 3) produced the second best predictor. These three equations all explained 65 % of the variance (Table 16.2). Kloiber et al. (2002a, b) and Olmanson et al. (2002) also found similar results in the USA. However, both TM1 and TM2, as single bands, were poor predictors of SDT in this study.

Remotely sensed reflectance in the red and infrared wavelength regions was found suitable in estimating suspended sediment. In this study, we found a strong relationship with TM4 (Table 16.1); however,TM3 also produced a reasonable correlation. The multiple regression results for TSS are shown in Table 16.2. These show that the independent variable combination of TM3 with TM1/TM3, TM1 with TM3/TM1, and TM1 with TM1/TM3 explained 56–57 % of the variance in TSS. Surprisingly, TM4 did not produce good model fit when used in the regression equation despite strong positive relationship being found in the Pearson correlation analysis (Table 16.1). Only 49 % of the variance was accounted for by TM1 and TM4 (Table 16.2). The best model for TSS prediction can be found as TSS = −59.66 (TM3) – 1386.2(TM1/TM3) + 1386.2 that explained 57 % of the variance. The second best model was TSS = −21.17(TM1) – 834.4(TM1/TM3) + 3933.849, followed by TSS = 20.17(TM1) + 6157.1(TM3/TM1) – 671.3, each of later two models explaining 56 % of the variance. The model using TM3 and the combined bands TM(1 + 2 + 3) explained 54 % of the variance. However, the ratio of other visible bands such as TM1/TM2 and TM2/TM3 was a poor predictor of TSS concentrations in water. Note that TM4 used either as a single band or in a band

Table 16.2 Coefficient of determination (r^2) and multiple regression equations for LN(SDT) and TSS

WQ[a]	Regression equations	Multiple r^2	Adjusted r^2
SDT	LN(SDT) = 0.073(TM1) + 4.129(TM1/TM3) − 13.493	0.66	0.57
	LN(SDT) = 0.067(TM1) − 30.32(TM3/TM1) + 9.276	0.65	0.56
	LN(SDT) = 0.215(TM3) + 6.121(TM1/TM3) − 19.341	0.65	0.56
	LN(SDT) = 0.26(TM2) + 80.582(TM1/(TM1 + 2 + 3)) − 50.025	0.65	0.56
	LN(SDT) = 0.05(TM1) + 49.448(TM1/(1 + 2 + 3)) − 28.747	0.54	0.42
	LN(SDT) = 0.458(TM1) – 0.23(TM11 + 2 + 3) − 0.427	0.53	0.42
	LN(SDT) = 0.05(TM1) + 49.448(TM1/(1 + 2 + 3))	0.53	0.41
	LN(SDT) = −0.516(TM3) + 0.109(TM1 + 2 + 3) + 2.601	0.51	0.30
	LN(SDT) = −0.219(TM2) – 7.002(TM3/TM2) + 15.193	0.40	0.25
	LN(SDT) = −0.21(TM2) + 5.77(TM2/TM3) + 2.446	0.38	0.23
	LN(SDT) = −0.003(TM1) + 2.47(TM1/TM2) − 3.283	0.21	0.01
	LN(SDT) = −0.007(TM1) − 15.94(TM2/TM1) + 9.075	0.20	0.00
	LN(SDT) = −0.077(TM2) – 0.031(TM1 + 2 + 3) + 8.621	0.16	0.00
TSS	TSS = −59.66(TM3) − 1386.2(TM1/TM3) + 1386.2	0.57	0.46
	TSS = −21.17(TM1) − 834.4(TM1/TM3) + 3933.849	0.56	0.45
	TSS = −20.17(TM1) + 6157.1(TM3/TM1) − 671.3	0.56	0.45
	TSS = 129.879(TM3) − 34. 205(TM1 + 2 + 3) + 1067.909	0.54	0.43
	TSS = −79.037(TM2) − 17591.98(TM1/(1 + 2 + 3)) + 12331.7	0.53	0.42
	TSS = −27.997(TM1) + 55.489(TM4) + 1090.015	0.49	0.36
	TSS = 66.87(TM4) − 14.77(TM1 + 2 + 3) + 742.53	0.47	0.33
	TSS = −37.207(TM2) + 60.35(TM4) + 86.868	0.47	0.34
	TSS = −1.104(TM3) + 42.126(TM4) − 551.856	0.36	0.20
	TSS = 37.076(TM4) − 1054.94(TM1/(1 + 2 + 3)) + 113.521	0.36	0.20
	TSS = 41.34(TM4) − 56.972	0.36	0.29
	TSS = 27.52(TM2) + 1775.83(TM3/TM2) − 2191.18	0.36	0.20
	TSS = 27.54(TM2) − 1478.1(TM2/TM3) − 1051.5	0.34	0.17
	TSS = −13.045(TM1) − 7992.243(TM1/(1 + 2 + 3)) + 5637.38	0.30	0.12
	TSS = 25.26(TM3) − 3928.6(TM1/(1 + 2 + 3)) + 1800.723	0.30	0.13
	TSS = −78.95(TM1) − 37.742(TM1 + 2 + 3) + 1059.35	0.29	0.12
	TSS = −18.928(TM2) + 12.274(TM1 + 2 + 3) − 764.73	0.06	0.00
	TSS = −2.68(TM1) − 238.29(TM1/TM2) + 993.02	0.04	0.00
	TSS = −2.1(TM2) + 1360.6(TM2/TM1) − 181.68	0.03	0.00

[a]*WQ* water quality parameter

combination or ratio did not contribute significantly to the prediction of suspended sediment concentration in water (Table 16.2). The best model fit for TM band 4 was TSS = −27.997(TM1) + 55.489(TM4) + 1090.015 that explained 49 % of the variance. As mentioned earlier, some earlier studies have found that TM3 is the best predictor for suspended sediment concentration, while others reported TM4 is suitable. Compared with other work in the literature, our study predicted TSS relatively poorly as demonstrated by low r^2. For instance, Zhou et al. (2006) reported that TM4 had the strongest single-band relationship with in situ suspended sediment measurement, and they found good coefficient of determination ($r^2 = 0.77$) for data collected in March; however, our study only explained 57 % of the variance in TSS estimation. Our low prediction could have been influenced by a variety of factors. For example, the presence of algae and phytoplankton in the river waters could inhibit more accurate prediction of suspended sediment (Zhou et al. 2006; Wang et al. 2006).

In addition, the prediction of water quality parameters using satellite data is dependent on the characteristics of area being analyzed as much as on the TM band or their ratio or band combination chosen (Brezonik et al. 2005). The lack of predictive power was also probably related to the small sample size (11 samples), though Pulliainen et al. (2001) suggest that only a few representative samples are adequate to calibrate remotely sensed data for water quality assessment. An important point to note is that in this study we used both rivers and lakes to calibrate remotely sensed brightness values, but most of the studies reported in the literature used either lakes or reservoirs to demonstrate the relationship. This may be another factor that produced low predictability. Despite the low prediction value in terms of coefficient of determination (r^2), we believe that the strength of measurement of TSS through remotely sensed data is highly variable in the tropical environment. While Pearson correlation coefficient showed TM band 4 had good relationship, regression analysis resulted in band 1 and the ratio of band 1 and band 3 or vice versa is a good predictor to estimate TSS in inland waterbodies.

The spatial distribution of Secchi disk transparency (SDT) and suspended sediment is shown in Figs. 16.1 and 16.2. In each case, we have used the best prediction equation to map the water quality expressed as SDT and TSS. Generally, water appeared to have lower clarity in the areas where human activities are most intense, and wetlands have been filled in for urban development. As noted by Wang et al. (2006), wetlands trap sediment and nutrition which have obvious effects on water quality. Since wetlands in the study area have been developed continuously to meet the increasing demand for urban land, this may have considerable effect on water clarity. Higher clarity was found in the downstream part of the Buriganga River where discharge from other tributaries is contributing to maintain a consistent flow in the river during the period of study, i.e., February of 2011. In contrast, the Turag River had the lowest clarity owing to the fact that point source pollution in these areas is highest (Sohel et al. 2003) during the satellite data acquisition period with the gradual decline of wetlands. In addition, the water color of the Turag River was found to be completely black during our field visits due to decomposition of organic matter and the discharge of sewage and domestic and industrial wastewater into that river. Lakes such as Gulshan and Banani had comparatively good water quality as both point and nonpoint source pollution are lower in these lakes.

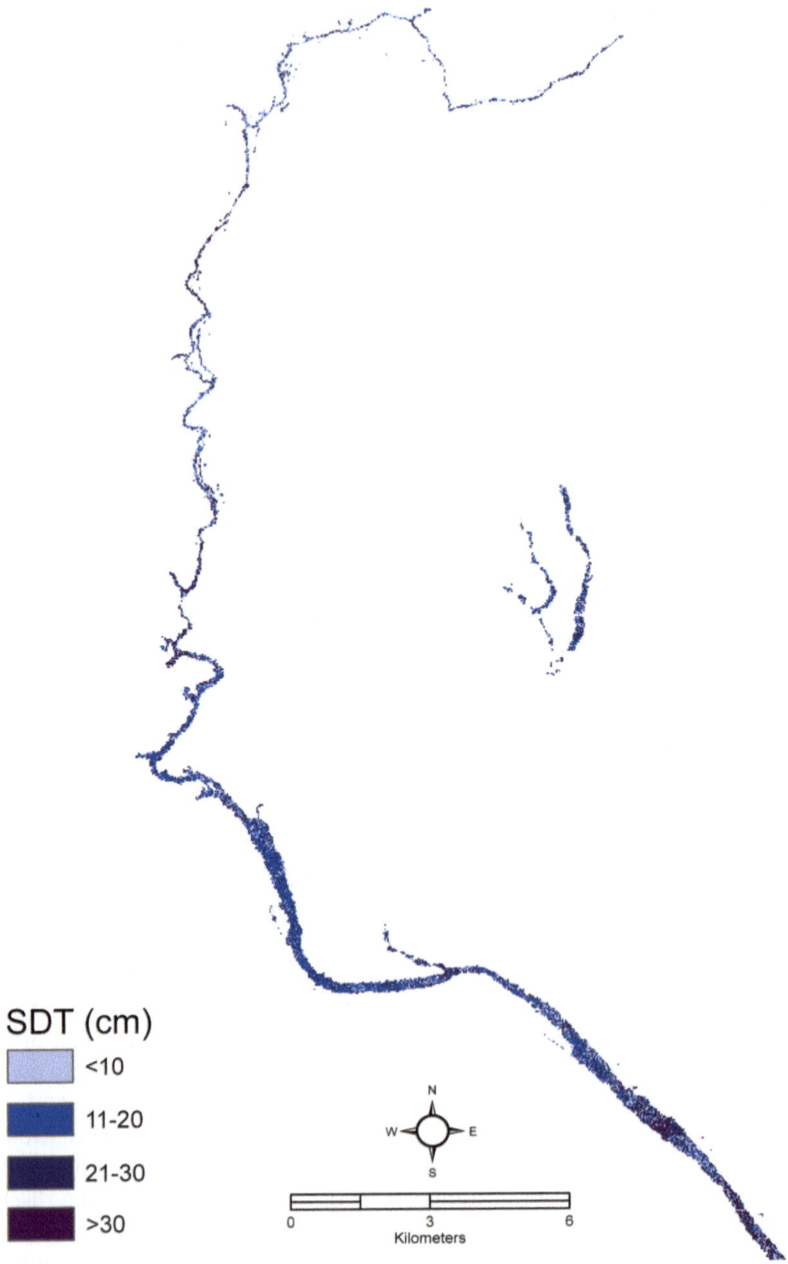

Fig. 16.1 Distribution of Secchi disk transparency (SDT) in the rivers and lakes on 2 February 2011

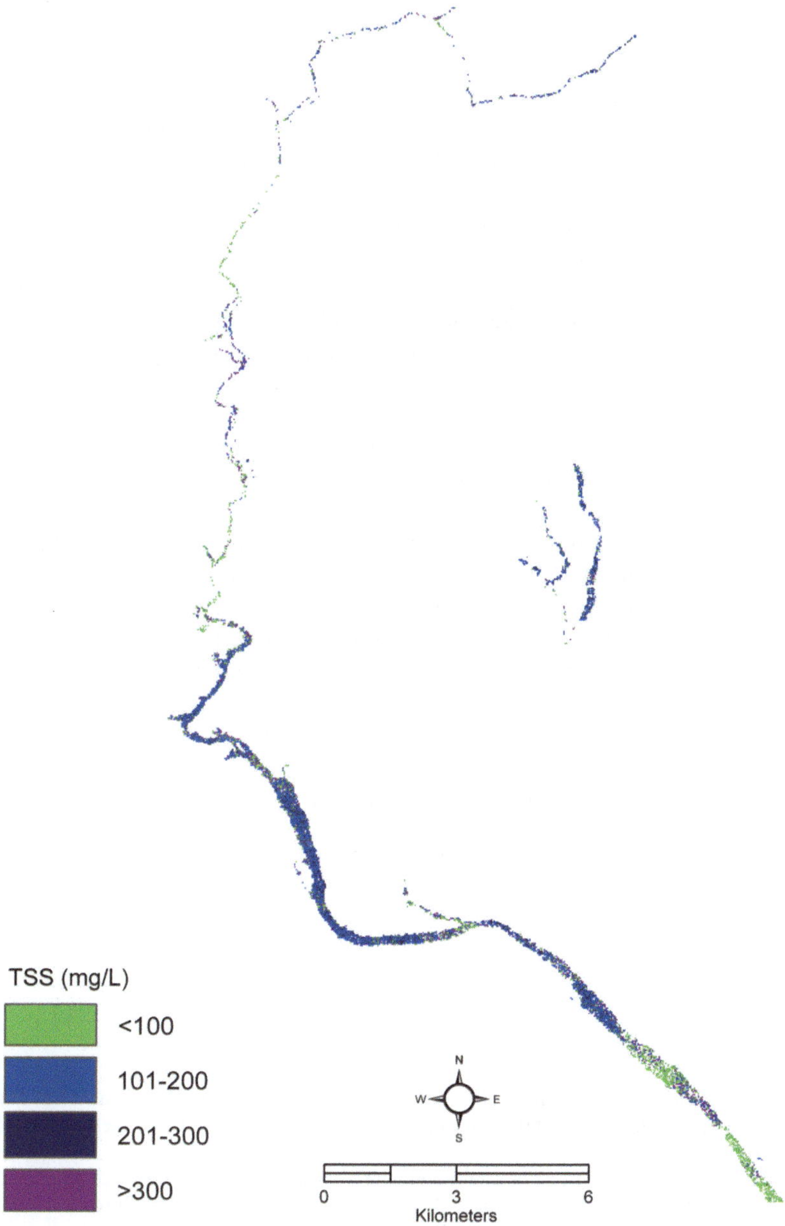

Fig. 16.2 Distribution of TSS in the rivers and lakes on 2 February 2011

16.4 Conclusions

This chapter has examined the spatial distribution of water quality using Landsat TM and in situ measurements. Correlation and regression analyses were performed to establish the relationships between satellite-based brightness values and ground truth data.

Correlation analysis revealed that the ratio of TM1 and TM3 had strongest relation with Secchi disk transparency. TSS analysis demonstrated that TM4 was highly correlated with ground measurement followed by the TM3. Multivariate regression showed that TM1 and the ratio of TM1/TM3 was the best predictor for water clarity as measured by SDT. Likewise, the best prediction for suspended sediment concentration was obtained with TM3 and the ratio of TM1 and TM3. Analysis of the spatial distribution of water quality revealed that areas where there is intense human activity have relatively poor water quality compared with areas that are away from point and nonpoint source pollution. The sediment load is higher in the rivers than lakes.

Despite the fact that this work was exploratory in nature, the results of this study could be of significant value for a data poor country where increasing human activities are consistently causing deterioration of surface water quality by the discharge of untreated domestic, agricultural, and industrial waste into the rivers and lakes. Further research is needed to understand the dynamics of water quality by using more sample points and other parameters. A study is currently underway to examine the temporal patterns of water quality of rivers and lakes in Dhaka.

Acknowledgment The first author of this chapter wishes to express her gratitude to Dr. Ashraf M. Dewan for his assistance with data, methods, and field work in carrying out this study. We also acknowledge the support of the International Foundation for Science (IFS), Sweden, for partial funding of this work under a project (Reference: W4656-1).

References

Aktar Z, Fakhruddin ANM, Ahmed MSU, Nasreen T, Parveen S (2009) Assessment and treatment of the supplied pipeline water of selected areas of Dhaka City. Bangladesh J Sci Ind Res 44(3):335–340

Allee RJ, Johnson JE (1999) Use of satellite imagery to estimate surface chlorophyll a and Secchi disc depth of Bull Shoals Reservoir, Arkansas, USA. Int J Remote Sens 20(6):1057–1072

Almanza E, Melack JM (1985) Chlorophyll differences in Mono Lake (California) observable on Landsat imagery. Hydrobiologia 122(1):13–17

Aranuvachapun S, LeBlond PH (1981) Turbidity of coastal water determined from Landsat. Remote Sens Environ 84:113–132

Aranuvachapun S, Walling DE (1988) Landsat-MSS radiance as a measure of suspended sediment in the lower Yellow River (Hwang Ho). Remote Sens Environ 25(2):145–165

Baban SMJ (1995) The use of Landsat imagery to map fluvial sediment discharge into coastal waters. Mar Geol 123(3–4):263–270

Baban SMJ (1996) Trophic classification and ecosystem checking of lakes using remotely sensed information. Hydrol Sci J 41(6):939–957

Baban SMJ (1997) Environmental monitoring of estuaries; estimating and mapping various environmental indicators in Breydon water estuary, U.K., using Landsat TM imagery. Estuar Coast Shelf Sci 44(5):589–598

Bangladesh Bureau of Statistics (BBS) (2010) Compendium of environment statistics of Bangladesh 2009, Bangladesh Bureau of Statistics. Ministry of Planning, Dhaka

Begum DA, Tanvir AKM (2010) Water quality aspects in and around Dhaka city. In: Proceedings of international conference on environmental aspects of Bangladesh (ICEABO10), Japan, 2010, pp 175–178

Bilge F, Yazici B, Dogeroglu T, Ayday C (2003) Statistical evaluation of remotely sensed data for water quality monitoring. Int J Remote Sens 24(24):5317–5326

Brezonik P, Menken KD, Bauer M (2005) Landsat-based remote sensing of lake water quality characteristics, including chlorophyll and colored dissolved organic matter (CDOM). Lake Reserv Manag 21(4):373–382

Brivio PA, Giardino C, Zilioli E (2001) Determination of chlorophyll concentration changes in Lake Garda using an image-based radiative transfer code for Landsat TM images. Int J Remote Sens 22(2–3):487–502

Carlson RE (1977) A trophic state index for lakes. Limnol Oceanogr 22(2):361–369

Carpenter DS, Carpenter SM (1983) Modeling inland water quality using Landsat data. Remote Sens Environ 13(4):345–352

Chander G, Markham B (2003) Revised Landsat-5 TM radiometric calibration procedures and postcalibration dynamic ranges. IEEE Trans Geosci Remote Sens 41(11):2674–2677

Chavez PS (1996) Image-base atmospheric corrections – revisited and improved. Photogramm Eng Remote Sens 62:1025–1036

Choubey VK (1997) Monitoring turbidity with IRS-1A. Hydrol Process 11(15):1907–1915

Clark GL, Ewing C, Lorenzen CJ (1970) Spectra of backscattered light from the sea obtained from aircraft as a measure of chlorophyll concentration. Science 167:1119–1121

Curran PJ, Novo EMM (1988) The relationship between suspended sediment concentration and remotely sensed spectral radiance: a review. J Coast Res 4(3):351–368

Curran PJ, Hansom JD, Plummer SE, Pedley MI (1987) Multispectral remote sensing of nearshore suspended sediments: a pilot study. Int J Remote Sens 8(1):103–112

Dekker AG, Peters SWM (1993) The use of the Thematic Mapper for the analysis of eutrophic lakes: a case study in the Netherlands. Int J Remote Sens 14(5):799–821

Dekker AG, Vos RJ, Peters SWM (2002) Analytical algorithms for lake water TSM estimation for retrospective analysis of TM and SPOT sensor data. Int J Remote Sens 23(1):15–35

Dewan AM, Kabir MH, Nahar K, Rahman MZ (2012) Urbanization and environmental degradation in Dhaka Metropolitan Area of Bangladesh. Int J Environ Sustain Dev 11(2):118–146

Dewan AM, Corner RJ, Hashizume M, Ongee ET (2013) Typhoid fever and its association with environmental factors in the Dhaka Metropolitan Area of Bangladesh: a spatial and time-series approach. PLoS Negl Trop Dis 7(1):e1998

Dwivedi RM, Narain A (1987) Remote sensing of phytoplankton: an attempt from the Landsat Thematic Mapper. Int J Remote Sens 8(10):1563–1569

Ekercin S (2007) Water quality retrievals from high resolution Ikonos multispectral imagery: a case study in Istanbul, Turkey. J Water Air Soil Pollut 183(1–4):239–251

Fraser RN (1998) Multispectral remote sensing of turbidity among Nebraska Sand Hills Lakes. Int J Remote Sens 19:3011–3016

Hadjimitsis DG, Hadjimitsis MG, Toulios L, Clayton CRI (2010) Use of space technology for assisting water quality assessment and monitoring of inland water bodies. Phys Chem Earth 35(1–2):115–120

Harrington JA Jr, Schiebe FR, Nix JF (1992) Remote sensing of lake Chicot, Arkansas: monitoring suspended sediments, turbidity, and secchi depth with Landsat MSS data. Remote Sens Environ 39(1):15–27

Hellweger FL, Schlosser P, Lall U, Weissel JK (2004) Use of satellite imagery for water quality studies in New York harbor. Estuarine, Coast Shelf Sci 61(3):437–448
Hellweger FL, Miller W, Oshodi KS (2007) Mapping turbidity in the Charles River, Boston using a high-resolution satellite. Environ Monit Assess 132(1–3):311–320
Islam MA, Gao J, Ahmad W, Neil D, Bell P (2003) Image calibration to like-values in mapping shallow water quality from multi-temporal data. Photogramm Eng Remote Sens 69(5):567–575
Jensen JR (2000) Remote sensing of the environment: an earth resources perspective. Prentice Hall, Upper Saddler River
Johnson RW, Harriss RC (1980) Remote sensing for water quality and biological measurements in coastal waters. Photogramm Eng Remote Sens 46(1):77–85
Julien RY (1995) Erosion and sedimentation. Cambridge University Press, New York
Kamal MM, Malmgren-Hansen A, Badruzzaman ABM (1999) Assessment of pollution of the river Buriganga, Bangladesh using a water quality model. Water Sci Technol 40(2):129–136
Karn SK, Harada H (2001) Surface water pollution in three urban territories of Nepal, India and Bangladesh. Environ Manag 28(4):483–496
Khorram S (1981) Water quality mapping from Landsat digital data. Int J Remote Sens 2(2):145–153
Khorram S, Cheshire HM (1985) Remote sensing of water quality in the Neuse river estuary, North Carolina. Photogramm Eng Remote Sens 51:329–341
Klemas V, Borchardt VJF, Treasure WM (1973) Suspended sediment observations from ERTS-1. Remote Sens Environ 2:205–221
Klemas V, Borchardt JF, Treasure WM (1974) Coastal and estuarine studies with ERTS-1 and Skylab. Remote Sens Environ 3(3):153–174
Kloiber SM, Brezonik PL, Bauer ME (2002a) Application of Landsat imagery to regional-scale assessment of lake clarity. Water Res 36(17):4330–4340
Kloiber SM, Brezonik PL, Olmanson LG, Bauer ME (2002b) A procedure for regional lake water clarity assessment using Landsat multispectral data. Remote Sens Environ 82(1):38–47
Lambrix JM, Naugle BI (2003) Linear regression prediction models for turbidity in Kentucky Lake utilizing Landsat 7 ETM+ and KLMP in situ data. Geol Soc Am Abstr Progr 35(6):168
Lathrop RG Jr (1992) Landsat thematic mapper monitoring of turbid inland water quality. Photogramm Eng Remote Sens 58(4):465–470
Lathrop RG Jr, Lillesand TM (1986) Utility of thematic mapper data to assess water quality in southern Green Bay and West-Central Lake Michigan. Photogramm Eng Remote Sens 52:671–680
Lillesand TM, Johnson WL, Deuell RL, Lindstrom OM, Meisner DE (1983) Use of Landsat data to predict the trophic state of Minnesota lakes. Photogramm Eng Remote Sens 49(2):219–229
Lindell LT, Steinvall O, Jonsson M, Thcalesson T (1985) Mapping coastal water turbidity using Landsat imagery. Int J Remote Sens 16(5):629–642
Lo CP, Hutchinson WT (1991) Determination of turbidity patterns of the Zhijiang estuarine region, South China, using satellite images and a GIS approach. Geocarto Int 6(3):27–38
Lodhi MA, Runquist DC, Han L, Kuzila MS (1998) Estimation of suspended sediment concentration in water integrated surface reflectance. Geocarto Int 13(2):11–15
Lyon JC, Bedford KW, Yen JC, Lee DH, Mark DJ (1988) Determination of suspended sediments from multiple day Landsat and AVHRR data. Remote Sens Environ 25(1):107–115
Ma R, Dai J (2005) Investigation of chlorophyll-*a* and total suspended matter concentrations using Landsat ETM and field spectral measurement in Taihu lake, China. Int J Remote Sens 26 (13):2779–2787
Mayo M, Gitelson A, Yacobi YZ, Ben-Avraham Z (1995) Chlorophyll distribution in Lake Kinneret determined from Landsat Thematic Mapper data. Int J Remote Sens 16(1):175–182
McFeeters SK (1996) The use of normalized difference water index (NDWI) in the delineation of open water features. Int J Remote Sens 17(7):1425–1432

Mertes LAK, Smith MO, Adams JB (1993) Estimating suspended sediment concentrations in surface waters of the Amazon river wetlands from Landsat images. Remote Sens Environ 43(3):281–301
Moore GK (1980) Satellite remote sensing of water turbidity. Hydrol Sci 25(4):407–421
Nas B, Karabork H, Ekercon S, Berktay A (2009) Mapping chlorophyll-*a* through in-situ measurements and Terra Aster satellite data. Environ Monit Assess 157(1–4):375–382
Nas B, Ekercin S, Karabork H, Berktay A, Mulla DJ (2010) An application of Landsat-5TM image data for water quality mapping in Lake Beyşehir, Turkey. J Water Air Soil Pollut 212 (1–4):183–197
Nelson SAC, Soranno PA, Cheruvelil KS, Batzli SA, Skole DL (2003) Regional assessment of lake water clarity using remote sensing. J Limnol 62(Suppl 1):27–32
Olmanson LG, Kloiber SM, Bauer ME, Brezonik PL (2001) Image processing protocol for regional assessments of lake water quality. Water Resources Center and Remote Sensing Laboratory, University of Minnesota, St. Paul
Olmanson LG, Bauer ME, Brezonik PL (2002) Use of Landsat imagery to develop a water quality atlas of Minnesota's 10,000 lakes. In: Pecora 15/Land Satellite Information IV/ISPRS Commission I/FIEOS 2002 conference proceedings, Denver, 10–14 Nov 2002
Östlund C, Flink P, Strombeck N, Pierson D, Lindell T (2001) Mapping of the water quality of lake Erken, Sweden, from imaging spectrometry and Landsat thematic mapper. Sci Total Environ 268(1–3):139–154
Pattiaratchi C, Lavery P, Wyllie A, Hick P (1994) Estimates of water quality in coastal waters using multidate Landsat Thematic Mapper data. Int J Remote Sens 15(8):1571–1584
Peckham SD, Lillesand T (2006) Detection of spatial and temporal trends in Wisconsin lake water clarity using Landsat-derived estimates of Secchi depth. Lake Reserv Manag 22(4):331–341
Potes M, Costa MJ, Da Silva JCB, Silva AM, Morais M (2011) Remote sensing of water quality parameters over Alqueva reservoir in the south of Portugal. Int J Remote Sens 32(12):3373–3388
Pulliainen J, Kallio K, Eloheimo K, Koponen S, Servomaa H, Hannonen T (2001) A semi-operative approach to lake water quality retrieval from remote sensing data. Sci Total Environ 268(1–3):79–93
Rahman S, Hossain F (2008) Spatial assessment of water quality in peripheral rivers of Dhaka city for optimal location of water intake point. Water Resour Manag 22(3):377–391
Reddy MA (1997) A detailed statistical study on selection of optimum IRS LISS pixel configuration for development of water quality models. Int J Remote Sens 18(12):2559–2570
Ritchie JC, Cooper CM (1987) Comparisons of Landsat MSS pixel array sizes for estimating water quality. Photogramm Eng Remote Sens 53(11):1549–1553
Ritchie JC, Cooper CM (1988) Comparisons of measured suspended sediment concentrations with suspended sediment concentrations estimated from Landsat MSS data. Int J Remote Sens 9(3):379–387
Ritchie JC, Schiebe FR, Cooper CM (1986) Surface water quality measurements of Lake Chicot, Arkansas using data from Landsat satellites. J Freshw Ecol 3(3):391–397
Ritchie JC, Cooper CM, Yongqing J (1987) Using Landsat multispectral scanner data to estimate suspended sediments in Moon Lake, Mississippi. Remote Sens Environ 23(1):65–81
Ritchie JC, Schiebe FR, Cooper CM (1989) Landsat digital data for estimating suspended sediment in inland water, IAHS Pub. No. 182. International Association of Hydrological Sciences, pp 151–158
Ritchie JC, Cooper CM, Schiebe FR (1990) The relationship of MSS and TM digital data with suspended sediments, chlorophyll, temperature in Moon Lake, Mississippi. Remote Sens Environ 33(2):137–148
Sawaya KE, Olmanson LG, Heinert NJ, Brezonik PL, Bauer ME (2003) Extending satellite remote sensing to local scales: land and water resource monitoring using high-resolution imagery. Remote Sens Environ 88(1–2):144–156
Scarpace FL, Holmquist KW, Fisher LT (1979) Landsat analysis of lake quality. Photogramm Eng Remote Sens 45(5):623–633

Schiebe FR, Ritchie JC (1986) Suspended sediment monitored by satellite. In: Proceedings of fourth Federal Interagency Sedimentation conference, Las Vegas, 1986, vol 1. pp 70–78
Schneider WJ (1968) Color photographs for water resources studies. Photogramm Eng 34(3):257–262
Slonecker ET, Jennings DB, Garofalo D (2001) Remote sensing of impervious surface: a review. Remote Sens Rev 20(3):227–235
Sohel KMA, Chowdhury MAI, Ahmed MF (2003) Surface water quality in and around Dhaka City. J Water Supply: Res Technol 52(2):141–153
Stefouli M, Dimitrakopoulos D, Papadimitrakis J, Charou E (2005) Monitoring and assessing internal waters (Lakes) using operational space borne data and field measurements. Eur Water 9(10):25–33
Svab E, Tyler AN, Preston T, Presing M, Balogh KV (2005) Characterizing the spectral reflectance of algae in lake waters with high suspended sediment concentrations. Int J Remote Sens 26(5):919–928
Tan Q, Shao Y, Yang S, Wei Q (2003) Landsat TM optimal bands selection for freshwater lake international importance wetland interpretation and monitoring. IEEE Geosci Remote Sens Proc 2003:3299–3301
Tassan S (1997) A numerical model for the detection of sediment concentration in stratified river plums using Thematic Mapper data. Int J Remote Sens 18(12):2699–2705
Tyler AN, Svab E, Preston T, Présing M, Kovács WA (2006) Remote sensing of the water quality of shallow lakes: a mixture modelling approach to quantifying phytoplankton in water characterized by high-suspended sediment. Int J Remote Sens 27(8):1521–1537
Usali N, Ismail MH (2010) Use of remote sensing and GIS in monitoring water quality. J Sustain Dev 3(3):228–238
Verdin JP (1985) Monitoring water quality conditions in a large western reservoir with Landsat imagery. Photogramm Eng Remote Sens 51(3):343–353
Volpe V, Silvestri S, Marani M (2011) Remote sensing retrieval of suspended sediment concentration in shallow waters. Remote Sens Environ 115(1):44–54
Wang F, Luoheng H, Hsiang-Te K, Roy BVA (2006) Applications of Landsat-5 TM imagery in assessing and mapping water quality in Reelfoot Lake, Tennessee. Int J Remote Sens 27(23):5269–5283
Wang J, Lu X, Zhou Y (2007) Retrieval of suspended sediment concentrations in the turbid water of the upper Yangtze River using Landsat ETM+. Chin Sci Bull 52(suppl II):273–280
Wen-yao L, Klemas V (1988) Quantitative analysis of distribution of suspended sediments in the Yellow river estuary from MSS data. Geocarto Int 3(1):51–62
Witzig AS, Whitehurst CA (1981) Current use and technology of Landsat MSS data for lake trophic classification. Water Resour Bull 17(6):962–970
Wu C, Wu J, Qi J, Zhang L, Huang H, Lou L, Chen Y (2010) Empirical estimation of total phosphorus concentration in the mainstream of the Qiantang River in China using Landsat TM data. Int J Remote Sens 31(9):2309–2324
Zhang Y, Pulliainen J, Koponen S, Hallikainen M (2002) Application of an empirical neural network to surface water quality estimation in the Gulf of Finland using combined optical data and microwave data. Remote Sens Environ 81(2–3):327–336
Zhang Y, Pulliainen J, Koponen S, Halikainen M (2003a) Empirical algorithms for Secchi disk depth using optical and microwave remote sensing data from the Gulf of Finland and the archipelago sea. Boreal Environ Res 8(3):251–261
Zhang Y, Pulliainen J, Koponen S, Hallikainen M (2003b) Water quality retrievals from combined Landsat TM data and ERS-2 SAR data in the Gulf of Finland. IEEE Trans Geosci Remote Sens 41(3):622–629
Zhou W, Wang S, Zhou Y, Troy A (2006) Mapping the concentration of total suspended matter in Lake Taihu, China using Landsat-5 TM data. Int J Remote Sens 27(6):1177–1191

Chapter 17
Emissions from the Brick Manufacturing Industry

Sarath K. Guttikunda

Abstract Brick manufacturing is the fastest-growing industrial sector in Bangladesh and among the major contributors to the air pollution and health problems in Dhaka, along with vehicle exhaust, resuspended road dust, and domestic fuel burning. There are about 1,000 brick kilns in Dhaka region from the districts of Dhaka, Gazipur, Manikganj, and Narayanganj. Brick manufacturing is confined to the non-monsoonal months and produces 3.5 billion bricks per year, using energy-inefficient fixed-chimney Bull trench kiln technology fuelled with coal and agricultural waste. The total annual emissions are estimated as 23,300 tons of $PM_{2.5}$, 15,500 tons of SO_2, 302,000 tons of CO, 6,000 tons of black carbon, and 1.8 million tons of CO_2. The associated health impacts largely fall on the densely populated districts of Dhaka, Gazipur, and Narayanganj. Using the ATMoS dispersion model, the impact of brick kiln emissions was estimated over Dhaka region – ranging from 7 to 99 $\mu g/m^3$ (5th and 95th percentile concentration per model grid) at an average of 38 $\mu g/m^3$ – and provincial cluster contributions of 27 % originating from Narayanganj (to the south with the highest kiln density), 30 % from Gazipur (to the north with equally large cluster spread along the river and canals), and 23 % from Savar of Dhaka district. The modelling results were validated using evidence from receptor modelling studies conducted in Dhaka region. An introduction of emerging vertical shaft combustion technology and a possible relocation of the northern clusters to the southeast can provide faster benefits for public health and reduce climate precursor emissions.

Keywords Particulate pollution • Emissions inventory • Air quality • Brick kiln clusters • Modelling

---

S.K. Guttikunda (✉)
Division of Atmospheric Sciences, Desert Research Institute, Reno, NV 89512, USA
e-mail: sguttikunda@gmail.com

A. Dewan and R. Corner (eds.), *Dhaka Megacity: Geospatial Perspectives on Urbanisation, Environment and Health*, Springer Geography,
DOI 10.1007/978-94-007-6735-5_17,

17.1 Introduction

The air quality in the Dhaka city has deteriorated over the last decade due to a rapid change in the vehicular fleet, increased congestion, and a large increase in industrial activity, resulting in mortality and morbidity impacts (WHO 2011; World Bank 2006, 2007). The World Health Organization (WHO) studied publicly available air quality data from 1,100 cities and ranked Dhaka within the top 20 cities with the worst air pollution (WHO 2011). The economic costs associated with mortality and morbidity due to poor air quality is estimated at US$500 million per year for Dhaka city (World Bank 2006). An overview of the measured monthly average $PM_{2.5}$ concentrations at a monitoring station near the Sangsad Bhaban (Parliament house) (SB – urban site) and another station located at the Atomic Energy Centre (Dhaka) (AECD – campus site) is presented in Fig. 17.1 ($PM_{2.5}$ refers to the particulate matter (PM) with aerodynamic diameter less than 2.5 μm). The average concentration of $PM_{2.5}$ due to all sources is 100 μg/m^{3}at SB and 30 μg/m^{3}at AECD, both well above the health guidelines of 10 μg/m^3 prescribed by the World Health Organization and the Bangladesh national ambient air quality standard of 15 μg/m^3.

In Dhaka Metropolitan Area (DMA), emissions from vehicle exhaust, resuspended road dust (due to vehicular movement), and industries are dominant sources of air pollution (Begum et al. 2006, 2008, 2011; Azad and Kitada 1998; Hasan and Mulamoottil 1994). Among the industries, brick kiln emissions form a major source. Source apportionment studies conducted in 2001–2002, 2005–2006, and 2007–2009 estimated that brick kilns contribute, on average, ca. 30–40 % of total measured $PM_{2.5}$ pollution (Begum et al. 2006, 2008, 2011). Simulations using the CMAQ chemical transport dispersion modelling system over Dhaka and Bangladesh estimated that at

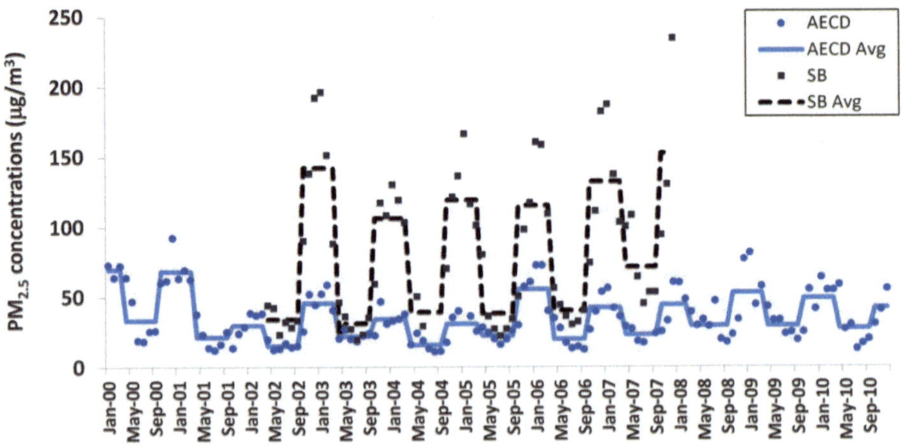

Fig. 17.1 Monthly average ambient $PM_{2.5}$ concentrations (mg/m^3) measured near Sangsad Bhaban (SB – traffic junction) and at the Atomic Energy Centre (AECD – campus); the average concentrations are segregated between brick manufacturing season (October to March) and the remaining months

least 35 % of ambient PM_{10} and at least 15 % of ambient $PM_{2.5}$ in DMA are associated with brick kiln emissions (Billah Ibn Azkar et al. 2012).

There is limited understanding of the movement of the emissions and the contributions attributable to various brick kiln clusters available to support effective urban air quality management in Dhaka. Given the uncertainties and deficiencies in source apportionment studies (Johnson et al. 2011), which assign contributions to a certain location, qualitatively and quantitatively, results from emissions and dispersion modelling can further support and compliment the requirements of integrated urban air quality management (Hidy and Pennell 2010). In this chapter, we present the impact of emissions from brick kilns surrounding Dhaka region via dispersion modelling of particulate pollution, with supporting information from ongoing source apportionment studies, and discuss possible interventions to benefit Dhaka's air quality.

17.2 Study Area

The Dhaka region, spread over an area of 1,500 km^2, includes the districts of Dhaka, Gazipur, Manikganj, and Narayanganj and contains approximately 1,000 brick kilns. The location of the brick kilns presented in Fig. 17.2 has been digitised using visual images from Google Earth and physically verifying the location of some clusters. Most of these kilns are located along the intra-city canals linking the rivers, which also serve as the arteries for transporting raw material to the kilns and delivery of the finished product to distribution and construction sites. The study domain extends from 89.7°E to 90.8°E in longitude and 23.3°N to 24.4°N in latitude. This area has been subdivided into 55 × 55 grid cells at a grid resolution of 0.02° for the purposes of this study.

17.2.1 Brick Kiln Clusters

The largest clusters are located in Gazipur (~320) and Narayanganj (~270), followed by Savar, Dhamrai in Dhaka, and Kaliganj in Gazipur, each with over 100 kilns, and Rupganj in Narayanganj with less than 100 kilns. Brick manufacturing in Bangladesh is dominated by small individual operators, each employing 200–300 daily wage workers per kiln, on a seasonal basis. Most of these are conventional fixed-chimney Bull trench kilns (FCBTK), which are relatively more polluting and energy-inefficient when compared to the newer, cleaner technologies such as Hoffmann kilns, high-draught kilns, or vertical shaft brick kilns (World Bank 2010; CAI-Asia 2008). A summary of various kiln technologies and comparison to the FCBTK currently used is presented in Table 17.1. Of the technologies listed in Table 17.1, FCBTK is the most common technology, and vertical shaft and zigzag technology kilns are in a pilot stage (World Bank 2010).

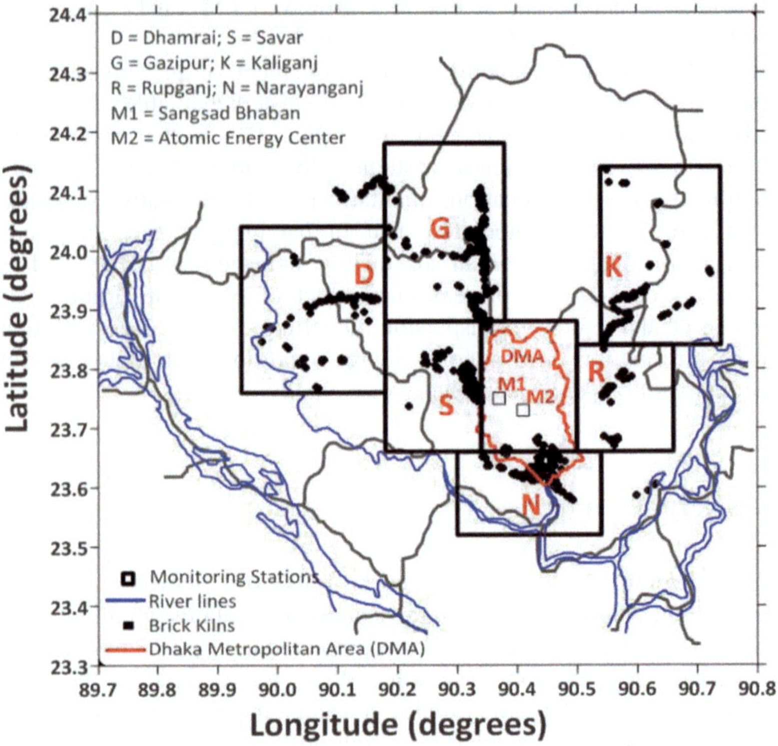

Fig. 17.2 Modelling domain covering Dhaka region, along with the location of the brick kiln clusters and monitoring stations (*small boxes* within DMA)

Table 17.1 Comparison of current and alternative brick manufacturing technologies available in Bangladesh[a, b]

Technology	Fuel consumed per 100,000 bricks	Average tons of CO_2 produced per 100,000 bricks	Average reduction in PM emissions compared to FCBTK (%)
FCBTK	20–22 tons coal	50	
Zigzag[c]	16–20 tons coal	40	40
Hoffmann[d]	15,000–17,000 m^3 NG	30	90
Hoffmann[e]	12–14 tons coal	30	60
VSBK[f]	10–12 tons coal	25	60

[a]The table is a summary of information presented in the World Bank (2007)
[b]*FCBTK* fixed-chimney Bull trench kiln, *NG* natural gas, *VSBK* vertical shaft brick kiln
[c]Some zigzag pilot kilns are in operation, listed as poor to medium performance. Any improvement in the efficiency of operations can lead to further reductions in coal consumption
[d]Manufacturing period for Hoffmann kilns is round the year, compared to the 6-month operations for the other kilns, thus increasing the land and raw material requirements; link to natural gas grid and continuous supply is a major constraint
[e]Initial investments are higher for Hoffmann kilns
[f]Operational models are available in India and Kathmandu (CAI-Asia 2008)

Due to a lack of information on the brick manufacturing rates by individual kiln, an average production rate of 20,000 bricks per day is assumed per kiln (World Bank 2007). Since all the kilns follow the same design framework for heating and stacks, the assumption of an average production rate per stack is reasonable for estimating the final emissions.

17.2.2 Emissions Inventory

Brick kilns are primarily associated with PM, carbon monoxide (CO), sulphur dioxide (SO_2), volatile organic compounds (VOCs), nitrogen oxides (NO_x), and heavy metals depending on the type of fuel burnt. In this study, a combination of local surveys (World Bank 2007) and recent measurement campaigns (Maithel et al. 2012) are utilised to estimate activity-based combustion emissions. Maithel et al. (2012) is the first set of measurements in South Asia to quantify the emission factors for brick kilns in the Indo-Gangetic plains. While these factors cannot be translated and used directly for the brick kilns in Dhaka, due to variations in the fuel mix between India and Bangladesh, they do provide a baseline for comparison, given that the technologies in use are similar between the regions.

At the brick kilns in Dhaka region, the energy required for baking 100,000 bricks is estimated as being 20 tons of coal with a calorific value of 22 MJ/kg. The majority (~80 %) of the energy needs are provided by coal with occasional use of biomass (~10 %) and heavy fuel oil (~10 %) (World Bank 2007). For comparative purposes, in India, most of the kiln clusters are located closer to the agricultural areas, which results in a higher portion of field residue available for combustion at the kilns. The majority of the coal used at the kilns in Dhaka is imported from India's northeastern states, with an ash content of 20–25 % and sulphur content of 1.0 % by weight. The use of biomass is limited to the harvest season, when residue from neighbouring agricultural land can be acquired at a price lower than that of coal (per MJ). However, this is not a sustainable option for the entire season. The heavy fuel oil is used as a replacement fuel during the occasional rainy days to avoid the necessity of using wet coal. No specific pollution control measures are implemented at the kilns, either to control SO_2 emissions or to settle the total suspended particulates. The kiln's chimney extending up to 50 m is designed to settle most of the heavier particles and release most of the PM_{10} and $PM_{2.5}$ into the flue gas.

The World Bank (2007) presents an emissions inventory for the kilns located north of Dhaka, in the Gazipur district, including a measured emissions factor of 44.0 g/s for total suspended particulates, which, for an average production rate of 20,000 bricks per day, translates to 190 g per brick. Gains (2010) suggests an average PM_{10} emissions factor for brick kilns in Asia as 1,750 tons per million tons of bricks, which, for an average weight of 3 kg per brick, translates to 5.85 g per brick. Le and Oanh (2010) measured $PM_{2.5}$ emissions of 0.64–1.4 g per brick produced in kilns smaller in design and production capacities than those available in Dhaka. For the base year 2010, based on fuel consumption and fuel

Table 17.2 Estimated emissions from brick kiln clusters in Dhaka region

Kiln clusters	No. of kilns[a]	Select box area (km^2)[b]	Total $PM_{2.5}$ emissions (tons/year)[a,c]
Dhaka (DMA)	0	400	
Savar (S)	120	400	2,850 (12 %)
Dhamrai (D)	80	750	1,900 (8 %)
Gazipur (G)	320	675	8,500 (37 %)
Kaliganj (K)	90	675	2,150 (9 %)
Rupganj (R)	50	325	1,250 (5 %)
Narayanganj (N)	270	380	6,700 (27 %)

[a]Numbers are rounded and all the kilns are tagged to cluster regions
[b]For the boxes drawn in Fig. 17.4 over each district
[c]Numbers in the bracket indicate the percentage of district emissions to the total domain emissions from brick kilns

characteristics data, we estimated the emission factors in this study, in grams per brick, produced as 6.8 for $PM_{2.5}$, 9.7 for PM_{10}, 4.6 for SO_2, 4.7 for NO_x, 90.0 for CO, 520 for CO_2, and 2.8 for black carbon (BC).

The total emissions from the brick kilns located in the six districts are estimated at 33,100 tons of PM_{10}, 23,300 tons of $PM_{2.5}$, 15,500 tons of SO_2, 16,100 tons of NO_x, 302,000 tons of CO, and 6,000 tons of BC per year. The total emissions for each district cluster are presented in Table 17.2. The percentage contribution by district cluster to total emissions is same for all the pollutants. For convenience, only $PM_{2.5}$ is presented in the table. Brick production rates varied from 10,000 to 45,000 per day per kiln, depending on their manufacturing capacity and on the availability of raw material and labour. For the clusters around DMA, during the operational season, the total manufactory averages 3.5 billion bricks per year, consuming 1.0 million tons of coal and producing 1.8 million tons of CO_2. The spatial spread of the emissions is shown on the density map presented in Fig. 17.2, with highest densities being observed over Gazipur and Narayanganj. We acknowledge that there is spatial uncertainty associated with the production rates and related emissions by region and by kiln. However, since all the kilns in the region are designed in similar fashion, this allows for averaging of the energy consumed per brick produced and thus the emissions by grid cell.

17.3 Particulate Pollution Modelling

We utilised the Atmospheric Transport Modelling System (ATMoS) dispersion model – a forward trajectory Lagrangian Puff-transport model (Calori and Carmichael 1999) – to characterise the movement of brick kiln emissions. The model has previously been utilised to study regional- and urban-scale pollution management in Asia for sulphur, nitrogen, and PM pollutants (Guttikunda and Calori 2013; Guttikunda and Jawahar 2012; Li et al. 2004; Guttikunda et al. 2003; Holloway et al. 2002). The horizontal resolution of the model is 0.02° for

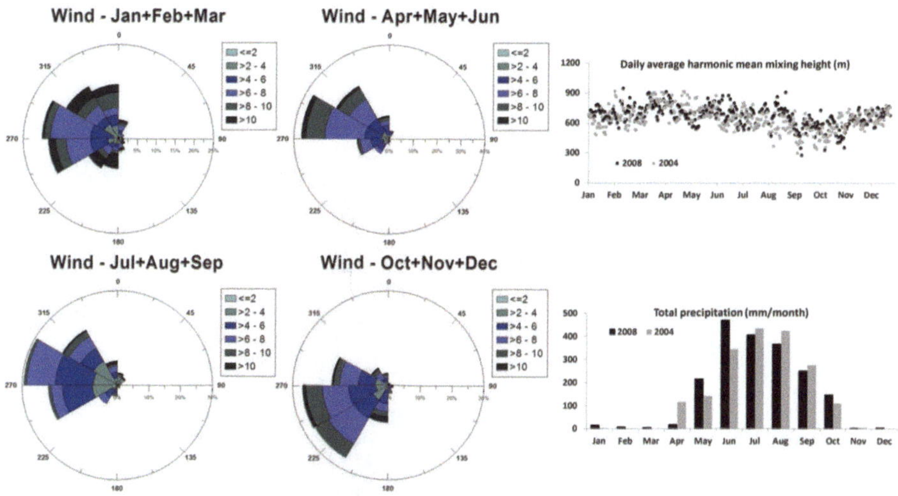

Fig. 17.3 Wind speed and wind direction for year 2008; monthly precipitation totals (mm/h) and daily average harmonic mean mixing height (metres) estimated from the National Centers for Environmental Prediction (NCEP) reanalysis data for years 2004 and 2008 for the grid covering Dhaka region

both latitude and longitude, equivalent in the study area to about 2 km. The vertical resolution includes a surface layer, boundary layer (designated as the mixing layer height), and a top layer. The multiple layers allow the model to differentiate the contributions of diffused area sources like transport and domestic combustion emissions and point source emitters such as brick kilns and power plants. The model includes a simplified chemical mechanism to convert SO_2 to sulphates and NO_x emissions to nitrate concentrations, which are added to the $PM_{2.5}$ totals as part of the secondary contributions. In this study modelling was conducted for $PM_{2.5}$ fractions only, given the propensity of this fraction to penetrate further into the human lungs and result in exacerbated respiratory and cardiovascular diseases and in some cases lead to premature mortality (HEI 2010).

The dispersion model was run for a simulation period of 6 months, corresponding to the brick manufacturing season. The manufacturing season corresponds to the non-monsoonal season, when there is less precipitation. The precipitation patterns, wind speeds and wind directions, and daily average harmonic mean mixing heights in Fig. 17.3 are obtained from the National Centers for Environmental Prediction reanalysis (NCEP 2012).

The modelled brick manufacturing seasonal average $PM_{2.5}$ concentrations are shown in Fig. 17.4. It is important to note that the modelled concentrations are those attributable to brick kiln emissions only and do not include emissions from vehicle exhaust, road dust, other industries, or domestic fuel combustion. The modelled concentrations of $PM_{2.5}$ over DMA ranged from 7 to 99 $\mu g/m^3$ over the 6-month period (the range indicates the 5th and 95th percentile concentration per model grid) with an average of 38 $\mu g/m^3$.

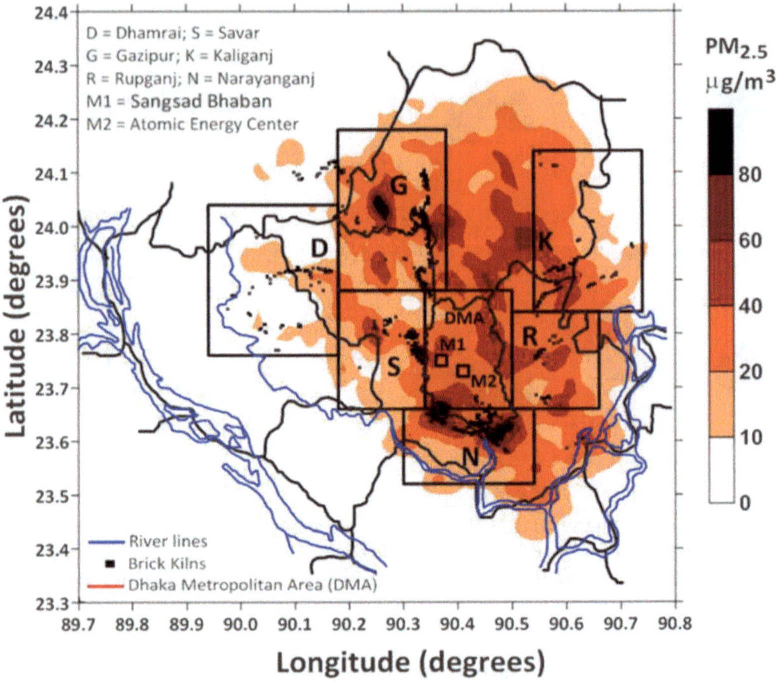

Fig. 17.4 Modelled $PM_{2.5}$ (mg/m^3) concentrations averaged over the brick manufacturing season for Dhaka region

Table 17.3 Model average $PM_{2.5}$ concentrations due to brick kiln emissions in Dhaka region

District	Seasonal 5th and 95th percentile (average) (µg/m^3) concentrations	Dec–Jan 5th and 95th percentile (average) (µg/m^3) concentrations	Percentage contributions to seasonal averages over DMA (%)
Dhaka (DMA)	7–99 (38)	8–161 (56)	
Savar (S)	3–59 (19)	3–85 (27)	23
Dhamrai (D)	1–30 (8)	1–39 (12)	10
Gazipur (G)	3–91 (28)	4–134 (40)	30
Kaliganj (K)	3–55 (21)	4–63 (28)	3
Rupganj (R)	8–79 (31)	12–96 (43)	7
Narayanganj (N)	4–134 (34)	4–186 (41)	27

The concentration ranges modelled for the area covering the boxed regions in Fig. 17.2 are presented in Table 17.3. The variation in the concentrations is primarily due to the selection of a large area around each district cluster. For the boxed regions, the modelled data is extracted by grid (2 × 2 km^2) to estimate the range of modelled concentrations. The contribution of the brick kilns in the respective regions is proportional to the number of the kilns located in the vicinity, except for DMA, which is surrounded by regions with kilns and experiences dispersed pollution. The Narayanganj (N) area with the highest density of kilns and closest to DMA experiences averages of $PM_{2.5}$ as high as 120 μg/m^3 (reaching 180 μg/m^3 in December and January) from brick kilns emissions alone, which is already well in excess of the WHO guidelines for exposure risks.

The source apportionment studies estimated an average contribution of 30–40 % originating from the brick kilns (Begum et al. 2006, 2008, 2011). This was used as proxy to compare the modelled concentrations in Fig. 17.4 to the measured $PM_{2.5}$ concentrations in Fig. 17.1. The $PM_{2.5}$ monitoring data is also split between the brick manufacturing season and the remaining months, which present a distinct change in the pollution trends, coinciding with the brick manufacturing cycles. On the basis of the source apportionment results, the seasonal average measured $PM_{2.5}$ concentration of 140 μg/m^3 at the Sangsad Bhaban translates to 42–56 μg/m^3. The site at the Atomic Energy Centre in Dhaka located on campus, which is deemed a background site for monitoring activities, experienced a seasonal average $PM_{2.5}$ concentration of 50 μg/m^3, with limited exposure to vehicle exhausts. For the boxed region covering DMA, we modelled an average concentration of 38 μg/m^3 due to the emissions from brick kiln clusters (Table 17.2). We did not make point to point comparison between measured and modelled concentrations, due to limited monitoring activities carried out in DMA. In Fig. 17.1, although we split the averages to brick manufacturing and non-manufacturing months, this in no way suggests that emissions from other sources are contributing any less to the growing PM pollution in the city.

The average $PM_{2.5}$ concentrations for December and January are at least twice the seasonal averages. The monthly averages of $PM_{2.5}$ (due to all sources) measured at the two monitoring stations also observed peaks in December or January. At the Sangsad Bhaban, $PM_{2.5}$ concentrations averaged 168 $\pm$ 35 μg/m^3 between 2002 and 2008, peaking at 230 μg/m^3 in January 2008. The peak concentrations also increased from 130 μg/m^3 in 2004 to 230 μg/m^3 in 2008, an indication of the growing demand for fired bricks, an increase in the operational kilns, and a booming demand in the construction industry. The peaks in concentrations are also partly due to slow-moving winds in December and January, which are known to enhance unfavourable dispersion characteristics and exacerbate the ground-level concentrations.

17.4 Provincial Apportionment of Pollution

Emissions from low-lying sources like vehicle exhaust and domestic fuel use are fairly constant over all months of a year, and being confined to the city limits thus contributes more to the locally attributable pollution levels. However, the brick kilns, spread across Dhaka region, operate with a stack height of 50 m each, and their emissions are capable of being transported for larger distances, as it is evident in the monitoring data presented in Fig. 17.1 and the modelled concentrations presented in Fig. 17.4.

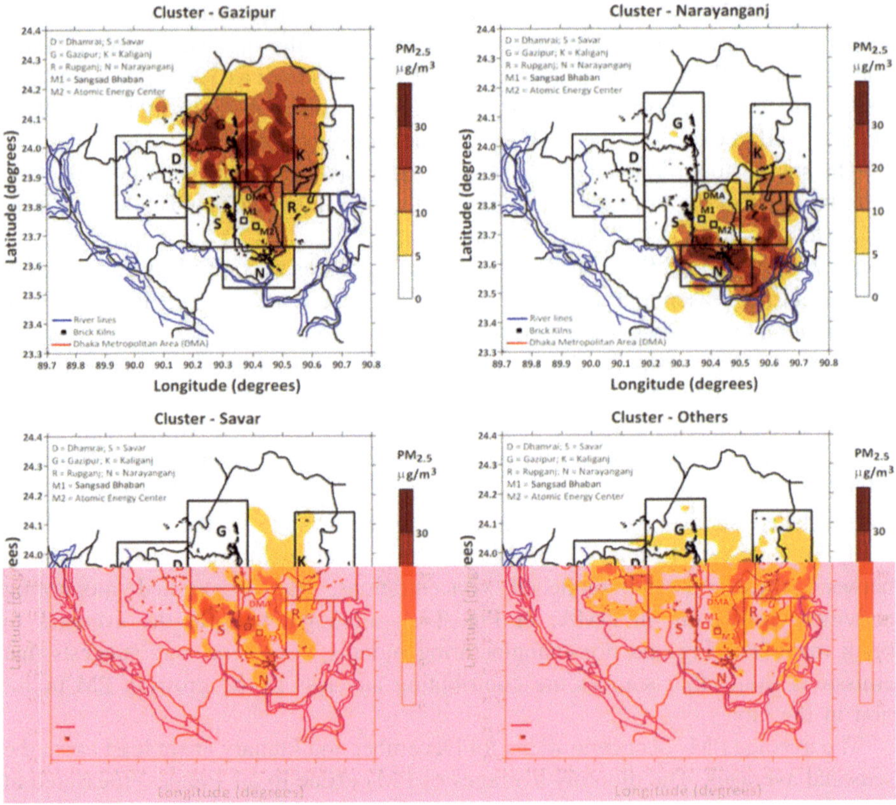

Fig. 17.5 Modelled $PM_{2.5}$ (mg/m^3) concentrations averaged over the brick manufacturing season for the emissions originating from individual brick kiln clusters in Dhaka region

While source apportionment is a resource for estimating the contribution of brick kilns to a particular location (where the monitoring is conducted), it fails to assess the spatial origin of the pollution. This is especially difficult when all the kilns (irrespective of their location) are operating with similar fuel characteristics and the receptor modelling depends on the fuel characteristics to apportion pollution loads. An added advantage of dispersion modelling is the spatial allocation of the emissions and tagging of the sources to estimate contributions to a select location. For analytical purposes, the emissions from kilns are tagged by source for the six districts surrounding DMA (for the boxes presented in Fig. 17.4). A summary of the modelled concentrations due to these emissions is presented in Table 17.3, and the contribution of each cluster is presented in Fig. 17.5.

The highest contributions to DMA originate from the kilns in Narayanganj (N) and Gazipur (G), being 27 and 30 %, respectively. Though the density of kilns in Savar (S) is not as large, being only 11 % of the total, due to the proximity of these kilns to Dhaka and favourable winds (Fig. 17.3), the contribution of these kilns is also substantial at 23 %. In the various districts, the contribution of in situ kilns is the highest, as expected, followed by those originating from the neighbouring district clusters. Gazipur was identified as the hotspot which contributed the most to Dhaka's poor air quality. Similar to DMA, Gazipur is also densely populated and experiences very high pollution exposure problems (World Bank 2007).

17.5 Implications

Brick kilns are thought to be a major source of rural and urban air pollution throughout South Asia. In Bangladesh, supporting the booming infrastructure and construction industry, the brick kiln industry is one of the fastest-growing sectors, with a manufacturing capacity of 12 billion bricks a year from 4,500 brick kilns surrounding the major cities of Dhaka, Khulna, Rajshahi, and Chittagong. This industry is expected to grow at least 50 % by 2020 (UNDP 2011; World Bank 2007). Few studies have rigorously analysed this issue, and fewer have measured the emissions factors and/or modelled their contribution to the ambient particulate pollution levels in and around the cities in South Asia. This study is one of the first to do so – in both mapping the kiln clusters and modelling the particulate pollution for Dhaka region. Given the unorganised nature of this manufacturing sector, we believe that uncertainties in estimating the fuel consumption patterns, production cycles, and emission rates are unavoidable; yet it is important to raise the necessary scientific awareness of a growing industrial sector and highlight its contributions to air quality and climate change precursors.

The energy and environmental consequences are already evident in the daily monitoring data in DMA. An important conclusion of this study is the spatial apportionment of brick kiln pollution in DMA, with the majority originating from three clusters – Narayanganj (to the south with the highest kiln density), Gazipur (an equally large cluster to the north spread along the river and canals), and Savar (to the west with winds favouring the movement of pollution towards the city). The introduction of emerging technologies and a possible relocation of the kilns from these clusters could provide rapid benefits to air quality and public health by being targeted to the most polluting clusters in the region.

There is an acute need to leapfrog from the outdated and inefficient FCBTK technology and introduce newer and cleaner technologies like zigzag (high and natural) draught kilns, Hoffmann kilns, tunnel kilns, and vertical shaft brick kilns (World Bank 2010; Zhihong 1997). From Table 17.1, at least a 40 % reduction in PM emission rates is possible by shifting away from the currently used technology, which, based on the average concentrations in Fig. 17.4, translates to a reduction of 20 $\mu g/m^3$ of ambient $PM_{2.5}$. The World Bank (2010) and UNDP (2011) are working with the regional bodies and kiln owners in Bangladesh to pilot and promote one of the emerging technologies (vertical shaft brick kilns) in Dhaka. The World Bank (2011) estimated health damages ranging from 2.5 to 14.0 million taka per kiln under the current FCBTK technology which could be reduced to from 0.9 to 5.7

million taka per kiln for vertical shaft brick kilns (1US$ = 80.5 taka in 2012). The combination of energy savings and reduction in pollution is a win-win situation, in which the industry benefits because of savings in energy costs and the city benefits because of reduction in adverse health impacts.

The potential for combined benefits for both health and climate by controlling the climate change precursors like CO_2 and BC is an emerging science. Since anthropogenic climate change is largely driven by fossil fuel combustion, this analysis of emissions from the brick kilns also provides insight into climate forcing and associated mitigation strategies.

Acknowledgements I would like to thank Dr. Bilkis Begum from the Atomic Energy Centre (Dhaka) for the synthesis of their source apportionment results and Dr. Zia Wadud from the Bangladesh University of Engineering and Technology (Dhaka) for the monitoring data.

References

Azad AK, Kitada T (1998) Characteristics of the air pollution in the city of Dhaka, Bangladesh in winter. Atmos Environ 32(11):1991–2005

Begum BA, Biswas SK, Hopke PK (2006) Temporal variations and spatial distribution of ambient PM2.2 and PM10 concentrations in Dhaka, Bangladesh. Sci Total Environ 358(1–3):36–45

Begum B, Biswas S, Hopke P (2008) Assessment of trends and present ambient concentrations of $PM_{2.2}$ and PM_{10} in Dhaka, Bangladesh. Air Qual Atmos Health 1(3):125–133

Begum BA, Biswas SK, Hopke PK (2011) Key issues in controlling air pollutants in Dhaka, Bangladesh. Atmos Environ 45(40):7705–7713

Billah Ibn Azkar MAM, Chatani S, Sudo K (2012) Simulation of urban and regional air pollution in Bangladesh. J Geophys Res 117(D7), D07303

CAI-Asia (2008) Clean brick making technology – success of VSBK in Kathmandu. Clean Air Initiative for Asia Cities, Manila

Calori G, Carmichael GR (1999) An urban trajectory model for sulfur in Asian megacities: model concepts and preliminary application. Atmos Environ 33(19):3109–3117

GAINS (2010) Greenhouse gas and air pollution interactions and synergies – South Asia program. International Institute of Applied Systems Analysis, Laxenburg

Guttikunda SK, Calori G (2013) A GIS based emissions inventory at 1 km × 1 km spatial resolution for air pollution analysis in Delhi, India. Atmos Environ 67:101–111

Guttikunda SK, Jawahar P (2012) Application of SIM-air modeling tools to assess air quality in Indian cities. Atmos Environ 62:551–561

Guttikunda SK, Carmichael GR, Calori G, Eck C, Woo J-H (2003) The contribution of megacities to regional sulfur pollution in Asia. Atmos Environ 37(1):11–22

Hasan S, Mulamoottil G (1994) Environmental problems of Dhaka City: a study of mismanagement. Cities 11(3):195–200

HEI (2010) Outdoor air pollution and health in the developing countries of Asia: a comprehensive review, Special Report 18. Health Effects Institute, Boston

Hidy GM, Pennell WT (2010) Multipollutant air quality management. J Air Waste Manag Assoc 60(6):645–674

Holloway T, Levy Ii H, Carmichael G (2002) Transfer of reactive nitrogen in Asia: development and evaluation of a source-receptor model. Atmos Environ 36(26):4251–4264

Johnson TM, Guttikunda SK, Wells G, Bond T, Russell A, West J, Watson J (2011) Tools for improving air quality management. A review of top-down source apportiontment techniques and their application in developing countries, ESMAP publication series. The World Bank, Washington, DC

Le H, Oanh N (2010) Integrated assessment of brick kiln emission impacts on air quality. Environ Monit Assess 171(1–4):381–394

Li J, Guttikunda SK, Carmichael GR, Streets DG, Chang Y-S, Fung V (2004) Quantifying the human health benefits of curbing air pollution in Shanghai. J Environ Manag 70(1):49–62

Maithel S, Uma R, Bond T, Baum E, Thao VTK (2012) Brick kilns performance assessment, emissions measurements, & a roadmap for cleaner brick production in India. Study report prepared by Green Knowledge Solutions, New Delhi

NCEP (2012) NCEP/NCAR global reanalysis products, 1948-continuing, distributed by National Center for Atmospheric Research, Colorado, USA. http://rda.ucar.edu/datasets/ds090.0/index.html

UNDP (2011) Improving kiln efficiency in the brick making industry. GEF-United National Development Program, UNDP, Dhaka

WHO (2011) Outdoor air pollution in the world cities. World Health Organization, Geneva

World Bank (2006) Country environmental assessment, Bangladesh. The World Bank, Washington, DC

World Bank (2007) Small study on air quality of impacts of the North Dhaka brickfield cluster by modeling of emissions and suggestions for mitigation measures including financing models. Consultant report prepared by Bangladesh University of Engineering and Technology for the World Bank, Washington, DC

World Bank (2010) Vertical shaft brick kiln – design manual, ESMAP publication series. The World Bank, Washington, DC

World Bank (2011) Introducing energy-efficient clean technologies in the brick sector of Bangladesh, Report No. 60155-BD. Environment, Climate Change, and Water Resources Unit, South Asia Region, The World Bank, Washington, DC

Zhihong Z (1997) Energy efficiency and environmental pollution of brickmaking in China. Energy 22(1):33–42

Chapter 18
Rainfall Dependence of Hospital Visits of *Aeromonas*-Positive Diarrhoea

Masahiro Hashizume, Abu S.G. Faruque, and Ashraf M. Dewan

Abstract The numbers of patients with cholera and *Aeromonas*-positive diarrhoea show similar bimodal seasonality in Dhaka, Bangladesh. Considering the association between these two pathogens, our previous finding that the number of cholera cases increased following the period of high rainfall led us to investigate the potential role of rainfall on the transmission of *Aeromonas*-positive diarrhoea. This study quantifies the impact of rainfall on the number of cases of *Aeromonas*-positive diarrhoea, to gain a deeper insight into the mechanisms of the seasonality of the disease. We examined a time series of the number of hospital visits due to *Aeromonas*-positive diarrhoea per week in relation to weekly rainfall from 1996 to 2000, using Poisson regression models and adjusting for seasonal and between-year variation, public holidays and temperature. The weekly number of cholera cases increased by 20.7 % (95 % confidence interval, 10.6–31.6) for a 10-mm increase in average rainfall over lags of 0–16 weeks. There was no clear relationship between the number of cases and river level or temperature.

Keywords *Aeromonas* • Climate • Diarrhoea • Rainfall • Time-series analysis

---

M. Hashizume (✉)
Institute of Tropical Medicine (NEKKEN), Nagasaki University, 1-12-4, Sakamoto, Nagasaki 852-8523, Japan
e-mail: hashizum@nagasaki-u.ac.jp

A.S.G. Faruque
Department of Clinical Medicine, International Centre for Diarrhoeal Disease Research (ICDDR), Bangladesh, GPO Box No. 128, Dhaka, Bangladesh
e-mail: gfaruque@icddrb.org

A.M. Dewan
Department of Spatial Sciences, Curtin University, Kent Street, Bentley, Perth 6102, WA, Australia
e-mail: A.Dewan@curtin.edu.au

A. Dewan and R. Corner (eds.), *Dhaka Megacity: Geospatial Perspectives on Urbanisation, Environment and Health*, Springer Geography,
DOI 10.1007/978-94-007-6735-5_18,

18.1 Introduction

Diarrhoea is one of the principal causes of mortality and morbidity in developing countries. Many diarrhoeal diseases show clear seasonality, suggesting that weather factors could play a role, either directly or indirectly through other intermediate pathways. Clarifying the potential effect of weather on the transmission of diarrhoeal diseases could give a deeper insight into the mechanisms of seasonality and therefore has the potential to improve disease control. Quantitative estimates of weather effects on diarrhoeal transmission, which can be applied to estimate population-attributable risk of climate exposure, could also contribute to assess the attributable burden of climate change. However, this has been little studied, probably because it is difficult to isolate the contribution of other seasonally varying factors.

Many different bacterial, viral and parasitic agents have been associated with diarrhoea. The main burden of diarrhoeal diseases falls on children; therefore, most studies have focused on this age group. A review of 24 community-based studies from 13 developing countries showed that the most common cause of diarrhoeal episodes was enterotoxigenic *Escherichia coli* (ETEC), with a median of 14 %, followed by *Giardia lamblia* (10.5 %), rotavirus (6 %) and *Campylobacter* spp. (6 %) (Black 2001). A review of 73 studies in health facilities, in which patients were frequently seeking care because of more severe symptoms, showed that the most common cause of diarrhoeal episodes was rotavirus with a median of 20 %, followed by ETEC (11 %) and *Campylobacter* spp. (7 %). The review found that >30 % of cases of diarrhoea were caused by bacteria.

Aeromonas spp. are ubiquitous organisms found in aquatic environments, and there is convincing evidence for a role of a waterborne route for diarrhoea, although the pathogenicity and epidemiology remain uncertain. These bacteria are members of the family Vibrionaceae, to which *Vibrio cholerae* also belongs. The numbers of patients with cholera and *Aeromonas*-positive diarrhoea show similar bimodal seasonality in Dhaka, Bangladesh (Hashizume et al. 2008; Pascual et al. 2002). Considering the association between these two pathogens, our previous finding that the number of cholera cases increased with high rainfall (Hashizume et al. 2008) led us to investigate the potential role of rainfall on the transmission of *Aeromonas*-positive diarrhoea. This study quantifies the impact of rainfall on the number of cases of *Aeromonas*-positive diarrhoea, to gain a deeper insight into the mechanisms of the seasonality of the disease. To gain some insight into possible causal pathways linking rainfall to the occurrence of *Aeromonas*-positive diarrhoea, the association between river levels and diarrhoea incidence was also investigated. Population factors potentially affecting vulnerability to the effect of rainfall on the incidence of *Aeromonas*-positive diarrhoea were also investigated.

18.2 Methods

18.2.1 Data

The primary outcome variable for this study was the weekly number of *Aeromonas*-positive stools from patients with diarrhoea who visited the International Centre for Diarrhoeal Disease Research, Bangladesh (ICDDR,B) hospital in Dhaka. The ICDDR,B hospital serves an urban population of approximately ten million and provides free treatment for more than 100,000 cases of diarrhoea each year. Every 50th patient was enrolled in the surveillance system. For each diarrhoea patient at the ICDDR,B hospital, stool samples were examined microbiologically to identify common enteric pathogens including *Aeromonas*. The details of the laboratory procedures have been described in a previous study (Stoll et al. 1982). We retrieved information on the date of hospital visit and the pathogens identified from each stool specimen during a 5-year period (January 1996–December 2000). Information on age, sex, socioeconomic status (educational level and roof structure of the house) and hygiene and sanitation practices (drinking water source, distance to the water source and type of toilet) for each patient was also compiled to investigate potential vulnerable groups. These were selected as potential markers of vulnerability because socioeconomic status (Ali et al. 2002; Emch 1999) and hygiene and sanitation practices (Hughes et al. 1982; Sommer and Woodward 1972) have been identified as important determinants of diarrhoea in Bangladesh. Patients were classified as *Aeromonas* positive when *Aeromonas* spp. were identified from the stool specimens, regardless of the presence of other pathogens. From these data, the weekly *Aeromonas*-positive cases were counted and used for analysis.

We obtained data on daily rainfall and maximum and minimum temperatures in Dhaka from the Bangladesh Meteorological Department (BMD). The daily level of Buriganga River in Dhaka was recorded by the Bangladesh Water Development Board (BWDB). The weekly mean temperature, river level and the total weekly rainfall were calculated from the daily records.

18.2.2 Statistical Analysis

We examined the relationship between the number of the *Aeromonas*-positive cases per week and rainfall using generalised linear Poisson regression models allowing for overdispersion (McCullagh and Nelder 1989). To account for seasonality in the incidence of *Aeromonas* positivity that was not directly due to the weather, Fourier terms up to the sixth harmonic, for which the time span was approximately 2 months, were included in the model. Indicator variables for the years of the study were incorporated into the model to allow for long-term trends and other variations between years. To allow for autocorrelation, an autoregressive term at order one was incorporated into the models (Brumback et al. 2000). From exploratory analyses, we considered lags (delays in effect) of up to 16 weeks for rainfall. In our initial analyses designed to identify the broad shape of any association, we fitted a natural cubic spline (3 df) to the average rainfall over lags of 0–16 weeks. We also included temperature as a natural cubic spline (3 df) in the model to control confounding, with lags of 0–4 weeks. These analyses suggested a linear positive association between the number of cases and rainfall; we therefore fitted a linear model. An increase in the number of *Aeromonas*-positive cases that were associated with a 10-mm increase in a given measure of rainfall, estimated from the regression model, was reported as a percentage change. Using the simple linear model, we then examined lag effects in more detail, by fitting linear unconstrained distributed lag models, which were comprised of terms for rainfall at each lag up to the previous 16 weeks.

For the models for river level, we considered lags of up to 4 weeks. Temperature terms with natural cubic splines (3 df) were included in all models to control confounding of temperature over lags of 0–4 weeks. We also examined the relationship with temperature adjusting for potential confounding by rainfall using the same methods of analysis.

The same analyses were conducted for *Aeromonas*-positive diarrhoea excluding the cases coinfected with *V. cholerae* because its seasonality was similar to that of *Aeromonas*-positive diarrhoea. Sensitivity of estimates to the degree of seasonal control (3 and 12 harmonics) was also examined.

18.3 Results

The distribution of diarrhoeal pathogens by age group is shown in Table 18.1. There were 1,222 hospital visits (in the 2 % sample) at which *Aeromonas* was isolated from stools from 1996 to 2000; 30 % of patients were <1 year old and 31 % were aged 1–14 years. The mean weekly number of hospital visits was 4.7 (SD 2.9). Seasonal variation in the number of cases showed bimodal seasonality, peaking before and at the end of the monsoon (Fig. 18.1), with a trough in the middle.

The relationships between rainfall and the number of cases of *Aeromonas*-positive diarrhoea, after adjusting for season, between-year variations and temperature, are shown in Fig. 18.2. The risk–response curves for rainfall at lags 0–8, 9–16 and 0–16 weeks showed a positive slope with high rainfall, and it seemed to be a linear relationship, especially for rainfall at lag 0–16 weeks. No 'low rainfall' effect was observed at any lags. There was strong evidence for a linear association at all lags. For a 10-mm increase in average rainfall in the previous 16 weeks, the number of *Aeromonas*-positive cases increased by 20.7 % [95 % confidence interval (CI), 10.6–31.6] by using a model that assumed a log-linear increase in risk. The detailed analysis of lag structure showed that the rainfall effect was not observed in the index week (week 0), while the positive effect was observed afterwards up to 16 weeks (Fig. 18.3).

Table 18.1 Distribution of diarrhoea by pathogens by age in ICDDR,B Dhaka Hospital, 1996–2000

Pathogens[a]	<1 year		1–14 years		15–29 years		>30 years		Total	
Cholera	278	(5.4)	1,298	(24.6)	778	(34.8)	718	(29.6)	3,072	(20.4)
Rotavirus	1,702	(33.2)	942	(17.8)	73	(3.3)	98	(4.0)	2,815	(18.7)
Shigella	167	(3.3)	321	(6.1)	146	(6.5)	137	(5.6)	771	(5.1)
Salmonella	62	(1.2)	60	(1.1)	32	(1.4)	45	(1.9)	199	(1.3)
Campylobacter	474	(9.2)	452	(8.6)	177	(7.9)	130	(5.4)	1,233	(8.2)
E. coli	1,194	(23.3)	812	(15.4)	257	(11.5)	302	(12.4)	2,565	(17.0)
Aeromonas	366	(7.1)	377	(7.1)	202	(9.0)	277	(11.4)	1,222	(8.1)
Other diarrhoea[b]	889	(17.3)	1,020	(19.3)	572	(25.6)	722	(29.7)	3,203	(21.2)
Total	5,132	(100.0)	5,282	(100.0)	2,237	(100.0)	2,429	(100.0)	15,080	(100.0)

[a]A patient was diagnosed with cholera, rotavirus, *Shigella, Salmonella, Campylobacter, E. coli* and *Aeromonas* when the respective pathogen was identified from the stool specimen

[b]Patients from whom pathogens other than *V. cholerae*, rotavirus, *Shigella, Salmonella, Campylobacter, E. coli* or *Aeromonas* were identified, or none of pathogens were identified

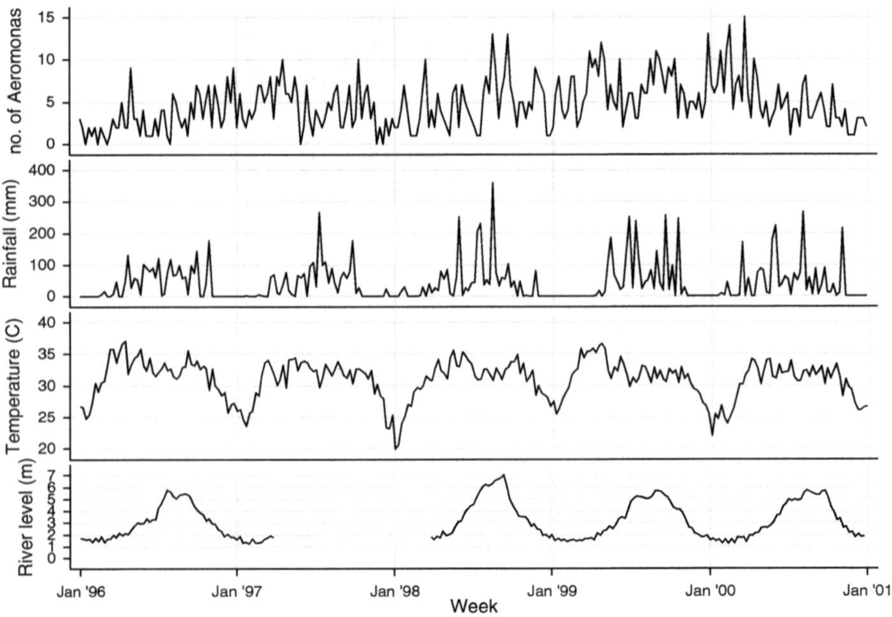

Fig. 18.1 Seasonal variations in the number of cases of *Aeromonas*-positive diarrhoea per week and meteorological and river level data in Dhaka, 1996–2000. The river level data are missing from April 1997 to March 1998

The higher point estimates of the risk in those living in houses with a concrete roof, drinking tap water, using water sources at close distance and using sanitary toilet facilities were observed, but the risks could be easily explained by chance (Table 18.2).

There was no clear relationship between the number of cases and river level and temperature.

Repeating the analyses excluding cases coinfected with *V. cholerae* left patterns of the effects of rainfall largely unchanged. When in sensitivity analyses the degree of seasonal control was halved (3 harmonics) or doubled (12 harmonics), the estimates of the effect of rainfall changed little.

18.4 Discussion

This study shows that there was a strong positive and linear (i.e. increasing over the entire rainfall range) association between hospital visits due to *Aeromonas*-positive diarrhoea and high rainfall in Dhaka.

This study showed that the number of cases of *Aeromonas*-positive diarrhoea increased with high rainfall in previous weeks after controlling for potential confounding by seasonal and between-year differences and temperature. This

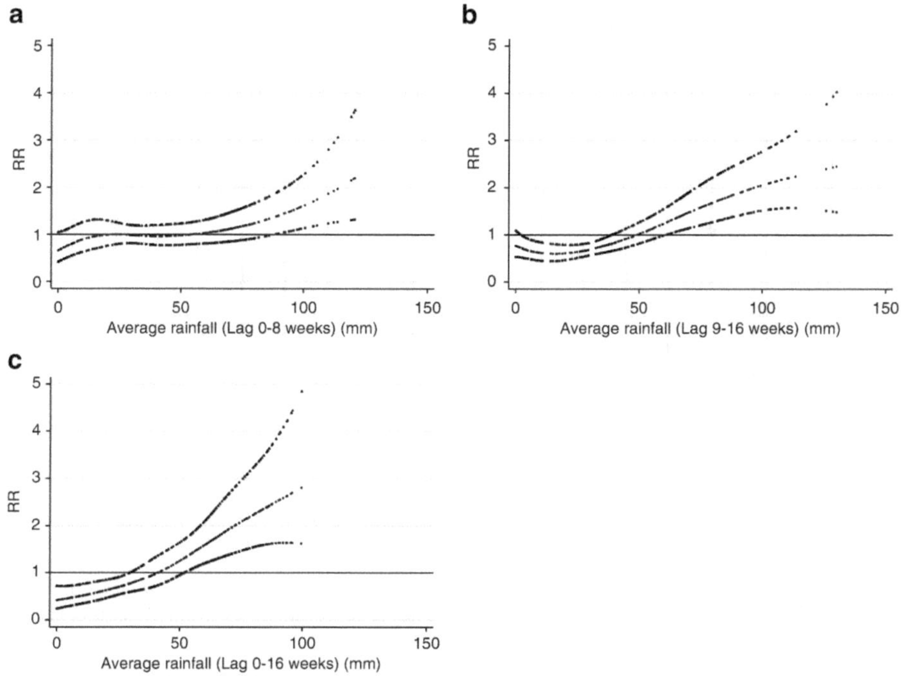

Fig. 18.2 Relationship between the number of cases of *Aeromonas*-positive diarrhoea and average rainfall over lags of (**a**) 0–8 weeks, (**b**) 9–16 weeks and (**c**) 0–16 weeks (shown as a 3 df natural cubic spline) adjusted for seasonal variation, between-year variations and temperature. *RR* represents the relative risk of *Aeromonas*-positive diarrhoea (scaled against the mean weekly number of cases). The *centre line* shows the estimated spline curve, and the *upper* and *lower lines* represent the 95 % CIs

means that high rainfall can explain the departure of the number of cases from the usual seasonal pattern, but it cannot explain the usual seasonal pattern itself. These findings are broadly in accordance with the limited number of previous studies quantifying the relationship between rainfall and pathogen-non-specific diarrhoea, although most of the studies were conducted in developed countries or Pacific Islands where the climate and geography are different from those in Bangladesh. Some mechanisms of the relationship between rainfall and diarrhoea might work in such different circumstances, but careful interpretation is needed when the relevance of the results of the current study is discussed with reference to previous studies.

A time-series analysis in Fiji reported that monthly diarrhoea incidence in infants increased by 2 % (95 % CI, 1.5–2.3) for each 1 kg/m^2/min increase in rainfall above the median value, after allowing for the effects of long-term trends and seasonal patterns (Singh et al. 2001). In that study, high rainfall was associated with significant increases in diarrhoea in the same month but decreased in the following month. The authors hypothesised that the initial effect of high rainfall is

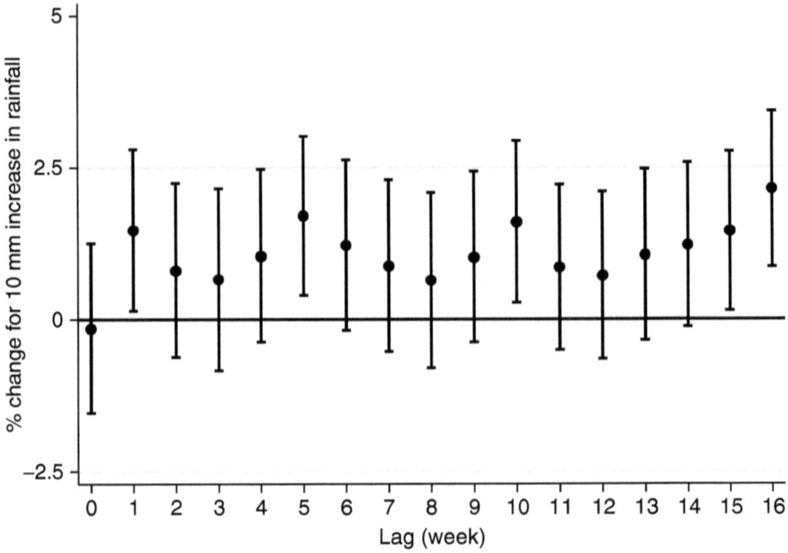

Fig. 18.3 Percentage change (and 95 % CIs) in the number of cases of *Aeromonas*-positive diarrhoea for rainfall (per 10-mm increase) at each lag (unconstrained distributed lag models)

to flush faecal contaminants from pastures and dwellings into water supplies, but continued rain can lead to subsequent improvement in water quality. In the present study, however, no protective effects of high rainfall were observed in any lag period by detailed analysis of lag structure.

However, this previous study in Fiji had some methodological differences from the present investigation. The case definition of diarrhoea was not obvious and the unit of analysis was monthly rather than weekly. Furthermore, the causative agents of diarrhoea in Fiji are likely to have been different from those in Bangladesh (Watson 2001; Bennish and Wojtyniak 1991).

The current study is also broadly in accordance with a US study reporting that approximately half of waterborne disease outbreaks were preceded by heavy rainfall within a 2-month lag (Curriero et al. 2001). According to the US study, outbreaks due to surface water contamination were most significant for extreme rainfall in the same month as the outbreak, while outbreaks due to groundwater contamination showed highest significance for extreme rainfall 2 months prior to the outbreak. A positive association with high rainfall and lag within 2 months to the occurrence of diarrhoea were also observed in the current study, although there were many differences in study design. Firstly, the outcome of the previous study was waterborne disease, and the events are not all necessarily a disease causing diarrhoea. Secondly, the previous study analysed only outbreaks rather than incidence of overall waterborne diseases, while the current study included the outbreak as well as sporadic cases. The pathways of infection and therefore the size of effect and lag time between heavy rainfall events and outbreaks may have differed from those between continuous measurements of high rainfall and numbers of diarrhoea patients. Finally, causative agents and general hygiene and sanitation practices in

Table 18.2 Percentage change (and 95 % CIs) in weekly number of *Aeromonas*-positive diarrhoea for 10-mm increase in rainfall at lags of 0–16 weeks (1996–2000: ICDDR,B Dhaka Hospital)

			High rainfall[a]		
	Mean[b]	SD	% change	95 % CI	*p* value[c]
Total morbidity	4.7	(2.9)	20.7	(10.6, 31.6)	
Characteristics					
Sex					
Female	2.0	(1.6)	20.9	(6.6, 37.1)	0.986
Male	2.7	(2.0)	21.0	(8.9, 34.4)	
Age (years)					
<15	2.9	(2.1)	20.4	(8.6, 33.5)	0.677
15–29	0.8	(0.9)	13.7	(−3.5, 33.9)	
≥30	1.1	(1.1)	26.8	(5.9, 51.7)	
Socio-economic status					
Education					
No education	2.3	(1.8)	16.8	(4.2, 31.0)	0.625
Informal or <6 years	0.8	(0.9)	16.6	(−4.5, 42.3)	
≥6 years	1.6	(1.4)	27.1	(10.6, 46.0)	
Roof structure					
Non concrete	3.6	(2.5)	20.3	(8.9, 32.7)	0.464
Concrete	1.0	(1.0)	29.4	(9.1, 53.5)	
Hygiene and sanitation					
Drinking water source					
Tube well	1.9	(1.6)	17.1	(3.0, 33.3)	0.555
Tap water	2.8	(2.1)	23.1	(11.1, 36.4)	
Distance to water source					
More than 5 m	3.0	(2.2)	18.1	(6.3, 31.1)	0.477
5 m or less	1.7	(1.5)	25.8	(9.4, 44.6)	
Type of toilet					
Non sanitary	2.0	(1.6)	20.7	(7.2, 35.8)	0.991
Sanitary	2.7	(1.9)	20.8	(8.9, 34.0)	

[a]A linear model was used to quantify the effect
[b]Mean weekly count of cases due to *Aeromonas* ($n = 260$ weeks)
[c]*p* value for test for heterogeneity between subgroups

the US population are likely to be different from those in Bangladesh. Nevertheless, the lag between rainfall exposure and occurrence of waterborne disease observed in the previous study (within 2 months) was broadly similar to that in the current study.

In communities where there are no water treatment and distribution systems, as in many rural areas in developing countries, an association between rainfall and contamination of well water and groundwater has been reported. For example, water samples from boreholes are consistently positive for coliform bacteria after heavy rainfall in Lagos, Nigeria (Egwari and Aboaba 2002). Shallow wells are more contaminated than deep wells and boreholes. Rainfall is also associated with the presence of *Pseudomonas aeruginosa*, *Aeromonas hydrophila* and coliform bacteria in underground water (Stukel et al. 1990).

Aeromonas spp. are ubiquitous organisms found in aquatic environments, and there is convincing evidence for a role of a waterborne route for diarrhoea, although the pathogenicity and epidemiology remain uncertain (Hunter 1997). These bacteria are members of the family *Vibrionaceae*, to which *V. cholerae* also belongs, which means that *Aeromonas* spp. and *V. cholerae* are taxonomically similar. The number of cases of cholera and *Aeromonas*-positive diarrhoea showed similar bimodal seasonality in Bangladesh: the first peak being in March to May (hot humid summer) and the second in September (during the monsoon) (Hashizume et al. 2008). If this relationship is considered to fit the Bradford-Hill criteria of causality, the finding that the number of cholera cases also increased with higher rainfall may help to evaluate whether the observed association of high rainfall with these pathogens is likely to be causal through a waterborne route. However, the positive relationship of *Aeromonas* with river level, proposed as a potential intermediate of the association between rainfall and diarrhoea, was not observed, unlike for cholera. Further research on the association of concentration of *Aeromonas* spp. in the aquatic environment or drinking water source in relation to rainfall and river level may provide more direct evidence for the pathways.

18.5 Limitations

Time-series regression studies measure the associations over time between an outcome indicator (diarrhoea) and an exposure (rainfall). The underlying hypothesis is that the fluctuations of the health outcome over time are associated with the fluctuations of the exposure indicator over time, after considering other potential confounders that also change over time. The covariates that vary between people but not over time are not potential confounders in time-series studies. Therefore, time-fixed covariates, such as sex, socioeconomic status or any other factors whose distribution does not vary over the short- to midterm in the study population, do not confound the short- to midterm association between rainfall and diarrhoea.

In the rainfall study, confounding refers to a situation in which an observed relative risk of diarrhoea apparently due to exposure to rainfall is in part due to a third factor that is associated with rainfall and independently affects the risk of diarrhoea. Two main groups of variables fall into this category: variables that have a causal effect on diarrhoea and that vary in time in a similar manner to rainfall (e.g. temperature) and those that are not causally related to diarrhoea, but are indicators for unmeasured causal factors that change slowly or seasonally (e.g. seasonal and between-year differences). These potential confounders were taken into account in the present study. The temperature was controlled by incorporating linear and smoothed terms into the models. Seasonal and between-year differences were in particular considered by incorporating Fourier terms (up to six harmonics) and indicator variables for the years, respectively, into the models. Nevertheless, residual confounding could have been an issue in the study.

Selection bias occurs when there is a difference in the characteristics related to the exposure or outcome of interest between people who are selected for a study and those who are not. In the present study, counts of hospital visits were used to assess the potential effects of exposure to rainfall (or temperature) on incidence of diarrhoea, and selection of cases (under-ascertainment) would have caused bias only if associated with rainfall or temperature.

The ICDDR,B hospital maintains a surveillance system in which data from every 50th patient presenting to the hospital are collected. The main advantage of these data is the resulting extended coverage of the population in Dhaka. Originally established as the Cholera Research Laboratory Hospital in 1962, the ICDDR,B hospital is well known among people in Dhaka as a hospital for diarrhoea. The hospital now serves an urban population of approximately ten million individuals (BBS 2003) and provides free treatment for more than 100,000 cases of diarrhoea each year. Therefore, the data used in this study were, to some extent, representative of underlying diarrhoeal morbidity of the population in Dhaka where there are constraints on surveillance systems at the population level.

However, there may be concerns for possible limitations in the capacity of the hospital to receive the patients, and it could be a problem especially during epidemics of diarrhoeal diseases. This could influence the number of cases as well as the timing of hospital visits. However, in principle, the hospital accepts all patients visiting the hospital and never refuses patients on the grounds of capacity. Most individuals presenting with diarrhoea to the hospital are treated in the outpatient treatment pavilion and released within 24 h of presentation; therefore, the number of beds in the inpatient ward is not a critical problem. Thus, the capacity of the hospital should not have been a threat to this study.

Although treatment was free, there was possible under-representation of poorer patients because of constraints of access to the hospital due to, for example, a limited transportation budget for a trip to the hospital or higher costs of time for leaving work to go to the hospital. However, as described before, under-ascertainment will cause bias only if associated with rainfall or temperature. If poorer individuals are more likely to be affected by rainfall or temperature, it could introduce some degrees of selection bias, but there is no reason to believe that this was the case in the present study.

18.6 Conclusions

The presence of *Aeromonas* in the stools of patients with diarrhoea showed a strong positive and linear association with rainfall. Rainfall leading to flooding is thought to overwhelm water and sanitation systems, resulting in intake of contaminated drinking water. However, the positive relationship of *Aeromonas* with river level, proposed as a potential intermediate of the association between rainfall and diarrhoea, was not observed, unlike for cholera. Analysis of the change in drinking water quality in relation to rainfall may provide more direct evidence for the pathways, although that information was not available for this study.

Acknowledgements This study was supported by the Global COE Program, Nagasaki University, Japan. ICDDR,B acknowledges with gratitude the commitments of all donors to the Centre's research efforts.

References

Ali M, Emch M, Donnay JP, Yunus M, Sack RB (2002) The spatial epidemiology of cholera in an endemic area of Bangladesh. Soc Sci Med 55(6):1015–1024

Bangladesh Bureau of Statistics (BBS) (2003) National population census 2001, Preliminary report. Bangladesh Bureau of Statistics, Ministry of Planning, Dhaka

Bennish ML, Wojtyniak BJ (1991) Mortality due to shigellosis: community and hospital data. Rev Infect Dis 13(Suppl 4):S245–S251

Black RE (2001) Diarrheal diseases. In: Nelson KE, Williams CM, Graham NMH (eds) Infectious disease epidemiology: theory and practice. Aspen, Gaithersburg, pp 497–517

Brumback B, Burge HA, Neas LM, Ryan LM, Schwartz JD, Stark PC (2000) Transitional regression models, with application to environmental time series. J Am Stat Assoc 95(449):16–27

Curriero FC, Patz JA, Rose JB, Lele S (2001) The association between extreme precipitation and waterborne disease outbreaks in the United States, 1948–1994. Am J Public Health 91(8):1194–1199

Egwari L, Aboaba OO (2002) Environmental impact on the bacteriological quality of domestic water supplies in Lagos, Nigeria. Rev Saude Publica 36(4):513–520

Emch M (1999) Diarrheal disease risk in Matlab, Bangladesh. Soc Sci Med 49(4):519–530

Hashizume M, Armstrong B, Hajat S, Wagatsuma Y, Faruque AS, Hayashi T, Sack DA (2008) The effect of rainfall on the incidence of cholera in Bangladesh. Epidemiology 19(1):103–110

Hughes JM, Boyce JM, Levine RJ, Khan M, Aziz KM, Huq MI, Curlin GT (1982) Epidemiology of eltor cholera in rural Bangladesh: importance of surface water in transmission. Bull World Health Organ 60(3):395–404

Hunter PR (1997) Waterborne disease: epidemiology and ecology. Wiley, Chichester

McCullagh P, Nelder J (1989) Generalized linear models. Chapman & Hall, London

Pascual M, Bouma MJ, Dobson AP (2002) Cholera and climate: revisiting the quantitative evidence. Microbe Infect 4(2):237–245

Singh RB, Hales S, de Wet N, Raj R, Hearnden M, Weinstein P (2001) The influence of climate variation and change on diarrheal disease in the Pacific Islands. Environ Health Perspect 109(2):155–159

Sommer A, Woodward WE (1972) The influence of protected water supplies on the spread of classical-Inaba and El Tor-Ogawa cholera in rural East Bengal. Lancet 2(7785):985–987

Stoll BJ, Glass RI, Huq MI, Khan MU, Holt JE, Banu H (1982) Surveillance of patients attending a diarrhoeal disease hospital in Bangladesh. Br Med J (Clin Res Ed) 285(6349):1185–1188

Stukel TA, Greenberg ER, Dain BJ, Reed FC, Jacobs NJ (1990) Longitudinal study of rainfall and coliform contamination in small community drinking water supplies. Environ Sci Technol 24(4):571–575

Watson C (2001) Death from multi-resistant shigellosis in Fiji Islands. Pac Health Dialog 8(1):99–102

Chapter 19
Modelling Spatiotemporal Patterns of Typhoid Cases Between 2005 and 2009 Using Spatial Statistics

Ashraf M. Dewan, Robert J. Corner, and Masahiro Hashizume

Abstract The objectives of this study were to analyse spatiotemporal patterns of reported typhoid cases between 2005 and 2009 and to model the spatial relationship of demographic and socioeconomic factors associated with the occurrence of typhoid in Dhaka. The lowest level census units were used as the scale of analysis. Data in relation to typhoid was collected from 11 major hospitals by scanning individual medical reports, while demographic and socioeconomic variables were encoded in GIS from the population censuses of 2001 and 2011. Global (Moran's *I*) and local models (G_i*) were used to test how census districts were dispersed or clustered over space. The spatial relationships were modelled through ordinary least square (OLS) and geographically weighted regression (GWR) techniques. Spatial pattern analysis as measured by Moran's *I* demonstrated that the distribution of the affected communities with typhoid was spatially autocorrelated across the study period, 2005–2009. Hotspot analysis using local G_i* indicated large variation in the locations and sizes of clusters. The demographic model outperformed the socioeconomic and demographic + socioeconomic models in predicting the occurrence of typhoid in the study area. The results of this study are of great aid to identify spatial risk factors, essential to develop the control and prevention measures to specific areas.

Keywords Typhoid • Water-borne disease • Spatial regression • Space-time • Spatial patterns • Hotspot

---

A.M. Dewan (✉) • R.J. Corner
Department of Spatial Sciences, Curtin University, Kent Street, Bentley, Perth 6102, WA, Australia
e-mail: A.Dewan@curtin.edu.au; R.Corner@curtin.edu.au

M. Hashizume
Institute of Tropical Medicine (NEKKEN), Nagasaki University, 1-12-4, Sakamoto, Nagasaki 852-8523, Japan
e-mail: hashizum@nagasaki-u.ac.jp

A. Dewan and R. Corner (eds.), *Dhaka Megacity: Geospatial Perspectives on Urbanisation, Environment and Health*, Springer Geography,
DOI 10.1007/978-94-007-6735-5_19,

19.1 Introduction

Typhoid infection, caused by *Salmonella enterica* serotype Typhi (*S. typhi*), remains a serious public health concern in many parts of the world (Ochiai et al. 2008; Bhan et al. 2005). Available records show that about 22 million new cases of typhoid disease occur annually around the world resulting in 200,000 deaths (Patel et al. 2010), and the incidence of typhoid is higher in certain age groups such as children and young adults (Karkey et al. 2010). The burden of disease is extremely high in developing countries, primarily due to inadequate provision of public health infrastructure such as safe water and sanitation (Nagashetty et al. 2010; Kanungo et al. 2008). However, the infection is occasionally found in developed countries due to importation via travellers and migrants (Patel et al. 2010). Previous studies have demonstrated that South Central and Southeast Asia exhibit the highest incidence of typhoid fever (>100 cases per 100,000 persons per year) with high fatality rates (Nagashetty et al. 2010; Parry 2005; Crump et al. 2004).

In common with other endemic settings in South Asia, Bangladesh has a high typhoid prevalence rate – an incidence of 2,000 per 100,000 per year (WHO 2003). Although detailed studies on typhoid in the country are unavailable, a number of studies conducted primarily on micro-level urban populations indicate that typhoid remains a serious health concern in Dhaka that is largely affecting a large number of middle- and lower-income groups (Rahman et al. 2011; Naheed et al. 2010; Ram et al. 2007; Brook et al. 2005; Saha et al. 2001; Butler et al. 1991; Roy et al. 1985; Stoll et al. 1983). As a result, the socioeconomic impact of typhoid is also very high (Punjabi 1998 cited in WHO 2003). The incidence rate is higher in the youngest population, and the major risk factors include contaminated water and food (see Ram et al. 2007) and poor quality of life (see Corner et al. 2013; Rahman et al. 2011). These studies also pointed out that the burden of typhoid infection could be higher than anticipated previously.

Use of geospatial techniques, such as Geographic Information Systems (GIS) and spatial statistics, is increasingly being recognised as a powerful tool for mapping spatial patterns of disease and for modelling causative factors (Maantay and McLafferty 2011; Rezaeian et al. 2007; Kazmi and Usery 2001). Common geospatial techniques in health research include disease mapping, detection of clustering, identification of risk factors and regression analysis (Gesler 1986). Currently, spatiotemporal pattern analysis of disease has gained renewed popularity with the rapid advancement of spatial statistics as it allows quantitative models for the prediction of risk areas to be developed (Unwin 1996), which is essential to minimise the effects of disease on humans (Khormi and Kumar 2011; Ali et al. 2002). A number of methods have now become available to detect clustering of disease, particularly for data encoded in areal units. These include join-count statistics (Upton and Fingleton 1985), global Moran's *I* (Pfeiffer et al. 2008), local indicators of spatial autocorrelation (LISA) (Lai et al. 2009; Anselin 1995) and local and global G_i* (Getis and Ord 1992). Numerous studies have utilised either of these techniques to identify spatial clustering across a range of health issues such as dengue (Hsueh et al. 2012; Castillo et al. 2011;

Lin and Wen 2011; Wen et al. 2010), malaria (Haque et al. 2012; Zhang et al. 2008), chlamydia (Owusu-Edusei and Owens 2009), Barmah Forest virus (Naish et al. 2011), La Crosse virus (Haddow and Odoi 2009), gonorrhoea (Yin et al. 2012), childhood cancer (Rainey et al. 2006), avian influenza (Loth et al. 2010), typhoid (Dewan et al. 2013; Hinman et al. 2006), scrub typhus infection (Kuo et al. 2011), cardiovascular disease (Bertazzon et al. 2010), traumatic brain injury (Colantonio et al. 2011) and other leading causes of deaths (Tsai et al. 2009). A review of these studies indicates that there are two broad lineages of spatiotemporal pattern analysis. They are: (1) identification of spatial patterns through global and local techniques and (2) the assessment of causal factors accountable for disease occurrence by using spatial and temporal statistics. Though spatial clustering analysis enables the mapping of "hot and cold" spots, modelling the factors influencing the occurrence of a disease is no less important in developing appropriate measures for effective health-care delivery.

The megacity of Dhaka is one of the fastest growing cities in the world, with growth driven primarily by rural–urban migration (Islam 2005). The rate of urban expansion is extremely high and is mostly being carried out in an unplanned manner which is causing severe environmental degradation (Dewan et al. 2012). Consequently, both waterborne and vector-borne diseases have become acute, posing as yet unknown threats to the inhabitants (BBS 2010). As rapid urbanisation is known to alter social and cultural practices of people (Kanungo et al. 2008), inadequate provision of public health infrastructures in the event of ill-structured urban expansion could put more people at risk of disease (Wu et al. 2009), which is already evident in Dhaka (BBS 2010). As mentioned earlier, case control studies in Dhaka assert that the burden of disease could be higher than commonly believed (see Naheed et al. 2010; Saha et al. 2001). However, there is no information on the spatiotemporal distribution of typhoid at a regional scale that could assist in planning preventive measures against typhoid fever. Considering this fact, this chapter aims to: (1) analyse spatiotemporal patterns of typhoid cases between 2005 and 2009 and (2) to model the socioeconomic factors influencing the occurrences of disease by using spatial statistical techniques.

19.2 Materials and Methods

19.2.1 Study Area and Data

The study area is a subset of the Dhaka Metropolitan Development Plan (DMDP) area that includes three municipalities, namely, Dhaka City Corporation (DCC), Savar Pourashava and Tongi Pourashava, and a number of other communities (see Chap. 1 for detailed description).

Water supply and provision of safe sanitation for the population in the study area are inadequate. While most of the inhabitants in the municipal areas have access to piped water, drinking and domestic water sources outside of these areas vary substantially – from ponds to rivers. Due to the flat topography, the area is subject

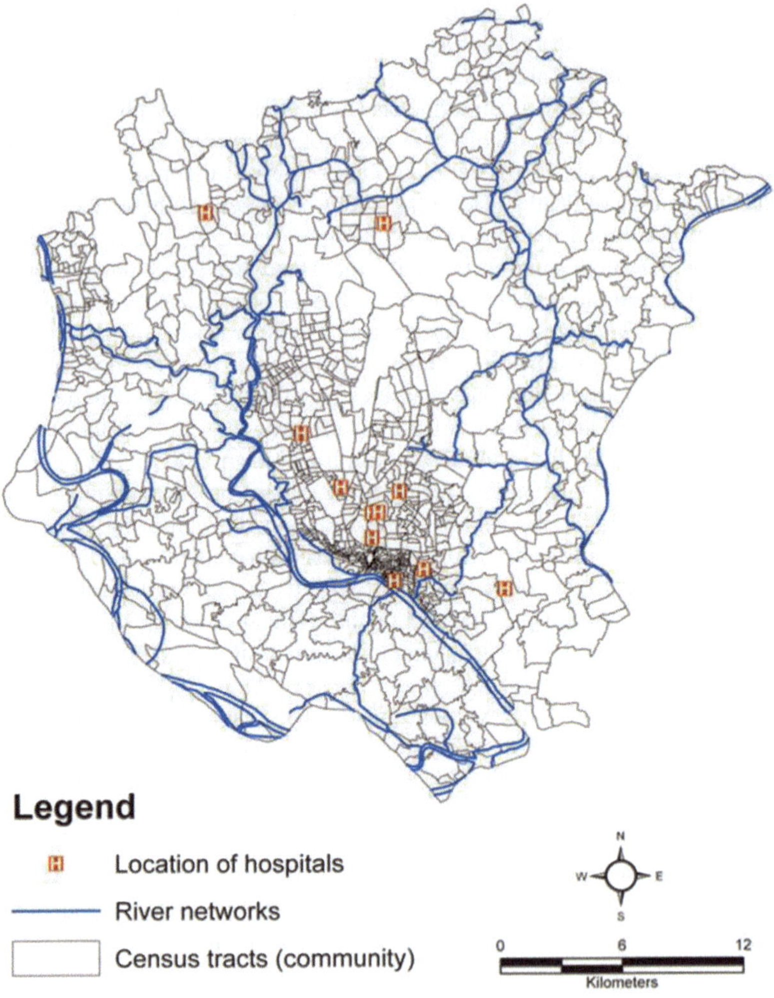

Fig. 19.1 Location of hospitals studied

to monsoonal flooding almost every year, which increases the risk of surface water contamination with major implications for waterborne and food-borne diseases such as typhoid (BBS 2010).

The study is based on passive surveillance typhoid data which was collected between April and December 2009 from 11 major health service providers (Fig. 19.1). A detailed description of the data collection methods for reported typhoid cases and the creation of the geographic data, including encoding procedures, case definition and logistics, are provided in Dewan et al. (2013). For this study, individual typhoid cases were aggregated by community for each year and then summed to obtain the total cases pertaining to each census tract in the study area throughout 2005–2009. The final database contains the case data for each tract for each of the individual years.

The population and socioeconomic data were obtained from Bangladesh Bureau of Statistics community series (BBS 2003, 2012) which represented both the 2001 and 2011 population censuses. Since these data were in tabular format, they were first encoded in a spreadsheet and then linked with the geographic unit by using a unique ID. Eleven demographic and socioeconomic variables were obtained from the decennial censuses. These were population counts, total literacy rate, percent of male literacy, percent of female literacy, water sources (including tap, tube well and others) and toilet facility (namely, water-sealed sanitary, nonwater-sealed sanitary, non-sanitary and no sanitation). Apart from these variables, the percent of population under the poverty line and mean housing value for each census tract were computed from various sources. The procedures for deriving these two variables can be found elsewhere (Dewan 2013). It should be noted that mean housing value was used as a surrogate parameter due to the absence of household income data for each community and that it was log-transformed because of high skewness. Population density was calculated as total population in each census tract divided by its area. To determine the relationship between number of reported typhoid cases in each individual year and the demographic variables, population counts were estimated for each year assuming a uniform rate of population growth during the decade in each census tract. The following equation was used to estimate the population for the years 2005–2009:

$$P_{\mathrm{i}} = P_{2001}(1 + x)^n \tag{19.1}$$

where P_{i} is population for year of interest, P_{2001} is the population of 2001, x is the annual growth rate between 1991 and 2001 and n is the number of years between 2001 and the year of interest.

For the aggregated data covering the entire period 2005–2009 typhoid cases, the 2011 population census was used with an assumption that the geographic variation in the covariates did not change significantly between years. However, the study did not attempt to interpolate other socioeconomic variables. Based on an extensive literature review, a total of 14 variables were identified that were categorised as either demographic, socioeconomic or demographic + socioeconomic (Table 19.1) in order to model the spatial relationships between typhoid infection and causal factors.

19.2.2 Analysing Spatial Patterns

To understand the degree of spatial clustering among neighbouring census tracts, the global spatial autocorrelation statistic Moran's I was used for the five different annual typhoid datasets and the combined 2005–2009 data. This was to measure

Table 19.1 Explanatory variables used in the study

Variable	Category
Total population	Demographic
Population density	Demographic
Total literacy rate (%)	Socioeconomic
Male literacy rate (%)	Socioeconomic
Female literacy rate (%)	Socioeconomic
Mean housing value ('000 BDT[a])	Socioeconomic
Percentage of households with piped water	Socioeconomic
Percentage of households with tube well	Socioeconomic
Percentage of households with other water sources (e.g. ponds, rivers)	Socioeconomic
Percentage of households with water-sealed sanitation	Socioeconomic
Percentage of households with nonwater-sealed sanitation	Socioeconomic
Percentage of households with non-sanitation	Socioeconomic
Percentage of households with no sanitation facility	Socioeconomic
Percentage of households under poverty line	Socioeconomic
All variables	Demographic + socioeconomic

[a]*BDT* Bangladeshi currency (TAKA)

and test the way in which cases in census districts were dispersed or clustered over space. Due to the irregular shape and size of the census tracts, conceptualisation of the spatial adjacency relationship was based on the first-order queen's case contiguity rule with Euclidian distance. The values of Moran's I range from +1 (extremely positive autocorrelation) to −1 (extremely negative relationship), and a value close to 0 indicates complete spatial randomness (Wong and Lee 2005). Global Moran's I can be expressed as

$$I = \frac{n\sum_{i=1}^{n}\sum_{j=1}^{n} w_{ij}(x_i - \bar{x})(x_j - \bar{x})}{\left(\sum_{i\neq j}\sum w_{ij}\right)\left(\sum_{i=1}^{n}(x_i - \bar{x})^2\right)} \tag{19.2}$$

where n is the total number of census tracts in the study, i and j represented different census units, x_i is the case incidence in unit i, and $\bar{x}$ is the mean case incidence. w_{ij} is a measure of the spatial proximity of pairs i and j.

19.2.3 Hotspot Analysis

After analysing the spatial pattern of reported typhoid cases in the study area, the local G_i* (Ord and Getis 1995) statistic was used to assess both the statistical significance of clusters and their spatial extent. The local G_i* statistic compares the local mean infection rate with the global mean rate and produces two statistics – *z-score* and *p-value* for each census tract – indicating the location of statistically significant high or low values. A positive *z-score* indicates hotspots with high rates and surrounded by high values. By contrast, a statistically significant negative *z-score* reflects clustering of low rates or "cold spots" (ESRI 2012; Getis and Ord 1992). To determine hot and cold spots, total typhoid cases for the six different periods were used, and the first-order queen's case polygon contiguity was used to conceptualise the spatial adjacency relationships among features.

19.2.4 Modelling Spatial Relationships

Since both global and local measures indicated high clustering of typhoid cases in the study area, the next step was to model the relationship of disease occurrences at the local level with the demographic and socioeconomic factors listed in Table 19.1. This was accomplished by using both global and local regression techniques.

Ordinary least squares (OLS) regression, a global method, was first used to model the spatial relationship. This is particularly useful to determine whether the explanatory variable set is free from multicollinearity that the coefficients are statistically significant and that the residuals are not spatially correlated (ESRI 2011; Poole and O'Farrell 1971). However, OLS regression examines variables globally and can give misleading results when applied to phenomena that vary over space (Clement et al. 2009).

Reported typhoid case numbers were used as the dependent variable, and the demographic and socioeconomic parameters were used as explanatory variables for each year and for the aggregated data of 2005–2009. The results were then checked to see the suitability of individual variables to the model fit. Multicollinearity was assessed through the variance inflation factor (VIF) values, and variables having a VIF of >7.5 were removed from the regression analysis (ESRI 2011). Similarly, stationarity was evaluated using the Koenkar's studentised Breusch-Pagan statistic which indicated statistical heteroscedasticity and nonstationarity in the dataset with a confidence level of $p < 0.05$. Further, the residuals from the OLS model were tested for spatial autocorrelation using Moran's *I* statistic. The results of this test suggested the presence of spatial dependency in the data, and therefore, a local model was required to investigate the spatial relationship at the local level.

Geographically weighted regression (GWR) was used to examine the spatial dependency and highlight local variability. GWR is a local model and is effective in determining the underlying local factors for particular spatial patterns (Fotheringham et al. 2001). It performs regressions on local subsets of features within a user-defined bandwidth. As the spatial configuration of features being analysed was inhomogeneous, an adaptive kernel, whose bandwidth was set to minimise the Akaike information criterion (AICc) (ESRI 2010), was adopted for the regression analysis. In order to understand the model fit and compare the outcomes of both global and local regressions, the adjusted coefficient of determination (Adjusted R^2) and the AICc were used (Fotheringham et al. 2002). Further, local collinearity, independency and normality of residuals of GWR were evaluated by inspection of the condition number of the design matrix of the regression.

As discussed above, OLS results were assessed using VIF and other statistics. Explanatory variables that presented high VIF value were excluded. A total of eight explanatory variables were finally used in the GWR model. These variables were population count, population density, total literacy rate, non-sanitary, no sanitary, other water sources, poverty and mean household value. They were categorised as demographic (total population and density), socioeconomic (total literacy rate, non-sanitary, no sanitation, other water sources, poverty and mean household value) and the demographic + socioeconomic category that included all eight of the variables. These three groupings were used as explanatory variables in three separate models in which total cases of typhoid were the dependent variable. The GWR model was tested for individual years as well as for the aggregated data.

19.3 Results

The result of global autocorrelation analysis using Moran's I is shown in Table 19.2. This demonstrates that the distribution of communities affected by typhoid was spatially autocorrelated for all years (2005–2009), implying that the infection is neither uniformly nor randomly distributed over the study area. Moran's I statistic was highest (0.457) in 2008 followed by 2009 ($I = 0.343$). Aggregated data on the total cases of typhoid from 2005 to 2009 also showed significant clustering of the disease in Dhaka ($I = 0.392$).

The spatial extent of hot and cold spots of typhoid infection as obtained from the local G_i* is shown in Fig. 19.2. This shows a large variation in the locations and sizes of clusters among the years studied which is particularly useful when examined spatially. The locations of the two highest categories in the figure were indications of statistically significant clusters (>1.96 std. dev) – that is, high values surrounded by high. Close examination of these maps clearly shows a particular spatial pattern of clustering between 2005 and 2009. For instance, typhoid clusters were located in the southern, central and eastern part of the study area in 2005. However, an apparent shift in typhoid hotspots was observed in 2007 and 2009

Table 19.2 Global autocorrelation analyses of typhoid data (2005–2009)

Year	Moran's I	Z-score	Pattern
2005	0.297	17.924	Clustered
2006	0.198	12.767	Clustered
2007	0.297	18.047	Clustered
2008	0.457	27.611	Clustered
2009	0.343	20.287	Clustered
2005–2009	0.392	24.119	Clustered

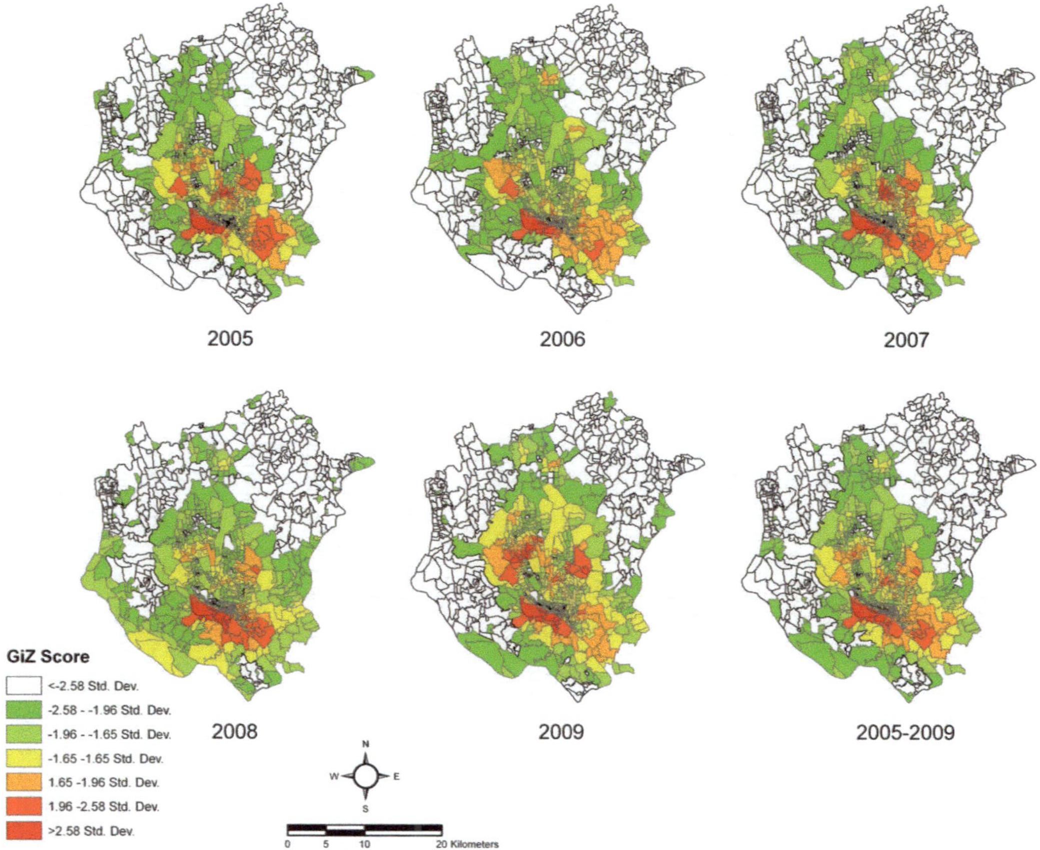

Fig. 19.2 Spatial clusters (*hotspots*) of typhoid cases in Dhaka (2005–2009). Hot and cold spots are measured by local G_i* for each year as well as for the entire period

when new clusters extended to the central and west-central locations of the city. An interesting pattern of clustering was that specific semi-urban communities such as Keraniganj, Zinzira and Kaliganj along the river Buriganga exhibited persistent hotspots since 2005 (Fig. 19.2).

In this study, we compared the AICc and adjusted R^2 values from GWR with those from OLS to investigate the strength of the relationship between typhoid occurrence and demographic and socioeconomic factors over different parts in Dhaka. Fourteen explanatory variables, categorised into three broad headings, were tested separately and used to assess the spatial dependency.

Table 19.3a, b, c showed the OLS results for three explanatory categories. It reveals that the demographic variables were free from multicollinearity as indicated by low VIF values. By contrast, the collinearity among the socioeconomic variables was high, particularly for male literacy rate, female literacy rate, piped and tube well water sources, water-sealed sanitation and nonwater-sealed sanitation (Table 19.3b). Further, the OLS result of the combined demographic + socioeconomic parameters was also not free from multicollinearity as the VIF values were quite high for a few variables (e.g. piped water). Variables with high VIF were removed from the local regression analysis since they make no significant contribution to the models.

It is apparent from Table 19.4a, b, c that demographic variables explained the spatial variation of typhoid case incidence far better than the socioeconomic and demographic + socioeconomic variables. For example, for demographic variables for the aggregated 2005–2009 period, the higher adjusted R^2 and lower AICc values show that GWR outperformed the OLS models. The highest adjusted R^2 was achieved for the aggregated typhoid data for 2005–2009, with 66 % of the variance being explained by the model. By contrast, the socioeconomic parameter group poorly explained the typhoid case incidence both using OLS and GWR. Although GWR provided some small improvements in AICc values across the years studied, adjusted R^2 values were extremely low (Table 19.4b) – the highest adjusted R^2 value was obtained using the GWR method with aggregated data of 2005–2009 (adjusted $R^2 = 0.18$).

When demographic and socioeconomic parameters were used in combination, there is an improvement in the model fit compared with socioeconomic variables alone. However, the results were still lower than those for demographic variables alone. The highest adjusted R^2 value was obtained by GWR for the year 2008, explaining 50 % of the total variance in the model.

Figure 19.3 shows the distribution of standardised residuals from GWR modelling of the aggregated 2005–2009 typhoid cases as a function of demographic, socioeconomic and demographic + socioeconomic variables. A comparison of the three model outputs clearly indicates that the demographic variables explained the largest amount of variance. The strength of the GWR is highly spatially variable, and the results are extremely useful to discern which parts of the city have such stronger relationships than other parts.

Table 19.3 Ordinary least square (OLS) results

Parameter	Estimated value	Standard error	VIF
(a) *Demographic*			
Intercept	−1.639	0.9200	–
Total population	0.0001	0.0000	1.07
Population density	0.3772	0.0946	1.07
(b) *Socioeconomic*			
Intercept	−6.105	4.522	–
Total literacy	0.012	0.011	1.80
Male literacy	−0.062	0.036	7.55
Female literacy	−0.005	0.031	7.37
San_water	0.040	0.036	41.93
San_nw	0.028	0.036	33.68
NonSan	0.019	0.039	12.90
NoSan	−0.031	0.049	1.55
Pipe_water	0.032	0.055	124.67
Tube well	0.000	0.055	117.67
Other_sources	0.005	0.060	5.35
Poverty	16.913	5.44	2.52
Mean housing value	0.563	0.105	2.12
(c) *Demographic + socioeconomic*			
Intercept	−3.638	4.189	–
Total population	0.000	0.000	1.06
Population density	−0.000	0.000	1.08
Total literacy	0.012	0.100	1.80
Male literacy	−0.061	0.033	7.56
Female literacy	0.009	0.029	7.59
San_water	0.037	0.033	41.95
San_nw	0.025	0.034	33.71
NonSan	0.019	0.036	12.95
NoSan	−0.012	0.045	1.55
Pipe_water	0.007	0.051	124.93
Tube well	−0.015	0.051	117.68
Other_sources	0.003	0.056	5.35
Poverty	10.957	5.069	2.55
Mean housing value	0.399	0.098	2.16

San_water water-sealed sanitation, *San_nw* nonwater-sealed sanitation, *NonSan* non-sanitation, *NoSan* no sanitation, *Other_sources* other water sources (e.g. ponds, rivers)

19.4 Discussion and Conclusions

A GIS together with spatial statistics are valuable tools that allow for an examination of the degree of clustering of cases of infection and the investigation of the spatial relationship of causal factors. Though these tools have been used in a variety of health studies, their application to typhoid has been limited. Hence, this study could greatly benefit public health communities by providing a deeper understanding of disease distribution over space and factors associated with particular spatial patterns.

Table 19.4 Comparison of OLS and GWR results (2005–2009)

	OLS			GWR		
Year	R^2	Adjusted R^2	AICc	R^2	Adjusted R^2	AICc
(a) *Demographic variables*						
2005	0.243	0.242	4,232.74	0.605	0.506	3,881.95
2006	0.216	0.215	4,010.29	0.646	0.589	3,333.04
2007	0.226	0.224	4,670.88	0.636	0.543	4,196.06
2008	0.200	0.199	4,625.41	0.677	0.611	3,880.07
2009	0.231	0.229	3,968.74	0.505	0.424	3,721.79
2005–2009	0.168	0.166	7,912.94	0.739	0.659	7,034.25
(b) *Socioeconomic variables*						
2005	0.057	0.051	4,510.72	0.192	0.132	4,445.59
2006	0.033	0.027	4,276.55	0.062	0.045	4,258.95
2007	0.047	0.041	4,933.93	0.090	0.073	4,899.54
2008	0.049	0.043	4,847.46	0.102	0.084	4,800.51
2009	0.055	0.048	4,230.73	0.092	0.074	4,204.00
2005–2009	0.059	0.053	8,073.94	0.242	0.180	7,944.99
(c) *Demographic + socioeconomic variables*						
2005	0.264	0.259	4,213.56	0.351	0.338	4,081.85
2006	0.033	0.027	4,278.06	0.341	0.328	3,835.16
2007	0.238	0.232	4,666.86	0.473	0.431	4,345.33
2008	0.221	0.215	4,608.70	0.542	0.505	4,091.83
2009	0.248	0.242	3,957.44	0.398	0.349	3,814.31
2005–2009	0.200	0.193	7,880.83	0.486	0.450	7,453.44

This study used the global Moran's *I* statistic to investigate the spatial patterns of typhoid for each year from 2005 to 2009 and found strong evidence of cases being spatially autocorrelated. This information can guide public health professionals in their search for possible interventions.

The study also identified hotspots of typhoid infection using the local G-statistic. These maps are important particularly for two reasons: firstly, the evaluation of spatial distribution of disease could assist in understanding the aetiology (Moore and Carpenter 1999), and secondly, the correlation between hotspots of a particular disease and its aetiology provides valuable information in detecting risk factors from a spatial viewpoint (Tsai et al. 2009). In addition, the spatial extent of hotspots enables the recognition of diffusion patterns of disease over space and time, an important step to prevent further spread of disease (Hsueh et al. 2012; Wallace 1994). Although the study did not set out to analyse diffusion patterns of the disease, examination of spatial clustering of typhoid obtained from the G-statistic may provide information on how typhoid disease has evolved from our benchmark year of 2005. Four general types of spatial diffusion have been identified as expansion, contiguous, relocation and hierarchical (Hornsby, no date). The assessment of multi-temporal hotspot maps suggests that relocation and contiguous types of diffusion have been prominent in the study area since 2005. For instance, the spread of typhoid fever was close to contiguous diffusion in 2005. The 2006 map,

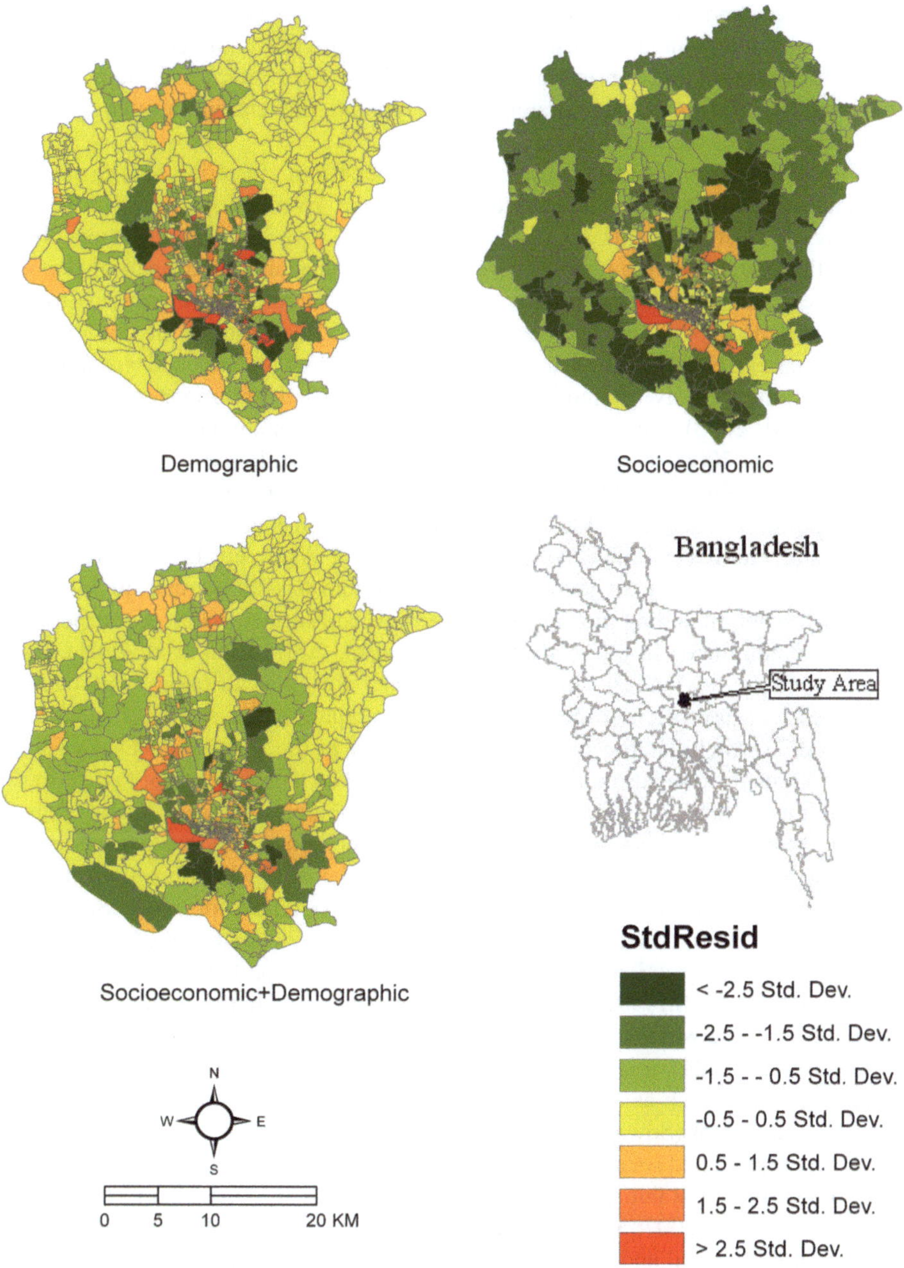

Fig. 19.3 The distribution of standardised residuals of typhoid cases from GWR as a function of demographic, socioeconomic and demographic + socioeconomic variables

on the other hand, did not follow any specific diffusion pattern, but a single focus pattern was dominant. The hotspot maps of 2007 and 2009 could be best explained as both expansion and contiguous diffusions. Thus, these maps may be of significant value for formulating strategies for typhoid prevention through a better understanding of the aetiology.

Further analysis of hotspot maps indicates persistent clustering among specific communities in Dhaka mostly located along the river Buriganga. The reason for having clusters in this area may be attributed to a number of factors. First of all, these areas are the residence of marginal groups and the density of population is extremely high (Rahman et al. 2011), and therefore, the combination of overcrowdedness with low socioeconomic status may account for the clustering of typhoid cases (Nagashetty et al. 2010). Secondly, both surface and groundwater have become heavily contaminated due to intensified anthropogenic activities (Zahid et al. 2006) which give rise to higher faecal contamination load in rivers (Hörman et al. 2004) with major implication on the transmission and distribution of typhoid (Mermin et al. 1999). Thus, typhoid in that area may be associated with water pollution for which the level of contamination has already exceeded the permissible limit (Dewan et al. 2012; BBS 2010; Aktar et al. 2009; Karn and Harada 2001). Risk factor identification for typhoid has shown that all source of drinking water, including piped water, tube wells and surface water, are perpetually highly contaminated in the study area (Ram et al. 2007; Saha et al. 2001). As a result, people relying on surface waterbodies for their drinking and bathing purposes will have elevated risk of contracting typhoid (Sur et al. 2007; Mermin et al. 1999) which has recently been investigated in the same area (Dewan et al. 2013). Another explanation for the clustering could be the existence of flood control embankments that have created multifaceted problems in recent years (Dewan 2013). Many argue that water resources development often leads to dramatic changes in ecology and demography in a particular area which has considerable impact on the transmission and distribution of water-related diseases (Sow et al. 2002; Emch 2000; Singh et al. 1999; Molyneus 1997; Hunter et al. 1982). For instance, the construction of the Bargi Dam in India resulted in an epidemic of malaria (Singh et al. 1999). Since the flood control project in that area of Dhaka has resulted in severe waterlogging during monsoon in recent times (Alam and Rabbani 2007; Tawhid 2004), it is not surprising that there is a higher occurrence of water-related diseases such as typhoid due to ecological changes which are linked with flood embankment (Sow et al. 2002).

Demographic and socioeconomic factors are known to impact typhoid incidence and transmission (Rahman et al. 2011; Sharma et al. 2009; Kothari et al. 2008; Kelly-Hope et al. 2007; Sur et al. 2007; Gasem et al. 2001; Luby et al. 2008). The application of global and local regressions has provided useful information on the causation of typhoid incidence in the study area.

The predictive model based on demographic indicators comprising of population counts and density clearly provided superior results in terms of both coefficient of determination (adjusted R^2) and AICc values. For example, the AICc values from the GWR model resulted in considerable improvement compared with the

AICc values from OLS model during the study period, with the reduction in AICc values being greater than the three points suggested by the literature as being significant (Fotheringham et al. 2002). Furthermore, the maximum value of the condition number of the regression design matrix for the demographic variables was 7.8, which is much smaller than a typical value of 30, meaning that our model was free from statistical concerns. This result indicates that GWR can serve as a new approach to the exploration of spatially varying relationships and can minimise spatial autocorrelation. Since humans are the only host of typhoid fever (Bhan et al. 2005), overcrowding in Dhaka may have significant impact on the incidence of the disease. Overcrowded populations are of especial risk of harbouring or spreading diseases (Hewitt 1997) through close contact with the cases or carriers (Tran et al. 2005) and sharing of food from the same plate (Vollaard et al. 2004). Apart from these, poor lifestyle (Kothari et al. 2008) and larger household size (Sur et al. 2007) have been found to have an impact on the transmission of disease. The result of this study is in agreement with previous studies, conducted in similar endemic settings, revealing that crowdedness is conducive to the occurrence of typhoid fever (Rahman et al. 2011; Sur et al. 2007).

Compared to the demography-based model, the socioeconomic model performed very poorly in explaining the spatial nonstationarity. As shown in Table 19.4b, the adjusted R^2 was very low in both regression models, though there was some improvement in AICc with the local model. To determine the cause of such low adjusted R^2, the study tested condition number and the spatial autocorrelation (Moran's *I*) of standardised residuals from OLS and GWR models. The analysis revealed that the socioeconomic model was spatially autocorrelated (Moran's $I = 0.320$ and 0.216 from OLS and GWR, respectively). The condition number from the GWR model ranged from 21.37 to 27.93, indicating the data was not fully free from statistical concerns. These two measures clearly showed spatial dependency in the socioeconomic parameters that resulted in the poor adjusted R^2. As noted elsewhere (Karkey et al. 2010), the local environment may have an important role to play in the distribution and transmission of typhoid. Moreover, there may be some other covariates that confound the socioeconomic model. For example, cultural and individual food handling practices, hygiene routines and environmental differences may put some people at elevated risk of contracting typhoid (Tran et al. 2005). Our model only included variables that can be determined in the census reports, so data on these additional factors were unavailable. Therefore, further study could include additional variables acquired through detailed surveys to develop a more spatially explicit relationship. Nevertheless, poverty, lack of education, lower-income status, poor housing, living under unsanitary conditions, drinking untreated water and inadequate water supply are believed to contribute substantially to the occurrence of enteric fever (Sharma et al. 2009; Sur et al. 2007; Gasem et al. 2001), which are quite widespread in the study area.

Compared to the socioeconomic model, the combined model using both demographic and socioeconomic variables provided a slightly better prediction as shown by the higher adjusted R^2 and the decrease in AICc from the OLS to the GWR models (Table 19.4c). This was due to multicollinearity among the socioeconomic variables which also exhibited spatial dependency as indicated by Moran's I values for the standardised residuals from both OLS and GWR (Moran's $I = 0.310$ and 0.100 from OLS and GWR, respectively). Large-sized study areas such as that under consideration here with highly dissimilar characteristics between neighbourhoods can cause multicollinearity among socioeconomic variables leading to spatial nonstationarity. In other words, GWR tends to predict better where the gradient of change in explanatory variable values is not high (Ogneva-Himmelberger et al. 2009). Since GWR is sensitive to kernel type and bandwidth (Cho et al. 2010; Wheeler and Tiefelsdorf 2005), local coefficients that vary spatially may have impacted the predictors (Griffith 2008), as marginal coefficients of GWR have a tendency to be degraded substantially by collinearity (Wheeler and Calder 2007). Moreover, absence of actual information such as household income could also play a crucial role in establishing the best predictors for a demographic + socioeconomic model.

Cluster maps developed in this study could assist planners to evaluate spatial risk factors and help in developing appropriate health-care measures to reduce the burden of disease in Dhaka. Though vaccination is available to prevent typhoid infection, it cannot be an alternative to sound environmental health infrastructure (Ivanoff et al. 1994). While the presence of spatial nonstationarity led to poor results, except for the demographic model, GWR still can be applied as an exploratory technique for investigating relationships in multivariate spatial datasets (Harris et al. 2011).

Rapid urbanisation in Dhaka with little provision of public health infrastructure could trigger large-scale outbreaks of typhoid in the years ahead. This could further be aggravated by overcrowding which could act as an agent of disease transmission. Furthermore, Dhaka is likely to receive more intense rainfall driven by climatic change. These changes may all put more people at risk of typhoid. This study therefore underscores the need for appropriate policies as well as critical public health infrastructure to curb the further spread of waterborne infection, particularly typhoid.

There are a few limitations to this study. Firstly, the disease data that were acquired from hospitals may have either underestimated or overestimated the number of typhoid cases. Because the data were historical records documented from the record room of each hospital, there was no valid method to ascertain repeated hospitalisations of an individual patient. In addition, hospital-based surveillance may underestimate the actual population infected, since only severely sick people tend to get admitted for treatment. Secondly, we only considered 11 major health service providers, the majority of which were public hospitals. The study could be improved by incorporating data from private clinics where most of the affluent people seek health services. Thirdly, it was not possible to separate cases into those caused by *S. typhi* and those caused by the paratyphoid serovars. Separation of these two types would allow us to estimate the disease dynamics and identify the most prevalent typhoid types in Dhaka. Finally, new methods are needed to overcome some of the problems associated with GWR such as the mixed geographically weighted regression proposed by Mei et al. (2006).

Acknowledgement We acknowledge the support of the International Foundation for Science (IFS), Sweden, for funding this work under a project (Reference: W4656-1) on which Ashraf M. Dewan was the Principal Investigator.

References

Aktar Z, Fakhruddin ANM, Ahmed MSU, Nasreen T, Parveen S (2009) Assessment and treatment of the supplied pipeline water of selected areas of Dhaka City, Bangladesh. J Sci Ind Res 44 (3):335–340

Alam M, Rabbani MG (2007) Vulnerabilities and responses to climate change for Dhaka. Environ Urban 19(1):81–97

Ali M, Emch M, Donnay JP, Yunus M, Sack RB (2002) The spatial epidemiology of cholera in an endemic area of Bangladesh. Soc Sci Med 55(6):1015–1024

Anselin L (1995) The local indicators of spatial association – LISA. Geogr Anal 27(2):93–115

Bangladesh Bureau of Statistics (BBS) (2003) Population census – 2001, Community series. Bangladesh Bureau of Statistics, Ministry of Planning, Dhaka

Bangladesh Bureau of Statistics (2010) Compendium of environment statistics of Bangladesh 2009. Bangladesh Bureau of Statistics, Ministry of Planning, Dhaka

Bangladesh Bureau of Statistics (BBS) (2012) Population and housing census – 2011, Community report, Dhaka, Narayanganj and Gazipur Zila. Bangladesh Bureau of Statistics, Ministry of Planning, Dhaka

Bertazzon S, Olson S, Knudtson M (2010) A spatial analysis of the demographic and socio-economic variables associated with cardiovascular disease in Calgary (Canada). Appl Spat Anal Policy 3(1):1–13

Bhan M, Bahl R, Bhatnagar S (2005) Typhoid and paratyphoid fever. Lancet 366(9487):749–762

Brook WA, Hossain A, Goswami D, Sharmeen AT, Nahar K, Alam K, Ahmed N, Naheed A, Nair GB, Luby S, Breiman RF (2005) Bacteremic typhoid fever in children in an urban slum, Bangladesh. Emerg Infect Dis 11(2):326–329

Butler T, Islam A, Kabir I, Jones PK (1991) Patterns of morbidity and mortality in typhoid fever dependent on age and gender: review of 552 hospitalized patients with diarrhea. Rev Infect Dis 13(1):85–90

Castillo KC, Körbl B, Stewart A, Gonzalez JF, Ponce F (2011) Application of spatial analysis to the examination of dengue fever in Guayaquil, Ecuador. Proc Environ Sci 7:188–193

Cho S-H, Lambert DM, Chen Z (2010) Geographically weighted regression bandwidth selection and spatial autocorrelation: an empirical example using Chinese agricultural data. Appl Econ Lett 17(8):767–772

Clement F, Orange D, Williams M, Mulley C, Epprecht M (2009) Drivers of afforestation in northern Vietnam: assessing local variations using geographically weighted regression. Appl Geogr 29(4):561–576

Colantonio A, Moldofsky B, Escobar M, Vernich L, Chipman M, McLellan B (2011) Using geographical information systems mapping to identify areas presenting high risk for traumatic brain injury. Emerg Themes Epidemiol 8:7

Corner RJ, Dewan AM, Hashizume M (2013) Modeling typhoid risk in Dhaka Metropolitan Area in Bangladesh. Int J Health Geogr 12:13

Crump JA, Luby SP, Mintz ED (2004) The global burden of typhoid fever. Bull World Health Organ 82(5):346–353

Dewan A (2013) Floods in a megacity: geospatial techniques in assessing hazards, risk and vulnerability. Springer, Dordrecht

Dewan AM, Kabir MH, Nahar K, Rahman MZ (2012) Urbanisation and environmental degradation in Dhaka Metropolitan Area of Bangladesh. Int J Environ Sustain Dev 11(2):118–146

Dewan AM, Corner RJ, Hashizume M, Ongee ET (2013) Typhoid fever and its association with environmental factors in Dhaka Metropolitan Area: a spatial and time-series approach. PLoS Negl Trop Dis 7(1):e1998

Emch M (2000) Relationships between flood control, kala-azar, and diarrheal disease in Bangladesh. Environ Plan A 32(6):1051–1063

ESRI (2010) Interpreting GWR results. http://webhelp.esri.com/arcgisdesktop/9.3/index.cfm?TopicName=Interpreting_GWR_results. Accessed 12 Oct 2012

ESRI (2011) Common regression problems, consequences and solutions in regression analysis basics. http://resources.esri.com/help/9.3/arcgisdesktop/com/gp_toolref/spatial_statistics_toolbox/regression_analysis_basics.htm. Accessed 12 Oct 2012

ESRI (2012) How hot spot analysis (Getis-Ord Gi* works). http://help.arcgis.com/en/arcgisdesktop/10.0/help/index.html#/How_Hot_Spot_Analysis_Getis_Ord_Gi_works/005p00000011000000/. Accessed 12 Oct 2012

Fotheringham AS, Charlton ME, Brunsdon C (2001) Spatial variations in school performance: a local analysis using geographically weighted regression. Geogr Environ Model 5(1):43–66

Fotheringham SA, Brunsdon C, Charlton M (2002) Geographically weighted regression: the analysis of spatially varying relationships. Wiley, Chichester

Gasem MH, Dolamans WM, Keuter MM, Djokomoeljanto RR (2001) Poor food hygiene and housing as risk factors for typhoid fever in Semarang, Indonesia. Trop Med Int Health 6 (6):484–490

Gesler W (1986) The uses of spatial analysis in medical geography: a review. Soc Sci Med 23 (10):963–973

Getis A, Ord JK (1992) The analysis of spatial association by use of distance statistics. Geogr Anal 24(3):189–206

Griffith DA (2008) Spatial-filtering-based contribution to a critique of geographically weighted regression (GWR). Environ Plan A 40(11):2751–2769

Haddow AD, Odoi A (2009) The incidence risk, clustering, and clinical presentation of La Crosse virus infections in the Eastern United States, 2003–2007. PLoS One 4(7):e6145

Haque U, Scott LM, Hashizume M, Fisher E, Haque R, Yamamoto T, Glass GE (2012) Modelling malaria treatment in Bangladesh using spatial statistics. Malar J 11:63

Harris P, Brunsdon C, Fotheringham AS (2011) Links, comparisons and extensions of the geographically weighted regression model when used as a spatial predictor. Stoch Environ Res Risk Assess 25(2):123–138

Hewitt K (1997) Regions of risk: a geographical introduction to disasters. Addison-Wesley Longman Limited, Harlow

Hinman SE, Blackburn JK, Curtis A (2006) Spatial and temporal structure of typhoid outbreaks in Washington D.C., 1906–1909: evaluating local clustering with the G_i* statistic. Int J Heal Geogr 5:13

Hörman A, Rimhanen-Finne R, Maunula L, von Bonsdorff C-H, Torvela N, Heikinheimo A, Hanninen M-L (2004) Campylobacter spp., Giardia spp., Cryptosporidium, spp., Noroviruses, and indicator 372 organisms in surface water in south-western Finland, 2000–2001. Appl Environ Microbiol 70(1):87–95

Hornsby K (no date) Spatial diffusion: conceptualizations and formalizations [Online]. Available: www.spatial.maine.edu/khornsby/KHI21.pdf

Hsueh Y-H, Lee J, Beltz L (2012) Spatio-temporal patterns of dengue fever cases in Kaoshiung City, Taiwan, 2003–2008. Appl Geogr 34:587–894

Hunter JM, Rey L, Scott D (1982) Man-made lakes and man-made diseases. Soc Sci Med 16 (11):1127–1145

Islam N (2005) Dhaka now: contemporary urban development. The Bangladesh Geographical Society, Dhaka

Ivanoff B, Levine MM, Keuter PH (1994) Vaccination against typhoid fever: present status. Bull World Health Organ 72(6):957–971

Kanungo S, Dutta S, Sur D (2008) Epidemiology of typhoid and paratyphoid fever in India. J Infect Dev Ctries 2(6):454–460

Karkey A, Arjyal A, Anders KL, Boni MF, Sabina D, Koirala S, My PVT, Nga TVT, Clements ACA, Holt K, Duy PT, Day JN, Campbell JI, Dougan G, Dolecek C, Farrar J, Basnyat B, Baker S (2010) The burden and characteristics of enteric fever at a healthcare facility in a densely populated area of Kathmandu. PLoS One 5(11):e13988

Karn SK, Harada H (2001) Surface water pollution in three urban territories of Nepal, India and Bangladesh. Environ Manag 28(4):483–496

Kazmi SJH, Usery EL (2001) Application of remote sensing and GIS for the monitoring of diseases: a unique research agenda for geographers. Remote Sens Rev 20(1):45–70

Kelly-Hope LA, Alonso WJ, Theim VD, Anh DD, Canh DG, Lee H, Smith DL, Miller MA (2007) Geographical distribution and risk factors associated with enteric diseases in Vietnam. Am J Trop Med Hyg 76(4):706–712

Khormi HS, Kumar L (2011) Modeling dengue fever risk based on socioeconomic parameters, nationality and age groups: GIS and remote sensing based case study. Sci Total Environ 409 (22):4713–4719

Kothari A, Pruthi A, Chugh TD (2008) The burden of enteric fever. J Infect Dev Ctries 2 (4):253–259

Kuo C-C, Huang J-L, Ko C-Y, Lee P-F, Wang H-C (2011) Spatial analysis of scrub typhus infection and its association with environmental and socioeconomic factors in Taiwan. Acta Trop 120(1–2):52–58

Lai P-C, So F-M, Chan K-W (2009) Spatial epidemiological approaches in disease mapping and analysis. CRC Press, Boca Raton

Lin C-H, Wen T-H (2011) Using geographically weighted regression (GWR) to explore spatial varying relationships of immature mosquitoes and human densities with the incidence of dengue. Int J Environ Res Public Health 8(7):2798–2815

Loth L, Gilbert M, Osmani MG, Kalam AM, Xiao X (2010) Risk factors and clusters of highly pathogenic avian influenza H5N1 outbreaks in Bangladesh. Prev Vet Med 96(1–2):104–113

Luby SP, Gupta SK, Sheikh MA, Johnston RB, Ram PK, Islam MS (2008) Tubewell water quality and predictors of contamination in three flood-prone areas in Bangladesh. J Appl Microbiol 105(4):1002–1008

Maantay JA, McLafferty S (2011) Environmental health and geospatial analysis: an overview. In: Maantay JA, McLafferty S (eds) Geospatial analysis of environmental health. Springer, Dordrecht, pp 3–37

Mei CL, Wang N, Chang WX (2006) Testing the importance of the explanatory variables in a mixed geographically weighted regression model. Environ Plan A 38(3):587–598

Mermin JH, Villar R, Carpenter J, Roberts L, Samaridden A, Gasanova L, Lomakina S, Bopp C, Hutwagner L, Mead P, Ross B, Mintz ED (1999) A massive epidemic of multidrug-resistant typhoid fever in Tajikistan associated with consumption of municipal water. J Infect Dis 179 (6):1416–1422

Molyneus DH (1997) Patterns of change in vector borne diseases. Ann Trop Med Parasitol 91 (7):827–829

Moore DA, Carpenter TE (1999) Spatial analytical methods and geographic information systems: use in health research and epidemiology. Epidemiol Rev 21(2):143–161

Nagashetty K, Channappa ST, Gaddad SM (2010) Antimicrobial susceptibility of Salmonella Typhi in India. J Infect Dev Ctries 4(2):070–073

Naheed A, Ram PK, Brooks WA, Hossain MA, Parsons MB, Talukder KA, Mintz E, Luby S, Breiman RF (2010) Burden of typhoid and paratyphoid fever in a densely populated urban community, Dhaka, Bangladesh. Int J Infect Dis 14(S3):e93–e99

Naish S, Hu W, Mengersen K, Tong S (2011) Spatio-temporal patterns of Barmah Forest virus disease in Queensland, Australia. PLoS One 6(10):e25688

Ochiai RL, Acosta CJ, Danovaro-Holliday MC, Baiqing D, Bhattacharya SK, Agtini MD, Bhutta ZA, Canh DG, Ali M, Shin S, Wain J, Page A-L, Albert MJ, Farrar J, Abu-Elyazeed R, Pang T, Galindo CM, von Seidlein L, Clemens JD, Domi Typhoid Study Group (2008) A study of

typhoid fever in five Asian countries: disease burden and implications for controls. Bull World Health Organ 86(4):260–268
Ogneva-Himmelberger Y, Pearsall H, Rakshit R (2009) Concrete evidence & geographically weighted regression: a regional analysis of wealth and the land cover in Massachusetts. Appl Geogr 29(4):478–487
Ord JK, Getis A (1995) Local spatial autocorrelation statistics: distributional issues and an application. Geogr Anal 27(4):286–306
Owusu-Edusei K Jr, Owens C (2009) Monitoring county-level Chlamydia incidence in Texas, 2004–2005: application of empirical Bayesian smoothing and exploratory spatial data analysis (ESDA) methods. Int J Health Geogr 8:12
Parry C (2005) Typhoid fever and cholera. Medicine 33(7):34–36
Patel TA, Armstrong M, Morris-Jones SD, Wright SG, Doherty T (2010) Imported enteric fever: case series from the hospital for tropical diseases, London, United Kingdom. Am J Trop Med Hyg 82(6):1121–1126
Pfeiffer D, Robinson T, Stevenson M, Stevens K, Rogers D, Celments A (2008) Spatial analysis in epidemiology. Oxford University Press, New York
Poole MA, O'Farrell PN (1971) The assumptions of the linear regression model. Trans Inst Br Geogr 52(March):145–158
Rahman AKMM, Ahmad M, Begum RS, Hossain MZ, Hoque SA, Matin A, Yeasmin L, Mamun MGS (2011) Prevalence of typhoid fever among the children in a semi urban area of Bangladesh. J Dhaka Med Coll 20(1):37–43
Rainey JJ, Omenah D, Sumba PO, Moormann AM, Rochford R, Wilson ML (2006) Spatial clustering of endemic Burkitt's lymphoma in high-risk regions of Kenya. Int J Cancer 120 (1):121–127
Ram PK, Naheed A, Brooks WA, Hossain MA, Mintz ED, Breiman RF, Luby SP (2007) Risk factors for typhoid fever in a slum in Dhaka, Bangladesh. Epidemiol Infect 135:458–465
Rezaeian M, Dunn G, Leger SS, Appleby L (2007) Geographical epidemiology, spatial analysis and geographical information systems: a multidisciplinary glossary. J Epidemiol Community Health 61(2):98–102
Roy SK, Speelman P, Butler T, Nath S, Rahman H, Stoll BJ (1985) Diarrhea associated with typhoid fever. J Infect Dis 151(6):1138–1143
Saha SK, Baqui AH, Hanif M, Darmstadt GL, Ruhulamin M, Nagatake T, Santosham M, Black RE (2001) Typhoid fever in Bangladesh: implications for vaccination policy. Paediatr Infect Dis J 20(5):521–524
Sharma PK, Ramakrishnan R, Hutin Y, Manickam P, Gupte MD (2009) Risk factors for typhoid in Darjeeling, West Bengal, India: evidence for practical action. Trop Med Int Health 14 (6):696–702
Singh N, Mehra RK, Sharma VP (1999) Malaria and the Narmada-river development in India: a case study of the Bargi dam. Ann Trop Med Parasitol 93(5):477–488
Sow S, de Vlas SJ, Engels D, Gryseels B (2002) Water-related disease patterns before and after the construction of the Diama dam in northern Senegal. Ann Trop Med Parasitol 96(6):575–586
Stoll BJ, Glass RI, Banu H, Alam M (1983) Enteric fever in patients to a diarrhoeal disease hospital in Bangladesh. Trans R Soc Trop Med Hyg 77(4):548–551
Sur D, Ali M, Seidlein LV, Manna B, Deen JL, Acosta CJ, Clemens JD, Bhattacharya SK (2007) Comparisons of predictors for typhoid and paratyphoid fever in Kolkata, India. BMC Publ Health 7:289
Tawhid KG (2004) Causes and effects of water logging in Dhaka City, Bangladesh. TRITA-LWR Master's thesis, Department of Land and Water Resource Engineering, Royal Institute of Technology, Sweden
Tran HH, Bjune G, Nguyen BM, Rottingen JA, Grais RF, Guerin PJ (2005) Risk factors associated with typhoid fever in Son La Province, northern Vietnam. Trans R Soc Trop Med Hyg 99 (11):819–826

Tsai P-J, Lin M-L, Chu C-M, Perng C-H (2009) Spatial autocorrelation of health care hotspots in Taiwan in 2006. BMC Publ Health 9:464

Unwin DJ (1996) GIS, spatial analysis and spatial statistics. Prog Hum Geogr 20(4):540–551

Upton GJG, Fingleton B (1985) Spatial data analysis by example, point pattern and quantitative data volume I. Wiley, New York

Vollaard AM, Ali S, van Asten HAGH, Widjaja S, Visser LG, Surjadi C, van Dissel JT (2004) Risk factors for typhoid and paratyphoid fever in Jakarta. JAMA 291(21):2607–2615

Wallace D (1994) The resurgence of tuberculosis in New York City: a mixed hierarchically and spatially diffused epidemic. Am J Public Health 84(6):1000–1002

Wen T-H, Lin NH, Chao D-Y, Hwang K-P, Kan C-C, Lin KC-M, Wu JT-S, Huang SY-J, Fan I-C, King C-C (2010) Spatial-temporal patterns of dengue in areas at risk of dengue hemorrhagic fever in Kaohsiung, Taiwan, 2002. Int J Infect Dis 14(4):e334–e343

Wheeler D, Calder C (2007) An assessment of coefficient accuracy in linear regression models with spatially varying coefficients. J Geogr Syst 9(2):145–166

Wheeler D, Tiefelsdorf M (2005) Multicollinearity and correlation among local regression coefficients in geographically weighted regression. J Geogr Syst 7(2):161–187

Wong DWS, Lee J (2005) Statistical analysis of geographic information with ArcView GIS and ArcGIS. Wiley, Hoboken

World Health Organization (WHO) (2003) Background document: the diagnosis, treatment and prevention of typhoid fever. Department of Vaccines and Biologicals. Available at: www.who.int/vaccines-documents/

Wu PC, Lay JG, Guo HR, Lin CY, Lung SC, Su HJ (2009) Higher temperature and urbanization affect the spatial patterns of dengue fever transmission in subtropical Taiwan. Sci Total Environ 407(7):2224–2233

Yin F, Feng Z, Li X (2012) Spatial analysis of county-based gonorrhoea incidence in mainland China, from 2004 to 2009. Sex Health 9(3):227–232

Zahid A, Balke KD, Hassan MQ, Flegr M (2006) Evaluation of aquifer environment under Hazaribagh leather processing zone of Dhaka city. Environ Geol 50(4):495–504

Zhang W, Wang L, Fang L, Ma J, Xu Y, Jiang J, Hui J, Liang S, Yang H, Cao W (2008) Spatial analysis of malaria in Anhui province, China. Malar J 7:206

Chapter 20
Spatiotemporal Analysis of Dengue Infection Between 2005 and 2010

Sarwa Ali, Robert J. Corner, and Masahiro Hashizume

Abstract The high incidence of dengue fever in Dhaka is a constant threat to the population and a recurring problem for the health authorities. This chapter investigates the spatial and temporal epidemiology of dengue fever between 2005 and 2010. This epidemiological analysis provided important information about the pattern of the virus cases with standard deviation ellipses being used for directional examination of the incidences. To investigate spatial dependencies and examine the occurrence pattern for clustering, Moran's *I* and Local Indicators of Spatial Association (LISA) analysis were utilised. Results showed that there was obvious spatial autocorrelation as well as significant clustering of dengue cases in Dhaka, revealing that the virus is concentrated around the heart of the city.

Keywords Dengue fever • Spatial analysis • Spatial epidemiology • Clustering

20.1 Introduction

Dengue is an arbovirus (an arthropod-borne virus), not dissimilar to yellow fever and malaria (Gubler 1998). The virus infects 50–100 million people worldwide a year, leading to approximately 500,000 severe case hospitalisations (many of whom are children); out of which about 2.5 % (12,500 people) result in death (WHO

S. Ali (✉) • R.J. Corner
Department of Spatial Sciences, Curtin University, Kent Street,
Bentley, Perth 6102, WA, Australia
e-mail: sarwa210@gmail.com; R.Corner@curtin.edu.au

M. Hashizume
Institute of Tropical Medicine (NEKKEN), Nagasaki University,
1-12-4, Sakamoto, Nagasaki 852-8523, Japan
e-mail: hashizum@nagasaki-u.ac.jp

A. Dewan and R. Corner (eds.), *Dhaka Megacity: Geospatial Perspectives on Urbanisation, Environment and Health*, Springer Geography,
DOI 10.1007/978-94-007-6735-5_20,

2012). It is one of Dhaka's most significant prevailing viral diseases (Choudhury et al. 2008).

The name dengue covers two different manifestations of the disease: classic dengue fever and dengue haemorrhagic fever/dengue shock syndrome (Erickson et al. 2010). The more common of the two, classic dengue fever, is often asymptomatic or similar to "the flu" but can sometimes cause headaches and muscle and joint pains (Erickson et al. 2010). More severe cases can progress to dengue haemorrhagic fever and dengue shock syndrome (Kalayanarooj et al. 1997).

Dengue virus infection is regarded as being emerging or re-emerging diseases for three main reasons. Firstly, the number of reported cases worldwide is increasing (through either increased infection rate or better diagnosis). Secondly, the virus has a possibly fatal manifestation if left untreated (dengue haemorrhagic fever and dengue shock syndrome), and finally, the areas endemic to the disease are expanding (Igarashi 1997).

The consequence of simple dengue fever is loss of workdays for communities dependent on wage labour. The consequence of severe illness (resulting in haemorrhage) is high mortality rates, as late stages of the virus require tertiary level management/care, which is beyond the reach of most of the people at risk (Guha-Sapir and Schimmer 2005).

Another reason for the virus's current significance is the lack of vaccine. Despite much effort, currently there is no vaccine or specific therapy for the treatment of the virus. Much of the focus has, therefore, been directed to vector control as being the main measure for virus management. Consequently, an analysis of the recorded dengue cases and an identification of risk patterns could improve decision-making for controlling the disease in an endemic region (Castillo et al. 2011).

The disease is transmitted through the bite of the *Aedes aegypti* and *Ae. albopictus* mosquitoes. These mosquitoes bite primarily during the day, especially in the hours just after dawn and before sunset, and an infection can be acquired via a single bite. A female mosquito that takes a blood meal from a dengue-infected person becomes infected itself with the virus in the cells lining its gut. About 8–10 days later, the virus spreads to other tissues including the mosquito's salivary glands and is subsequently released into its saliva. The virus seems to have no detrimental effect on the mosquito, which remains infected for life. Dengue can also be transmitted via infected blood products and through organ donation (Wilder-Smith et al. 2009).

The mosquito breeding cycle involves the laying of eggs on water, preferably in shaded and dark locations. The eggs hatch into larvae which feed on material suspended in water, eventually hatching into adult mosquitoes. The cycle from egg to adult can take as little as 7–8 days, whilst an adult mosquito lives for about 3 weeks (CDC 2012).

Many factors have been linked to the recent increase of dengue virus transmission; the main focuses being on increased urbanisation, inadequate water supply and storage and new trends in population movements (Hsueh et al. 2012; Wu 2009; Nakhapakorn and Tripathi 2005; Ali et al. 2003). In recent decades, the expansion of villages, towns and cities in endemic areas and the increased mobility of humans have increased the number of epidemics and circulating viruses. Furthermore, migration of people from lower socioeconomic groups to the cities has created slums which have poor sanitation and a deteriorating environment. Other factors affecting transmission rate include the improper dumping of rubbish such as used tyres, empty tin cans or food containers which provides breeding sites for *Aedes aegypti*.

Using GIS (Geographic Information Systems), the spatial distribution of dengue virus has been investigated in many regions of the world such as in Taiwan (Hsueh et al. 2012; Wu 2009; Wen et al. 2006), Thailand (Nakhapakorn and Jirakajohnkool 2006; Vanwambeke et al. 2006; Nakhapakorn and Tripathi 2005), Sri Lanka (Pathirana et al. 2009), Bangladesh (Ali et al. 2003), Brazil (Mondini et al. 2005; Braga 2003), Puerto Rico (Morrison et al. 1998), Saudi Arabia (Khormi and Kumar 2011), India (Bohra and Andrianasolo 2001; Bhandari 2008), French Guinea (Tran et al. 2004), Ecuador (Castillo et al. 2011) and in the United States (Erickson et al. 2010). Apart from mapping disease distribution, these studies have shown that GIS also provides a useful range of spatial analytical tools that can yield valuable information for the study of public health issues and enable health officials to plan for informed decision-making (Rezaeian et al. 2007).

The World Health Organization (Martinez 2007) has explained how GIS and related geospatial technology has the potential to aid in worldwide dengue prevention and control programmes. It gives the relevant personnel the ability to organise and link datasets from different sources. This enables them to access data from GPS receivers and digital imagery from satellites and aerial photos. Remote sensing can provide up-to-date information on soil moisture, vegetation type, land cover/use, urban planning, crop monitoring, forestry and water and air quality that influence the vector-borne disease occurrences. It also provides the capabilities for authorities to synthesise and visualise information in maps.

The aim of this chapter is to present a straightforward approach using spatial techniques to investigate and evaluate the spatial pattern of dengue virus in Dhaka between the years of 2005 and 2010. Our intention was to determine whether dengue virus cases in Dhaka are clustered or conform to the pattern known as complete spatial randomness.

20.2 Data and Methods

20.2.1 Data

Data on dengue cases for the period of 2005–2009 and the first half of 2010 was obtained from the record rooms of 11 major health service providers in Dhaka megacity (see Chap. 19) with a standardised patient abstraction form that includes date of admission, location of patient's residence, demographic and clinical data and date of discharge and outcome (dead/alive). Only those admitted to hospital with dengue fever were included in the database, and outpatients were excluded. The diagnosis of dengue was made by physicians at the respective hospitals, and some, but not all, were confirmed by laboratory investigation. To avoid data duplication, we first matched data in the case records using all the demographic variables and then cross-checked the data against the corresponding day/year in the logbooks of the hospitals. If a case occurred in both these records, then we included it in the database. We excluded cases residing outside of the study area (see Chap. 1) along with duplicates which resulted in a total of 3,169 dengue cases being available for analysis. Census tracts, the lowest level of census geography in Bangladesh, were used as a geographic data. Case mapping, creation of geographic feature dataset and encoding method were performed using procedures described elsewhere (Dewan et al. 2013).

Population data from the 2001 census, including a breakdown into male and female categories within different age groups, were obtained from the Bangladesh Bureau of Statistics community series (BBS 2003). Since the data were in tabular format, they were first encoded in a spreadsheet and then linked with the geographic unit by using a unique ID.

20.2.2 Analytical Techniques

GIS modelling has been used to investigate disease patterns in a number of different areas using a number of different methodologies. Morrison et al. (1998), for example, performed a space-time analysis of reported dengue cases during an outbreak in Florida and Puerto Rico in 1991–1992. Pratt (2003, p. 2) discussed how "incorporating traditional epidemiological statistical techniques into a GIS interface allows researchers to gain a greater insight into the spatial aspect of the spread of disease".

A study by Nakhapakorn and Tripathi (2005) followed two main processes for the analysis of dengue. Firstly, the relationship between cases and the areas' climatic variables was evaluated through multiple regressions, and then an Information Value approach was used to determine which physical and environmental factors are more crucial in dengue incidences. Mondini et al. (2008) examined the spatial correlation between dengue incidents and socioeconomic, demographic and environmental factors in a city in Brazil, whilst Haddow et al. (2009) claimed to be "the first use of smoothing techniques, the global Moran's I, and the Local Indicators of Spatial Association (LISA) to detect spatial clustering of La Crosse virus infections at a national level in the United States". More recently, Hsueh et al. (2012) examined the spatiotemporal patterns of dengue fever in Kaohsiung City, Taiwan. Their study focused on three main variables (density, transportation arteries and water bodies) and confirmed to some degree the importance of these variables in the spread of dengue fever.

The methodology used in this study comprises three major parts: data processing and inspection through visualisation, statistical testing to perform epidemiological analysis and spatial analysis employing autocorrelation techniques and cluster pattern identification.

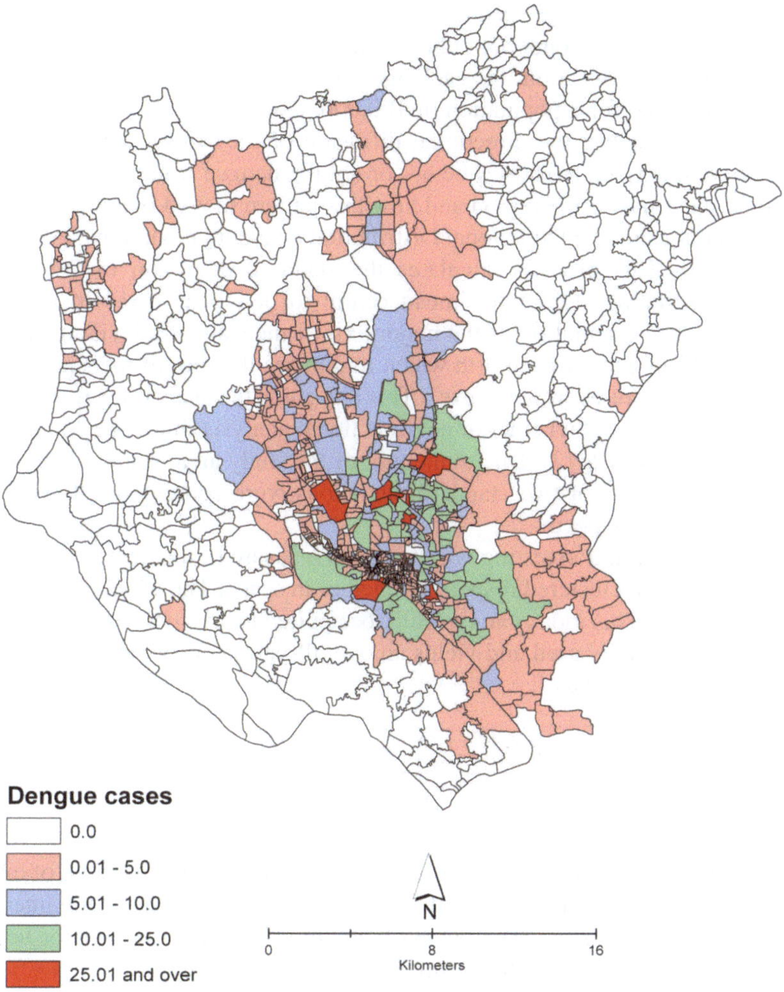

Fig. 20.1 Frequency of dengue cases by census tract, 2005–2010

Visualisation is an important tool for showing the change in disease patterns over time. An initial infection occurrence map was produced using ArcGIS 10 software (ESRI 2011), with appropriate symbolisation and a suitable number of classes. The resulting map (Fig. 20.1) portrays the frequency of virus occurrence over the region for the study period. The main data file comprised 3,169 dengue cases over a five-and-a-half-year period, attributed according to sex, season and age group. Using the SPSS software (SPSS Inc 1999), the data were initially analysed by year and season, to determine any obvious temporal pattern, and then by the patient's sex and age group, to determine any socio-statistical relationships. The dependent variables used for this section were the recorded frequency counts for the virus throughout the study period. The one exception to this is that in the case of the standardised age-specific incidence analysis – the independent variable used was not frequency count but rather the age-standardised incidences, computed as the ratio of the virus count for each age group to population count of each age group, standardised for a population of 100,000 people.

In order to visualise the directional diffusion of the virus, standard deviation ellipsoids were derived to show whether or not the dengue occurrences followed a directional pattern over the region and how this pattern moved over the years. This was carried out using the Directional Distribution (standard deviational ellipse) tool in ArcGIS 10. This process calculates the standard deviation of the x and y coordinates differences from the mean centre to define the axes of the ellipse (ESRI 2012). The orientation of the major axis of the standard deviation ellipse is that rotation from geographic north which minimises the sum of the squares of the deviation of the features from the axes. This can guide the analyst in terms of which regions to focus on, and whether or not directional analysis should be considered in later studies (Blewitt 2012). Due to the size of the study area, standard deviational ellipses with a radius of two standard deviations were computed, allowing for a wider directional perspective.

Spatial autocorrelation testing was used as a measure of the degree to which the occurrence data are clustered/dispersed together in space. The Moran's I index was used in this study. This index can typically be applied to area units where numerical ratio or interval data is used and yields an overall value for the whole dataset.

Moran's I is defined as

$$I = \frac{n \sum_{i=1}^{n} \sum_{j=1}^{n} w_{ij}(x_i - \bar{x})(x_j - \bar{x})}{\left(\sum_{i \neq j} \sum w_{ij}\right)\left(\sum_{i=1}^{n} (x_i - \bar{x})^2\right)} \tag{20.1}$$

where n is the number of spatial units indexed by i and j, x is the variable of interest, w is an element of a matrix of spatial weights and x_i is the value of the interval or ratio variable in area unit i. The value of Moran's I ranges from -1 for negative spatial autocorrelation to $+1$ for positive spatial correlation.

The weights matrix can be constructed based on the contiguity (adjacency) of the polygon boundaries or calculated from the distance between polygons centroids. For this analysis, queen's case contiguity was used, with a weight of 1 implying that polygons are adjacent and a weight of 0 implying non-contiguity.

Finally, the geographical pattern of the occurrence points was examined in more detail using the Local Indicators of Spatial Association (LISA) measure. LISA is a local autocorrelation measure proposed by Anselin (1995) to assess spatial autocorrelation and identify regions with disease rates statistically similar to and dissimilar from their neighbours. LISA analysis yields a measure of spatial autocorrelation for each individual location and allows us to identify high-high clusters (hotspots) in an area indicating the area's high values of a variable that are surrounded by high values on the neighbouring areas, as well as the low-low clusters (cold spots) which are areas of low values of a variable surrounded by low values. The procedure is implemented in the GeoDa software (Anselin et al. 2006).

The definition of the LISA measure is given below:

$$I(i) = \frac{x_i - \overline{x}}{\delta} \times \sum_{j=1}^{n} \left[W_{ij} \times \frac{(x_j - \overline{x})}{\delta} \right] \tag{20.2}$$

where $I(i)$ is the LISA index for region i, w_{ij} is weight describing the proximity of region i to region j, $x_i - \overline{x}$ is the deviation of region i with respect to the mean, $x_j - \overline{x}$ is the deviation of the region j with respect to the mean, δ is the standard deviation and n is the total number of the regions to be evaluated. The weights w_{ij} are set so that if region i is adjacent to region j, a value of 1 is assigned; a value of 0 is assigned otherwise. Adjacency may be assigned in a number of ways, usually in relation to the hypothetical movements of chess pieces (Anselin 1995).

In this study, as for the Moran's I calculations, the dependent variable used was total frequency of dengue cases for the study area with adjacency defined as being "queen's case". The GeoDA implementation of LISA uses a randomisation technique to infer significance of the results (Anselin 2003, 2004). Inference for significance of both global and local Moran's I was based on 499 permutations with an alpha level of 0.01 to test the statistical significance.

20.3 Results

20.3.1 Visual Inspection

Figure 20.1 is a choropleth map showing the total dengue frequency in the study period for each census tract with five classes determined by using a natural breaks algorithm. The map shows that the dengue cases are most common closer to the centre of the city with fewer occurrences towards the outer limits of the city.

Table 20.1 suggests that over the study period, there has been a general downward trend in dengue cases, with the exception of 2008, when recorded occurrences increased by 63 cases.

20.3.2 Epidemiology of Dengue

The monsoon season in Bangladesh runs from July to October, and the post-monsoon season (winter) defines from November to February, and the pre-monsoon season from March to June. Figure 20.2 shows the difference in seasonal distribution of the dengue virus. It is evident that the dengue virus is most active during the monsoon, with a few residual cases before and after. During

Table 20.1 Annual dengue occurrences, 2005–2010

Year	Number of cases	Percent (%) of total	Cumulative percent (%)
2005	727	22.9	22.9
2006	584	18.4	41.4
2007	497	15.7	57.1
2008	560	17.7	74.7
2009	421	13.3	88.0
2010	380	12.0	100.0
Total	3,169	100.0	

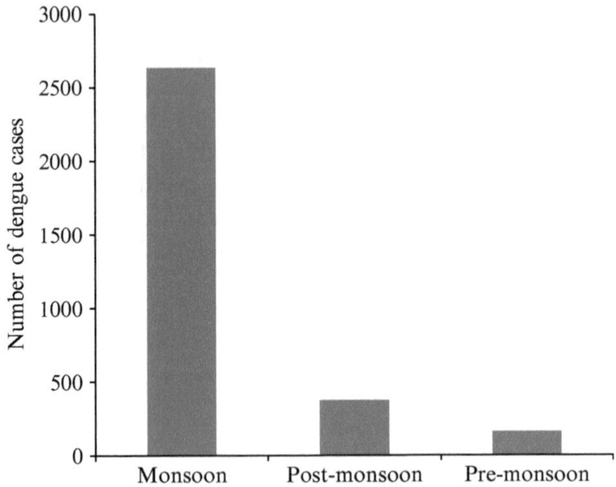

Fig. 20.2 Seasonal distribution of dengue occurrences

this monsoon period, very heavy rainfall lashes the city as well as the entire country, providing excellent damp breeding sites for the *Aedes aegypti* mosquito to thrive in.

Further descriptive analysis shows that there is a considerable difference between the number of male and female dengue cases over the years. From 3,169 recorded dengue patients, 72.6 % of total (2,301) were male, giving a male/female ratio of 2.65:1.

Table 20.2 shows that the age group of people most affected are between 18 and 34. This may be attributable to the fact that they are the most mobile and likely to be involved in the workforce. The next most affected age group is 35–59, closely followed by infants 0–4. The mean age of the dengue cases is within the 18–34 age group, whilst the median age (26) also lies in the same category. Surprisingly, our data does not show any cases for the age group of 5–9. This is similar to the result found for the prevalence of typhoid case in the same area (Dewan et al. 2013).

In analysing the number of deaths recorded in the data, we found that out of the 3,169 patients, almost 8 % (251 instances) of dengue cases resulted in death from the haemorrhagic complications of the disease, whilst 80 % (2,538 instances) of all cases reported being nonfatal classic dengue fever (the outcome is unknown for 2 % of the cases). Examining this in more detail, we found that the largest number of

Table 20.2 Standardised age- and gender-specific incidences

Age	Sex	Cases (%)	Population	Annual incidence rate (per 100,000)	Annual incidence rate for both sex
Total	Male	23,011 (72.6)	4,548,189	506.94	289.60
Total	Female	868 (27.4)	3,697,393	23.48	
0–4	Male	244	401,545	60.77	47.70
	Female	119	359,439	33.11	
10–14	Male	43	482,972	8.90	6.16
	Female	15	458,050	3.27	
15–17	Male	140	279,568	50.08	36.15
	Female	50	246,056	20.32	
18–34	Male	1,223	1,752,085	69.80	52.03
	Female	432	1,428,841	30.23	
35–59	Male	579	1,048,331	55.23	47.38
	Female	209	614,674	34.00	
60+	Male	67	183,561	36.50	34.52
	Female	40	126,443	31.63	

deaths from dengue is amongst the working age group (18–34) with 132 (53 % of total deaths) cases, followed by the 35–59 age groups with 65 total cases (25 %). However, the 0–4 age group had the highest number of deaths relative to the number of recorded cases in that age group, with 31 (12 %) out of 52 cases resulting in death. The 10–14 age group only made up 1 % of deaths, whilst the 15–17 and 60+ age groups together made up the remaining 9 %.

20.3.3 Spatial Analysis

A frequently used method of visualising the spatial trend, through time, of the attributes of a set of points or areas is to calculate the standard deviation of the points for each year. Figure 20.3 shows the overlapping standard deviational ellipses of the dengue occurrences per census tract over the years 2005–2009 (half-year data for 2010 excluded), each year being represented by a different colour. This shows that the dengue occurrences follow a diagonal South-South Easterly to North-North Westerly pattern with little change over the years.

Figure 20.4 shows the scatterplot results for the autocorrelation test carried out using GeoDa for each year of the dengue occurrence data. The Moran's I spatial statistic is visualised as the slope of the scatterplot with the spatially lagged variable on the vertical axis and the original variable on the horizontal axis. The slope of the regression line is Moran's I statistic and is shown at the top of each window.

In all instances, adjustment for outliers (highlighted with yellow in each diagram) was made. However, this had little overall impact as outliers were few and

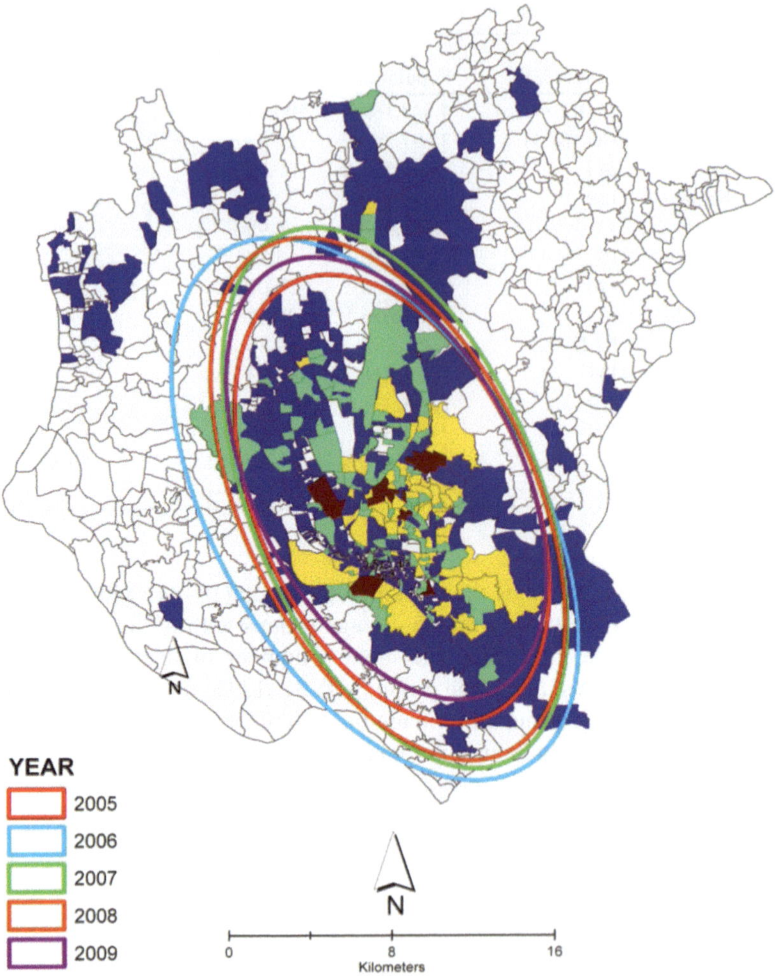

Fig. 20.3 Standard deviation ellipsoids of dengue occurrences, 2005–2009

insignificant. The scatter plots show that the actual dengue cases (*x*-axis) were positively correlated with the spatially lagged cases (*y*-axis) and follow a clustered pattern.

To assess the significance of the Moran's *I* statistic against a null hypothesis of no spatial autocorrelation, GeoDa uses a permutation procedure; in this case 499 permutations were used. Since each set of permutations is based on a different randomisation, the results will not be exactly replicable.

Results for the statistical significance testing is shown in Table 20.3, which shows significant positive spatial autocorrelation in dengue occurrences for the first 5 years recorded, with Moran's *I* statistics of 0.36 ($p = 0.01$) in 2005, 0.25 ($p = 0.01$) in 2006, 0.23 in 2007 ($p = 0.01$), 0.31 ($p = 0.01$) in 2008 and 0.42

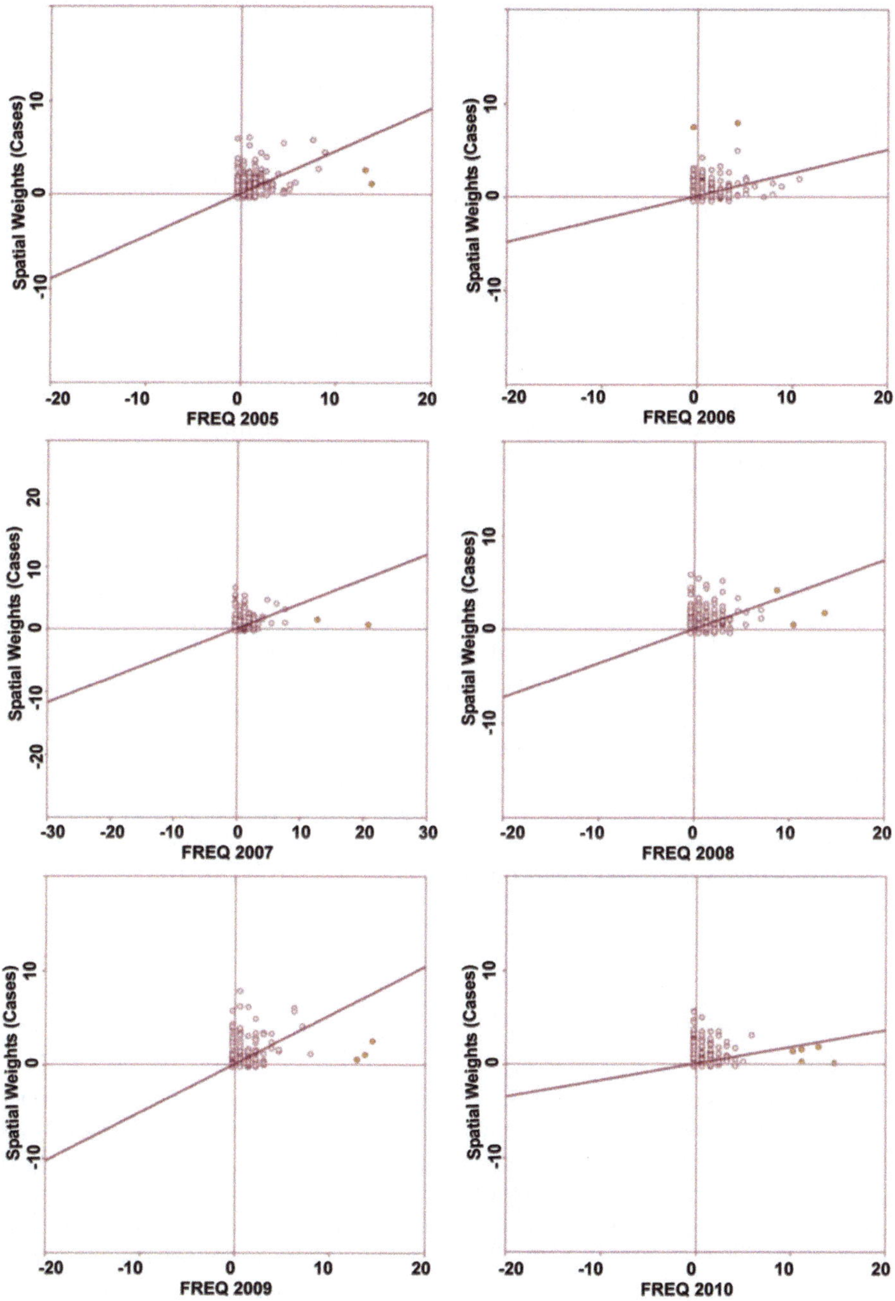

Fig. 20.4 Moran's *I* scatterplots for total dengue frequency, 2005–2010

Table 20.3 Spatial autocorrelation for dengue, 2005–2010 (non-outlier adjusted)

Period	Moran's I	Mean	SD	$E[I]$	p-value
2005	0.3578	−0.0018	0.0158	−0.0008	0.01
2006	0.2463	−0.0016	0.0195	−0.0008	0.01
2007	0.2259	−0.0027	0.0165	−0.0008	0.01
2008	0.3056	−0.0029	0.0197	−0.0008	0.01
2009	0.3204	−0.0005	0.178	−0.0008	0.01
2010	0.1751	−0.0018	0.0166	−0.0008	0.01

($p = 0.01$) in 2009. Whilst the 2010 data does not follow this trend, results for this year are deemed unreliable since they are from an incomplete year.

Since the Moran's I global spatial autocorrelation statistic indicated a clustered pattern of dengue cases in Dhaka, the analysis proceeded to investigate that pattern further. The LISA analysis produces two maps: a cluster map and a significance map. The combination of the two allows us to see which locations are contributing most strongly to the local outcome and in which direction.

The cluster map for dengue occurrences is shown in Fig. 20.5. The map distinguishes between clusters of high values, shown in red, which also have neighbours of high values (HH); clusters of low values, shown in blue, with low-value neighbours (LL); outliers, in pink, where a high value is surrounded primarily by low values (HL); and outliers, in pale blue, where low value is surrounded primarily by high values (LH). The strongly coloured regions on the map are therefore those that contribute significantly to a positive global spatial autocorrelation outcome, whilst paler colours contribute significantly to a negative autocorrelation outcome.

Figure 20.6 shows the statistical significance level of each region's contribution to the local autocorrelation outcome. This was determined using an automated complex Monte Carlo randomisation procedure (O'Sullivan 2012).

20.4 Discussion

The data visualisation provides a perspective on the nature of the disease in Dhaka megacity. Hanafi-Bojd et al. (2012) discuss how the maps produced by their study provided a visual tool for decision-making about initiating and focusing control programmes for malaria in Iran. However, whilst visualisation may be a powerful tool for providing a "bigger picture" perspective, it is still only a stepping stone to further analysis. As Wen et al. (2006) explained in a similar dengue study, visualisation cannot definitively confirm clustering of cases or spatial correlations.

The results for the epidemiological analysis show that dengue incidences had a yearly decreasing trend except in 2008. This may be due to higher virus awareness, changing environmental conditions or even more use of control measures in publicly accessible mosquito breeding sites. However, reporting on dengue in

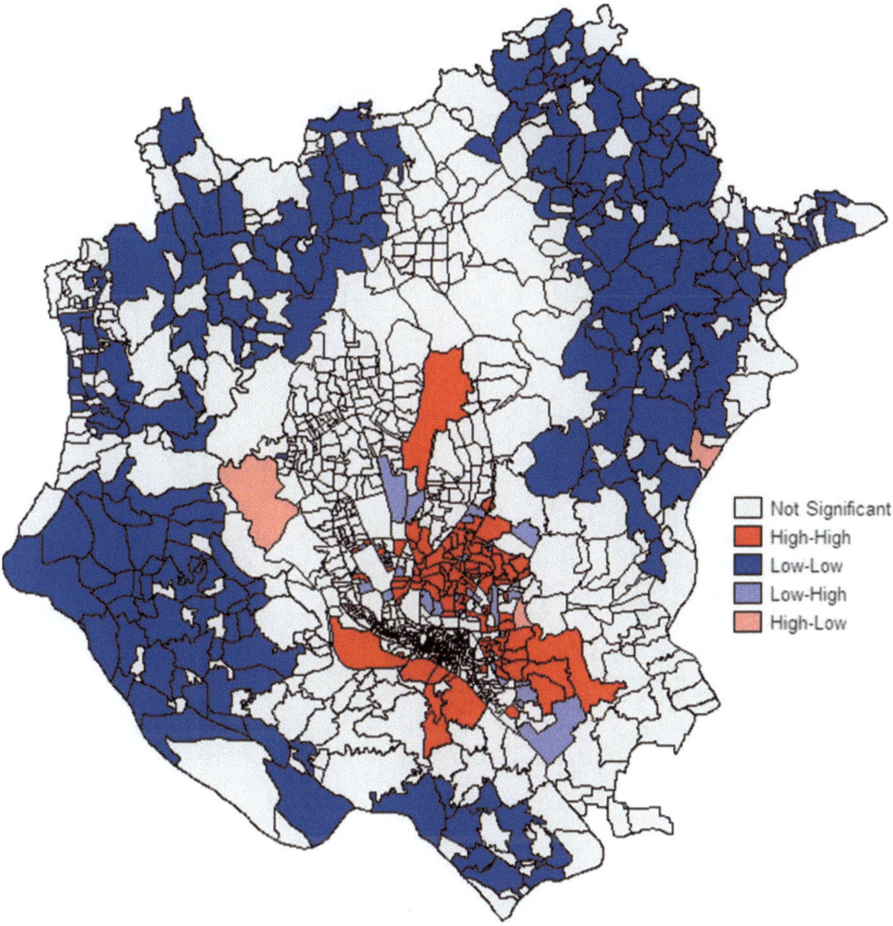

Fig. 20.5 Spatial clustering of dengue, 2005–2010

Dhaka, the United Nations Environmental Programme (2006) stated, "Even though the incidence of dengue fever has decreased, the mosquito problem prevails" (UNEP 2006, p. 62).

In the seasonal analysis, a significant difference in seasonal distribution of the dengue virus is noted, with cases being concentrated during the monsoon season where rainfall is significantly heavy in the region. This concurs with findings by Ahmed et al. (2007, p. 209) which showed that "the seasonal pattern of the mosquitoes was fairly close to variations in rainfall [...] the highest rainfall indicated the highest larval population except in May, which is the starting time of the rainy season in Bangladesh". Other studies related to the *Aedes aegypti* mosquito have shown similar seasonal patterns. Vezzani et al. (2004) found that the highest *Aedes aegypti* density was associated with accumulated rainfall above

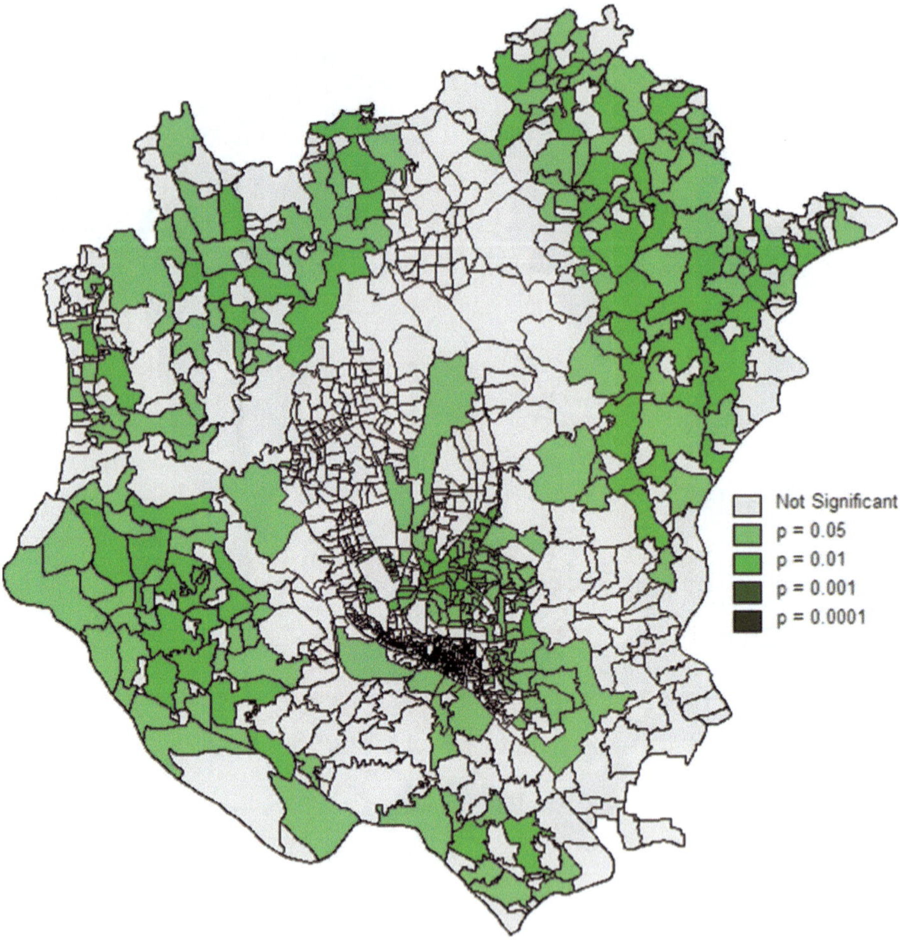

Fig. 20.6 Dengue significance map

150 mm. Micieli and Campos (2003) observed the close relationship of the highest peak of *Aedes aegypti* population with high rainfall and the fact that the mosquito population decreased during months with less rainfall. Additionally, Hashizume et al. (2012) presented strong evidence indicating that dengue fever increased with river levels, proving that factors associated with both high and low river levels increase the hospitalisations of dengue fever cases in Dhaka.

In regard to the significantly higher percentage of male patients than female patients, we propose two possible explanations. One is that men are more exposed to dengue-carrying mosquitoes during the daytime either at the workplace or whilst travelling to and from work. The other is that adult men are more likely to seek health care than adult women. A similar finding has been reported for typhoid incidence in the same area (Dewan et al. 2013). Anker and Arima (2011) have elaborated further on this issue in a paper looking at gender differences in dengue cases in six Asian countries.

It is widely recognised that in many of the Asian communities, lower disease incidence in women may be a statistical artefact related to lower reporting and to women seeking care from traditional practitioners who do not report to public surveillance systems. By the same token, women are less likely to be taken for care at a hospital when ill or are taken at late stages of disease, when no other options are available. Determining gender differences, both in infection and severity of disease, requires well-designed and targeted studies to capture both the biological and social factors that drive disease patterns in a community (Guha-Sapir and Schimmer 2005). Furthermore, dengue is typically regarded as a childhood disease and is an important cause of paediatric hospitalisation in Southeast Asia. Severe disease in Southeast Asia is also common in babies and young children, as found in this study, due to their low immunity (Ranjit and Kissoon 2011).

Whilst it is possible to get a general sense of the spatial orientation of cases by plotting a simple choropleth map such as Fig. 20.1, the use of standard deviation ellipses makes the trend clear. By looking at the orientation and size of standard deviation ellipses for several years, it is possible to predict which areas should prepare for a rise in incidence of that disease (Blewitt 2012), or whether the pattern remains largely static over time.

The global spatial autocorrelation analysis using Moran's I showed that the distribution of dengue virus was spatially clustered, for all years with complete data. The scatter plots indicated the presence of spatial dependence across the years. The highest indices were observed for the years 2005 and 2009 (0.45 and 0.5 ($p = 0.01$), respectively (after outlier elimination)), with the rest of the years remaining around the 0.3 value ($p = 001$). It is not unusual for data of an infectious disease, like dengue, to have a strong clustered spatial pattern due to the method of propagation of the disease involving proximity and neighbourhood (Jeefoo et al. 2011). This information corresponds with the public health opinion that dengue mainly occurs in clusters and does not spread regularly or randomly throughout the area (WHO 2009). There are many reasons as to why dengue occurrences appear to be strongly clustered around the heart of Dhaka City: urbanisation and population density (Wu 2009; Bhandari et al. 2008; Hsueh et al. 2012; Khormi and Kumar 2011; Ali et al. 2003), land use (Pathirana et al. 2009; Vanwambeke et al. 2006), presence of standing water and impervious surfaces in a high rainfall area (Pathirana et al. 2009), and socioeconomic factors relating to literacy and correct management of water awareness (Mondini and Neto 2008).

It is instructive to compare the land use maps shown in Chap. 5 with the cluster map in Fig. 20.5. This shows an apparent correspondence between the areas with a dense combination of both built-up and vegetation land use areas and the disease case count. However, this relationship requires further examination as this apparent causality may be due to multicollinearity of several external variables and needs to be tested further.

20.5 Conclusions

The incidence of dengue fever in Dhaka is a constant threat to the population and a recurring problem for the health authorities. Through spatial autocorrelation, clustering and epidemiological analysis, this chapter examines the spatial and temporal distribution of dengue cases in the study area. It identifies potential dengue risk areas in the city based on recorded virus frequency for 2005–2010. We have shown that there is a clear pattern of clustering to dengue virus occurrences in Dhaka.

The spatial distribution of dengue occurrences was observed to follow a geographically clustered pattern. This was confirmed by statistical testing using Moran's *I* which indicated strong autocorrelation within the data. The largest clusters of dengue cases were present around the centre of the city and around the heavily urbanised regions of the city. Whilst this may seem to suggest that the highest rate of dengue illness in Dhaka occurs in areas of high urbanisation, this is an incomplete picture.

Overall, the spatial analyses that were carried out were capable of identifying simple relationships within the data and will need to be taken further in order to effectively bring about more census tract-specific findings. This would require consideration of both the effects of urbanisation and other socioeconomic and biophysical factors in the region. Further analysis could assist in focusing and implementing precautionary and preventive strategies to more effectively monitor and control the incidence of dengue.

Acknowledgement We acknowledge the support of the International Foundation for Science (IFS), Sweden, for funding part of this work under a project (Reference: W4656-1) on which Ashraf M. Dewan was the Principal Investigator.

References

Ahmed MTU, Rahman GMS, Bashar K, Shamsuzzaman M, Samajpati S, Sultana S, Hossain MI, Banu NN, Rahman MS (2007) Seasonal prevalence of dengue vectors mosquito in Dhaka city, Bangladesh. Bangladesh J Zool 35(2):205–212

Ali M, Wagatsuma Y, Emch M, Breiman RF (2003) Use of geographic information system for defining spatial risk for dengue transmission in Bangladesh: role for *Aedes Albopictus* in an urban outbreak. Am J Trop Med Hyg 69(6):634–640

Anker M, Arima Y (2011) Male-female differences in the number of reported incident dengue fever cases in six Asian countries. West Pac Surveill Response J 2(2):17–23

Anselin L (1995) Local indicators of spatial association – LISA. Geogr Anal 27(2):93–115

Anselin L (2003) GeoDa 0.9 user's guide, Urbana-Champaign, Spatial Analysis Laboratory (SAL), Department of Agricultural and Consumer Economics, University of Illinois, Illinois

Anselin L (2004) GeoDa 0.95i release notes, Urbana-Champaign, Spatial Analysis Laboratory (SAL), Department of Agricultural and Consumer Economics, University of Illinois, Illinois

Anselin L, Syabri I, Kho Y (2006) GeoDa: an introduction to spatial data analysis. Geogr Anal 38(1):5–22

Bangladesh Bureau of Statistics (BBS) (2003) Community series-2001. http://www.bbs.gov.bd/PageWebMenuContent.aspx?MenuKey=142. Accessed June 2011

Bhandari KP, Raju PLN, Sokhi BS (2008) Application of GIS modelling for dengue fever prone area based on socio-cultural and environmental factors – a case study of Delhi city zone. Int Arch Photogramm Remote Sens Spat Inf Sci 37(B8):165–170

Blewitt M (2012) Multiple sclerosis geographics, investigating aetiology through space/time epidemiology. http://www.msgeographics.com/index_files/Page660.htm. Accessed 12 Dec 2012

Bohra A, Andrianasolo H (2001) Application of GIS in modeling of dengue risk based on sociocultural data: case of Jalore, Rajasthan, India. Dengue Bull 25:92–102

Braga A (2003) Serious youth gun offenders and the epidemic of youth violence in Boston. J Quant Criminol 19(1):33–54

Castillo KC, Korbl B, Stewart A, Gonzalez JF, Ponce F (2011) Application of spatial analysis to the examination of dengue fever in Guayaquil, Ecuador. Proc Environ Sci 7:188–193

CDC (2012) Dengue and the *Aedes aegypti* mosquito. United States Center for Disease Control, Atlanta. http://www.cdc.gov/dengue/resources/30jan2012/aegyptifactsheet.pdf. Accessed 25 Jan 2013

Choudhury Z, Banu S, Islam A (2008) Forecasting dengue incidence in Dhaka, Bangladesh: a time series analysis. Dengue Bull 32:29–36

Dewan AM (2013) Floods in a megacity: geospatial techniques in assessing hazards, risk and vulnerability. Springer, Dordrecht

Dewan AM, Corner RJ, Hashizume M, Ongee ET (2013) Typhoid fever and its association with environmental factors in the Dhaka Metropolitan Area of Bangladesh: a spatial and time-series approach. PLoS Negl Trop Dis 7(1):e1998

Environmental System Research Institute (ESRI) (2012) http://resources.esri.com/help/9.3/ArcGISengine/java/Gp_ToolRef/spatial_statistics_tools/directional_distribution_standard_deviational_ellipse_spatial_statistics_.htm. Accessed 15 Dec 2012

Erickson RA, Presley SM, Allen LJS, Long KR, Cox SB (2010) A dengue model with a dynamic *Aedes albopictus* vector population. Ecol Model 221(24):2899–2908

ESRI (2011) ArcGIS desktop: release 10. Environmental Systems Research Institute, Redlands

Gubler DJ (1998) Resurgent vector-borne diseases as a global health problem. Emerg Infect Dis 4(3):442–450

Guha-Sapir D, Schimmer B (2005) Dengue fever: new paradigms for a changing epidemiology. Emerg Themes Epidemiol 2(1):1–10

Haddow AD, Odoi A (2009) The incidence risk, clustering, and clinical presentation of La Crosse virus infections in the Eastern United States, 2003–2007. PLoS ONE 4(7):1–8

Hanafi-Bojd AA, Vatandoost H, Oshagi MA, Charrahy Z, Haghdoost AA, Zamani G, Abedi F, Sedaghat MM, Soltani M, Shahi M, Raeisi A (2012) Spatial analysis and mapping of malaria risk in an endemic area, south of Iran: a GIS based decision making for planning of control. Acta Trop 122(1):132–137

Hashizume M, Dewan AM, Sunahara T, Rahman MZ, Yamamoto T (2012) Hydroclimatological variability and dengue transmission in Dhaka, Bangladesh: a time-series study. BMC Infect Dis 12:98

Hsueh Y-H, Lee J, Beltz L (2012) Spatio-temporal patterns of dengue fever cases in Kaohsiung City, Taiwan 2003–2008. Appl Geogr 34:587–594

Igarashi A (1997) Impact of dengue virus infection and its control. FEMS Immunol Med Microbiol 18(4):291–300

Jeefoo P, Tripathi NK, Souris M (2011) Spatio-temporal diffusion pattern and hotspot detection of dengue in Chachoengsao province, Thailand. Int J Environ Res Public Health 8:51–74

Kalayanarooj S, Vaughn DW, Nimmannitya S, Green S, Suntayakorn S, Kunentrasai N, Viramitrachai W, Ratanachu-eke S, Kiatpolpoj S, Innis BL, Rothman AL, Nisalak A, Ennis FA (1997) Early clinical and laboratory indicators of acute dengue illness. J Infect Dis 176(2):313–321

Khormi HM, Kumar L (2011) Modeling dengue fever risk based on socioeconomic parameters, nationality and age groups: GIS and remote sensing based case study. Sci Total Environ 409(22):4713–4719

Martinez R (2007) Geographic information system for dengue prevention and control. World Health Organization report of the scientific working group on dengue-on behalf of the special programme for research and training in tropical diseases, WHO Regional Office for the Americas (AMRO)/PanAmerican Health Organization (PAHO), pp 134–139

Micieli MV, Campos RE (2003) Oviposition activity and seasonal pattern of a population of Aedes (Stegomyia) aegypti in Subtropical Argentina. Mem Inst Oswaldo Cruz 98(5):659–663

Mondini A, Neto FC (2008) Spatial correlation of incidence of dengue with socioeconomic, demographic and environmental variables in a Brazilian city. Sci Total Environ 393(2–3):241–248

Mondini A, Neto FC, Sanches MGY, Lopes JCC (2005) Spatial analysis of dengue transmission in a medium-sized city in Brazil. Revista De Saude Publica 29(3):444–451

Morrison AC, Getis A, Santiago M, Rigau-Perez JG, Reiter P (1998) Exploratory space-time analysis of reported dengue cases during an outbreak in Florida, Puerto Rico, 1991–1992. Am J Trop Med Hyg 58(3):287–298

Nakhapakorn K, Jirakajohnkool S (2006) Temporal and spatial autocorrelation statistics of dengue fever. Dengue Bull 30:171–183

Nakhapakorn K, Tripathi NK (2005) An information value based analysis of physical and climatic factors affecting dengue fever and dengue haemorrhagic fever incidence'. Int J Health Geogr 4(13):177–183

O'Sullivan D (2012) Local indicators of spatial association. Penn State College of Earth and Mineral Sciences. https://www.e-education.psu.edu/geog586/l9_p11.html. Accessed 12 Dec 2012

Pathirana S, Kawabata M, Goonetilake R (2009) Study of potential risk of dengue disease outbreak in Sri Lanka using GIS and statistical modelling. J Rural Trop Publ Health 8:8–17

Pratt M (2003) Down-to-earth approach jumpstarts GIS for dengue outbreak, The magazine for ESRI software users. http://www.esri.com/library/reprints/pdfs/arcuser_dengue-outbreak.pdf. Accessed 8 Aug 2012

Ranjit S, Kissoon N (2011) Dengue hemorrhagic fever and shock syndromes. Paediatr Crit Care Med 12(1):90–100

Rezaeian M, Dunn G, Leger SS, Appleby L (2007) Geographical epidemiology, spatial analysis and geographical information systems: a multidisciplinary glossary. J Epidemiol Community Health 61(2):98–102

SPSS Inc (1999) SPSS Base 10.0 for Windows user's guide. SPSS Inc., Chicago

Tran A, Deparis X, Dussart P, Morvan J, Rabarison P, Remy F, Polidori L, Gardon J (2004) Dengue spatial and temporal patterns, French Guiana, 2001. Emerg Infect Dis 10(4):615–621

United Nations Environmental Programme (UNEP) (2006) Regional resource centre for Asia and the Pacific State of the Environment report. Asian Institute of Technology, Dhaka

Vanwambeke SO, Van-Benthem BHB, Khantikul N, Burghoorn-Mass C, Panart K, Oskam L (2006) Multi-level analyses of spatial and temporal determinants for dengue infection. Int J Health Geogr 5(5)

Vezzani D, Velazquez SM, Schweigmann N (2004) Seasonal pattern of abundance of *Aedes aegypti* in Buenos Aires City, Argentina. Mem Inst Oswaldo Cruz 99(4):351–356

Wen T, Lin N, Lin C, King C, Su M (2006) Spatial mapping of temporal risk characteristics to improve environmental health risk identification: a case study of a dengue epidemic in Taiwan. Sci Total Environ 367(2–3):631–640

Wilder-Smith A, Chen LH, Massad E, Wilson ME (2009) Threat of dengue to blood safety in dengue-endemic countries. Emerg Infect Dis 15(1):8–11

World Health Organisation (WHO) (2009) Dengue: Guidelines for diagnosis, treatment, prevention and control – New edn. WHO Press, Geneva

World Health Organisation (WHO) (2012) Dengue and severe dengue (fact sheet). http://www.who.int/mediacentre/factsheets/fs117/en/. Accessed 13 June 2012

Wu P-C (2009) Higher temperature and urbanization affect the spatial patterns of dengue fever transmission in subtropical Taiwan. Sci Total Environ 407(7):2224–2233

Index

A
Accessibility, 6, 11, 13, 15, 26, 30, 38, 47, 56, 112, 126–128, 130–131, 135, 188, 195, 198, 214, 242, 264, 343, 347, 369, 378
Accuracy
  error matrix, 81
  fuzzy kappa, 129, 138, 139
  kappa coefficient, 81, 139
  misclassification, 80
  overall, 81, 82
  producers, 82
  users, 82
Acid throwing, 260, 264, 267
Adam, J.S., 47
Adaptation measures, 178
Adaptive kernel
  bandwidth, 351
ADB. *See* Asian Development Bank (ADB)
Advanced spaceborne thermal emission and reflection radiometer (ASTER), 220, 221, 304
Aedes aegypti mosquitoes, 368, 369, 374, 379, 380
Aedes albopictus mosquitoes, 368
Aeromonas hydrophila, 341
Aeromonas-positive, 18, 333–343
Aerosol, 11
  apportion pollution loads, 327
  carbon monoxide, 323
  chronic obstructive pulmonary disease, 11
  combustion emissions, 323–325, 330
  diesel particulate filter, 286
  dispersed pollution, 326–327
  domestic fuel, 325
  domestic fuel burning, 324, 325, 327
  emission rates, 324, 329
  fossil fuel combustion, 216, 330
  Gaseous Emission of Industries, 286
  nitrogen oxides, 323
  oxidation catalyst, 286
  particulate, 11
  particulate matter (PM), 11, 320, 323, 327
  resuspended road dust, 11, 320
  trace gases, 11
  urban, 11
  vehicle exhaust, 11, 320, 325, 327
  volatile organic compounds, 323
Aetiology, 356
Africa, 4, 24, 208, 238, 259
Agent-based models, 76, 100, 141
Agricultural suitability, 148, 155
Agrochemical wastewater, 304
Ahmad, N., 262
Ahmed, M.T.U., 379
Ahmed, S.J., 15, 17, 101, 107
AHP. *See* Analytic hierarchy process (AHP)
Ahsanullah Engineering School, 30
Air conditioning demands, 206, 220
Air pollution, 11, 17, 117, 320, 329
Akaike information criterion, 351, 354–356, 358
Alam, M., 13
Algae
  blue-green, 306
Algeria, 208
Ali, S., 18
Alluvial deposits, 2, 12
Amazonian floodplain, 164
Ambient Air Quality Standards, 286, 320
American Standard Code for Information Interchange (ASCII), 128
Amin, Md.R., 17

A. Dewan and R. Corner (eds.), *Dhaka Megacity: Geospatial Perspectives on Urbanisation, Environment and Health*, Springer Geography,
DOI 10.1007/978-94-007-6735-5, © Springer Science+Business Media Dordrecht 2014

AMIP. *See* Atmospheric Model Intercomparison Project (AMIP)
Analytic hierarchy process (AHP), 150, 151, 155, 157, 185
Andersen, M.A., 259, 276
Anderson Scheme Level I, 79
Angel, S., 100, 104
Aniya, M., 80
Anker, M., 381
Annual rainfall, 2, 3, 9, 13, 63, 68–70
Anselin, L., 372
Anthropogenic activities, 11, 75, 229, 358
Anthropogenic climate change, 330
AOI. *See* Area of interest (AOI)
Apportionment study, 320, 321, 327
Apportion pollution loads, 328
Aquaculture development, 148
Aquatic ecosystems, 305
ArcGIS, 49, 128, 267, 371, 372
Area of interest (AOI), 78, 80, 307
Arima, Y., 381
Arthropod borne virus, 367
Arthur, S.T., 225
ASCII. *See* American Standard Code for Information Interchange (ASCII)
Asia
  south, 188, 323, 329, 346
  south-central, 258
  southeast, 27, 346, 381
Asian Development Bank (ADB), 32, 100, 293, 296
Asian population, 24, 100
Assaults, 260, 261, 264, 267, 268, 271, 274
ASTER. *See* Advanced spaceborne thermal emission and reflection radiometer (ASTER)
ATMoS. *See* Atmospheric Transport Modelling System (ATMoS)
Atmospheric CO_2, 76
Atmospheric Model Intercomparison Project (AMIP), 165
Atmospheric Transport Modelling System (ATMoS), 324
Atomic Energy Center, 320, 327
Australia, 259, 303
Autocorrelation, 127, 335, 370, 372, 375, 378, 382
Autoregressive term, 335
Avian influenza, 347

B
Bacterial, 334, 341, 342
Ballal Sen, 25
Banani, 30, 31, 35, 112, 311
Bandwidth, 351, 360
Bangalore, 4
Bangla Bazar, 27
Bangladesh
  Bangladesh Bureau of Statistics (BBS), 1–3, 6, 8–11, 15, 16, 26, 37, 47–49, 77, 102, 148, 149, 179, 183, 187, 197, 242, 243, 254, 265, 269, 304, 305, 343, 347–349, 358, 370
  Bangladesh Centre for Advanced Studies (BCAS), 215
  Bangladesh Legal Digest, 288
  Bangladesh Meteorological Department (BMD), 62, 68, 335
  Bangladesh National Building Code (BNBC), 12
  Bangladesh period, 30–31
  Bangladesh police, 260–262, 264
  Bangladesh Power Development Board (BPDB), 204–206
  Bangladesh Transverse Mercator (BTM), 78, 128, 305
  Bangladesh Water Development Board (BWDB), 179, 335
Baqee, M.A., 262
Bargi dam, 358
Baridhara, 31, 84–85, 112
Barmah forest virus, 347
Bartlett's test of sphericity, 244
Baseline climate, 63
Batty, M., 46
Beraid, 8
Berry, B.J.L., 47
Bhatara, 8
Bhatta, B., 110
Bhawal, 25
Bid rent model, 127, 130
Bimodal seasonality, 334, 336, 342
Biodiversity, 75, 100
Biogeochemical, 76
Biological oxygen demand (BOD), 11, 286, 304
Biomass, 232, 323
Biophysical, 91, 105, 127, 130, 148, 179, 181, 219, 221, 226, 230, 233, 238, 239
Bishop Heber, 28
Bivariate correlation matrix, 246
Bivariate regression, 267
Blackbody temperature, 225
BLHI. *See* Boundary layer heat island (BLHI)
Blood meal, 368
BOD. *See* Biological oxygen demand (BOD)
Boundary layer heat island (BLHI), 220
Bradford-Hill criteria, 342
Bramley, G., 15, 17

Brazil, 259, 369, 370
Brezonik, P., 306
Brick kiln
  clusters, 321–323, 324, 327, 328
  emissions, 320, 321, 323–327
  Hoffmann kilns, 321, 329
  tunnel kilns, 329
  vertical shaft, 321, 322, 329
  zigzag technology, 321
Brick manufacturing industry, 15, 319–330
Brightness values, 302, 304, 305, 307, 311, 314
British Crown, 28
British period, 26, 28–30
Building construction rules, 287, 297
Built-up surface, 85, 89, 102, 104–105, 107, 108, 110, 114–116, 234, 243
Burden of typhoid, 346
Burglary, 13, 261, 262, 267–269, 271, 276, 277
Bursik, R.J. Jr., 274

C
Cahill, M., 260
Calcutta, 28, 47, 52
Calibration, 124, 125, 129, 132, 136, 138–141, 169, 225, 304, 306, 311
Campos, R.E., 380
Campylobacter spp., 334, 337
Canopy layer heat island (CLHI), 220
Capital investment, 204, 213
Carbon monoxide, 323
Cardiovascular disease, 325, 347
Caretaker government, 206
Carlson, R.E., 303
Carlson, T.N., 225
Cartographic visualisation, 267
Catalytic converter, 286
Catastrophe theory, 100
CBD. *See* Central business district (CBD)
Ceccato, V., 259
CEGIS. *See* Center for Environmental and Geographic Information Services (CEGIS)
Cellular automata, 76, 78, 81, 83, 89–91, 100, 132
Celsius, 225
Census boundaries, 48, 49, 102, 149, 190, 243
Center for Earth Observation and Digital Earth (CEODE), 305
Center for Environmental and Geographic Information Services (CEGIS), 48, 149, 183, 265
Central business district (CBD), 11, 27, 30, 46–50, 52, 53, 56, 102, 103, 133–135
Central place theory, 127
Centroids, 372
CEODE. *See* Center for Earth Observation and Digital Earth (CEODE)
CETP. *See* Common effluent treatment plant (CETP)
Chakma, S., 17
Chander, G., 305
Chandnighat, 26
Chaos theory, 100
Chauk, 27
Chennai, 4, 260
Chicago School of Sociology, 258
Childhood cancer, 347
Children abuse, 13, 261, 264, 268, 271
China, 47, 52, 177, 221, 304, 305, 309
Chittagong, 9, 329
Chlamydia, 347
Chlorophyll concentration, 303, 304
Cholera, 334, 337, 342, 343
Choropleth mapping, 53, 266, 269, 277, 373, 381
Chowdhury, A., 24
Chowdhury, M.H., 262
Chowdhury, R.A., 17
CHRI. *See* Commonwealth Human Rights Initiative (CHRI)
Chronic obstructive pulmonary disease, 11
CI. *See* Consistency index (CI)
CIR. *See* Crime incidence rate (CIR)
City planning and management, 30, 39, 41, 238
Clark, C., 46, 51
Clarke, K.E., 127
Clark, G.L., 302
Classic dengue fever, 368, 374
CLHI. *See* Canopy layer heat island (CLHI)
Climate change
  anthropogenic, 330
  baseline climate, 63
  El Niño–Southern Oscillation (ENSO), 73
  exposure, 334
  fourth assessment report, 63
  global warming, 3, 73
  Intergovernmental Panel on Climate Change (IPCC), 63
  Kyoto Protocol, 216
  non-climatic factor, 13, 149
  precursors, 329
Climate exposure, 334
Climatic stationarity, 178
Cluster maps
  cold spots, 271, 347, 350, 351, 372
  hot spots, 271, 347–349, 353
CMAQ, 320

CNG. *See* Compressed natural gas (CNG)
Coefficient of determination, 267, 274, 276, 302, 309–311, 351, 358
Cold spots, 271, 274, 347, 351–353, 372
Coliform bacteria, 341
Colombo Plan, 32
Combustion emissions, 323–325, 330
Common effluent treatment plant (CETP), 295
Commonwealth Human Rights Initiative (CHRI), 262
Community, 17, 49, 149, 180–184, 186–191, 193–198, 213, 214, 242–243, 254, 258, 271, 296, 298, 334, 341, 347–349, 352, 354, 359, 368, 370, 381
Community series, 9, 48, 49, 102, 254, 349, 370
Comparison matrix, 151–153, 185, 186
Complete conversion, 164, 169–174
Compressed natural gas (CNG), 37, 286
Concentric circle, 100
Concentric rings, 100
Conceptualisation, 46, 165, 166, 180–182, 266, 350, 351
Condition number, 352, 359
Confounding, 303, 336, 338, 342, 359
Confusion matrix, 101
Consistency index (CI), 153, 185, 186, 336, 339
Consistency ratio, 153, 154, 185
Constrain maps, 151
Contaminated water and food, 346, 348
Contiguous diffusion, 356, 358
Control and prevention, 287, 345
Conversion matrix, 125, 136–137, 226
Conversion of land use and its effects at small regional extent (CLUE-s) land use modelling framework, 124, 128, 140
Cooper, C.M., 303
Coping capacity, 181, 182, 184, 189–190, 193, 194, 198
Corner, R.J., 17, 18
Correlation matrix, 239, 243–246, 254
Crime
  acid throwing, 260, 264, 267
  assaults, 260, 261, 267, 271, 274
  burglary, 13, 261, 262, 267, 269, 271, 276, 277
  children abuse, 13, 261, 264, 268
  homicide, 259, 260, 267, 271
  incidence, 17, 258, 260, 262–264, 266–277
  juvenile delinquents, 258, 262
  murder, 13, 260–262, 267, 268, 271, 277
  narcotics-related, 260, 261, 264, 267, 268, 271
  property, 259, 260, 264, 269, 271, 276, 277
  racially motivated, 258, 259
  ransom, 260, 264, 267
  rape, 260, 264, 267
  rescue-related, 268, 271, 277
  robbery, 13, 261, 262, 264, 267, 268, 271
  violence against women, 260–262, 264, 267, 268, 277
  violent, 258–260, 267, 268, 271, 277
Crime incidence rate (CIR), 266, 267, 269, 271, 272, 274–277
Criminal syndicates, 13, 262–263
Criminological conditions, 265
Culpin Planning Ltd, 33
Cutter, S.L., 179, 181
Cyclone, 3, 62, 182, 262
Cyprus, 339

D
Dacca Improvement Trust (DIT), 30, 31, 286–287
Dacca Master Plan 1959, 31, 35–38
Dacoity
  gang robbery, 13, 267, 268, 271
Daily mean temperature, 62, 63
Dakshinkhan, 8
DAP. *See* Detailed Area Plan (DAP)
Dasymetric mapping, 243
Daudpur, 13
Dawes, C., 28
Decision making, 8, 77, 91, 141, 150, 215, 258, 298, 368, 369, 378
Deficiencies, 36, 204, 217, 297, 321
Degradation
  environmental, 2, 3, 34, 77, 78, 91, 101, 108, 149, 239, 284, 294, 347
  land, 2, 76, 91
  soil, 75
Delhi, 4, 11
Demand and allocation analysis, 125, 149
Demographic, 8, 9, 15, 16, 49, 102, 128, 239, 254, 258, 263, 265, 349, 350, 352–354, 358, 359, 369, 370
  composition, 258
  and socioeconomic, 48, 128, 183, 242, 243, 254, 259, 260, 263, 277, 349–356, 358–360, 370
Demra, 8, 274

Dengue
classic, 368, 374
fever, 368–370, 374, 379, 380, 382
hemorrhagic fever, 368
occurrences, 372, 374–376, 378, 381, 382
prevention and control programmes, 369, 378
shock syndrome, 368
Densifying areas, 112, 114
Density
housing, 242, 243, 245–247, 249, 253
map, 104, 110, 116, 324
medium, 104, 110, 112, 151
population, 8, 11, 13, 15, 31, 46, 48–53, 56, 77, 102, 104–105, 112–116, 140, 149, 151–154, 178, 186, 188, 189, 193, 196, 198, 221, 242, 245–246, 249, 253, 254, 274, 275, 349, 350, 352, 355, 381
urban, 46–49, 56
Department of Energy (DOE), 165
Department of Environment (DOE), 284–291, 293, 294
Detailed Area Plan (DAP), 33, 35, 38, 49, 149, 183, 242
Developing countries, 4, 151, 157
Dewan, A.M., 17, 18, 77, 101, 179, 181, 209, 285, 348
Dhaka
Dhaka City Corporation (DCC), 6, 8, 13, 15–17, 33, 35, 39, 40, 209–211, 213, 239–241, 244, 249, 254, 287, 347
Dhaka College, 28, 29
Dhaka Electricity Supply Authority (DESA), 205, 206
Dhaka Electric Supply Company (DESCO), 205
Dhaka megacity (DM), 4, 6–13, 15, 41, 48, 78, 90, 91, 110, 158, 205–207, 209, 213, 222, 276, 285, 298, 369, 378
Dhaka Metropolitan Area (DMA), 8, 13, 15–17, 34, 37, 128, 134, 209, 264–267, 271–273, 276, 320–321, 324–329, 349
Dhaka Metropolitan Area Integrated Urban Development Plan (DMAIUDP), 31–33, 36
Dhaka Metropolitan Development Plan (DMDP), 6, 8, 13, 33–36, 38–40, 131–132, 148, 149, 157, 182, 347
Dhaka Metropolitan Police (DMP), 264, 275
Dhaka Municipal Corporation Ordinance, 1993, 287
Dhaka-Narayangonj-Demra (DND), 85
Dhaka Paurashava, 287
Dhaka Statistical Metropolitan Area (DSMA), 6
Dhaka University, 30, 207, 208
Dhaleswari River, 192
Dhamsana, 8
Dhanmondi, 30, 31, 85, 112, 269
Dholai khal, 24, 26
Diagonalisation, 81, 243, 345
Diarrhoea, 18, 333–343
Diarrhoeal transmission, 334
Diesel particulate filter, 286
Diffusion
contiguous, 356
expansion, 356
hierarchical, 356
relocation, 356
Digital elevation model, 149, 165, 179, 183, 184
Digital number (DN), 225, 305
Dilapidated drainage systems, 178
Disaster
management, 178, 179
mitigation, 181
preparedness, 181
risk, 178, 180
risk reduction, 178, 180
Disaster Management Bureau (DMB), 178, 183, 187, 188
Disaster resilience of place (DROP), 180–182
Discharge, 11, 164–169, 285, 293–295, 304, 311, 314, 369
Discharge of sewage, 311
Discriminant function, 239
Disease
Aeromonas-positive, 18, 333–343
arthropod borne virus, 367
avian influenza, 347
Barmah Forest virus, 347
childhood cancer, 347, 381
chlamydia, 347
cholera, 334, 342
dengue, 18
diarrhoea, 334, 339, 340, 342, 343
distribution, 305, 357, 369
epidemics, 343, 368, 373–375
food-borne disease, 305, 348
harbouring or spreading diseases, 359

Disease (*cont.*)
headaches, 368
immunity, 381
incidence, 18, 239, 346, 354, 358, 370, 379, 381, 382
malaria, 347, 358, 367, 378
muscle pains, 368
parasitic agents, 334
respiratory and cardiovascular diseases, 325
transmission, 334, 358, 360
typhoid, 305, 345–360, 374, 380–381
Vibrio cholera, 334, 336–338, 342
Vibrionaceae, 334, 342
water-borne disease, 117, 183, 189, 340, 341, 347, 348
water-related diseases, 13, 358
yellow fever, 367
Dispersed pollution, 327
Dispersion model (ATMoS), 320–321, 324
Dissolve oxygen (DO), 11, 286, 304
DIT. *See* Dacca Improvement Trust (DIT)
DMB. *See* Disaster Management Bureau (DMB)
DN. *See* Digital number (DN)
DO. *See* Dissolve oxygen (DO)
Domestic fuel, 325, 327
Domestic fuel burning, 323
Downgrading, 263
Dowry, 262
Drainage system
dilapidated, 178
facilities, 149, 178
Drinking water, 11, 15, 183, 335, 338, 341–343, 347, 358
DROP. *See* Disaster resilience of place (DROP)
Dutt, A.K., 10
Dynamic pressures, 180

E

Earth filling, 12
Earth Observation-1, 77
Earthquake, 12, 182
Earth Resources Technology Satellite-1 (ERTS-1), 101, 302
Earth system, 75, 100, 209, 220
East India Company, 27, 28
East Pakistan, 4, 30
Ecological changes, 358
Econometric modelling, 52, 56, 78, 104
Economic development, 10, 174, 205, 219–220, 258, 297
Ecosystem
aquatic, 305
services, 76
Ecuador, 369
Edge effect, 265
Egypt, 115
Eigenvalue, 153, 185, 186, 244, 247, 254
Eigenvector, 185
Electricity, 15, 28, 38, 204–207, 211, 213–217, 290–291
Elevation, 9, 13, 30, 130, 134, 140–141, 150–155, 164, 168, 184, 186, 187, 195, 196, 246
El-Nino Southern Oscillation (ENSO), 73
Emergency management, 179
Emission rates, 329
Emissivity, 225
EMP. *See* Environmental Management Plan (EMP)
Empirical method, 302
Empirical-statistical models, 76, 132
Endlicher, W., 17
Enforcement, 33, 262, 264, 288, 294, 295, 297, 298
Enhanced Thematic Mapper Plus (ETM+), 302, 304
ENSO. *See* El-Nino Southern Oscillation (ENSO)
Enteric pathogens, 335
Entropy theory, 100, 104, 110
Environmental
auditing, 290
awareness programs, 294
clearance, 289–291
Clearance Certificate (ECC), 285, 290–291, 294
Conservation Act (ECA)1995, 284, 287–289, 296–297
Conservation Rules (ECR) 1997, 284, 288, 289
constraints, 85, 107, 151
Court Act, 288, 293
degradation, 2–3, 77, 78, 91, 101, 108, 149, 239, 284, 294, 347
governance, 17, 283–298
hazards, 12, 182
human-related, 77, 220
impact assessment (EIA), 17, 185, 284, 285, 287–290, 292–298
impacts, 46, 62, 76, 91, 100, 101, 232, 285, 288, 295, 296
impact statement (EIS), 288, 290, 294, 296
planning, 91, 147, 174, 293

pollution, 238, 285–287
processes, 148
quality of city, 239
SEA, 296
urban, 91, 101, 158, 181, 219, 220, 232, 233, 238, 239, 243, 249, 287, 297
Environmental Management Plan (EMP), 290
Environmental System Research Institute (ESRI), 105, 351, 371, 372
Environment Research and Technology Development Fund, 73
ENVISAT, 304
Epidemics, 117, 343, 358, 368
Epidemiology, 334, 342, 370, 373–375, 378, 382
Epistemological orientations, 180
EPZ. *See* Export processing zone (EPZ)
EREC. *See* European Renewable Energy Council (EREC)
Error matrix, 81
Error propagation, 80, 81
ERS. *See* European remote sensing (ERS)
ERTS-1. *See* Earth Resources Technology Satellite-1 (ERTS-1)
Escherichia coli (ETEC), 334
ESDA. *See* Exploratory spatial data analysis (ESDA)
ESRI. *See* Environmental System Research Institute (ESRI)
Estimated value, 355
Ethnic composition, 258
ETM+. *See* Enhanced Thematic Mapper Plus (ETM+)
Euclidean distance, 49, 151, 186
European remote sensing (ERS), 304
European Renewable Energy Council (EREC), 216
European traders, 27, 28
Eusuf, A.Z., 47
Evans, D.J., 258
Evapotranspiration, 165, 220
Expansion diffusion, 356, 358
Explanatory Report, 35
Exploratory analysis, 301–314
Exploratory spatial data analysis (ESDA), 244–246, 259, 266
Exponential, 46, 49–52, 267, 275
Export processing zone (EPZ), 83, 114
Exposure risk, 327, 334
Extension growth, 109, 110, 116

F

Faber, B.J., 80, 243
Factor analysis, 239, 243, 244, 246–249, 254
Family disputes, 262
Farmlands, 31
Faruque, A.S.G., 17
Faruqui, S., 24
Fatality rates, 346
FCBTK. *See* Fixed chimney bull trench kilns (FCBTK)
Feature dataset, 183, 242, 265, 370
FEM. *See* Finite element method (FEM)
Female mosquito, 368
Field survey, 48, 79, 183, 311
Fiji, 339, 340
Finite element method (FEM), 164
Finland, 304
Fixed chimney bull trench kilns (FCBTK), 321, 322, 329
Flood
affected-frequency, 184, 185, 197
control project, 358
control works, 178
floodwater depth, 184, 185, 197
fluvial, 12, 127, 148, 183
forecasting, 83, 117
pluvial, 12, 127, 148, 183
policy, 178
regulation, 169
risk, 17, 35, 174, 178, 179, 188, 195, 197, 198
risk management, 178, 188
water depth, 184, 185, 197
Flooding, 3, 12, 13, 17, 25, 31, 34, 36, 108, 112, 117, 148, 155, 157, 164, 169, 174, 178, 182–184, 343, 348
Fluvial flooding, 12, 127, 148, 183
Food-borne disease, 305, 348
Food containers, 369
Fossil fuel combustion, 330
Fourth Assessment Report, 63
Fractals theory, 100
Fractional vegetation cover (FVC), 221, 226
Fraser, R.N., 303
French, 27, 28
French Guinea, 369
Frenkel, A., 100
Fuzzy kappa, 129, 138, 140
FVC. *See* Fractional vegetation cover (FVC)

G

Galerkin, 168
Galster, G., 100
Ganges, 2
Garbage, 28, 343
Gaseous Emission of Industries, 286

Gazipur, 6, 8, 33, 321, 323, 324, 328–329
GCPs. *See* Ground control points (GCPs)
GDP. *See* Gross domestic product (GDP)
Geddes, Sir Patrick, 29
GEF. *See* Global Environmental Facility (GEF)
Gentrification, 263
GeoDa, 266, 372, 373, 375, 376
GEOEYE-1, 79
Geographically weighted regression (GWR), 260, 277, 351–356, 358–360
Geographic information science (GIS), 239–244, 249, 254, 259
Geo-Informatics and Space Technology Development Agency (GISTDA), 224
Geologically, 186
Geophysical, 204, 207–213, 217
Geospatial techniques, 17, 46, 76, 77, 101, 148, 149, 179, 180, 182, 197, 259, 264, 346, 369
Germany, 214, 216, 259
'Get-rich-quick'syndrome, 263
GHG emission, 204, 216
Giardia lamblia, 334
Gillies, R.R., 226
GI Science, 147
GISTDA. *See* Geo-Informatics and Space Technology Development Agency (GISTDA)
Global carbon cycle, 76
Global Environmental Facility (GEF), 205, 208
Global horizontal irradiance (GHI), 204, 205, 207–209, 217
Global model, 346
Global positioning system (GPS), 49, 79, 306, 369
Global warming, 3, 13, 73
GoB. *See* Government of Bangladesh (GoB)
Google Earth, 149, 265, 321
Governance quality, 215–216, 293
Government of Bangladesh (GoB), 32, 33, 102, 149, 151, 178, 283, 288, 296–298
GPS. *See* Global positioning system (GPS)
Grameen Shakti (GS), 206, 214
Grandeur, 28
Grasmick, H.G., 274
Green, N.E., 239
Griffiths, P., 101
Gross domestic product (GDP), 2, 10, 77
Ground-based SDT, 307
Ground control points (GCPs), 78, 305
Groundwater, 2, 11, 285, 340, 341, 358
GS. *See* Grameen Shakti (GS)
Gulshan, 30, 31, 35, 84, 112, 205, 269, 271, 311
Gurjar, B.R., 11
Guttikunda, S.K., 17
GWR. *See* Geographically weighted regression (GWR)

H

Haddow, A.D., 370
Haider, Md.R., 17
Hails, 62
Hai, P.T., 169
Hanafi-Bojd, A.A., 378
Handcraft cotton, 28
Haphazard growth, 149, 240
Harirampur, 8
Harmonics, 325, 335, 336, 338, 342
Harries, K., 276
Hasan, M.R., 149
Hashizume, M., 18, 380
Hayashi, T., 17
Hazards
  catastrophic events, 178
  earthquake, 12, 182
  flood, 12, 178, 179, 182–185, 187, 190–193, 195, 197, 198, 285
  hails, 62
  nor'wester, 12
  storm surge, 3
  thunderstorms, 2, 62
  tornado, 12, 62
  water-related, 178
Hazards-of-place (HOP), 180–182
Hazratpur, 13
Health service providers, 348, 360, 369
Heat island
  boundary layer, 220
  canopy layer, 220
  surface urban, 220
Heteroscedasticity, 351
Heywood, I., 157
Hierarchical decision model, 151, 185
Hierarchical diffusion, 356
HIES. *See* Household Income and Expenditure Survey (HIES)
High-high, 271, 372
High-low, 271
Hiramatsu, K., 17
Hoaglin, D., 49
Hoffmann kilns, 321, 322, 329
Holt, J.B., 243

Homicide, 258–260, 264, 267, 271
HOP. *See* Hazards-of-place (HOP)
Hoque, S., 17
Hospitalization, 360, 367, 380, 381
Hot spots, 124, 259, 273, 274, 329, 350–354, 356, 372
Household Income and Expenditure Survey (HIES), 2, 183, 243, 265
Households
  general, 8, 16
  Income and Expenditure Survey (HIES), 2, 183, 243, 265
  institutional, 8
  size, 8, 9, 246, 247, 249, 359
Housing
  census, 2, 3, 15, 56, 183, 242, 265, 349
  density, 242–243, 246, 247, 249, 253
  Katcha, 15, 188, 189, 196–198, 242, 243, 265
  pucca, 15, 188, 189, 242, 243, 245–247, 265
  semi-pucca, 15, 188, 189, 242, 243
  types, 188, 189, 242
Hsueh, Y.-H., 370
Human activity, 100, 222, 226, 305, 311, 314
Human-induced, 76, 77
Human–nature interaction, 180
Human-related environmental changes, 77, 220
Hungary, 303, 304, 309
Hybrid classification technique, 79
Hyderabad, 4
Hydrological cycles, 76
Hygiene and sanitation practices, 335, 340–341

I

ICDDR,B. *See* International Centre for Diarrhoeal Disease Research, Bangladesh (ICDDR,B)
IGBP. *See* International Geosphere-Biosphere Project (IGBP)
Iglewicz, B., 49
IHDP. *See* International Human Dimensions Program (IHDP)
IKONOS, 79, 304
Illegal development, 157, 284
Illuminated roof-tops, 204, 205, 209–211, 217
Image combination, 125, 185, 303
Image ratio, 307, 309, 311, 314
Immunity, 381
Incidence, 2, 10, 12, 17, 18, 241, 258, 260, 262–264, 266, 268–271, 273–278, 334, 335, 339, 343, 346, 354, 358, 359, 370–372, 375, 378–382
Independent Power Producers (IPPs), 206
India, 1, 4, 11, 25, 28, 30, 47, 177, 221, 259, 304, 322, 323, 358, 369
Indian Ocean dipole mode, 73
Indian Remote Sensing (IRS), 78, 79, 304
Industrial effluents, 11
Industrial wastewater, 311
Inequality, 10, 13, 258, 260, 277
Infection rate, 350, 368
Infill growth, 108
Information Value approach, 370
Informed decision making, 77, 91, 258, 369
Inhomogeneous, 351
Inland water systems, 304, 311
Inorganic components, 11, 304
Insolation, 208
Institutional households, 8
Intergovernmental Panel on Climate Change (IPCC), 63
International Centre for Diarrhoeal Disease Research, Bangladesh (ICDDR,B), 335, 337, 341, 343
International Geosphere-Biosphere Project (IGBP), 76
International Human Dimensions Program (IHDP), 76
IPCC. *See* Intergovernmental Panel on Climate Change (IPCC)
IPPs. *See* Independent Power Producers (IPPs)
Iran, 221, 378
IRS. *See* Indian Remote Sensing (IRS)
%ISA. *See* Percent impervious surface (%ISA)
Islam Khan Chisti, 26
Islam, M.A., 303
Islam, M.M., 184
Islam, N., 6, 8, 10, 24
Islam, T., 3, 262
ISODATA. *See* Iterative self-organizing data analysis (ISODATA)
Iterative self-organizing data analysis (ISODATA), 79

J

Jafarabad, 27
Jahangirnagar, 26–27
Jaishankar, K., 259, 260
Jankowski, P., 157
Japan, 216, 221
Jensen, I.R., 80
Jiang, F., 100
Joint pains, 368
Juvenile delinquents, 258

K
Kabir, M.H., 17, 205, 210–212
Kabir, S.M.Z., 17
Kadamrasul, 6
Kaiser-Meyer-Olkin (KMO), 244
Kalatia, 8
Kaliganj, 8, 321, 324, 326, 354
Kappa coefficient, 83, 89, 129, 138–140
Kasarda, J., 47
Kashimpur, 8
Katcha, 15, 188, 189, 194–198, 242, 243, 265
Kaufmann, R.K., 76
Kawahara, M., 168
Kawamura, Y., 17
Kernel, 104, 351, 360
Key driving factors, 123–141
Key variables, 126–128
Khals, 34
Khan, I., 25–27
Khan, S., 27
Khan, T., 17
Khulna, 329
Kidnapping, 13, 261, 267
Kiguchi, M., 17
Kim, K.S., 46
Kinney, J.B., 260
Kloiber, S.M., 306, 307, 309
KMO. *See* Kaiser-Meyer-Olkin (KMO)
Knox, P.L., 263
Kolkata, 4
Konabari, 8
Korea, 221
Kumartuli, 25–26
Kutubpur, 13
Kyoto Protocol, 216, 217

L
Lack of education, 359
Lagos, 341
Lagrangian Puff-transport model, 324
Lags, 335–336, 339, 341
Lake, 11, 79, 130, 169, 285–286, 303–306, 311–314
Lakhya, 9, 285
Lambert, E.G., 260
Land
  conversions, 85, 90, 125, 129, 136, 148, 169–174, 226–228
  degradation, 76, 91
  disputes, 262
  LULC, 17, 75–91, 101, 102, 124, 128, 149, 151, 165–166, 169, 183, 219–233
  regulations, 287
  surface temperature (LST), 17, 219–233, 243
  values, 31, 56, 112, 125, 128, 140, 141, 242
Landfalling, 3
Landlessness, 2–3, 77
Landsat
  ERTS-1, 101, 302
  ETM+, 221, 302, 304
  MSS, 302–304
  TM, 78, 79, 183, 191, 221–226, 301–314
  TM 5, 17, 78, 183, 224, 225, 232, 243, 305
Landscape
  evaluation and planning, 148
  metrics, 101, 110, 116
Land surface temperature (LST), 17, 219–233, 243
Land use and land cover (LULC), 17, 75–91, 101, 102, 124, 128, 149, 151, 165–166, 169, 183, 219–233
Land use prediction map, 125
Land use regulations, 287
Large metropolis, 77
Larvae, 368
Lauritesen, J.L., 258
Lau, S.S.Y., 110
Leapfrogging development, 141
Le, H., 323
Li, G., 244, 249, 253
Lightning, 62
Lillesand, T.M., 307
Linear
  association, 336, 343
  regression, 127, 274, 275, 277–278, 335
    logarithmic, 50
    negative exponential, 52
    power, 52
  relationship, 336
  unconstrained distributed lag models, 336
LISA. *See* Local indicators of spatial autocorrelation (LISA)
Literacy rate, 2, 9, 10, 15, 16, 189, 190, 196, 198, 242, 245–247, 249, 253, 265, 349, 350, 352, 354
Living under unsanitary conditions, 359
Loading matrix, 247
Local and global G_i*, 346
Local collinearity, 351–352
Local indicators of spatial autocorrelation (LISA), 266–267, 271, 277, 346, 372, 373, 378

Local models, 351, 359
Locational vulnerability, 186–187, 190, 191, 193, 194, 197–198
Lo, C.P., 80, 243
Logarithmic, 50, 52, 132, 267, 275, 308
Logistic regression, 125–127, 129, 133–135
Log-linear, 336
Low density development, 110, 116
Lower income groups, 10, 346
Lower income status, 359
Low-high, 271
Low-low, 372
Lowry model, 130
LST. *See* Land surface temperature (LST)
Lubunda, 9
LULC. *See* Land use and land cover (LULC)

M

Macfarlane, P.W., 31–32
Madhupur, 9, 152, 154, 164, 187
Mahtab-uz-Zaman, Q.M., 110
Maithel, S., 323
Malaria, 346–347, 358, 367, 378
Malaysia, 221
Management
- disaster, 179
- environmental, 158, 295–297
- flood risk, 178, 179, 188
- resource, 77

Manikganj, 321
Manning's roughness coefficient, 168, 170
Mann-Kendall rank, 63
Manual techniques, 148, 258
Manufacturing industries, 11, 15, 319–330
Map
- choropleth, 53, 266, 271, 277, 373, 381
- density, 104, 110, 116
- edge effect, 265
- modifiable areal unit problem (MAUP), 265
- significance, 378, 380
- urban suitability, 155

Marginalised groups, 31, 178
Marginal population, 9, 10
Markham, B., 305
Markov-CA modelling, 89–91
Markov-cellular model, 76, 89
Markov chain analysis (MCA), 76, 78, 81, 89
Masuya, A., 17
Matuail, 8
Maximum likelihood classification (MLC), 80
Maximum rainfall, 68
MCA. *See* Markov chain analysis (MCA)
McDonald, J.F., 47
MCE. *See* Multi-criteria evaluation (MCE)
McKay, H.D., 258–259
McMillen, D.P., 47
Mean centre, 50–51, 372
Mean housing value, 242, 245–247, 249, 253, 349, 350, 355
Medical waste, 12
Medium-density, 110, 112
Medium Resolution Imaging Spectrometer (MERIS), 304
Megacity, 4–13, 15, 23–41, 77, 85, 99–101, 117, 148, 149, 206, 211
Meghna, 2, 9
Mei, C.L., 360
Meng, Y., 46, 47
MERIS. *See* Medium Resolution Imaging Spectrometer (MERIS)
Meteorological
- disasters, 62
- disturbances, 62

Methods and techniques
- Akaike information criterion (AICc), 351
- analytic hierarchy process (AHP), 150–151, 155, 157, 185, 191
- ATMoS, 324
- autoregressive term, 335
- Bartlett's Test of Sphericity, 244
- bimodal seasonality, 334, 336, 342
- buffer, 105, 107, 114–117, 183
- calibration, 129, 136, 138–141, 169, 225, 304, 306
- cellular automata (CA), 76, 78, 81, 83, 100, 132
- CLUE-s land use modeling framework, 123–141
- discriminant function, 241
- econometric modelling, 52, 56, 78, 104
- empirical method, 302
- empirical-statistical models, 76, 132
- Euclidean distance, 49, 151, 186
- exploratory spatial data analysis (ESDA), 244–246, 259, 266
- factor analysis, 241, 243, 244, 246–249, 254
- finite element method (FEM), 164
- geographically weighted regression (GWR), 260, 277, 351–352, 354, 356–360
- geospatial, 17, 46, 76–78, 101, 148, 149, 179, 180, 182, 197, 259, 264, 346
- hybrid classification, 79, 80
- Information Value approach, 370

Methods and techniques (*cont.*)
iterative self-organizing data analysis (ISODATA), 79
local and global G_i*, 346
local indicators of spatial autocorrelation (LISA), 266–267, 271, 274, 277, 346, 372, 373, 378
Mann-Kendall rank, 63
manual, 148, 258
Markov-CA modelling, 89–91
Markov-cellular model, 76, 89
Markov Chain Analysis (MCA), 76, 78, 81
maximum likelihood classification (MLC), 80
mixed geographically weighted regression, 360
mono-window routine, 224
Monte Carlo randomization, 267, 378
Moran's *I*, 127, 266–267, 271, 346, 349–352, 356, 359–360, 370, 372, 373, 375–378, 381, 382
multi-criteria evaluation (MCE), 150, 157, 177–198, 241
Nash-Sutcliffe evaluation criteria, 168–170
object based image analysis (OBIA), 205, 209
Principal Axis Factoring, 244
principal component (PC), 243
semiempirical, 302
single-channel algorithm, 224–225
space-time analysis, 370
standard deviational ellipse (SDE), 50–53, 372, 375, 376, 381
transform divergence (TD), 79–80
weighted linear combination (WLC), 150, 155, 157, 185, 186, 190, 191, 243
weighted overlay, 241
zonal operations, 244
Metronomica, 127
Micieli, M.V., 380
Middle income groups, 37, 263
Migration
population, 3, 10, 77, 91, 258, 297
rural to urban, 3, 10, 77, 258, 276–277, 284, 297, 347
rural to urban exodus, 3, 10, 77, 258, 284, 297
Minimum rainfall, 68
Ministry of Environment and Forest (MoEF), 287, 290, 293
Minoprio Spensely, 31–32
Mirpur, 8, 27, 30, 205, 274
Mirzapur, 8
Misclassification, 80
Mitchell, J.K., 4
Mitigation, 181, 285, 287, 290, 294, 298, 330
Mixed geographically weighted regression, 360
MLC. *See* Maximum likelihood classification (MLC)
Model
Bid Rent, 127, 130
global, 346
hierarchical decision, 151, 185
local, 351, 359
Lowry, 130
Metronomica, 127
spatially explicit, 76
tank, 165–168
Moderate resolution image spectroradiometer (MODIS), 220, 221
Modifiable areal unit problem, 265
MODIS. *See* Moderate resolution image spectroradiometer (MODIS)
MoEF. *See* Ministry of Environment and Forest (MoEF)
Mohammadpur, 30
Momtaz, S., 17
Mondini, A., 370
Mono-window routine, 224
Monsoon, 2, 9, 13, 31, 62, 63, 65, 66, 68, 71, 112, 127, 130, 140, 148, 155, 157, 178, 183, 224, 265, 268, 336, 342, 347–348, 358, 373–374, 379
Monte Carlo randomization, 267, 378
Monthly mean temperature, 63
Moran's *I*
global autocorrelation, 352, 378, 381
spatial autocorrelation, 349, 351, 359, 372, 376, 378, 381
Morbidity, 11, 320, 334, 341, 343
Morrison, A.C., 370
Mortality, 320, 325, 334
Mosquito
female, 368
larvae, 368
Motijheel, 30, 49, 102, 274
Motor vehicle thefts, 260
Mott Mac Donald Ltd., 33
MSS. *See* Multispectral scanner (MSS)
Mughal period, 26–28
Mulligan, G., 260
Multicollinearity, 241, 254, 351, 354, 359–360, 381
Multicontextual opportunity theory, 259, 260

Multi-criteria evaluation (MCE), 150, 157, 177–198, 241
Multiple nuclei theory, 100
Multispectral scanner (MSS), 302–304
Multivariate regression, 125, 267, 274, 276, 277, 314
Mumbai, 4
Mundia, C.N., 80
Municipal committee, 28
Municipal wastewater, 285, 304
Murata, F., 17
Murder, 13, 260–262, 264, 267, 268, 271, 277
Murshidabad, 28
Muscle pains, 368
Muslim dynasty, 25
Myanmar, 1
Mymensingh, 28

N
Nahar, K., 17
Nahiduzzaman, Kh. Md., 15, 37
Nakhapakorn, K., 370
Narayanganj, 6, 8, 28, 33, 36, 37, 206, 321, 324, 327–329
Narcotics-related, 260
Nash-Sutcliffe evaluation criteria, 168–170
National Centers for Environmental Predictions (NCEP), 165, 325
Natural breaks algorithm, 191, 249, 373
Natural hazards, 12, 179–183, 190, 285
Natural resources, 2, 77, 78, 85, 91, 117, 148, 155–158, 228, 233
NCEP. *See* National Centers for Environmental Predictions (NCEP)
NDVI. *See* Normalised difference vegetation index (NDVI)
NDWI. *See* Normalized difference water index (NDWI)
Negative relationship, 231, 246, 350
Neighbourhood
  deprivation, 271
  effect, 125–127, 132, 137, 140, 141
  inequality, 258
  quality, 241
Nelson, S.A.C., 304
Nephelometric turbidity unit (NTU), 304
Netherlands, 309
Newling, B.E., 46
Nichol, J.E., 225
Nigeria, 341
Nitrogen oxides (NO_x), 323–325
NOAA–AVHRR (advanced very high resolution radiometer), 220
Noble, A.G., 10
Non-biodegradable components, 11, 304
Non-climatic factor, 13, 149, 178
Non-linear regression, 50
Non-monsoonal, 325
Non-point source pollution, 76, 305, 311, 314
Non-spatial module, 124
Nonwater-sealed sanitation, 349, 350, 354, 355
Normalised difference vegetation index (NDVI), 221, 226, 229, 230, 232, 243–247, 249, 253
Normalized difference water index (NDWI), 306
Nor'wester, 12
NTU. *See* Nephelometric turbidity unit (NTU)
NVivo analysis, 292

O
Oanh, N., 323
OBIA. *See* Object based image analysis (OBIA)
Object based image analysis (OBIA), 205, 209
Objective indicators, 240, 254
Olmanson, L.G., 303, 306, 309
OLS. *See* Ordinary least square (OLS)
Ongee, E.T., 17
Onishi, T., 17
Opportunity theory, 259, 260, 269
Optimal ecosystem functioning, 100
Optimum site location, 148
Ordinary least square (OLS), 50, 351, 352, 354–356, 358–359
Orenstein, D., 100
Organ donation, 368
Organisational framework, 290–292, 295
Other urban area (OUA), 6, 8, 13, 15, 16, 18
OUA. *See* Other urban area (OUA)
Outcome variable, 277, 335
Overall accuracy, 81, 82
Overcrowding, 117, 359, 360
Overdispersion, 335
Oxidation catalyst, 286

P
Pakistan, 3, 4, 10, 30, 36, 47, 124, 262
Pakistan period, 30
PAR. *See* Pressure and release model (PAR)
Paradigm shift, 178
Parasitic agents, 334

Paratyphoid, 360
Parr, J.B., 46
Particulate matter (PM), 11, 320, 323, 324, 327, 329
Particulate pollution, 321, 324–327, 329
Pathogenicity, 334, 342
Pathogens
  enteric, 335
  pathogenicity, 334, 342
Patient diagnosis, 337, 369
Pattiaratchi, C., 303
PC. *See* Principal components (PC)
PCA. *See* Principal component analysis (PCA)
PCI. *See* Per capita income (PCI)
Pearson correlation, 232, 307, 309, 311
Per capita income (PCI), 10, 77
Percent impervious surface (%ISA), 226
Permutations, 267, 373, 376
Peterson, R.E., 3
PGCB. *See* Power Grid Company of Bangladesh Limited (PGCB)
Phytoplankton, 311
Phytoplankton algae concentrations, 309
Piecemeal development, 12, 148, 149, 158, 241
Piped water, 11, 28
Pixel window, 307
Planning Commission, 33
Plastic waste, 12
Pleistocene terraces, 9, 157
Pluvial flooding, 12, 127, 148, 183
PM. *See* Particulate matter (PM)
Point source pollution, 311
Poisson regression models, 335
Police
  district, 264, 266, 269, 271, 274
  jurisdiction, 8, 271, 277
Policy making, 8, 46, 48, 240, 260
Pollution
  control, 287, 323
  loads, 230, 328
Polyethylene, 306
Ponds, 11, 15, 34, 131, 155, 347, 355
Poor
  capacity, 13
  housing, 359
  life style, 359
  quality of life, 346
Poorer housing estates and mixed society, 258
Population
  Asian, 99–100
  census, 2, 3, 6, 8, 10, 15, 49–51, 53–56, 102, 105, 116, 179, 183, 242, 265, 349, 370
  density, 3, 8, 11, 13, 15, 31, 46–53, 56, 77, 102, 104–105, 112–117, 140, 149, 151, 178, 186, 188, 189, 193, 221, 242, 245, 246, 249, 253, 254, 274, 275, 349, 352, 358, 381
  dependent, 33, 46, 50, 128, 274, 352
  explosion, 9, 77
  marginal, 9, 10, 31, 77, 254, 358
  movements, 368
  overcrowding, 117, 358–360
  pressure, 4, 35, 38, 53, 85, 148, 174, 204
  transient, 258, 263, 271
  urban, 3, 4, 10, 24, 46, 47, 49–50, 52, 56, 77, 99–100, 233, 240, 258, 335, 343, 346
Population-attributable risk, 334
Portuguese, 27, 28
Positive autocorrelation, 350
Post-classification, 80
Post-monsoon, 2, 63, 65, 67, 68, 72, 265, 268, 269, 277, 373
Postogola, 27
Potes, M., 304
Pourashava Ordinance, 1977, 18
Poverty
  line, 10, 187, 194, 214, 243, 249, 274–276, 349
  lower income groups, 10, 346
  lower income status, 359
  marginalised groups, 31, 178
  marginal population, 10, 178
  rate, 2, 183, 188, 274
Power, 28, 49, 50, 129, 138, 203–217, 263, 267, 275, 287, 293, 303, 311, 325
Power Grid Company of Bangladesh Limited (PGCB), 206
Power supply, 205–206, 213, 215, 216
Pratt, M., 370
Precipitation, 100, 148–149, 165, 183, 220, 222, 325
Pre-monsoon, 2, 62, 63, 65, 66, 68, 71, 265, 268–269, 277, 373
Pre-mughal period, 24–26
Preparedness, 178, 181, 190
Pressure and release model (PAR), 180, 182
Prevalence rate, 346
PreventionWeb, 178
Principal Axis Factoring, 244
Principal component analysis (PCA), 241, 243
Principal components (PC), 243
Principal factors, 153, 244
Priori knowledge, 152
Project planning, 33

Propensity, 25, 259, 271, 276, 325
Property crimes, 259, 260, 269, 271, 276, 277
Pseudomonas aeruginosa, 341
Psychology, 240
Public
  health, 11, 12, 305, 329, 346, 347, 355, 356, 360, 369, 381
    diarrhoeal, 334, 336, 343
    epidemiology, 334, 342, 378, 382
    morbidity, 11, 320, 334, 343
    mortality, 320, 325, 334
  services, 10
Pucca, 15, 188, 242, 243, 245–247, 265
Puerto Rico, 369, 370
Pulliainen, J., 311
Purana Paltan, 30

Q
QOL. *See* Quality of life (QOL)
Quadrated disk, 306
Quadratic, 267, 274, 275
Quality of life (QOL), 17, 239–254
Quantile classification scheme, 269, 271
Queen's polygon contiguity, 351
Questionnaire survey, 240
Quickbird, 205, 209–211

R
Rabbani, M.G., 13
Racially motivated crimes, 258, 259
Radarsat, 183
Rahimafrooz, 214
Rainfall
  annual, 9, 13, 63, 70
  maximum, 68
  minimum, 68
  seasonal, 63, 69–70
Rajdhani Unnayan Kartripakkha (RAJUK), 6, 8, 32–37, 39, 49, 130–132, 148, 149, 158, 183, 242, 286–287
Rajshahi, 329
RAJUK. *See* Rajdhani Unnayan Kartripakkha (RAJUK)
Ramna, 27–30
Ramsey, E.W., 80
Random
  index, 153, 186
  model, 139
Randomizations, 267, 373, 376, 378
Randomness, 266, 350, 369
Ransom, 260, 264, 267
Rao, P.K., 220
Rape, 260, 264, 267
Rashed, T., 179
Rashid, T., 17
Rate
  fertility, 91
  infection, 350, 368
  poverty, 2, 183, 188, 198, 274
  prevalence, 346
  unemployment, 242, 245–247, 258, 268–269
Rational choice perspective, 258–259
REB. *See* Rural Electrification Board (REB)
Regional planning, 148, 260
Regression
  adjusted R^2, 310, 351, 354, 358–360
  bivariate correlation matrix, 246
  coefficient of determination, 274, 276, 302, 309–311, 351, 357
  geographically weighted, 260, 277, 351–360
  linear, 49, 50, 274, 275, 277
  logarithmic, 50, 267, 275
  logistic, 125, 127, 129, 134, 135
  log-linear, 336
  modelling, 49, 76, 128, 129, 233, 241, 307, 309, 336, 359
  multivariate, 125, 267, 274, 276–277, 314
  negative exponential, 46, 49, 50
  non-linear, 50
  ordinary least square (OLS), 50, 351–355, 358–360
  Poisson, 335
  power, 49, 50, 267, 275
  stepwise, 267, 274, 307
Rehabilitation, 263
Relocation diffusion, 356
Remote sensing (RS), 76, 101, 104, 116, 179, 220, 221, 241–243, 254, 259, 302, 305
Renewable Energy Research Centre (RERC), 205, 208, 209, 217
RERC. *See* Renewable Energy Research Centre (RERC)
Rescue-related offences, 271, 277
Residuals, 127, 342, 350–352, 354, 357, 373
Resilience, 180, 181, 188
Respiratory and cardiovascular diseases, 325
Resuspended road dust, 11, 320
Return on investment (ROI), 38
Richardson, H.W., 110

Risk
exposure, 180, 327
factors, 346, 355–360
flood, 17, 35, 174, 178, 179, 188, 195, 197, 198
natural hazard, 12, 178–183, 285
population-attributable risk, 334
Ritchie, J.C., 303
River
Balu, 8, 164–166, 169–170, 174, 192, 193, 198, 285
Brahmaputra, 2
Buriganga, 8, 11, 24, 27, 32, 36, 83, 114, 192, 193, 198, 285, 311, 335, 354, 358
Dhaleswari, 192
flooding, 164
flow velocity, 167
Ganges, 2
Lakhya, 9, 285
Lubunda, 9
Meghna, 2, 9
Tongi khal, 9
transport, 32, 36
Turag, 9, 285, 311
Riverbanks, 26
River Transport (Mechanized) Emission Standard, 286
Robbery, 13, 262, 264, 267, 271
ROI. *See* Return on investment (ROI)
Root causes, 180
Rotated factor, 244, 253
Rotated factor loading matrix, 244, 247
Rotavirus, 334, 337
Roth, M., 220
Route selection, 148
Routine activities theory, 258–260, 271, 276
RS. *See* Remote sensing (RS)
Rural Electrification Board (REB), 205, 206, 215
Rural to urban migration, 3, 10, 77, 258, 284, 297
Rural–urban migration, 3, 10, 31, 91, 276–277, 347

S
Saaty, T.L., 150, 153
Sado, K., 184
Saint Venant equations, 164, 168
Saliva, 368
Salivary glands, 368
Salmonella enteric, 346
Salmonella typhi, 346, 360
Salt-and-pepper, 80
Sampling
random, 80, 81, 127, 128
systematic, 127
Sangsad Bhaban (SB), 320, 327
Sanitary toilet, 338
SAR. *See* Synthetic aperture radar (SAR)
Saralia, 8
Satarkul, 8
Satellite
advanced spaceborne thermal emission and reflection radiometer (ASTER), 220, 221, 304
Earth observation-1, 77
Earth Resources Technology Satellite-1 (ERTS-1), 101, 302
Enhanced Thematic Mapper Plus (ETM+), 221, 302, 304
ENVISAT, 304
European Remote Sensing (ERS), 304
GEOEYE-1, 79
IKONOS, 79, 304
Indian Remote Sensing (IRS), 304
Landsat, 78, 79, 101, 220–226, 243, 301–314
moderate resolution image spectroradiometer (MODIS), 220, 221
multispectral scanner, 302–304
Quickbird, 205, 209, 210
Radarsat, 183
remote sensing (RS), 76, 101, 179, 220, 221, 241–243, 253, 259, 305, 369
Shuttle Radar Topography Mission (SRTM), 165
SPOT, 259
synthetic aperture radar (SAR), 183, 304
Thematic Mapper (TM), 183, 224, 302–305, 311, 314
thermal infrared band (TIR), 224, 225, 233–234
thermal sensor, 220, 221
TIROS Operational Satellite (ITOS-1), 220
top-of-atmospheric (TOA), 225
Satellite-based LST, 221, 222
Satellite-based UHI, 220
Saudi Arabia, 241, 369
Savar, 6, 13, 15, 16, 33, 114, 321, 324, 326, 328, 329
SB. *See* Sangsad Bhaban (SB)
Scaled NDVI, 226
Scarpace, F.L., 303

Scatter plots, 375–377, 381
Scree plot, 244
Scrub typhus infection, 347
SDD. *See* Secchi disk depth (SDD)
SDE. *See* Standard deviational ellipse (SDE)
SDT. *See* Secchi disk transparency (SDT)
SEA. *See* Strategic environmental assessment (SEA)
Sea level, 3, 164, 216
Season
  dry, 2, 11, 304
  monsoon, 2, 9, 62, 63, 65, 66, 68, 71, 112, 127, 130, 148, 178, 223, 265, 268, 336, 342, 358, 373–374, 379
  post-monsoon, 2, 63, 65, 67, 68, 72, 265, 268, 269, 277, 373
  pre-monsoon, 2, 62, 63, 65, 66, 68, 71, 265, 268, 269, 277, 373
  winter, 2, 9, 62, 63, 65, 67, 68, 72, 78, 228, 229, 373
Seasonal average temperature, 63
Seasonality
  bimodal seasonality, 334, 336, 342
Seasonal rainfall, 62, 63, 69–70
Seasonal variations, 2, 83, 226, 262, 265, 268, 277, 303, 336, 338, 339
Seattle, 260
Secchi disk depth (SDD), 303
Secchi disk transparency (SDT), 302, 303, 305–312, 314
Sectoral plans, 33
Sector theory, 100
Segunbagicha, 30
Self-organising theory, 100
Semi-empirical method, 302
Semi-pucca, 15, 188, 242, 243
Sensitivity, 157, 198, 336, 338
Serotype, 346
Seto, K.C., 76
Shahbagh, 28
Shahid, S., 3
Shankharibazar, 25
Shankland Cox Partnership, 26, 32
Shannon's entropy, 104, 110, 116
Shantinagar, 30
Shape file, 49
Shaw, C.R., 258
Shimulia, 8
SHS. *See* Solar home systems (SHS)
Shuttle Radar Topography Mission (SRTM), 165
Shyampur, 8
Siddiqui, K., 13, 47
Siddirganj, 6
Signature editor, 79–80, 307
Simulation, 73, 83, 89, 91, 124, 128, 129, 131, 135–138, 140, 164–174, 320–321, 325
Single-channel algorithm, 224
Skewness, 349
Sky illumination, 208, 220
SLUETH, 127
Slum clusters, 11, 213
Slums, 11, 13, 31, 37, 38, 47, 51, 213, 269, 284, 285, 369
Smuggling, 13, 264
SO_2. *See* Sulphur dioxide (SO_2)
SOB. *See* Survey of Bangladesh (SOB)
Social cohesion, 263
Social disorganization theory, 276
Socially insecure, 13, 262–263
Socioeconomic
  factors, 127, 128, 148, 347, 351, 354, 358, 381
  strata, 269, 271
  variables, 130–131, 242, 246, 249, 274, 277, 349, 354–356, 359
Socio-political structure, 178
Soil degradation, 75
Solar energy, 205, 207, 208, 216
Solar home systems (SHS), 206, 213–215, 217
Solar PV systems, 203–217
Solid waste
  garbage, 11
  management services, 285
  medical, 12
Sonargaon, 25, 26
South Asia, 188, 323, 329, 346
South-central Asia, 258
Southeast Asia, 27, 346, 381
Sovereignty, 10, 40
Space Research and Remote Sensing Organization (SPARRSO), 48, 149, 224, 264
Space-time analysis, 370
Sparks, C.S., 259–260
SPARRSO. *See* Space Research and Remote Sensing Organization (SPARRSO)
Spatial
  adjacency, 350, 351
  analysis, 46, 56, 116, 125, 129, 149, 260, 264, 277, 285, 369, 370, 375–378, 382
  autocorrelation, 127, 128, 266, 271, 349, 351, 359, 372, 376, 378, 381, 382

Spatial (*cont.*)
  clusters, 18, 259, 266, 271, 273, 277, 346, 347, 349, 353, 355, 356, 370, 379
  database, 48, 79, 102, 157, 242, 243
  dependencies, 127, 128, 266, 351, 354, 359, 381
  diffusion, 356
  explicit models, 76, 124, 222
  module, 125
  multi-criteria assessment, 241
  non-spatial, 124
  pattern analysis, 277
  planning zones, 35
  risk factors, 346, 356, 360
  stationarity, 351
  statistics, 18, 102, 123–141, 259, 277, 345–, 375
  structure, 46, 101, 102, 116, 132
Spatially explicit models, 76, 124, 222
Spatially referenced population data, 17, 102, 116
Spatially targeted crime, 264
Spatiotemporal patterns, 45–57, 100, 258, 345–, 370
Spectral radiance, 225, 305
Speedy trial, 260, 264, 271
SPOT, 259
SPSS, 128, 267, 371
Sri Lanka, 369
SRTM. *See* Shuttle Radar Topography Mission (SRTM)
SSC. *See* Suspended sediment concentration (SSC)
Stable areas, 112
Standard deviational ellipse (SDE), 50, 53, 54, 372, 375
Standard error, 133
Standardised age-specific incidence analysis, 372
Standards, 10, 50, 51, 78, 83, 151, 168, 186, 215, 226, 240, 241, 254, 260, 286, 287, 293, 320, 354, 359, 360, 369, 372, 373, 375
Stationarity
  climate, 178
  spatial, 351
Step-wise regression, 267, 274, 307
Stewart, D.J., 115
Stool samples, 335
Storm surges, 3
Storm water runoff, 11
Strategic environmental assessment (SEA), 296
Street lights, 28
Structural failure, 12
Structure Plan, 33–36, 148
Subcontinent, 4
Subjective indicators, 198, 241, 254
Subtropical, 2, 241
SUHI. *See* Surface urban heat island (SUHI)
Suhrawardy Uddyan, 28
Suitability
  agricultural, 148
  aquaculture, 148
  landscape evaluation and planning, 148
  optimum site location, 148
  route selection, 148
  urban development, 17, 147–158
  urban land, 157
Sulphur dioxide (SO_2), 323–325
Sultanganj, 8
Sunshine duration, 204, 205, 208, 217
Surface meteorological data, 62, 68
Surface urban heat island (SUHI), 220, 221
Surface water, 11, 117, 167, 168, 182, 285, 301–314, 340, 348, 358
Surrogate variable, 253–254, 277
Survey, 37, 215, 224, 240, 254, 323, 359
Survey of Bangladesh (SOB), 79, 81, 149, 165, 183
Suspended sediment concentration (SSC), 303, 304, 310, 314
Sustainable development, 85, 91, 100, 158, 221
Sustainable urban planning and management, 240
Svab, E., 303
Sweden, 259
Symbolisation, 371
Synthetic aperture radar (SAR), 183, 304

T
Tank model, 165–168
Tantibazar, 25, 29
TD. *See* Transformed divergence (TD)
Technology supports, 204, 207, 213, 214
Tejgaon, 28, 30, 229
Temperature
  blackbody, 225
  gradient, 220, 231, 234
  land surface, 17, 219–234, 243
  monthly mean, 63
  satellite-based, 220–222, 314
  seasonal average, 63
  variability, 63–68

Terao, T., 17
Theft cases, 13, 260, 262, 267, 271
Thematic Mapper (TM), 183, 225, 302–305, 311, 314
Theory
catastrophe, 100
central place, 127
chaos, 100
disaster resilience of place (DROP), 180, 181
entropy, 100
fractals, 100
hazards-of-place (HOP), 180–182
multicontextual opportunity, 259, 260
multiple nuclei, 100
opportunity, 269
pressure and release (PAR), 180, 182
routine activities, 258–260, 271, 276
sector, 100
self-organising, 100
social disorganization, 276
Thermal infrared band (TIR), 224, 225, 233
Thermal sensor, 220, 221
Thinning out, 263
Thompson, M., 40
Three-hourly observations, 62
Thunderstorms, 2, 62
TIGER, 48
Tin cans, 369
TIR. *See* Thermal infrared band (TIR)
TIROS Operational Satellite (ITOS-1), 220
TM. *See* Thematic Mapper (TM)
TOA. *See* Top-of-atmospheric (TOA)
Tobler, W.R., 100, 127
Tongi, 6, 13, 15, 27, 32, 33, 36, 132, 206
Tongi khal, 8
Tonle Sap Lake, 169
Top-of-atmospheric (TOA), 225
Topographic maps, 78–81, 101, 305
Topography, 31, 108, 164, 347
Tornados, 12, 62
Total literacy rate, 245, 246, 249, 253, 265, 349, 352
Total suspended matter (TSM), 304
Total suspended sediment (TSS), 305–311, 313, 314
Town, 23–41, 85, 206, 209, 368
TPs. *See* Transition probabilities (TPs)
Trace gases, 11
Traditional epidemiological statistical, 370
Trajectories, 24–31, 324
Transformed divergence (TD), 80
Transient population, 258, 263, 271
Transition matrix, 81, 102, 224, 233, 234
Transition probabilities (TPs), 81, 83, 85, 87–89
Transmission of disease, 359, 360
Transport artery, 32
Tremor, 12
Tripathi, N.K., 370
Trophic state index (TSI), 303
Tropical
climate, 3
storms, 3
TSI. *See* Trophic state index (TSI)
TSM. *See* Total suspended matter (TSM)
TSS. *See* Total suspended sediment (TSS)
Tsunami wave propagation, 168
Tube wells, 11, 15, 188, 349, 354, 358
Tunnel kilns, 329
Turag, 8, 285, 311
Turbidity, 302–304
Turkey, 304, 309
Typhoid
burden of, 346
paratyphoid, 360
Salmonella enteric, 346
S. typhi, 346, 360
Typology, 105–109

U
UGAT. *See* Urban Growth Analysis Tool (UGAT)
UHI. *See* Urban heat island (UHI)
Unauthorised urban development, 148
Uncontrolled urban development, 178, 305
Unemployment rate, 242, 245–247, 258
United Nations
UNISDR, 180
United Nations Development Programme (UNDP), 32, 33, 329
United Nations Environment Program (UNEP), 12, 205, 208, 285, 286, 379
United Nations Framework Convention on Climate Change (UNFCCC), 216, 217
University of Dhaka, 207, 208
Unsafe conditions, 180
Untreated effluents, 231, 295
Urban
air pollution, 11, 117, 319, 329
Area Plan, 33, 35
built-up surface, 91, 102, 104–105, 107, 110, 114, 115, 234, 243
canyon effect, 220

Urban (*cont.*)
density, 47, 48
density gradient, 45, 46, 49, 50, 54, 56
development suitability, 17, 147–158
dwellers, 24, 178
environmental governance, 17, 284, 286–287, 293, 297
footprint, 31, 38, 124, 140
growth, 9, 12, 17, 35–38, 47, 77, 78, 91, 99–117, 123–141, 148, 151, 153, 155, 158, 178, 222, 240, 241, 264
growth analysis tool, 104, 108, 116
haphazard growth, 149, 240
heat island, 100, 218–220, 224, 228, 230, 231
illegal development, 284
infill growth, 108
infrastructures, 12, 149, 284
key driving factors, 123–141
land suitability, 148–150, 157
leapfrogging, 108, 109, 116, 141
low density, 110, 116, 210
megacity, 4–14, 23–41, 47, 48, 78, 85, 90, 91, 110, 117, 141, 158, 182, 204–207, 209, 213, 217, 222, 226, 276, 285, 298, 305, 347, 369, 378
metropolis, 158
non-climatic factor, 13, 149, 178
quality of life (QOL), 17, 239–254
services, 10, 112, 198, 254
sprawl, 75, 100–102, 104, 108–110, 115, 298
structure, 17, 37, 100, 263
suitability maps, 155
sustainability, 47, 156, 174, 240
systems, 148, 178
turn, 10, 24, 101
unauthorised, 148
uncontrolled, 178, 305
utilities, 38, 141, 204, 284, 285
vertical expansion, 51, 134, 151, 155, 222
vulnerability framework, 180–182
Urban Growth Analysis Tool (UGAT), 104, 108, 116
Urban heat island (UHI), 100, 220–222, 226, 230, 232, 233
Urbanization
gentrification, 263
USA, 48, 216, 221, 259, 264, 276, 303, 304, 309
Utility services, 290–291
UTM projection system, 78
Uttara, 30, 31, 36, 85, 205
Uttar Khan, 8

V
Validation, 129, 138–141, 169
Vandalism, 258
Variable
dependent, 50, 125, 128, 274, 351, 371, 373
explanatory, 125, 267, 274, 277, 350–352, 360
independent, 125, 127, 128, 133, 135, 140, 267, 275, 309, 372
key, 126–128
outcome, 277, 335
variance inflation factor (VIF), 351–352, 354
Variance inflation factor (VIF), 351–352, 354
Vedic period, 25
Vegetation
fractional vegetation cover (FVC), 221, 226
index, 221
NDVI, 221, 226, 229–231, 243, 244, 247, 249, 253
scaled NDVI, 226
Vehicle exhausts, 11, 320, 325, 327
Vehicular Exhaust Emission Standards, 286
Verburg, P.H., 17, 132
Vertical expansion, 51, 134, 151, 155, 222
Vertical shaft, 321, 322, 329
Vezzani, D., 379
Vibrio cholerae, 334, 336–338, 342
Vibrionaceae, 334, 342
Victoria Park, 29
Vietnam, 221
VIF. *See* Variance inflation factor (VIF)
Violence against women, 260, 261, 264, 267, 277
Violent crimes, 258–260, 267, 268, 270, 277
Viral, 334, 368
Visualisation, 228, 267, 369–372, 375, 378
VOCs. *See* Volatile organic compounds (VOCs)
Volatile organic compounds (VOCs), 323
Vulnerability
index, 186, 190
locational, 186–187, 190, 194
social, 181, 193
urban vulnerability framework, 180–182

W
Wang, F., 46, 114, 311
Wari, 28
WASA. *See* Water and Sewerage Authority (WASA)

Wastewater
  agrochemical, 304
  industrial, 285, 295, 304, 311
Water
  circulation, 34
  color, 311
  level, 164–166, 168–170, 184
  logging, 12, 34, 130, 183, 358
  pollution, 11, 12, 117, 231, 296, 302, 305, 358
  quality, 11, 17, 285, 286, 295, 301–314, 340, 343
  sources, 11, 189, 335, 338, 342, 347, 349, 352–355
  supply, 28, 128, 188, 340, 347, 359, 368
Water and Sewerage Authority (WASA), 285
Waterbodies, 79, 80, 82, 83, 89, 91, 151, 153, 155, 158, 305, 306, 311, 358
Water-borne disease, 117, 183, 189, 340, 341
Water pollution, 11, 12, 117, 231, 296, 302, 305, 358
  biological oxygen demand (BOD), 11, 286, 304
  chlorophyll concentration, 303, 304
  coliform bacteria, 341
  dissolve oxygen (DO), 11, 286, 304
  Escherichia coli (ETEC), 334
  ground-based SDT, 307
  industrial effluents, 11
  nephelometric turbidity unit (NTU), 304
  non-biodegradable components, 11, 304
  non-point source, 76, 305, 311, 314
  point source, 311, 325
  Secchi disk depth (SDD), 303
  Secchi disk transparency (SDT), 302, 303, 305–312, 314
  suspended sediment concentration (SSC), 303, 304, 310, 314
  total suspended matter (TSM), 304
  total suspended sediment (TSS), 305–311, 313, 314
  trophic state index (TSI), 303
  turbidity, 302–304
  untreated effluents, 231, 295
Water-related diseases, 358
Water-related hazards, 178
Water-sealed sanitation, 349, 354
Water sources, 11, 189, 335, 338, 342, 347, 349, 352–355
  piped, 11, 28
  ponds, 11, 15, 34, 131, 155, 347, 355
  rivers, 11, 335, 338, 342, 347, 354, 355
  tube wells, 11, 15, 188, 349, 354, 358
  wells, 11
Weather monitoring station, 62
Weeks, J., 179
Weighted linear combination (WLC), 150, 155, 157, 185, 186, 190, 243
Weighted overlay, 241
Well, 15, 38, 209, 244, 269, 271, 292, 341
Weng, Q., 225, 244, 249, 253
Wen, T., 378
Wetlands, 31, 35–36, 53, 78, 80, 83, 85, 91, 107, 117, 148, 155, 164, 169–171, 174, 178, 187, 209, 226, 297, 311
Whatman filter paper, 306
White, N.A.,258
WHO. *See* World Health Organization (WHO)
Winter, 2, 9, 62, 63, 65, 67, 68, 72, 78, 228, 229, 373
WLC. *See* Weighted linear combination (WLC)
Women and child abuse, 13, 264
World Bank, 2, 3, 24, 31, 37, 295, 296, 320–323, 329
World Health Organization (WHO), 320, 327, 346, 367–369, 381
Wu, L., 260

Y

Yamaguchi, Y., 101, 209
Yamane, Y., 17
Yellow fever, 367
Ye, X., 260

Z

Zhou, W., 311
Zhou, Y., 114
Zigzag technology, 321
Zonal operations, 244

MIX
Papier aus verantwortungsvollen Quellen
Paper from responsible sources
FSC® C105338

If you have any concerns about our products,
you can contact us on
ProductSafety@springernature.com

In case Publisher is established outside the EU,
the EU authorized representative is:
Springer Nature Customer Service Center GmbH
Europaplatz 3, 69115 Heidelberg, Germany

Printed by Libri Plureos GmbH
in Hamburg, Germany